INTRODUCTORY ALGEBRA: Programmed

THOMAS J. McHALE
ALLAN A. CHRISTENSON
KEITH J. ROBERTS
Milwaukee Area Technical College, Milwaukee, Wisconsin

ADDISON-WESLEY PUBLISHING COMPANY
Reading, Massachusetts
Menlo Park, California • London • Amsterdam • Don Mills, Ontario • Sydney

OTHER BOOKS OF INTEREST

ARITHMETIC MODULE SERIES

Programmed Modules

Module 1: Whole Numbers
Module 2: Fractions
Module 3: Decimal Numbers
Module 4: Percent, Ratio, Proportion
Module 5: Rounding, Estimation, Squares and Square Roots, Formula Evaluation, Exponents and Scientific Notation

Nonprogrammed Text

A single text with the content of all five modules in shorter, nonprogrammed form.

Reproduced by Addison-Wesley from camera-ready copy prepared by the authors.

ISBN 0-201-04747-0
CDEFGHIJ-AL-79

Preface

Introductory Algebra presents the basic skills of algebra to students who are learning algebra for the first time or who need an intensive review. To maximize the number of possible instructional strategies, the text is available in both a non-programmed form and a separate programmed form. The non-programmed form contains practice problem sets, section exercises, and unit review exercises, with all answers except the even-numbered answers for section exercises in the back of the text. The programmed form contains assignment self-tests with answers within the text and supplementary problems for each assignment with answers in the back of the text. Because the sections in the non-programmed form parallel the assignments in the programmed form, the two forms are perfectly coordinated with each other and with the set of tests provided.

The texts are accompanied by the book Tests For Introductory Algebra which contains a diagnostic test, fifty-four section or assignment tests, eighteen unit tests, nine multi-unit tests (every two units), and a comprehensive test. Three parallel forms are provided for unit tests, multi-unit tests, and the comprehensive test. A full set of answer keys for the tests and for even-numbered answers for the section exercises of the non-programmed text are included. Because of the large number of tests provided, various options are possible in using them. For example, an instructor can use the unit tests alone, the multi-unit tests alone, or some combination of the two types. Instructors desiring a multi-unit test covering more than two units can combine multi-unit tests or use the appropriate parts of the comprehensive test.

> Note: The test book is provided only to teachers. Copies of the tests for student use must be made by some copying process.

The following features make the objectives more achievable by students:

1. The content, which is limited to the simplest parts of a variety of topics, is presented in eighteen short units which are further divided into fifty-four sections or assignments.
2. The content includes a thorough review of topics from arithmetic (fractions, decimals, squares, square roots, proportions, percents, and geometric formulas) so that there is a natural transition from arithmetic to algebra.
3. The instruction, which is based on a task analysis, contains examples of all of the types of problems that appear in the exercises and tests.
4. Terminology is introduced only when necessary and is kept quite simple.
5. The applied problems are limited to fewer types but those types are discussed in detail.
6. The full and flexible set of tests can be used as a teaching tool to identify learning difficulties which can be remedied by class discussion or tutoring.

To make the content more relevant for students in career or pre-career programs as well as for students preparing for further mathematics courses, there is an initial emphasis on the more useful topics in algebra such as equation-solving, the use of formulas, and applied problems that are as realistic as possible. However, for a first contact with algebra, the full content is adequate for students with any goals.

The authors wish to thank Ms. Arleen D'Amore and Ms. Marylou Hansen who typed and proofread the camera-ready copy and Ms. Peggy McHale who prepared the drawings and graphs.

HOW TO USE THIS TEXT

This text and the tests available in Tests For Introductory Algebra can be used in various instructional strategies ranging from paced instruction with all students taking the same content to totally individualized instruction. The general procedure for using the materials is outlined below.

1. The diagnostic test can be administered either to simply get a measure of the entry skills of the students or as a basis for prescribing an individualized program.

2. Each unit is covered in a number of assignments. After the students have completed each assignment and the assignment self-test (in the text), the assignment test (from the test book) can be administered, corrected, and used as a basis for tutoring. The assignment tests are simply a teaching tool and need not be graded. The supplementary problems in the back of the text can be used at the instructor's discretion for students who need further practice.

   Note: Instead of using the assignment self-tests (in the text) as an integral part of the assignments, they can be used at the completion of a unit as a unit review exercise.

3. After the appropriate assignments are completed, either a unit test or a multi-unit test can be administered. Ordinarily, these tests should be graded. Parallel forms are provided to facilitate the test administration, including the retesting of students who do not achieve a satisfactory score.

4. After all desired units are completed in the manner above, the comprehensive test can be administered. Since the comprehensive test is a parallel form of the diagnostic test, the difference score can be used as a measure of each student's improvement.

Because of the flexibility of the tests provided, instructors who wish to modularize the course can do so.

ASSIGNMENTS FOR INTRODUCTORY ALGEBRA (PROGRAMMED)

Unit 1:	#1	(pp. 1-9)
	#2	(pp. 9-20)
	#3	(pp. 21-27)
	#4	(pp. 27-37)
Unit 2:	#5	(pp. 38-47)
	#6	(pp. 48-59)
	#7	(pp. 59-69)
	#8	(pp. 70-78)
Unit 3:	#9	(pp. 79-89)
	#10	(pp. 89-99)
	#11	(pp. 99-107)
Unit 4:	#12	(pp. 108-119)
	#13	(pp. 119-128)
	#14	(pp. 128-135)
Unit 5:	#15	(pp. 136-144)
	#16	(pp. 145-153)
	#17	(pp. 153-161)
Unit 6:	#18	(pp. 162-169)
	#19	(pp. 170-174)
	#20	(pp. 175-183)
Unit 7:	#21	(pp. 184-193)
	#22	(pp. 193-200)
	#23	(pp. 200-206)
Unit 8:	#24	(pp. 207-217)
	#25	(pp. 218-226)
Unit 9:	#26	(pp. 227-235)
	#27	(pp. 236-243)
	#28	(pp. 244-252)
	#29	(pp. 252-258)
Unit 10:	#30	(pp. 259-270)
	#31	(pp. 270-279)
Unit 11:	#32	(pp. 281-289)
	#33	(pp. 289-297)
	#34	(pp. 298-303)
Unit 12:	#35	(pp. 304-313)
	#36	(pp. 314-319)
	#37	(pp. 319-327)
Unit 13:	#38	(pp. 328-334)
	#39	(pp. 334-343)
	#40	(pp. 344-352)
Unit 14:	#41	(pp. 353-362)
	#42	(pp. 362-371)
	#43	(pp. 372-380)
Unit 15:	#44	(pp. 381-390)
	#45	(pp. 391-399)
	#46	(pp. 399-408)
Unit 16:	#47	(pp. 409-417)
	#48	(pp. 418-424)
	#49	(pp. 424-435)
Unit 17:	#50	(pp. 436-443)
	#51	(pp. 444-450)
	#52	(pp. 450-458)
Unit 18:	#53	(pp. 459-467)
	#54	(pp. 468-477)

CONTENTS

Unit 1 SIGNED WHOLE NUMBERS

In this unit, we will discuss addition, subtraction, multiplication, and division of signed whole numbers. The number line and comparisons of signed whole numbers are also discussed.

1-1 COMPARING WHOLE NUMBERS

In this section, we will discuss the various types of statements that are used to compare whole numbers. We will also show that comparisons of numbers are shown on the number line.

---

1. Since any whole number is equal to itself and is not equal to any other whole number, we can make statements like those below.

   5 is equal to 5.

   5 is not equal to 7.

   Instead of the words "is equal to" and "is not equal to", we ordinarily use the symbol "=" and the symbol "≠" since:

"=" means	"is equal to"
"≠" means	"is not equal to"

   Therefore: instead of "5 is equal to 5", we write $5 = 5$.

   instead of "5 is not equal to 7", we write ________.

$5 \neq 7$

2. Write each statement below with "=" or "≠" instead of the words.

   a) 38 is equal to 38 ________ b) 90 is not equal to 64 ________

---

a) $38 = 38$

b) $90 \neq 64$

3. Write each statement below with words instead of "=" or "≠".

   a) $27 \neq 31$ ________ b) $50 = 50$ ________

---

a) 27 is not equal to 31

b) 50 is equal to 50

4. Instead of simply saying that two different numbers are not equal, we frequently say that one "is greater than" or "is less than" the other. For example, we say:

9 is greater than 4

2 is less than 6

Instead of the words "is greater than" and "is less than", we ordinarily use the symbol ">" and the symbol "<" since:

">" means "is greater than"
"<" means "is less than"

Therefore: instead of "9 is greater than 4", we write 9 > 4.

instead of "2 is less than 6", we write ________.

---

2 < 6

5. Write each of these with ">" or "<" instead of the words.

a) 47 is greater than 34 ________ b) 51 is less than 69 ________

---

a) 47 > 34

b) 51 < 69

6. Write each of these with words instead of ">" or "<".

a) 5 < 9 ________ b) 12 > 8 ________

---

a) 5 is less than 9

b) 12 is greater than 8

7. On the number line, whole numbers are arranged according to their size with the "greater" numbers on the right. For example:

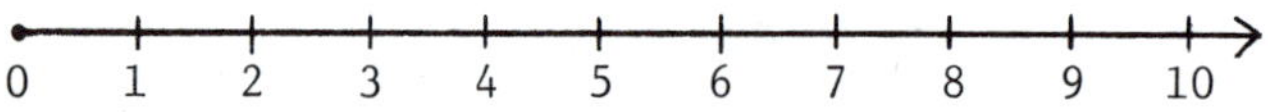

a) Since 5 > 2 , 5 is to the ________ (right/left) of 2 on the number line.

b) Since 3 < 8 , 3 is to the ________ (right/left) of 8 on the number line.

---

a) right

b) left

8. The part of the number line from 43 to 52 is shown below.

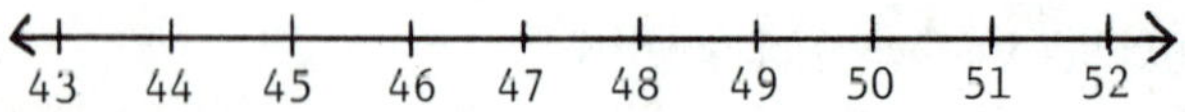

a) Since 45 < 51 , 45 is to the ________ (right/left) of 51 on the number line.

b) Since 50 > 48 , 50 is to the ________ (right/left) of 48 on the number line.

---

a) left

b) right

9. Use either ">" or "<" to complete these:

a) Since 68 is to the right of 57 on the number line, 68 ____ 57.

b) Since 72 is to the left of 85 on the number line, 72 ____ 85.

---

a) 68 > 57 b) 72 < 85

10. In the blanks below, write either ">" or "<".

a) 8 ____ 10 c) 46 ____ 19 e) 125 ____ 150

b) 19 ____ 4 d) 15 ____ 17 f) 500 ____ 300

---

a) < b) > c) > d) < e) < f) >

1-2 SIGNED WHOLE NUMBERS

In this section, we will introduce "signed" whole numbers. We will also discuss some situations in which "signed" whole numbers are useful.

---

11. Signed numbers are numbers that have either a "+" sign or a "-" sign in front of them. Those with a "+" in front of them are called "positive" numbers. Those with a "-" in front of them are called "negative" numbers. That is:

+5 and +13 are called "<u>positive</u>" numbers.

-7 and -10 are called "<u>negative</u>" numbers.

a) +3 is called a ________________ (positive/negative) number.

b) -8 is called a ________________ (positive/negative) number.

---

12. a) Does -30 mean "positive 30" or "negative 30"? ____________

b) Does +25 mean "positive 25" or "negative 25"? ____________

a) positive
b) negative

---

13. a) Does +2 or -2 mean "negative 2"? ______

b) Does +6 or -6 mean "positive 6"? ______

a) negative 30
b) positive 25

---

14. There are various applied situations in which signed numbers are useful. For example, they are useful in stating gains or losses. That is:

+\$10 can be used to represent a <u>gain</u> of \$10.

-\$10 can be used to represent a <u>loss</u> of \$10.

Write the signed number that can be used to represent these:

a) a loss of \$5 ______ b) a gain of \$50 ______ c) a loss of \$100 ______

a) -2
b) +6

---

15. Signed numbers can be used to represent deposits or withdrawals from a savings account. That is:

+\$25 can be used to represent a deposit of \$25.

-\$25 can be used to represent a withdrawal of \$25.

a) -\$5
b) +\$50
c) -\$100

(Continued on following page.)

15. (Continued)

Write the signed number that can be used to represent these.

a) a deposit of $40 _____ b) a withdrawal of $20 _____ c) a deposit of $100 _____

a) +$40
b) -$20
c) +$100

16. Signed numbers can be used to represent temperatures "above zero" and "below zero." That is:

+10° means "10° above zero".

-10° means "10° below zero".

Write the signed number than can be used to represent these:

a) 20° below zero _____ b) 15° above zero _____ c) 7° below zero _____

a) -20°
b) +15°
c) -7°

17. Signed numbers can also be used to represent distances "above sea level" and "below sea level". That is:

+500 feet means 500 feet above sea level.

-200 feet means 200 feet below sea level.

Represent each of these with a signed number.

a) 1,200 feet above sea level _____ b) 800 feet below sea level _____

a) +1,200 feet b) -800 feet

1-3 SIGNED NUMBERS AND THE NUMBER LINE

In this section, we will show how signed whole numbers are arranged on the number line.

18. On the number line, positive numbers are to the right of "0"; negative numbers are to the left of "0". For example:

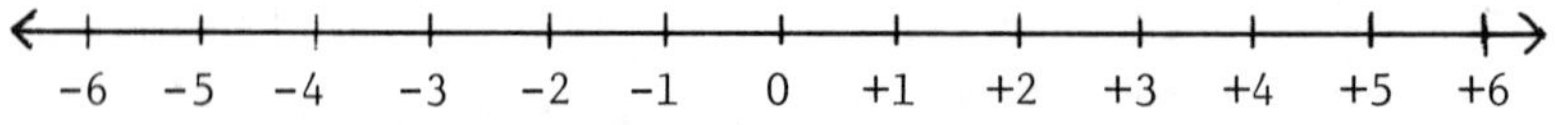

Notice this point: For every positive number, there is a negative number that is the same distance from "0". That is:

Both +2 and -2 are the same distance from "0".

Both +5 and _____ are the same distance from "0".

-5

19. Since "0" separates the positive and negative numbers, "0" is neither positive nor negative. That is, "0" has no sign.

Would it make sense to write either +0 or -0? _____

No

20. The numerical value of <u>positive</u> numbers increases as you move <u>to</u> <u>the</u> <u>right</u> on the number line.

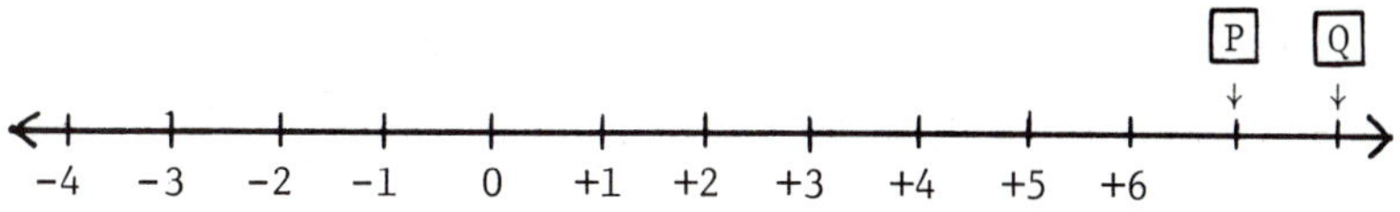

Therefore: a) What signed number is at point P above? +7

b) What signed number is at point Q above? +8

---

a) +7
b) +8

21. The part of the number line around +60 is shown at the right. What signed number is at:

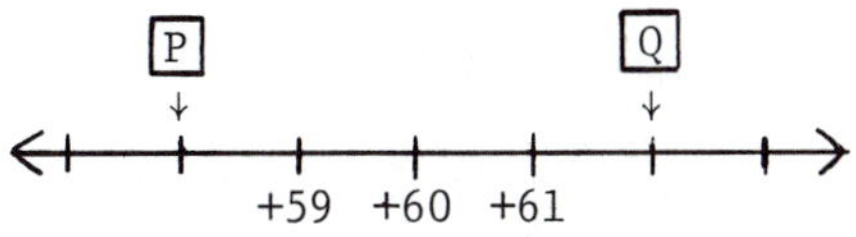

a) point P? +58 b) point Q? +62

---

a) +58
b) +62

22. On the number line at the right, what signed number is at:

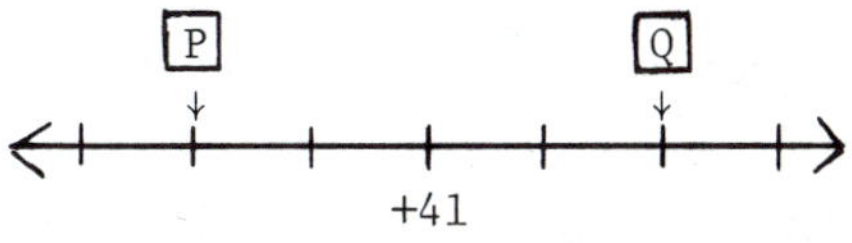

a) point P? +39 b) point Q? +43

---

a) +39
b) +43

23. The numerical value of <u>negative</u> numbers increases as you move <u>to</u> <u>the</u> <u>left</u> on the number line.

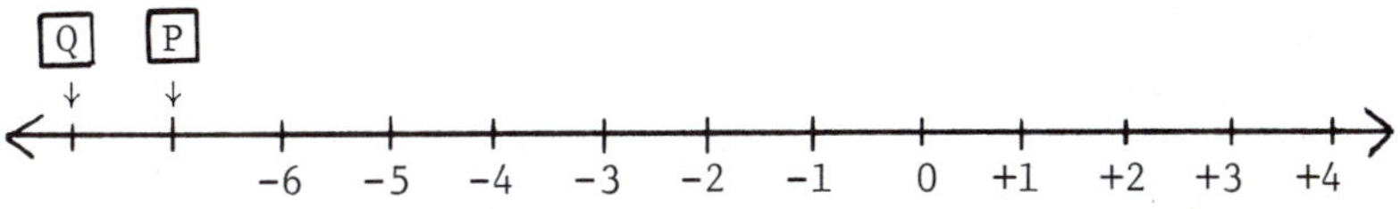

Therefore: a) What signed number is at point P above? -7

b) What signed number is at point Q above? -8

---

a) -7
b) -8

24. The part of the number line around -20 is shown at the right. What signed number is at:

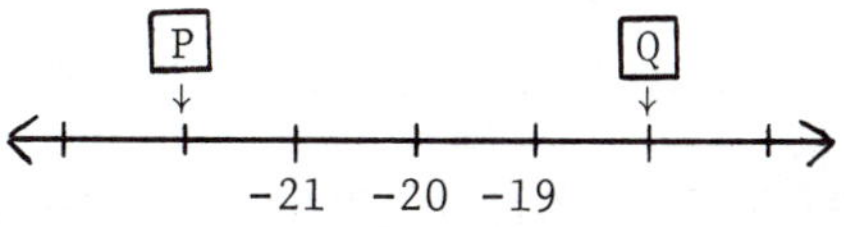

a) point P? -22 b) point Q? -18

---

a) -22
b) -18

25. On the number line at the right, what signed number is at:

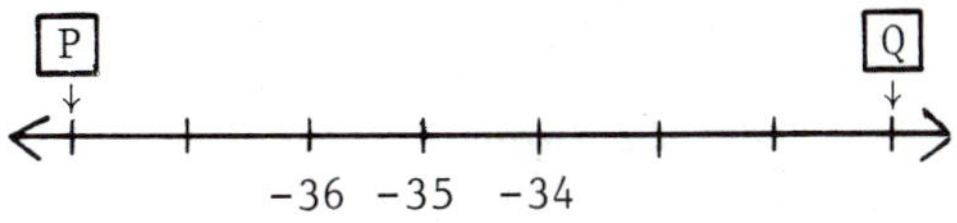

a) point P? -38 b) point Q? -31

---

a) -38
b) -31

26. On the number line at the right, what signed number is at:

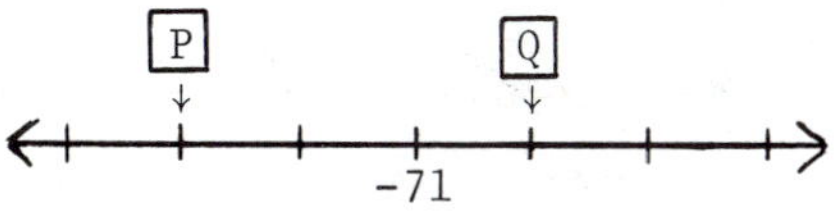

a) point P? -73 b) point Q? -70

---

a) -73 b) -70

1-4 COMPARING SIGNED WHOLE NUMBERS

Just as we can compare ordinary whole numbers, we can compare signed whole numbers. We will discuss comparisons of signed whole numbers in this section.

27. A part of the number line for signed numbers is shown below.

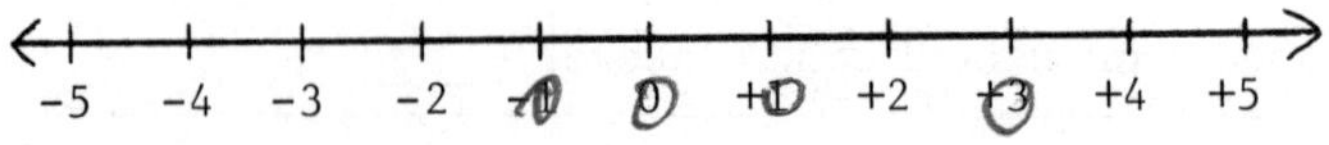

In each pair below, encircle the number that is "to the right" on the number line.

a) 0 or +3 b) +1 or -2 c) -4 or 0 d) -1 or -5

a) +3	c) 0
b) +1	d) -1

28. In each pair below, encircle the number that is "to the left" on the number line.

a) +10 or +27 b) 0 or -9 c) +12 or -15 d) -25 or -16

a) +10
b) -9
c) -15
d) -25

29. The definition of "is greater than" for signed numbers is given below.

> A first number is greater than a second number if the first is located to the right of the second on the number line.

Which number in each of the following pairs is the "greater" of the two because it is to the right on the number line?

a) +4 or +1 ______ c) -2 or +3 ______ e) -2 or -4 ______

b) +1 or 0 ______ d) -4 or 0 ______ f) -5 or -4 ______

a) +4	d) 0
b) +1	e) -2
c) +3	f) -4

30. The definition of "is less than" for signed numbers is given below.

> A first number is less than a second number, if the first is located to the left of the second on the number line.

Which number in each pair below is the "lesser" of the two because it is to the left on the number line?

a) +2 or +4 ______ c) 0 or -6 ______ e) -1 or -7 ______

b) +2 or 0 ______ d) -9 or +3 ______ f) -8 or -5 ______

a) +2	d) -9
b) 0	e) -7
c) -6	f) -8

31. Instead of the words "is greater than" and "is less than", we use the symbols ">" and "<". That is:

+5 > -3 means "+5 is greater than -3".

-6 < -3 means "-6 is less than -3".

In each blank below, write either ">" or "<".

a) +4 ______ -3 c) 0 ______ -1 e) -2 ______ -5

b) 0 ______ +5 d) -2 ______ +5 f) -7 ______ -4

a) >	d) <
b) <	e) >
c) >	f) <

32. To help you remember what the symbols ">" and "<" mean, you might note this:

">" points to the right on the number line and that is where the "greater" numbers are. ">" means "is greater than".

"<" points to the left on the number line and that is where the "lesser" numbers are. "<" means "is less than".

In each blank below, write either ">" or "<".

a) +69 _____ +75   c) +15 _____ -20   e) -14 _____ -12

b) +47 _____ 0   d) -10 _____ 0   f) -50 _____ -60

a) < b) > c) > d) < e) < f) >

1-5 ABSOLUTE VALUE

In this section, we will define what is meant by the "absolute value" of a signed number. Comparisons of the absolute values of signed numbers are also discussed.

---

33. The "absolute value" of a signed number is its distance from "0" on the number line. That is:

Since +7 is 7 units from "0", the absolute value of +7 is 7.

Since -5 is 5 units from "0", the absolute value of -5 is 5.

a) The absolute value of +9 is _____.   b) The absolute value of -12 is _____.

---

a) 9
b) 12

34. Any pair of signed numbers that are the same distance from "0" on the number line have the same absolute value. For example:

Both +3 and -3 have the same absolute value. It is 3.

Both +9 and -9 have the same absolute value. It is 9.

Name the pair of signed numbers whose absolute value is "12".

_____ and _____

+12 and -12

35. Since "0" is 0 units from "0" on the number line, the absolute value of "0" must be _____.

0

36. For practical purposes, we can think of the absolute value of a signed number as its numerical value with the sign ignored. Write the absolute value of each number below.

a) +9 _____   b) -11 _____   c) +50 _____   d) -79 _____

---

a) 9	c) 50
b) 11	d) 79

37. The symbol for absolute value is a pair of vertical lines around the signed number. For example:

$|+8|$ means "the absolute value of +8".

$|-5|$ means "the absolute value of -5".

Therefore: $|+8| = 8$ $\quad|-5| = 5$ $\quad$ a) $|+15| =$ ____ $\quad$ b) $|-27| =$ ____

---

38. a) $|+1| =$ ____ $\quad$ b) $|-1| =$ ____ $\quad$ c) $|0| =$ ____ $\quad$ d) $|-100| =$ ____

a) 15
b) 27

---

39. We can compare the absolute values of two numbers like +4 and -9. That is:

The absolute value of +4 is 4.

The absolute value of -9 is 9.

Therefore, -9 has a larger absolute value than +4.

Which number in each pair below has the larger absolute value?

a) +8 or -5 $\quad$ b) -4 or -7 $\quad$ c) +25 or -33 $\quad$ d) -65 or -59

a) 1
b) 1
c) 0
d) 100

---

40. Do not confuse the meaning of "greater than" and "less than" with that of absolute value.

Absolute value refers to the numerical value of a signed number with the sign ignored.

Greater than and less than refer to relative position on the number line.

When comparing two signed numbers:

1) Sometimes the one with the larger absolute value is "greater". For example:

+5 has a larger absolute value than +4.
+5 also is "greater" than +4 since it is to the right of +4 on the number line.

2) Sometimes the one with the larger absolute value is "less". For example:

-3 has a larger absolute value than -2.
-3, however, is "less" than -2 since it is to the left of -2 on the number line.

Of the numbers -2 and +1 : a) Which has the larger absolute value? ____

b) Which is "greater"? ____

a) +8
b) -7
c) -33
d) -65

---

41. Of -5 and -9 : a) Which has the larger absolute value? ____

b) Which is "greater"? ____

a) -2
b) +1

---

42. Of -35 , -25 , and +15 : a) Which has the largest absolute value? ____

b) Which is the "greatest"? ____

a) -9
b) -5

---

a) -35 $\quad$ b) +15

SELF-TEST 1 (pages 1-9)

Represent each of these with a signed number.

1. a deposit of \$50 ______
2. a loss of \$25 ______
3. 12° below zero ______
4. 100 feet above sea level ______

On the number line at the right, what signed number is at point:

5. A? ______ 6. B? ______ 7. C? ______

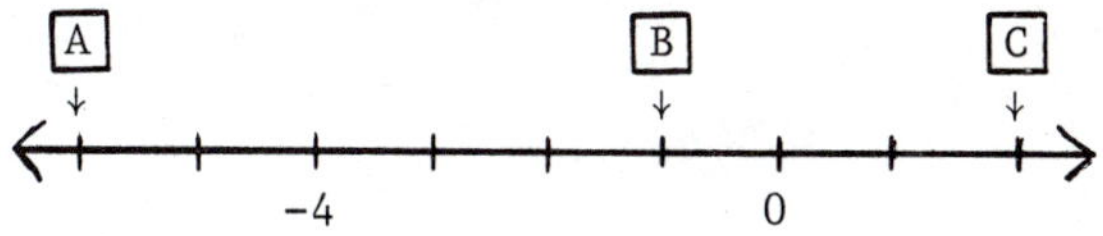

Write either ">" or "<" in each blank.

8. +7 ______ -5
9. -3 ______ +9
10. +6 ______ 0
11. -6 ______ 0
12. -10 ______ -4
13. -5 ______ -15

14. Name two signed numbers whose absolute value is 6. ______ and ______

Complete: 15. | +10 | = ______ 16. | 0 | = ______ 17. | -1 | = ______

Of -25 , -15 , and -5 : 18. Which has the largest absolute value? ______

19. Which is "greatest"? ______

ANSWERS:

1. +\$50	5. -6	8. >	11. <	14. +6 and -6	17. 1
2. -\$25	6. -1	9. <	12. <	15. 10	18. -25
3. -12°	7. +2	10. >	13. >	16. 0	19. -5
4. +100 feet					

1-6 TWO DIFFERENT USES OF THE "+" SYMBOL

The "+" symbol is used both as the sign of a number and as the symbol for addition. We will contrast the two different uses of the "+" symbol in this section.

43. In $(\overset{\downarrow}{+}5) + (\overset{\downarrow}{+}3)$: 1) The two "+" symbols under the arrows are signs of numbers.
2) The "+" symbol in the middle is the symbol for addition. It tells you to add the two numbers.

In (+10) + (+9) , draw an arrow over the one "+" symbol that tells you to add.

$(+10) \overset{\downarrow}{+} (+9)$

44. When signed numbers are involved in an addition, <u>we ordinarily write them in parentheses</u> so that the symbol for addition and the signs of numbers do not become confused. For example:

Instead of +5 + +6 , we write (+5) + (+6) .

Instead of -9 + -3 , we write (-9) + (-3) .

Instead of -7 + +4 , we write ________________ .

(-7) + (+4)

45. Though we <u>always write a negative sign (-)</u> before each negative number, <u>we frequently do not write a positive sign (+)</u> before each positive number. That is:

Both "+3" and "3" mean "positive 3".

Which two of the following mean "positive 7"? __________

+7 -7 7

+7 and 7

46. We frequently do not write a positive sign before the positive numbers in an addition. For example:

7 + 11 really means (+7) + (+11)

4 + (-2) really means (+4) + (-2)

(-6) + 3 really means ________________ .

(-6) + (+3)

47. Write each addition below so that the signs of both numbers are actually shown.

a) 5 + 10 ________________ b) (-7) + 2 ________________

---

a) (+5) + (+10)

b) (-7) + (+2)

48. 7 + 11 is actually a shorthand way of writing (+7) + (+11) .

Therefore, in 7 + 11 , is the "+" the sign of "11" or the symbol for addition? ________________

---

the symbol for addition

49. (-9) + 7 is actually a shorthand way of writing (-9) + (+7) .

Therefore, in (-9) + 7 , is the "+" the sign of "7" or the symbol for addition? ________________

---

the symbol for addition

50. In each addition below, only one "+" is actually shown. In each case, the one "+" is the symbol for ______________ .

7 + 3 (-10) + 1 2 + (-5)

---

addition

1-7 ADDING TWO POSITIVE OR TWO NEGATIVE NUMBERS

In this section, we will discuss the addition of two positive or two negative numbers. By showing some additions of both types on the number line, we will develop the "rule" that can be used for each type.

51. In an addition, the numbers added are called the "terms"; the answer is called the "sum". For example:

In (+3) + (+2) = +5 : "+3" and "+2" are the terms.
"+5" is the sum.

In (−1) + (−6) = −7 : a) "−1" and "−6" are the terms.
b) "−7" is the sum.

---

a) terms
b) sum

52. We performed (+5) + (+3) on the number line below.

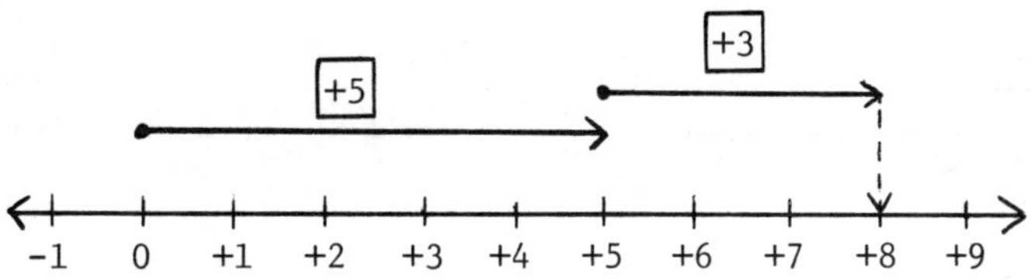

Note: 1) The bottom arrow represents the first term +5. It starts at "0" and moves 5 units to the right because +5 is positive.

2) The top arrow represents the second term +3. It starts at +5 and moves 3 units to the right because +3 is positive.

3) Reading down from the tip of the +3 arrow (see the dotted line), we find the sum +8.

Therefore: (+5) + (+3) = +8

+8 or 8

53. We performed 2 + 4 on the number line below. Notice that both arrows move to the right because both terms are positive.

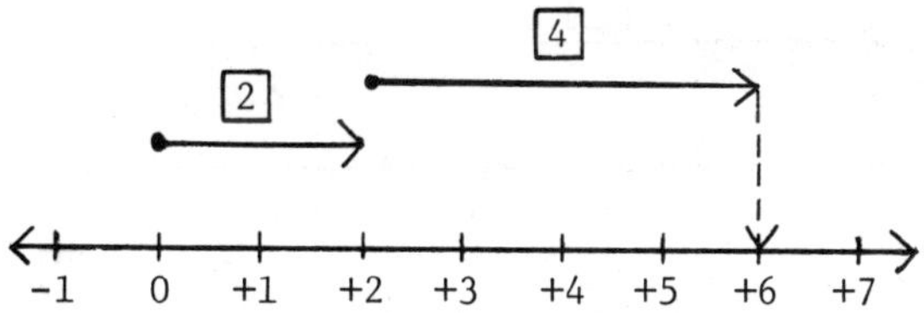

Reading down from the tip of the "4" arrow, we find the sum.

Therefore: 2 + 4 = 6

---

6 or +6

54. Here are the two additions we have done on the number line.

(+5) + (+3) = +8 or 8 2 + 4 = 6 or +6

By examining them, you can see the following "rule" that can be used.

> When adding two positive numbers:
> 1) The sign of the sum is always positive.
> 2) The absolute value of the sum is obtained by adding the absolute values of the two terms.

(Continued on following page.)

54. (Continued)

Let's use the rule to perform (+10) + (+5). Since both terms are positive:

a) The sign of the sum is ______________.

b) The absolute value of the sum is ______.

c) Therefore, (+10) + (+5) = ______.

---

a) positive (or "+")

b) 15

c) +15

55. We performed (-3) + (-2) on the number line below.

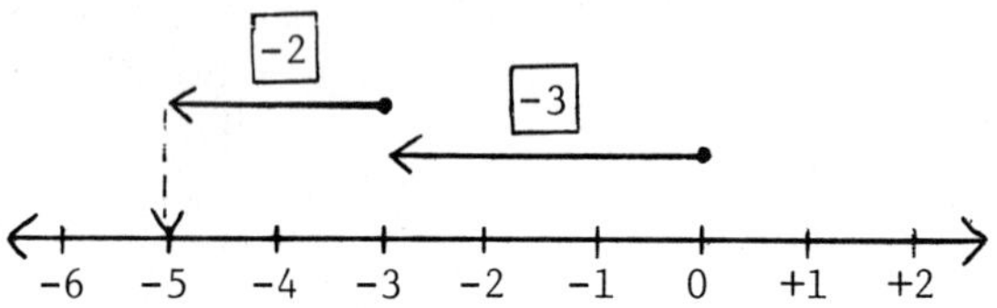

Note: 1) The bottom arrow represents the first term -3. It starts at "0" and moves 3 units to the left because -3 is negative.

2) The top arrow represents the second term -2. It starts at -3 and moves 2 units to the left because -2 is negative.

3) Reading down from the tip of the -2 arrow (see the dotted line), we find the sum -5.

Therefore: (-3) + (-2) = ______

-5

56. We performed (-1) + (-6) on the number line below. Notice that both arrows move to the left because both terms are negative.

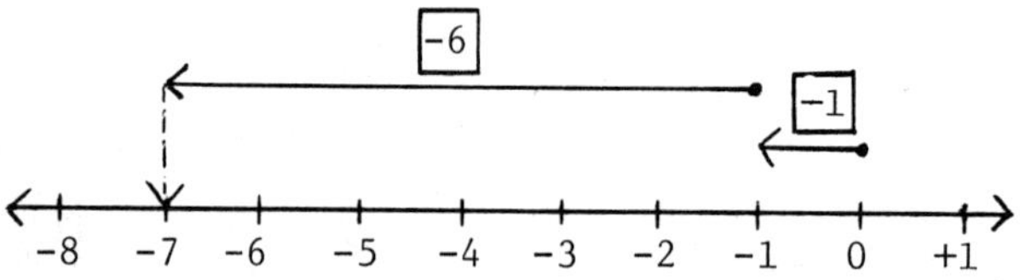

Reading down from the tip of the "-6" arrow, we find the sum.

Therefore: (-1) + (-6) = ______

-7

57. Here are the two additions we have done on the number line.

(-3) + (-2) = -5 (-1) + (-6) = -7

By examining them, you can see the following "rule" that can be used.

When adding two negative numbers:
1) The sign of the sum is always negative.
2) The absolute value of the sum is obtained by adding the absolute values of the two terms.

(Continued on following page.)

57. (Continued)

Let's use the rule to perform (-4) + (-5). Since both terms are negative:

a) The sign of the sum is ________.

b) The absolute value of the sum is ____.

c) Therefore, (-4) + (-5) = ____.

a) negative (or "-")
b) 9
c) -9

---

58. Complete each of the following additions.

a) (+7) + (+2) = ____ c) (-8) + (-1) = ____ e) (+20) + (+30) = ____

b) (-7) + (-2) = ____ d) (+4) + (+6) = ____ f) (-10) + (-15) = ____

a) +9 b) -9 c) -9 d) +10 e) +50 f) -25

1-8 ADDING ONE POSITIVE AND ONE NEGATIVE NUMBER

In this section, we will discuss the addition of one positive and one negative number. By showing some additions on the number line, we will develop the "**rule**" for additions of that type.

---

59. We performed (+5) + (-2) on the number line below.

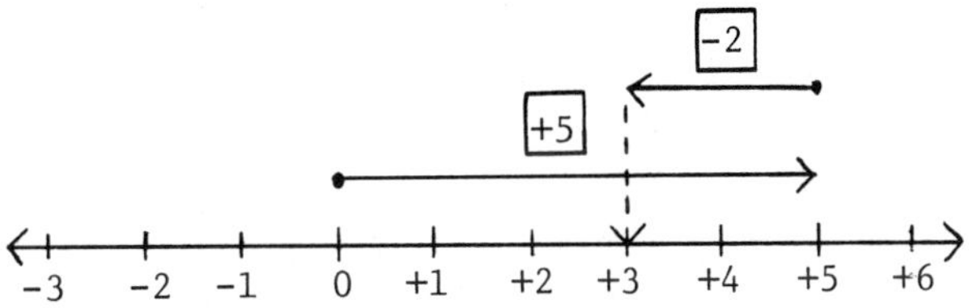

Note: 1) Since +5 is positive, the bottom arrow starts at "0" and moves 5 units to the right.

2) Since -2 is negative, the top arrow starts at +5 and moves 2 units to the left.

3) Reading down from the tip of the -2 arrow, we find the sum +3.

Therefore: (+5) + (-2) = ____

+3 or 3

---

60. We performed (-6) + (+4) on the number line below.

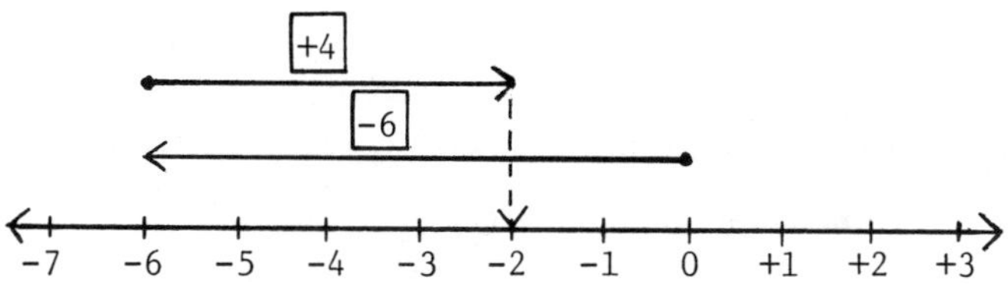

Note: 1) Since -6 is negative, the bottom arrow starts at "0" and moves 6 units to the left.

2) Since +4 is positive, the top arrow starts at -6 and moves 4 units to the right.

3) Reading down from the tip of the +4 arrow, we find the sum.

Therefore: (-6) + (+4) = ____

61. We performed (-3) + 7 on the number line below. Notice that the arrow for +7 moves past "0" to the positive side of the number line. -2

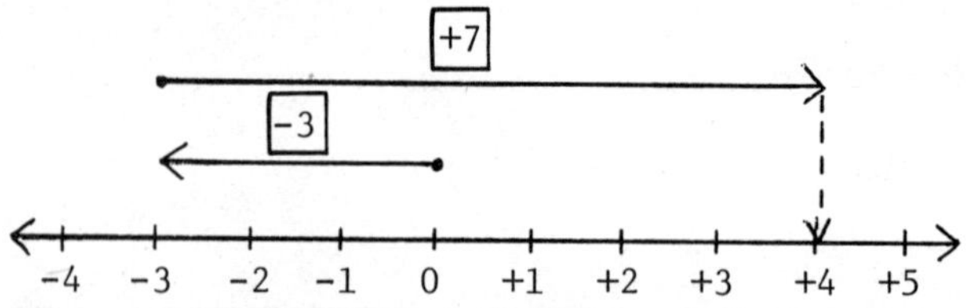

Reading down from the tip of the +7 arrow, we find the sum.

Therefore: (-3) + 7 = 4

---

62. We performed 1 + (-4) on the number line below. Notice that the arrow for -4 moves past "0" to the negative side of the number line. +4 or 4

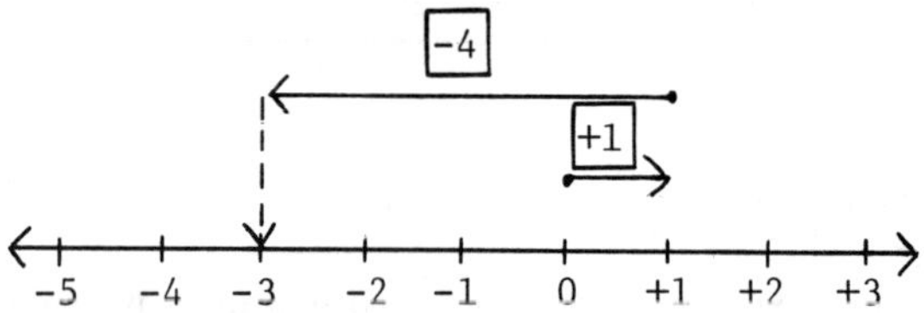

Reading down from the tip of the -4 arrow, we find the sum.

Therefore: 1 + (-4) = -3

---

63. The four additions done on the number line are shown below. -3

(+5) + (-2) = +3 (-3) + 7 = +4

(-6) + (+4) = -2 1 + (-4) = -3

In each case, the <u>sign of the sum</u> is the same as the sign of the term <u>with the larger absolute value</u>.

a) In (-15) + 8 , -15 has a larger absolute value than 8.
Since -15 is negative, the sign of the sum must be ____–____.

b) In 10 + (-3) , 10 has a larger absolute value than -3.
Since 10 is positive, the sign of the sum must be ____+____.

---

64. For each of the following, state whether <u>the sign of the sum</u> would be "+" or "-". a) negative (or "-") b) positive (or "+")

a) (-25) + (+11) ___–___ c) 37 + (-24) ___+___

b) (+17) + (-12) ___+___ d) (-50) + 20 ___–___

---

65. The four additions done on the number line are shown again below. a) - b) + c) + d) -

(+5) + (-2) = +3 (-3) + 7 = +4

(-6) + (+4) = -2 1 + (-4) = -3

(Continued on following page.)

65. (Continued)

In each case, the absolute value of the sum can be found by subtracting the smaller absolute value from the larger absolute value.

a) In 7 + (-9) , the two absolute values are 7 and 9.
Therefore, the absolute value of the sum is 9 - 7 = ______.

b) In (-8) + 5 , the two absolute values are 8 and 5.
Therefore, the absolute value of the sum is 8 - 5 = ______.

Answers to 65: a) 2 b) 3

66. The following "rule" can be used to add one positive and one negative number.

> When adding one positive and one negative number:
>
> 1) The sign of the sum is the same as the sign of the term with the larger absolute value.
>
> 2) The absolute value of the sum is obtained by subtracting the smaller absolute value from the larger absolute value.

Let's use the rule to perform 4 + (-9) .

a) Since -9 has the larger absolute value, the sign of the sum must be ______________.

b) Since the two absolute values are 4 and 9, the absolute value of the sum must be ______.

c) Therefore: 4 + (-9) = ______

Answers to 66: a) negative (or "-") b) 5, from 9 - 4 c) -5

67. Let's use the rule to perform (-7) + 11 .

a) Since 11 has the larger absolute value,
the sign of the sum must be ______________.

b) Since the two absolute values are 7 and 11,
the absolute value of the sum must be ______.

c) Therefore: (-7) + 11 = ______

Answers to 67: a) positive (or "+") b) 4, from 11 - 7 c) +4 or 4

68. Complete:

a) 10 + (-7) = ______ c) 2 + (-9) = ______

b) (-6) + 12 = ______ d) (-15) + 6 = ______

Answers to 68: a) +3 b) +6 c) -7 d) -9

69. Complete:

a) 8 + (-12) = ______ c) (-14) + 10 = ______

b) (-5) + 11 = ______ d) 25 + (-5) = ______

Answers to 69: a) -4 b) +6 c) -4 d) +20

1-9 ADDING A SIGNED NUMBER AND "0"

In this section, we will discuss the addition of either a positive number or a negative number and "0".

70. When one of the terms is "0", the sum equals the other term. For example:

(+6) + 0 = +6 a) 0 + (-1) = ______

(-9) + 0 = -9 b) 0 + (+1) = ______

---

a) -1
b) +1

71. Complete: a) (-5) + (-5) = ______ d) (-1) + (+1) = ______

b) 0 + (-8) = ______ e) (-2) + 0 = ______

c) (+4) + 0 = ______ f) 0 + (+10) = ______

---

a) -10 b) -8 c) +4 d) 0 e) -2 f) +10

72. Complete: a) 13 + (-33) = ______ e) (-19) + 49 = ______

b) (-12) + (-13) = ______ f) 70 + (-50) = ______

c) 50 + (-17) = ______ g) (-10) + (-20) = ______

d) 0 + (-99) = ______ h) (+60) + 0 = ______

---

a) -20 b) -25 c) +33 d) -99 e) +30 f) +20 g) -30 h) +60

1-10 ADDING THREE OR MORE SIGNED NUMBERS

In this section, we will discuss a method for additions containing three or more signed numbers.

73. To add three signed numbers, we can add "two at a time" as we have done at the left below. Using the same method, complete the other addition.

3 + (-5) + 7 → (-2) + 7 = +5

(-9) + (-2) + 6 → [ ] + 6 = ______

[-11] + 6 = -5

74. By adding "two at a time", find each sum below.

a) (-7) + 5 + (-8) = ______

b) 1 + 2 + (-7) = ______

c) (-3) + (-9) + 6 = ______

d) (-3) + (-4) + (-8) = ______

---

a) -10, from: (-2) + (-8)

b) -4, from: 3 + (-7)

c) -6, from: (-12) + 6

d) -15, from: (-7) + (-8)

75. To add four signed numbers, we again add "two at a time" as we have done at the left below. Using the same method, complete the other additions.

(-9) + 3 + (-4) + 12	2 + (-3) + 9 + (-5)
↓	↓
(-6) + (-4) + 12	☐ + 9 + (-5)
↓	↓
(-10) + 12 = +2	☐ + (-5) = ______

---

76. By adding "two at a time", find each sum below.

a) 10 + (-6) + (-8) + 1 = ______

b) (-1) + (-4) + 1 + 9 = ______

c) 12 + (-9) + 4 + (-5) = ______

d) (-6) + (-4) + 10 + (-1) = ______

Answer to 75: [-1] + 9 + (-5) ↓ [8] + (-5) = +3

---

77. The method of adding "two at a time" can be used when the addition contains more than four terms. Use that method to find each sum below.

a) (-9) + 8 + (-3) + 6 + (-7) = ______

b) (-5) + (-4) + (-1) + 12 + (-2) = ______

Answers to 76: a) -3 b) +5 c) +2 d) -1

Answers to 77: a) -5 b) 0

1-11 THE ORDER PROPERTY OF ADDITION

In this section, we will discuss the "order" property of addition. That is, we will show that the terms of an addition can be written in any order or added in any order without changing the sum. A shortcut for additions containing three or more terms is also discussed.

---

78. When adding two signed numbers, the terms can be written in either order without changing the sum. For example:

Both (-7) + (-3) and (-3) + (-7) = -10

Both (-5) + 9 and 9 + (-5) = 4

---

79. Using the "order" property of addition, write the correct term in each blank.

a) 10 + (-7) = (-7) + ______ b) (-1) + (-3) = (-3) + ______ c) 0 + (-5) = (-5) + ______

Answer to 78: +4

---

80. We have written the same three terms in different orders below.

(-4) + 8 + (-2) (-4) + (-2) + 8 (-2) + 8 + (-4)

Though the terms are in different orders, the sum for each addition is ______.

Answers to 79: a) (-7) + 10 b) (-3) + (-1) c) (-5) + 0

81. We have written the same four terms in different orders below. | +2

(-5) + (-1) + (-2) + 4 (-1) + (-5) + 4 + (-2) (-2) + (-1) + 4 + (-5)

Though the terms are in different orders, the sum for each addition is ______.

-4

82. No matter how many terms there are in an addition, we can write them in any order without changing the sum. Using this "order" property, write the correct term in each blank below.

a) (-10) + 7 + (-3) = (-3) + 7 + ________

b) 4 + (-2) + 9 + (-6) = ________ + 9 + (-2) + 4

---

a) (-3) + 7 + (-10)

b) (-6) + 9 + (-2) + 4

83. The terms of an addition can also be <u>added in any order</u> without changing the sum. For example:

The same three terms were added from <u>left to right</u> and from <u>right to left</u> below. In each case the sum is -4.

[(-7) + 5] + (-2)
↓
(-2) + (-2) = -4

(-7) + [5 + (-2)]
↓
(-7) + 3 = -4

The same four terms were added from <u>left to right</u> and from <u>right to left</u> below. In each case, the sum is ______.

[10 + (-3)] + 5 + (-9)
↓
[7 + 5] + (-9)
↓
12 + (-9) = 3

10 + (-3) + [5 + (-9)]
↓
10 + [(-3) + (-4)]
↓
10 + (-7) = 3

3

84. Because of the order property of addition, there is a shortcut that can be used for additions containing three or more terms. An example is discussed below.

To perform (-4) + 2 + (-5) + 3 , we can write all the negative and all the positive numbers together before adding. We get:

[(-4) + (-5)] + [2 + 3]
↓ ↓
(-9) + 5 = ______

-4

85. To perform 7 + (-8) + (-4) + 9 + 3 + (-1) , we grouped all the positive and all the negative terms together below.

[7 + 9 + 3] + [(-8) + (-4) + (-1)]
↓ ↓
19 + (-13) = ______

+6

86. It is even shorter to add all the positive and all the negative numbers mentally without actually rewriting the order of the terms.

Do the following mentally: (-7) + 9 + 6 + (-10)

a) The sum of the positive numbers is ______.

b) The sum of the negative numbers is ______.

c) Therefore, the total sum is ______.

---

a) 15
b) -17
c) -2

87. Do the following mentally: 5 + (-1) + 2 + (-6) + 7 + (-3)

a) The sum of the positive numbers is ______.

b) The sum of the negative numbers is ______.

c) Therefore, the total sum is ______.

---

a) 14
b) -10
c) +4 or 4

88. Use the shortcut to find each sum.

a) 4 + (-2) + 5 = ______

b) (-8) + 1 + (-4) = ______

c) (-2) + 5 + (-6) + 4 = ______

d) 7 + (-5) + 2 + (-6) = ______

e) (-7) + 8 + (-6) + 3 + (-6) = ______

f) 6 + (-5) + (-3) + 4 + 8 + (-9) = ______

---

a) +7 b) -11 c) +1 d) -2 e) -8 f) +1

1-12 OPPOSITES

Numbers with the same absolute value but different signs are called "opposites". We will discuss "opposites" in this section.

---

89. On the number line below, we have connected pairs of numbers with the same absolute value but different signs.

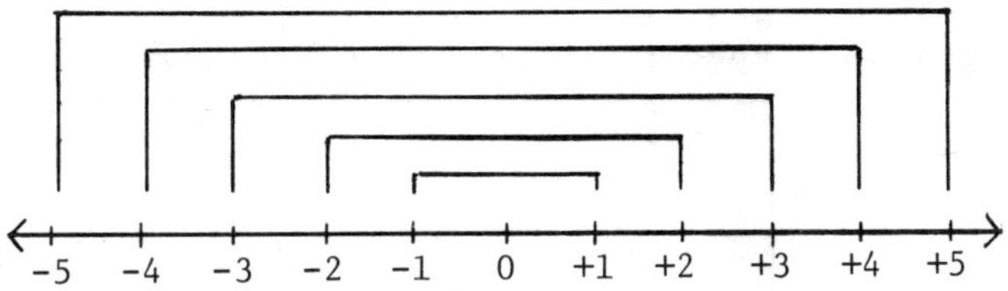

In each pair, the two numbers are the same distance from "0" but on "opposite" sides of "0". Therefore, they are called "opposites". That is:

+1 and -1 are called "opposites".

+3 and -3 are called "opposites".

+4 and ______ are called "opposites".

-4

90. Since +7 and -7 are a pair of opposites, we can make the following two statements:

The opposite of +7 is -7.

The opposite of -7 is +7.

Name the opposite of each of the following.

a) -3 ______ b) 9 ______ c) -15 ______ d) 1 ______

---

a) +3 b) -9 c) +15 d) -1

91. When a pair of opposites is added, the sum is "0". That is:

2 + (-2) = 0 a) (-4) + 4 = ______ b) 75 + (-75) = ______

---

a) 0 b) 0

92. There is one number that is its own opposite. You can see this fact by answering the questions below.

a) What number would you add to "0" to get "0" as a sum? ______

b) Therefore, the opposite of "0" is ______.

---

a) 0 b) 0

SELF-TEST 2 (pages 9-20)

Complete:

1. (-6) + (-5) = -11
2. 10 + (-3) = 7
3. (-12) + 1 = -11
4. 0 + (-1) = -1
5. (-7) + 8 = 1
6. 9 + (-9) = 0

Using the order property of addition, write the correct term in each blank.

7. 12 + (-8) = (-8) + 12
8. (-1) + 5 + (-9) = 5 + (-9) + (-1)

Complete:

9. (-4) + 7 + (-6) = -3
10. 9 + (-3) + 5 + (-7) = 4

11. The opposite of +10 is -10.
12. The opposite of -15 is 15.
13. The sum of any pair of opposites is always 0.

ANSWERS:

1. -11	4. -1	7. 12	9. -3	11. -10
2. +7	5. +1	8. -1	10. +4	12. +15
3. -11	6. 0			13. zero (or 0)

1-13 TWO DIFFERENT USES OF THE "-" SYMBOL

The "-" symbol is used both as the sign of a number and as the symbol for subtraction. We will contrast the two different uses of the "-" symbol in this section.

93. In a subtraction, the numbers subtracted are called "terms"; the answer is called the "difference". For example:

In 9 - 5 = 4 : "9" and "5" are the terms.
"4" is the difference.

In (-6) - (-1) = -5 : a) "-6" and "-1" are the ____________.

b) "-5" is the ______________________.

---

a) terms
b) difference

94. In $(\overset{\downarrow}{-}5) - (\overset{\downarrow}{-}7)$: 1) The two "-" symbols with the arrows over them are signs of numbers.

2) The "-" symbol in the middle is the symbol for subtraction.

a) In (-7) - (-6) , draw an arrow over the one "-" that tells you to subtract.

b) In 3 - (-8) , draw an arrow over the one "-" that is a sign of a number.

---

a) $(-7) \overset{\downarrow}{-} (-6)$
b) $3 - (\overset{\downarrow}{-}8)$

95. To perform a subtraction, you must be able to identify the sign of the second term. It is easy to do so when the second term is negative. For example:

In 5 - (-3) , the second term is -3.

In (-7) - (-1) , the second term is _____.

-1

96. When the second term is positive, however, there is a tendency to confuse the subtraction symbol with its sign because the positive sign is not usually shown. Remember:

6 - 9 really means 6 - (+9)

(-2) - 5 really means (-2) - (+5)

Write each of these so that the sign of the second term is actually shown.

a) 4 - 7 ____________________ b) (-10) - 3 ____________________

---

a) 4 - (+7)
b) (-10) - (+3)

97. a) Does 3 - 8 mean: 3 - (+8) or 3 - (-8) ? ____________________

b) Does (-5) - 9 mean: (-5) - (+9) or (-5) - (-9) ? ____________________

---

a) 3 - (+8)
b) (-5) - (+9)

98. In 7 - 9 : 1) The "-" is the symbol for subtraction.
2) The sign of 9 is "+". That is, 9 is a positive number.

In $(-3) \overset{\downarrow}{-} 4$: a) The "-" with the arrow over it is the symbol for ____________________.

b) Is 4 a positive or negative number? ____________________

---

a) subtraction b) positive

99. The second term of a subtraction is negative <u>only if the negative sign is actually shown</u>. That is:

In 5 - (-9) , the second term is <u>negative</u>.

In 5 - 9 , the second term is <u>positive</u>.

The second term is <u>negative</u> in which subtractions below? ____________

a) (-6) - 8 b) (-6) - (-8) c) 5 - (-10) d) 5 - 10

Both (b) and (c)

100. The second term is <u>positive</u> in which subtractions below? ____________

a) 1 - 3 b) 1 - (-3) c) (-7) - 2 d) (-7) - (-2)

Both (a) and (c)

1-14 CONVERTING SUBTRACTION TO ADDITION

Any subtraction of signed numbers is performed by converting it to an equivalent addition and then performing the addition. In this section, we will show how the conversion from subtraction to addition is made. In the next section, we will show the complete subtraction process. In the section after that, we will show that the method for subtraction makes sense.

---

101. Any subtraction can be converted to an equivalent addition by <u>ADDING THE OPPOSITE OF THE SECOND TERM TO THE FIRST TERM</u>. That is:

10 - (+6) = 10 + (the opposite of +6)

= 10 + (-6)

7 - (+8) = 7 + (the opposite of +8)

= 7 + ( )

---

7 + (-8)

102. Following the example below, complete the other two conversions.

5 - (-2) = 5 + (the opposite of -2)

= 5 + (+2)

a) (-4) - (+1) = (-4) + (the opposite of +1)

= (-4) + ( )

b) (-7) - (-10) = (-7) + (the opposite of -10)

= (-7) + ( )

---

a) (-4) + (-1)

b) (-7) + (+10)

103. Any subtraction can be converted to an equivalent addition by using the following definition:

The subtraction symbol "-" means:
ADD THE OPPOSITE OF THE SECOND TERM

(Continued on following page.)

103. (Continued)

By replacing the second term with its opposite, convert each subtraction below to an addition.

a) 5 - (+10) = 5 + (-10) c) 5 - (-10) = 5 + (10)

b) (-9) - (+3) = (-9) + (-3) d) (-9) - (-3) = (-9) + (3)

---

a) 5 + (-10)
b) (-9) + (-3)
c) 5 + (+10)
d) (-9) + (+3)

104. When the second term of a subtraction is positive, the positive sign is not usually shown. In such cases, remember that the second term is positive when converting to addition. For example:

2 - 7 = 2 - (+7) = 2 + (-7)

(-5) - 8 = (-5) - (+8) = (-5) + (-8)

Complete each of the following conversions to addition.

a) 9 - 6 = 9 - (+6) = 9 + ( )

b) (-1) - 3 = (-1) - (+3) = (-1) + ( )

---

a) 9 + (-6)
b) (-1) + (-3)

105. Complete the following conversions to additions.

a) 2 - 9 = 2 + ( ) c) (-3) - 5 = (-3) + ( )

b) 2 - (-9) = 2 + ( ) d) (-3) - (-5) = (-3) + ( )

---

a) 2 + (-9) c) (-3) + (-5)
b) 2 + (+9) d) (-3) + (+5)

106. Complete the following conversions to additions.

a) 4 - 19 = 4 + ( )

b) 15 - (-20) = 15 + ( )

c) (-9) - 25 = (-9) + ( ) d) (-40) - (-30) = (-40) + ( )

---

a) 4 + (-19) b) 15 + (+20) c) (-9) + (-25) d) (-40) + (+30)

1-15 SUBTRACTING SIGNED NUMBERS

After converting a subtraction to addition, we can complete the subtraction by performing the addition. We will discuss the method in this section.

---

107. Any subtraction can be performed by converting it to the equivalent addition and then performing the addition. For example:

4 - 7 = 4 + (-7) = -3

a) (-5) - 6 = (-5) + (-6) = ______ b) 1 - 9 = 1 + (-9) = ______

---

a) -11
b) -8

108. Following the example, complete the other two subtractions.

$8 - (-3) = 8 + 3 = 11$

a) $(-7) - (-5) = (-7) + 5 =$ ______ b) $1 - (-6) = 1 + 6 =$ ______

---

a) -2
b) 7

109. Complete each subtraction at the right by converting to addition and then performing the addition.

		ADDITION	
a)	$(-4) - 9 =$	____________	= ______
b)	$(-1) - (-3) =$	____________	= ______
c)	$3 - (-7) =$	____________	= ______
d)	$2 - 6 =$	____________	= ______

---

a) $(-4) + (-9) = -13$
b) $(-1) + 3 = 2$
c) $3 + 7 = 10$
d) $2 + (-6) = -4$

110. Using the same method, complete each subtraction at the right.

		ADDITION	
a)	$3 - 9 =$	____________	= ______
b)	$7 - (-7) =$	____________	= ______
c)	$(-3) - 4 =$	____________	= ______
d)	$(-8) - (-7) =$	____________	= ______

---

a) $3 + (-9) = -6$
b) $7 + 7 = 14$
c) $(-3) + (-4) = -7$
d) $(-8) + 7 = -1$

111. Use the same method to complete these:

a) $8 - 9 =$ ________________ c) $(-3) - 1 =$ ________________

b) $1 - (-1) =$ ________________ d) $(-5) - (-6) =$ ________________

---

a) -1	c) -4
b) 2	d) 1

112. Use the same method to complete these:

a) $10 - 15 =$ ________________ c) $(-17) - 13 =$ ________________

b) $25 - (-25) =$ ________________ d) $(-41) - (-41) =$ ________________

---

a) -5	c) -30
b) 50	d) 0

113. Earlier we showed that the terms of an addition can be written in either order without changing the sum. That is:

Both $(-7) + 2$ and $2 + (-7)$ equal -5.

However, if we change the order of the terms in a subtraction, we do not get the same difference. For example:

Though $9 - 6 = 3$, $6 - 9 = -3$

Though $(-3) - (-4) = 1$, $(-4) - (-3) =$ ______

-1

114. a) Does $8 + (-3) = (-3) + 8$? ________ b) Does $8 - (-3) = (-3) - 8$? ________

---

a) Yes b) No

1-16 JUSTIFYING THE PROCEDURE FOR SUBTRACTING SIGNED NUMBERS

In this section, we will justify the procedure for subtraction by showing that it makes sense. To do so, we will use the relationship between subtraction and addition.

---

115. When a subtraction that does not involve signed numbers is performed correctly, the sum of the second term and the difference equals the first term. For example:

$$8 - 3 = 5 \quad \text{and} \quad 3 + 5 = 8$$

$$7 - 1 = 6 \quad \text{and} \quad 1 + 6 = 7$$

If the method we are using to subtract signed numbers makes sense, the same fact should be true. That is, the sum of the second term and the difference should equal the first term. Let's check this fact for the subtraction below.

$$4 - 9 = -5 \qquad [\text{from } 4 + (-9) = -5]$$

a) At the right add 9 and -5 to see whether their sum is 4. $9 + (-5) =$ ____

b) Therefore, does the subtraction procedure make sense? ____

---

a) 4
b) Yes

116. Let's see whether the same fact holds for the subtraction below.

$$(-2) - 3 = -5 \qquad [\text{from } (-2) + (-3) = -5]$$

a) At the right add 3 and -5 to see whether their sum is -2. $3 + (-5) =$ ____

b) Therefore, does the subtraction procedure make sense? ____

---

a) -2
b) Yes

117. Let's see whether the same fact holds for the subtraction below.

$$7 - (-1) = 8 \qquad [\text{from } 7 + 1 = 8]$$

a) At the right add -1 and 8 to see whether their sum is 7. $(-1) + 8 =$ ____

b) Therefore, does the subtraction procedure make sense? ____

---

a) 7
b) Yes

118. Let's see whether the same fact holds for the subtraction below.

$$(-9) - (-2) = -7 \qquad [\text{from } (-9) + 2 = -7]$$

a) At the right, add -2 and -7 to see whether their sum is -9. $(-2) + (-7) =$ ____

b) Therefore, does the subtraction procedure make sense? ____

---

a) -9
b) Yes

1-17 SUBTRACTIONS AND EXPRESSIONS WITH MORE THAN TWO TERMS

In this section, we will discuss the procedure for evaluating expressions that contain more than two terms and involve subtraction.

---

119. When an expression containing more than two terms involves subtraction, we evaluate it by first converting each subtraction to addition and then finding the sum in the usual way. For example:

$$4 + 8 - (-5) = 4 + 8 + (+5) = 17$$

a) $2 - 6 + 7 = 2 + (-6) + 7 =$ ______

b) $5 - (-2) - 8 = 5 + (+2) + (-8) =$ ______

---

a) 3
b) -1

120. To evaluate the expressions below, we again convert each subtraction to addition and then find the sum in the usual way. That is:

$$10 - 7 + (-8) - (-1) = 10 + (-7) + (-8) + (+1) = -4$$

$$(-1) + 5 - (-9) - 6 = (-1) + 5 + (+9) + (-6) = \_\_\_\_$$

7

121. Evaluate each of these. Begin by converting each subtraction to addition.

a) $(-5) + (-3) - (-2) =$ ______

b) $4 - 9 + 8 =$ ______

c) $(-1) - 6 - (-7) =$ ______

d) $10 - (-7) - 1 =$ ______

---

a) -6
b) 3
c) 0
d) 16

122. Evaluate each of these. Begin by converting each subtraction to addition.

a) $(-3) - 1 - (-9) =$ ______

b) $10 - (-20) - 5 =$ ______

c) $9 - (-2) + (-8) - 5 =$ ______

d) $(-3) - 6 - (-10) - 1 =$ ______

---

a) 5
b) 25
c) -2
d) 0

123. We have already seen that the order property does not apply to simple subtractions. It also does not apply to more complex expressions involving subtractions. For example, we evaluated the expression below in the usual way and got "4" as the answer.

$$8 - 7 - (-3) = 8 + (-7) + (+3) = 4$$

The same three terms are written in two different orders below. Evaluate each expression to show that its value is not "4".

a) $7 - 8 - (-3) =$ ______

b) $(-3) - 8 - 7 =$ ______

124. When an expression contains more than two terms and <u>each operation is an addition</u>, we can write the terms in any order without changing its value. For example:	a) 2 b) -18

$$3 + (-5) + (-1) = (-5) + 3 + (-1)$$

However, when an expression contains more than two terms <u>and one or more operations is a subtraction</u>, we cannot write the terms in any order without changing its value. For example:

$$5 + (-1) - 7 \neq 7 + (-1) - 5$$

a) Does $(-4) + 8 + (-9) = (-9) + (-4) + 8$? ________

b) Does $12 - (-2) - 10 = 10 - (-2) - 12$? ________

a) Yes b) No

SELF-TEST 3 (pages 21-27)

Complete:

1. $4 - 7 =$ ______
2. $8 - (-2) =$ ______
3. $6 - (-6) =$ ______
4. $(-1) - 8 =$ ______
5. $(-10) - (-4) =$ ______
6. $(-2) - (-2) =$ ______

Complete:

7. $(-18) + 5 - (-7) =$ ______
8. $(-11) - 2 + 19 - (-4) =$ ______

True or False:

9. ______ $4 - 6 = 6 - 4$
10. ______ $(-3) + 5 + (-1) = (-1) + (-3) + 5$
11. ______ $10 - 7 - (-2) = 7 - (-2) - 10$

ANSWERS:

1. -3	4. -9	7. -6	9. False
2. 10	5. -6	8. 10	10. True
3. 12	6. 0		11. False

1-18 MULTIPLYING TWO POSITIVE FACTORS

In this section, we will discuss the basic meaning of multiplication and the procedure for multiplying two positive numbers.

---

125. There are various symbols that can be used to indicate a multiplication. For example, all six expressions below mean: <u>multiply 3 and 7</u>.

3 times 7	3(7)
3 x 7	(3)7
3 · 7	(3)(7)

a) 5 · 6 means multiply ______ and ______

b) (2)(9) means multiply ______ and ______

126. In a multiplication, the two numbers multiplied are called <u>factors</u>; the answer is called the <u>product</u>. For example:

In (6)(9) = 54 : the two <u>factors</u> are 6 and 9.
the <u>product</u> is 54.

a) In 5(3) = 15 , the product is ______.

b) In (7)(2) = 14 , the two <u>factors</u> are ______ and ______.

> a) 5 and 6
> b) 2 and 9

---

127. Any multiplication stands for an addition of identical terms. For example:

(4)(2) = 2 + 2 + 2 + 2   3(7) = 7 + 7 + 7

Notice that <u>the first factor tells how many times the second factor should be used as a term</u>. That is:

In (4)(2) , the "4" says that "2" should be used as a term <u>four</u> times.

In 3(7) , the "3" says that "7" should be used as a term ____________ times.

> a) 15
> b) 7 and 2

---

128. Write each of the following multiplications as an addition.

a) 2 · 5 = ____________________   b) 5(8) = ____________________

> three

---

129. When <u>both factors are positive</u>, each term in the corresponding addition is positive. Therefore, <u>the product is positive</u>. For example:

(3)(6) or (+3)(+6) = (+6) + (+6) + (+6) = +18

(5)(4) or (+5)(+4) = (+4) + (+4) + (+4) + (+4) + (+4) = +20

Using the fact that the product is positive when both factors are positive, complete these:

a) 7 · 8 = ________   b) 2(40) = ________   c) (6)(12) = ________

> a) 5 + 5
> b) 8 + 8 + 8 + 8 + 8

---

130. When multiplying two signed numbers, they can be written in either order without changing the product. For example:

4(3) = (+3) + (+3) + (+3) + (+3) = +12

3(4) = (+4) + (+4) + (+4) = +12

Using the "order" property of multiplication, write the correct factor in each blank.

a) 9(5) = 5( )   b) 6(71) = 71( )   c) (21)(47) = ( )(21)

> a) +56
> b) +80
> c) +72

> a) 5(9)   b) 71(6)   c) (47)(21)

1-19 MULTIPLYING ONE POSITIVE AND ONE NEGATIVE FACTOR

In this section, we will discuss the procedure for multiplying one positive and one negative factor.

131. When the first factor is positive and the second factor is negative, the multiplication stands for an addition of identical negative terms. That is:

Just as 4(6) means: (+6) + (+6) + (+6) + (+6)

4(-6) means: (-6) + (-6) + (-6) + (-6)

Write each of the multiplications below as an addition.

a) 2(-7) = ______ b) 5(-3) = ______

---

a) (-7) + (-7)

b) (-3) + (-3) + (-3) + (-3) + (-3)

132. When the first factor is positive and the second factor is negative, each term in the corresponding addition is negative. Therefore, the product is negative. For example:

3(-5) = (-5) + (-5) + (-5) = -15

4(-8) = (-8) + (-8) + (-8) + (-8) = -32

Using the fact that the product is negative when the first factor is positive and the second negative, complete these:

a) 2(-4) = ______ b) 7(-10) = ______ c) 5(-21) = ______

---

a) -8

b) -70

c) -105

133. When the first factor is negative and the second factor is positive, we can use the order property to get an equivalent multiplication in which the second factor is negative.

(-3)(7) = (7)(-3) (-9)(4) = (4)(-9)

Therefore, the product for any multiplication of that type is negative. That is:

(-3)(7) = (7)(-3) = -21 (-9)(4) = (4)(-9) = ______

-36

134. Write the product for each of these:

a) (-5)(2) = (2)(-5) = ______ b) (-10)(6) = (6)(-10) = ______

---

a) -10

b) -60

135. Whenever one factor is positive and one is negative, the product is negative. Therefore:

a) (-3)(3) = ______ b) (-4)(31) = ______ c) 20(-7) = ______

---

a) -9 b) -124 c) -140

1-20 MULTIPLICATIONS INVOLVING "1", "0", AND "-1"

In this section, we will discuss multiplications in which one factor is either "1", "0", or "-1".

---

136. When one of the factors is "1", the product is identical to the other factor. That is:

(1)(5) = 5 (1)(-3) = -3 a) (7)(1) = ______ b) (-9)(1) = ______

---

a) 7 b) -9

137. When one of the factors is "0", <u>the product is always "0"</u>. That is:

(7)(0) = 0 (-2)(0) = 0 a) (0)(8) = ____ b) (0)(-3) = ____

Answers: a) 0 b) 0

138. When one factor is "-1" and the other is positive, the product is <u>the opposite</u> of the other factor. That is:

6(-1) = -6 (-1)(10) = -10 a) 4(-1) = ____ b) (-1)(9) = ____

Answers: a) -4 b) -9

139. Complete: a) (1)(-1) = ____ b) (0)(1) = ____ c) (-1)(0) = ____

Answers: a) -1 b) 0 c) 0

1-21 MULTIPLYING TWO NEGATIVE FACTORS

In this section, we will discuss the procedure for multiplying two negative factors. The <u>rules for multiplying</u> signed numbers are also contrasted with the <u>rules for adding</u> signed numbers.

---

140. In the table at the right, the second factor is always (-5). Notice that each product increases by 5 as you read down the table. Complete the table by following that pattern.

(4)(-5) = -20
(3)(-5) = -15
(2)(-5) = -10
(1)(-5) = -5
(0)(-5) = 0
(-1)(-5) = ____
(-2)(-5) = ____
(-3)(-5) = ____

Answers: (-1)(-5) = <u>+5</u> (-2)(-5) = <u>+10</u> (-3)(-5) = <u>+15</u>

141. In the table at the right, the second factor is always (-2). Notice that each product increases by 2 as you read down the table. Complete the table by following that pattern.

(4)(-2) = -8
(3)(-2) = -6
(2)(-2) = -4
(1)(-2) = -2
(0)(-2) = 0
(-1)(-2) = ____
(-2)(-2) = ____
(-3)(-2) = ____

Answers: (-1)(-2) = +2 (-2)(-2) = +4 (-3)(-2) = +6

142. In the last two frames, we saw this fact: <u>When both factors are negative, the sign of the product is positive</u>.

Therefore: a) (-4)(-7) = ____ b) (-6)(-5) = ____

Answers: a) +28 b) +30

143. The three rules for multiplying signed numbers are listed below:

> (1) IF BOTH FACTORS ARE POSITIVE, THE PRODUCT IS POSITIVE.
>
> (2) IF BOTH FACTORS ARE NEGATIVE, THE PRODUCT IS POSITIVE.
>
> (3) IF ONE FACTOR IS POSITIVE AND ONE FACTOR IS NEGATIVE, THE PRODUCT IS NEGATIVE.

Using the rules above, write each product below.

a) (-3)(7) = ____ b) (2)(8) = ____ c) (-9)(-4) = ____ d) (2)(-7) = ____

---

a) -21

b) +16

c) +36

d) -14

144. Do not confuse the rules for adding signed numbers with the rules for multiplying signed numbers. For example:

When two negative numbers are ADDED, the sum is NEGATIVE.

When two negative numbers are MULTIPLIED, the product is POSITIVE.

Therefore: a) (-9) + (-3) = ____ b) (-9)(-3) = ____

---

a) -12

b) +27

145. Two more rules that should not be confused are given below.

When one positive and one negative number are ADDED, the sum can be either POSITIVE or NEGATIVE.

When one positive and one negative number are MULTIPLIED, the product is always NEGATIVE.

Be careful to get the correct signs for these:

a) 10 + (-5) = ____ c) (-10)(5) = ____

b) (-10) + 5 = ____ d) (10)(-5) = ____

---

a) +5 c) -50

b) -5 d) -50

146. Be careful to get the correct signs for these:

a) (-5) + (-20) = ____ c) (-5) + 20 = ____

b) (-5)(-20) = ____ d) (-5)(20) = ____

---

a) -25 b) +100 c) +15 d) -100

1-22 MULTIPLICATIONS WITH MORE THAN TWO FACTORS

In this section, we will discuss multiplications that contain three or more factors.

---

147. When a multiplication contains three factors, we can begin at the left and multiply two factors at a time. An example is shown at the left below. Complete the other multiplication.

(2)(-4)(5) → (-8) (5) = -40

(-4)(-5)(3) → ( ) (3) = ____

148.	When a multiplication contains four factors, we can also multiply "two factors at a time" from left to right. An example is shown at the left below. Complete the other multiplication. (-4)(3)(-1)(2) → (-12) (-1)(2) → (+12) (2) = 24 (-1)(5)(-3)(-2) → (-5) (-3)(-2) → ( ) (-2) = ______	(+20)(3) = 60
149.	Using the same method, complete each multiplication below: a) (-3)(2)(-7) b) (2)(-1)(-3)(-4)	(+15)(-2) = -30
150.	When one of the factors is "0", the product is always "0". For example: a) (6)(-2)(0) → (-12) (0) = ______ b) (-5)(0)(9)(-3) → (0) (9)(-3) → (0) (-3) = ______	a) +42 b) -24
151.	Always look for a "0" factor since you then know immediately that the product is "0". In which cases below is the product "0"? ____________ a) (-7)(-1)(0)(4)(-6) b) (85)(-15)(20)(-5) c) (125,000)(-375)(57,666)(0)(-1,789)	a) 0 b) 0
152.	The order property also applies to multiplications with more than two factors. For example, the same three factors are written in different orders below. In each case, the product is ______. (2)(-3)(4) (-3)(4)(2) (4)(2)(-3)	Both (a) and (c)
153.	Using the order property of multiplication, write the correct factor in each blank. a) (-1)(7)(5) = (5)( )(7) b) (-12)(15)(-11) = (-11)(15)( )	-24
154.	Do these: a) (-1)(-6) = ______ b) (-7)(2) = ______ c) (-5)(-1)(-2) = ______ d) (8)(0)(-7)(-3) = ______	a) (5)(-1)(7) b) (-11)(15)(-12)
155.	Do these: a) 5(-9) = ______ b) (-4)(-6) = ______ c) (-2)(-3)(10) = ______ d) (5)(-4)(-9)(0) = ______	a) +6 c) -10 b) -14 d) 0
		a) -45 c) +60 b) +24 d) 0

1-23 MULTIPLICATION WITH ADDITION OR SUBTRACTION

Some mathematical expressions involve both multiplication and either an addition or a subtraction. We will evaluate some expressions of that type in this section.

---

156. In expressions involving only addition and subtraction, each signed number is called a "term". For example, we have drawn a box around each term in each expression below.

[(-6)] - [5]   [10] + [(-4)] + [2]   [6] - [9] + [3] - [(-1)]

<u>Any multiplication is also one term</u>. That is:

In (5)(-4) + 7 , the two terms are (5)(-4) and 7 .

In 6 - (2)(-1) , the two terms are 6 and (2)(-1) .

In (8)(3) - (-2)(4) , the two terms are ______ and ______.

> (8)(3) and (-2)(4)

---

157. As you can see, any mathematical expression is made up of terms that are separated by either an addition symbol or a subtraction symbol. Draw a box around each term in the expressions below.

a) 7(-2) - 9   b) 10 + (-1)(-6)   c) (-3)(4) - (-9)(-7)

> Your answer should look like this:
>
> a) [7(-2)] - [9]
>
> b) [10] + [(-1)(-6)]
>
> c) [(-3)(4)] - [(-9)(-7)]

---

158. To evaluate an expression in which one term is a multiplication, <u>we perform the multiplication first</u>. An example is shown at the left below. Complete the other evaluation.

4(-2) + 5
↓
(-8) + 5 = -3

(-6)(3) + 10
↓
______ + 10 = ______

> (-18) + 10 = -8

---

159. Following the example at the left below, complete the other evaluation.

15 + (-2)(-5)
↓
15 + 10 = 25

(-6) + (-3)(4)
↓
(-6) + ______ = ______

> (-6) + (-12) = -18

---

160. Using the same steps, evaluate each of these.

a) (-3)(4) + 10 = ______   b) (-100) + (-5)(-10) = ______

> a) -2
>
> b) -50

---

161. In the expression at the left below, the two terms are connected by a subtraction symbol. <u>Notice that we performed the multiplication first and then converted the subtraction to addition</u>. Use the same steps to complete the other evaluation.

(4)(10) - (-8)
↓
40 - (-8) = 40 + (+8) = 48

(5)(-4) - 7
↓
______ - 7 = ______ + ______ = ______

162. To evaluate the expression at the left below, we multiplied first and then converted the subtraction to addition. Complete the other evaluation.

$$10 - \underbrace{(6)(-2)}_{\downarrow}$$
$$10 - (-12) = 10 + (+12) = 22$$

$$(-20) - \underbrace{(2)(5)}_{\downarrow}$$
$$(-20) - \_\_\_\_ = \_\_\_\_ + \_\_\_\_ = \_\_\_\_$$

Answer: $(-20) - 7 = (-20) + (-7) = -27$

---

163. Using the same steps, evaluate each of these.

a) $(8)(5) - 25 = \_\_\_\_$ b) $(-30) - (-3)(-10) = \_\_\_\_$

Answer: $(-20) - 10 = (-20) + (-10) = -30$

---

164. In the expression at the left below, both terms are multiplications. Notice that we performed both multiplications before adding. Complete the other evaluation.

Answer: a) 15 b) -60

$$\underbrace{(-5)(6)}_{\downarrow} + \underbrace{(-2)(-10)}_{\downarrow}$$
$$(-30) + 20 = -10$$

$$\underbrace{(-8)(-5)}_{\downarrow} + \underbrace{(3)(-10)}_{\downarrow}$$
$$\_\_\_\_ + \_\_\_\_ = \_\_\_\_$$

---

165. After performing both multiplications at the left below, we converted the subtraction to addition. Complete the other evaluation.

Answer: $40 + (-30) = 10$

$$\underbrace{(-4)(2)}_{\downarrow} - \underbrace{(-5)(-1)}_{\downarrow}$$
$$(-8) - 5 = (-8) + (-5) = -13$$

$$\underbrace{(3)(-7)}_{\downarrow} - \underbrace{(5)(2)}_{\downarrow}$$
$$\_\_\_\_ - \_\_\_\_ = \_\_\_\_ + \_\_\_\_ = \_\_\_\_$$

---

166. Complete each evaluation below.

Answer: $(-21) - 10 = (-21) + (-10) = -31$

a) $(6)(11) + (-7)(9) = \_\_\_\_$ b) $(-4)(-10) - (-1)(-60) = \_\_\_\_$

Answer: a) 3 b) -20

1-24 DIVIDING SIGNED NUMBERS

In this section, we will state and use the rules for dividing signed numbers. The rules are then justified in the next section.

---

167. In algebra, a division is always written as a fraction. That is:

Instead of $8 \div 4$, we write $\frac{8}{4}$. Instead of $35 \div 7$, we write $\frac{35}{7}$.

Write each division below as a fraction.

a) $50 \div 10 = \_\_\_\_$ b) $30 \div 2 = \_\_\_\_$ c) $45 \div 5 = \_\_\_\_$

168. There is a special name for each number in a fraction. That is:

In $\frac{15}{5}$: "15" is called the numerator. "5" is called the denominator.

a) In $\frac{28}{7}$, the numerator is ______. b) In $\frac{54}{6}$, the denominator is ______.

Answers: a) $\frac{50}{10}$ b) $\frac{30}{2}$ c) $\frac{45}{5}$

---

169. A fraction stands for a division in which the numerator is divided by the denominator. That is:

$\frac{10}{2} = 10 \div 2$ $\frac{32}{8} = 32 \div 8$ $\frac{49}{7} =$ ______ $\div$ ______

Answers: a) 28 b) 6

---

170. When a division is written as a fraction, the answer is still called the "quotient". That is:

In $\frac{30}{10} = 3$, "3" is called the quotient.   In $\frac{42}{3} = 14$, ______ is called the quotient.

Answer: $49 \div 7$

---

171. The rules for dividing signed numbers are similar to the rules for multiplying signed numbers. The three division rules are stated below:

> (1) IF BOTH NUMBERS ARE POSITIVE, THE QUOTIENT IS POSITIVE.
>
> For example: $\frac{30}{5} = +6$ $\frac{27}{9} = +3$
>
> (2) IF BOTH NUMBERS ARE NEGATIVE, THE QUOTIENT IS POSITIVE.
>
> For example: $\frac{-10}{-2} = +5$ $\frac{-35}{-5} = +7$
>
> (3) IF ONE NUMBER IS POSITIVE AND ONE NUMBER IS NEGATIVE, THE QUOTIENT IS NEGATIVE.
>
> For example: $\frac{-24}{6} = -4$ $\frac{32}{-4} = -8$

Using the rules above, write each quotient below.

a) $\frac{60}{3} =$ ______   b) $\frac{-20}{10} =$ ______   c) $\frac{-12}{-4} =$ ______   d) $\frac{40}{-10} =$ ______

Answer: 14

---

172. Complete each division below.

a) $\frac{-20}{5} =$ ______   b) $\frac{-12}{-2} =$ ______   c) $\frac{-80}{-10} =$ ______   d) $\frac{24}{-12} =$ ______

Answers: a) +20 b) -2 c) +3 d) -4

---

173. Complete each division.

a) $\frac{-8}{-4} =$ ______   b) $\frac{-7}{7} =$ ______   c) $\frac{40}{-5} =$ ______   d) $\frac{-60}{-20} =$ ______

Answers: a) -4 b) +6 c) +8 d) -2

Answers: a) +2 b) -1 c) -8 d) +3

174. When a number is divided by itself, the quotient is "+1". That is:

$\frac{5}{5} = +1$ $\frac{-9}{-9} = +1$ a) $\frac{10}{10} =$ ______ b) $\frac{-25}{-25} =$ ______

a) +1
b) +1

175. When a number is divided by "+1", the quotient is <u>identical to the numerator</u>. That is:

$\frac{4}{1} = 4$ $\frac{-7}{1} = -7$ a) $\frac{13}{1} =$ ______ b) $\frac{-22}{1} =$ ______

a) 13
b) -22

176. When a number is divided by "-1", the quotient is <u>the opposite of the numerator</u>. That is:

$\frac{3}{-1} = -3$ $\frac{-8}{-1} = +8$ a) $\frac{9}{-1} =$ ______ b) $\frac{-12}{-1} =$ ______

a) -9
b) +12

177. When "0" is divided by a signed number, the quotient is "0". That is:

$\frac{0}{5} = 0$ $\frac{0}{-6} = 0$ a) $\frac{0}{10} =$ ______ b) $\frac{0}{-12} =$ ______

a) 0
b) 0

178. Complete: a) $\frac{1}{1} =$ ______ b) $\frac{-1}{-1} =$ ______ c) $\frac{1}{-1} =$ ______ d) $\frac{0}{-1} =$ ______

a) +1 b) +1 c) -1 d) 0

1-25 JUSTIFYING THE PROCEDURE FOR DIVIDING SIGNED NUMBERS

In this section, we will justify the rules for dividing signed numbers. To do so, we will use the relationship between division and multiplication.

179. When a division is performed correctly in arithmetic, <u>the product of the denominator and quotient equals the numerator</u>. For example:

$$\frac{8}{4} = 2 \text{ and } 4(2) = 8$$

$$\frac{30}{5} = 6 \text{ and } 5(6) = 30$$

If the rules we are using to divide signed numbers make sense, the same fact should be true. That is, the product of the denominator and quotient should equal the numerator.

Let's check the above fact for $\frac{-10}{-2} = +5$.

a) Does (-2)(+5) equal -10 ? ______

b) Therefore, does the rule make sense? ______

a) Yes
b) Yes

180. Let's check the same fact for $\frac{-16}{8} = -2$.

a) Does $8(-2)$ equal -16 ? ________

b) Therefore, does the rule make sense? ________

---

a) Yes

b) Yes

181. Let's check the same fact for $\frac{35}{-5} = -7$.

a) Does $(-5)(-7)$ equal 35 ? ________

b) Therefore, does the rule make sense? ________

---

a) Yes b) Yes

SELF-TEST 4 (pages 27-37)

Complete: 1. $(4)(-7) =$ ______ 2. $(-6)(-3) =$ ______ 3. $(-5)(0) =$ ______

Complete: 4. $(-3) + 9 =$ ______ 6. $(-2) + (-6) =$ ______

5. $(-3)(9) =$ ______ 7. $(-2)(-6) =$ ______

Complete: 8. $(-1)(5)(-4) =$ ______ 9. $(-3)(2)(0)(-9) =$ ______

Evaluate: 10. $(-9) + 5(-4) =$ ______ 12. $(-2)(-3) + (-10)(-4) =$ ______

11. $(-10)(-3) - 5 =$ ______ 13. $(-1)(7) - (-2)(-8) =$ ______

Complete: 14. $\frac{-24}{6} =$ ______ 15. $\frac{42}{-7} =$ ______ 16. $\frac{-14}{-7} =$ ______ 17. $\frac{0}{-3} =$ ______

ANSWERS:

1. -28	4. 6	8. 20	10. -29	14. -4
2. 18	5. -27	9. 0	11. 25	15. -6
3. 0	6. -8		12. 46	16. 2
	7. 12		13. -23	17. 0

Unit 2 SOLVING EQUATIONS

In this unit, we will discuss simple equations and the algebraic principles that are used to solve them. Both the multiplication axiom and the addition axiom are introduced. The equations solved are limited to non-fractional equations with whole-number roots.

2-1 AN INTRODUCTION TO EQUATIONS

In this section, we will discuss what is meant by solving an equation. The meaning of the terms "satisfy", "solution", and "root" are also discussed.

---

1. Just like ordinary sentences, a mathematical sentence can be either true or false. For example:

   a) $5 + 4 = 9$ is a ______________ (true/false) sentence.

   b) $8 - 5 = 2$ is a ______________ (true/false) sentence.

---

2. A mathematical sentence with an equal sign "=" in it is called an equation. An equation may be either true or false.

   a) Is $4 + 8 = 11$ true or false? ______________

   b) Is $9 - 4 = 5$ true of false? ______________

Answers: a) true b) false

---

3. One of the numbers is missing in the equation below. A box is written in place of the missing number.

$$5 + \square = 7$$

As it stands, the equation is neither true nor false. But as soon as we write a number in the box, the equation becomes either true or false. That is:

   a) If we write a 4 in the box, we get $5 + 4 = 7$ which is ______________ (true/false).

   b) If we write a 2 in the box, we get $5 + 2 = 7$ which is ______________ (true/false).

Answers: a) false b) true

---

4. When writing in a number for a missing number in an equation, the number that makes the equation true is said to satisfy the equation.

$$\square + 2 = 8$$

What number satisfies the equation above? _____

Answers: a) false b) true

Answer: 6

5. Finding the missing number in an equation is called "solving the equation". Solve each equation below.

a) 7 + ☐ = 10 b) ☐ + 6 = 16 c) 14 - ☐ = 10

---

a) 3
b) 10
c) 4

6. A number that satisfies an equation is called the solution or root of the equation. Find the solution or root of each equation below.

a) 15 + ☐ = 20 b) 2 + ☐ = 10 c) ☐ - 4 = 6

---

a) 5
b) 8
c) 10

7. Sometimes a negative number is the solution or root of an equation.

5 + ☐ = 3

a) Is +2 the solution of the equation above? ________

b) Is -2 the root of the equation above? ________

---

a) No b) Yes

2-2 THE USE OF LETTERS IN EQUATIONS

Instead of using a box for the missing number in an equation, we ordinarily use a letter. That is:

Instead of 5 + ☐ = 8 , we write $5 + x = 8$.

Instead of ☐ - 2 = 7 , we write $y - 2 = 7$.

We will discuss the use of letters in equations in this section.

---

8. In the equation below, the letter "x" stands for a missing number.

$$7 + x = 11$$

a) If we substitute a 2 for the "x", is the equation true? ________

b) What number must we substitute for "x" to make the equation true? ______

---

a) No
b) 4

9. In the equation below, the letter "y" stands for a missing number.

$$y - 3 = 4$$

The equation above is true if we substitute 7 for the "y". Therefore, we say that the number ______ satisfies the equation.

7

10. Though mathematicians ordinarily use letters like "x", "y", and "z" in equations, any letter of the alphabet can be used. What number satisfies each of the equations below?

a) $V + 1 = 5$ b) $3 + d = 9$ c) $10 - p = 5$

---

a) 4 b) 6 c) 5

11. Finding the missing number in an equation is called "solving the equation". Solve each equation below.

a) $3 + t = 8$ b) $R + 7 = 8$ c) $m - 2 = 5$

---

12. A number that satisfies an equation is called the solution or root of the equation. Find the solution or root of each equation below.

a) $10 + x = 13$ b) $y + 1 = 9$ c) $5 - T = 1$

Answers: a) 5 b) 1 c) 7

---

13. The root of $x + 4 = 10$ is 6.

Instead of saying "the root is 6", we simply say $x = 6$.

The root of $7 - y = 6$ is 1.

Instead of saying "the root is 1", we simply say ______________.

Answers: a) 3 b) 8 c) 4

---

14. Write the root of each equation below.

a) $p + 2 = 4$ b) $4 + Q = 9$ c) $9 - x = 6$

p = ______ Q = ______ x = ______

Answer: $y = 1$

---

15. The root of an equation can be a negative number. For example:

The root of $x + (-2) = -5$ is -3.

The root of $(-1) + y = -9$ is ______.

Answers: a) $p = 2$ b) $Q = 5$ c) $x = 3$

Answer: -8

2-3 EXPRESSIONS LIKE "5x" AND "-3y"

In this section, we will discuss the meaning of expressions like "5x" and "-3y". The meaning of the word "coefficient" is also discussed.

---

16. To show a multiplication of a number and a letter, we could write parentheses around one or both factors. For example:

"5 times x" could be written: $5(x)$, $(5)x$, or $(5)(x)$

However, we ordinarily write the number and letter together without parentheses. That is:

"5 times x" is ordinarily written: $5x$

Similarly: a) Instead of $8(y)$, we ordinarily write __________.

b) Instead of $(-3)(t)$, we ordinarily write __________.

---

17. Remember that expressions like "4p" and "-9R" stand for a multiplication. That is:

"4p" means: 4 times p "-9R" means: ______________________

Answers: a) $8y$ b) $-3t$

		-9 times R

18. When $x = 3$, $7x = 7(3) = 21$.

a) When $y = 2$, $4y = 4(2) =$ ______ b) When $V = 8$, $-5V = -5(8) =$ ______

---

a) 8
b) -40

19. When $x = -4$, $9x = 9(-4) = -36$.

a) When $t = -2$, $6t = 6(-2) =$ ______ b) When $F = -9$, $-3F = -3(-9) =$ ______

---

a) -12
b) 27

20. In any multiplication of a number and a letter, the number is called the "coefficient" of the letter. For example:

In $9x$, 9 is called the coefficient of "x".

In $-3d$, ______ is called the coefficient of "d".

-3

21. When writing a multiplication of a number and a letter, we could write the letter first. However, we ordinarily write its coefficient first. That is:

Instead of "x(7)", we ordinarily write $7x$.

Instead of "y(-2)", we ordinarily write ________.

-2y

22. In each equation below, there is a multiplication of a number and a letter on the left side.

$7x = 56$ $-9y = 27$

Any multiplication of a number and a letter is only one term. Therefore, each equation above has only one term on each side.

In $7x = 56$: "7x" is called a letter-term.
"56" is called a number-term.

In $-9y = 27$: a) The letter-term is ________.

b) The number-term is ________.

---

a) -9y
b) 27

23. In the equation $6x = 30$, the left side is a multiplication. Therefore:

"6" and "x" are factors.

"30" is their product.

Since the letter "x" stands for a missing factor, solving the equation $6x = 30$ is the same as "finding the missing factor".

The missing factor is 5.

Therefore, the root of $6x = 30$ is ______.

5

24. a) Is +7 or -7 the root of $4m = 28$? ______

b) Is +3 or -3 the root of $2x = -6$? ______

---

a) +7, since: $4(+7) = 28$ b) -3, since: $2(-3) = -6$

2-4 RECIPROCALS

We ordinarily use the multiplication axiom to solve equations like $3y = 27$ and $-5t = 40$. However, before introducing the multiplication axiom, we must discuss what is meant by the reciprocals of both positive and negative whole numbers. We will discuss "reciprocals" in this section. Other types of multiplications needed for the multiplication axiom are also discussed.

---

25. Multiplying a whole number by a fraction whose numerator is "1" is the same as a division.

If both factors are positive, both numbers in the division are positive. Therefore, the product is positive. That is:

$\frac{1}{4}(8) = \frac{8}{4} = 2$ $\frac{1}{5}(35) = \frac{35}{5} = 7$

If both factors are negative, both numbers in the division are negative. Therefore, the product is positive. That is:

$-\frac{1}{3}(-6) = \frac{-6}{-3} = +2$ or 2 $-\frac{1}{7}(-56) = \frac{-56}{-7} = +8$ or 8

If one factor is positive and one factor is negative, the numbers in the division have different signs. Therefore, the product is negative. That is:

$\frac{1}{6}(-42) = \frac{-42}{6} = -7$ $-\frac{1}{8}(40) = \frac{40}{-8} = -5$

Using the facts above, complete each multiplication below.

a) $\frac{1}{9}(27) =$ ____ b) $-\frac{1}{4}(-20) =$ ____ c) $\frac{1}{8}(-32) =$ ____ d) $-\frac{1}{2}(10) =$ ____

---

a) +3
b) +5
c) -4
d) -5

26. The following rules can be used when multiplying two signed quantities.

1) If the two factors have like signs, the product is positive.
2) If the two factors have unlike signs, the product is negative.

Use the rules above to complete each multiplication below.

a) $\frac{1}{7}(-28) =$ ____ b) $-\frac{1}{5}(-15) =$ ____ c) $\left(-\frac{1}{2}\right)(14) =$ ____ d) $-\frac{1}{9}(-45) =$ ____

---

a) -4
b) +3
c) -7
d) +5

27. When the whole number and the denominator of the fraction are equal, the product is either "+1" or "-1". For example:

$\frac{1}{3}(3) = \frac{3}{3} = +1$ or 1 $-\frac{1}{7}(7) = \frac{7}{-7} = -1$

Write the product for each multiplication below.

a) $\frac{1}{5}(5) =$ ____ b) $\frac{1}{8}(-8) =$ ____ c) $-\frac{1}{2}(2) =$ ____ d) $-\frac{1}{6}(-6) =$ ____

---

a) +1 b) -1 c) -1 d) +1

28. Write the product for each multiplication below.

a) $\frac{1}{10}(-10) =$ ____ b) $-\frac{1}{2}(24) =$ ____ c) $\frac{1}{5}(-50) =$ ____ d) $-\frac{1}{12}(-12) =$ ____

29. When the product of two numbers is "+1", the two numbers are called a "pair of reciprocals".

| a) -1 | b) -12 | c) -10 | d) +1 |

Since $\frac{1}{4}(4) = +1$, $\frac{1}{4}$ and 4 are called a "pair of reciprocals".

We say: The reciprocal of $\frac{1}{4}$ is 4. The reciprocal of 4 is $\frac{1}{4}$.

Since $-\frac{1}{7}(-7) = +1$, $-\frac{1}{7}$ and -7 are called a "pair of reciprocals".

We say: a) The reciprocal of $-\frac{1}{7}$ is ______. b) The reciprocal of -7 is ______.

---

a) -7 b) $-\frac{1}{7}$

30. The reciprocal of any positive whole number is a positive fraction whose numerator is "1" and whose denominator is the whole number. For example:

The reciprocal of 6 is $\frac{1}{6}$. The reciprocal of 15 is $\frac{1}{15}$.

Write the reciprocal of each whole number. a) 2 ______ b) 8 ______ c) 20 ______

---

a) $\frac{1}{2}$ b) $\frac{1}{8}$ c) $\frac{1}{20}$

31. The reciprocal of any negative whole number is a negative fraction whose numerator is "1" and whose denominator is the absolute value of the whole number. That is:

The reciprocal of -7 is $-\frac{1}{7}$. The reciprocal of -12 is $-\frac{1}{12}$.

Write the reciprocal of each whole number. a) -2 ______ b) -8 ______ c) -15 ______

---

a) $-\frac{1}{2}$ b) $-\frac{1}{8}$ c) $-\frac{1}{15}$

32. The reciprocal of any positive whole number is a positive fraction. The reciprocal of any negative whole number is a negative fraction. Write the reciprocals of each whole number below.

a) 4 ______ b) -6 ______ c) 10 ______ d) -20 ______

---

a) $\frac{1}{4}$ b) $-\frac{1}{6}$ c) $\frac{1}{10}$ d) $-\frac{1}{20}$

2-5 THE MULTIPLICATION AXIOM FOR EQUATIONS

In this section, we will show how "the multiplication axiom for equations" is used to solve equations like $3x = 15$ and $-5y = 45$. A shortcut for the multiplication axiom is also shown.

---

33. The MULTIPLICATION AXIOM FOR EQUATIONS says this:

> If we multiply both sides of an equation that is true by the same quantity, the new equation is also true.

(Continued on following page.)

33. (Continued)

An example of the multiplication axiom for equations is shown below.

The equation below is true because both sides equal 12.

$$(3)(4) = 12$$

We multiplied both sides of the equation by 2 below.

$$2(3)(4) = 2(12)$$

The new equation is true because both sides equal ________.

24

34. When multiplying both sides of an equation by the same quantity, that quantity can be a fraction. An example is shown below.

The equation below is true because both sides equal 48.

$$(8)(6) = 48$$

We multiplied both sides of the equation by $\frac{1}{4}$ below.

$$\frac{1}{4}(8)(6) = \frac{1}{4}(48)$$

The new equation is true because both sides equal ________.

12

35. To solve $5x = 20$, we must isolate the "x". That is, we must get the "x" alone on one side of the equation. To do so, we can use the multiplication axiom to "get rid of" the coefficient "5" by multiplying both sides by $\frac{1}{5}$, the reciprocal of 5. We get:

$$5x = 20$$

$$\underbrace{\frac{1}{5}(5x)} = \underbrace{\frac{1}{5}(20)}$$

$$\downarrow \qquad \downarrow$$

$$1\,x = 4$$

$$x = 4$$

Is "4" the root of the original equation $5x = 20$? ________

Yes, since: $5(4) = 20$

36. When using the multiplication axiom to solve an equation like $5x = 20$, we isolate the letter BY MULTIPLYING BOTH SIDES BY THE RECIPROCAL OF THE COEFFICIENT OF THE LETTER. Multiplying by $\frac{1}{5}$ in the last frame worked because:

1) The product of a pair of reciprocals is "1". That is:

$$\frac{1}{5}(5) = 1$$

2) When any quantity is multiplied by "1", the product equals the original quantity. That is:

Just as $1(8) = 8$, $1x = x$

(Continued on following page.)

36. (Continued)

To isolate the "y" and solve the equation below, we must "get rid of" the -3 on the left. To do so, we multiply both sides by $-\frac{1}{3}$, the reciprocal of -3. We get:

$$-3y = 15$$

$$\underbrace{-\frac{1}{3}(-3}_{1}y) = \underbrace{-\frac{1}{3}(15)}_{-5}$$

$$1\ y = -5$$

$$y = -5$$

Is -5 the root of the original equation $-3y = 15$? __________

> Yes, since:
> $-3(-5) = 15$

---

37. To solve the equation at the left below, we "got rid of" the "2" by multiplying both sides by $\frac{1}{2}$. Using the same steps, solve the other equation by multiplying both sides by $-\frac{1}{10}$ to get rid of the "-10".

$$-16 = 2t \qquad\qquad\qquad -30 = -10V$$

$$\underbrace{\frac{1}{2}(-16)}_{-8} = \underbrace{\frac{1}{2}(2}_{1}t)$$

$$-8 = 1\ t$$

$$t = -8$$

> $$\underbrace{-\frac{1}{10}(-30)}_{3} = \underbrace{-\frac{1}{10}(-10}_{1}V)$$
> $$3 = 1\ V$$
> $$V = 3$$

---

38. It should be obvious that using the multiplication axiom is the same as <u>dividing the whole-number term by the coefficient of the letter</u>. Therefore, we can use a shortcut to solve equations like $4x = 32$ and $-5y = 30$. That is:

For $4x = 32$, $x = \frac{32}{4} = 8$ For $-5y = 30$, $y = \frac{30}{-5} = -6$

Use the shortcut to solve each equation below.

a) $9m = 45$ b) $-3p = 24$ c) $-6x = -18$

m = ______ p = ______ x = ______

> a) $m = 5$ $\left(\text{from } \frac{45}{9}\right)$
> b) $p = -8$ $\left(\text{from } \frac{24}{-3}\right)$
> c) $x = 3$ $\left(\text{from } \frac{-18}{-6}\right)$

---

39. Use the shortcut to solve each of these:

a) $7R = -21$ b) $-2y = 20$ c) $-10F = -40$

R = ______ y = ______ F = ______

> a) $R = -3$ $\left(\text{from } \frac{-21}{7}\right)$ b) $y = -10$ $\left(\text{from } \frac{20}{-2}\right)$ c) $F = 4$ $\left(\text{from } \frac{-40}{-10}\right)$

40. When the letter term is on the right side, <u>be sure to divide the whole-number term by the coefficient of the letter</u>. That is:

For $36 = 9x$, $x = \frac{36}{9} = 4$ For $-15 = 5t$, $t = \frac{-15}{5} = -3$

Use the shortcut to solve each of these.

a) $54 = 6y$ b) $-18 = 2d$ c) $24 = -8G$

$y =$ ______ $d =$ ______ $G =$ ______

---

a) $y = 9$ $\left(\text{from } \frac{54}{6}\right)$

b) $d = -9$ $\left(\text{from } \frac{-18}{2}\right)$

c) $G = -3$ $\left(\text{from } \frac{24}{-8}\right)$

41. We can switch the two sides of an equation without changing its root. For example:

The root of both $2x = 10$ and $10 = 2x$ is 5.

The root of both $-3y = 12$ and $12 = -3y$ is -4.

Therefore, if you find it difficult to solve an equation when the letter term is on the right side, you can simply switch the two sides. That is:

Instead of solving $-20 = 5t$, you can solve $5t = -20$.

Instead of solving $48 = -8x$, you can solve ______________.

$-8x = 48$

42. Solve each equation below.

a) $30 = -10x$ b) $-30 = 3t$ c) $-15 = -5R$

$x =$ ______ $t =$ ______ $R =$ ______

---

a) $x = -3$ $\left(\text{from } \frac{30}{-10}\right)$

b) $t = -10$ $\left(\text{from } \frac{-30}{3}\right)$

c) $R = 3$ $\left(\text{from } \frac{-15}{-5}\right)$

43. Solve each equation below.

a) $4x = 4$ b) $-7y = 7$ c) $-2 = 2d$ d) $-5 = -5P$

$x =$ ______ $y =$ ______ $d =$ ______ $P =$ ______

---

a) $x = 1$

b) $y = -1$

c) $d = -1$

d) $P = 1$

44. Remember, we can check a root by plugging it into the original equation to see if it produces a true statement.

a) Is -1 the root of the equation below? ________

$9x = -9$

b) Is -2 the root of the equation below? ________

$-14 = -7x$

---

a) Yes, since: $9(-1) = -9$

b) No, since: $-14 \neq -7(-2)$

45. When the number-term is "0", the root of the equation is "0". That is:

For $5x = 0$, $x = \frac{0}{5} = 0$ For $-9y = 0$, $y = \frac{0}{-9} = 0$

(Continued on following page.)

45. (Continued)

To show that "0" satisfies $5x = 0$, we checked it at the left below. Check "0" as the root of $-9y = 0$ at the right below.

$5x = 0$	$-9y = 0$
$5(0) = 0$	
$0 = 0$	

$-9(0) = 0$
$0 = 0$

46. Solve each equation.

a) $7x = 0$ b) $7x = 7$ c) $-3m = 0$ d) $-3m = 3$

x = ______ x = ______ m = ______ m = ______

---

a) $x = 0$ b) $x = 1$ c) $m = 0$ d) $m = -1$

SELF-TEST 5 (pages 38-47)

Find the root of each equation below.

1. $9 + x = 14$
2. $y + 5 = 15$
3. $8 - p = 5$

Solve each of the following equations.

4. $4x = 40$
5. $-18 = 9y$
6. $-2t = 24$
7. $-40 = -5F$
8. $9 = -9d$
9. $5p = 0$

ANSWERS:

1. $x = 5$	3. $p = 3$	5. $y = -2$	7. $F = 8$	9. $p = 0$
2. $y = 10$	4. $x = 10$	6. $t = -12$	8. $d = -1$	

2-6 THE DISTRIBUTIVE PRINCIPLE

To perform additions like $3x + 5x$ or $7y + (-2y)$, the "distributive principle of multiplication over addition" is used. We will introduce that principle in this section by showing how to multiply by it. Then in the next section, we will show how factoring by the distributive principle is used to add letter terms.

---

47. One of the factors in a multiplication can be an addition. To show that a factor is an addition, we enclose the addition in parentheses. For example:

In $(5 + 2)6$, the two factors are $(5 + 2)$ and 6 .

In $(4 + 1)8$, the two factors are $(4 + 1)$ and 8 .

To perform multiplications like those above, we can add before multiplying as we have done at the left below. Use the same method for the other multiplication.

$(5 + 2)6$ → $(7)\ 6 = 42$

$(4 + 1)8$ → $(\quad)8 =$ ______

Answer: $(5)8 = 40$

48. When the addition-factor contains one or two negative numbers, we usually enclose that factor in brackets [ ] to avoid a confusing series of parentheses. For example:

Instead of $((-2) + 5)9$, we write $[(-2) + 5]9$.

Instead of $((-7) + (-1))3$, we write $[(-7) + (-1)]3$.

To perform the multiplications above, we can also add before multiplying as we have done at the left below. Use the same method for the other multiplication.

$[(-2) + 5]9$ → $(3)\ 9 = 27$

$[(-7) + (-1)]3$ → $(\quad)\ 3 =$ ______

Answer: $(-8)3 = -24$

49. The two multiplications below were performed by the "adding before multiplying" method. The products are 20 and -10.

$(2 + 3)4$ → $(5)\ 4 = 20$

$[4 + (-6)]5$ → $(-2)\ 5 = -10$

We can perform the same multiplications by "multiplying before adding". The steps are:

$(2 + 3)4 = 2(4) + 3(4)$
$= 8 + 12 = 20$

$[4 + (-6)]5 = 4(5) + (-6)(5)$
$= 20 + (-30) = -10$

Did we get the same results using the second method? ______

Answer: Yes

50. In the "multiplying before adding" method, the first step uses the "DISTRIBUTIVE PRINCIPLE OF MULTIPLICATION OVER ADDITION". We diagrammed that principle below. The arrows show that the "7" is multiplied by both "8" and "5".

$$(8 + 5)7 = 8(7) + 5(7)$$

Following the pattern above, fill in the blanks below.

a) $(6 + 2)4 = 6(\quad) + 2(\quad)$ b) $(9 + 1)3 = \_\_\_(3) + \_\_\_(3)$

---

a) $6(4) + 2(4)$

b) $9(3) + 1(3)$

51. Another diagram of the "distributive principle" is shown below. Notice that the "9" is multiplied by both "4" and "-7".

$$[4 + (-7)]9 = 4(9) + (-7)(9)$$

Following the pattern above, fill in the blanks below.

a) $[(-2) + 8]5 = (-2)(\quad) + 8(\quad)$ b) $[(-7) + (-1)]6 = (\quad)(6) + (\quad)(6)$

---

a) $(-2)(5) + 8(5)$

b) $(-7)(6) + (-1)(6)$

52. The process we have been discussing is called "multiplying by the distributive principle". We can "multiply" by the distributive principle when a letter is involved. For example:

$$(7 + 4)x = 7(x) + 4(x) = 7x + 4x$$

$$(3 + 9)y = 3(y) + 9(y) = \_\_\_\_ + \_\_\_\_$$

$3y + 9y$

53. Following the example, write the product for each multiplication.

$$[5 + (-2)]x = 5(x) + (-2)(x) = 5x + (-2x)$$

a) $[(-3) + 8]t = (-3)(t) + 8(t) = \_\_\_\_ + \_\_\_\_$

b) $[(-7) + (-4)]m = (-7)(m) + (-4)(m) = \_\_\_\_ + \_\_\_\_$

---

a) $(-3t) + 8t$ b) $(-7m) + (-4m)$

2-7 ADDING LETTER-TERMS

In this section, we will review what is meant by "factoring" and then discuss "factoring by the distributive principle". We will show how the latter type of factoring can be used to add letter-terms.

---

54. Factoring is a process in which a product is broken up into the original factors. That is:

14 can be factored into 2 and 7 .

a) 30 can be factored into 6 and ____.

b) 72 can be factored into 8 and ____.

55. Here are two multiplications by means of the distributive principle:

$$(7 + 3)4 = 7(4) + 3(4)$$

$$(2 + 9)5 = 2(5) + 9(5)$$

By reversing the procedure, we can break up each product into the original factors. We get:

$$7(4) + 3(4) = (7 + 3)4$$

$$2(5) + 9(5) = (2 + 9)5$$

The process above is called "factoring by the distributive principle". That is:

7(4) + 3(4) was factored into (7 + 3) and 4 .

2(5) + 9(5) was factored into (   ) and ___.

a) 5
b) 9

---

56. Complete the following factorings by the distributive principle.

a) 7(3) + 6(3) = (7 + 6)(   )

b) 5(9) + 2(9) = (5 + 2)(   )

c) 3(2) + 8(2) = (___ + ___)(2)

d) 4(8) + 7(8) = (___ + ___)(8)

(2 + 9) and 5

---

57. Here are two more multiplications by the distributive principle.

$$(5 + 4)x = 5x + 4x$$

$$(2 + 8)t = 2t + 8t$$

By reversing the procedure, we can again break up each product into the original factors. That is:

5x + 4x can be factored to get (5 + 4)x .

2t + 8t can be factored to get ____________.

a) (7 + 6)(3)
b) (5 + 2)(9)
c) (3 + 8)(2)
d) (4 + 7)(8)

---

58. "Factoring by the distributive principle" can be used to add two letter terms. Two examples are shown below.

$$3x + 9x = (3 + 9)x = 12x$$

$$4y + 2y = (4 + 2)y = 6y$$

Following the steps above, add each pair of letter-terms below.

a) 8x + 7x = (8 + 7)x = ______

b) 6V + 4V = (6 + 4)V = ______

(2 + 8)t

---

59. Factor by the distributive principle to add each pair of letter-terms below.

a) 2t + 3t = (___ + ___)t = ______

b) 10x + 5x = (___ + ___)x = ______

a) 15x
b) 10V

---

a) (2 + 3)t = 5t
b) (10 + 5)x = 15x

60. We can factor by the distributive principle to add letter-terms with <u>negative</u> coefficients. For example:

$$7x + (-5x) = [7 + (-5)]x = 2x$$

$$(-8y) + 3y = [(-8) + 3]y = -5y$$

Using the same steps, add each pair of letter terms below.

a) $10t + (-6t) =$ ____________________

b) $(-5R) + (-4R) =$ ____________________

Answers: a) $[10 + (-6)]t = 4t$ b) $[(-5) + (-4)]R = -9R$

61. You should be able to see that <u>two letter-terms can be added by simply adding the two coefficients</u>. That is:

$$9y + 6y = 15y, \quad \text{since } 9 + 6 = 15$$

$$5x + (-2x) = 3x, \quad \text{since } 5 + (-2) = 3$$

Use the shortcut above to add each pair of letter-terms below.

a) $11m + 6m =$ ________ c) $(-4V) + 7V =$ ________

b) $6t + (-9t) =$ ________ d) $(-5x) + (-3x) =$ ________

Answers: a) 17m b) -3t c) 3V d) -8x

2-8 OPPOSITES AND SUBTRACTING LETTER-TERMS

We will begin this section by discussing the "opposites" of letter-terms. Then after showing how any subtraction of letter-terms can be performed by converting the subtraction to addition, we will show that any such subtraction can be performed by simply subtracting the coefficients.

---

62. When we add 5x and -5x, the coefficient of the sum is "0" because $5 + (-5) = 0$. That is:

$$5x + (-5x) = 0x$$

"0x" means "0 times x". And no matter what number we substitute for "x", "0 times x" = 0 .

Therefore: 0x equals what number? ______

Answer: 0

63. Two quantities are a pair of opposites if their sum is 0.

Since $5x + (-5x) = 0x$ or 0 : a) The opposite of 5x is ________.

b) The opposite of -5x is ________.

Answers: a) -5x b) 5x

64. Two letter-terms containing the same letter are a pair of opposites <u>if their coefficients are a pair of opposites</u>. That is:

$3y$ and $-3y$ are a pair of opposites, since $3 + (-3) = 0$

$7t$ and $-7t$ are a pair of opposites, since $7 + (-7) = 0$

Write the opposite of each letter-term below.

a) $10m$ ________ b) $-7S$ ________ c) $-6R$ ________ d) $14d$ ________

---

a) $-10m$ c) $6R$
b) $7S$ d) $-14d$

65. Any subtraction of letter-terms can be converted to an addition <u>by adding the opposite of the second term</u>. That is:

$5x - 8x = 5x + (\text{the opposite of } 8x) = 5x + (-8x)$

$7t - 2t = 7t + (\text{the opposite of } 2t) = 7t + (-2t)$

Convert each subtraction below to an addition.

a) $3R - 9R = 3R +$ ________ b) $10x - 5x = 10x +$ ________

---

a) $3R + \underline{(-9R)}$
b) $10x + \underline{(-5x)}$

66. Any subtraction of letter-terms can be performed by converting to addition and then performing the addition. That is:

a) $2x - 7x = 2x + (-7x) =$ ________ b) $12y - 4y = 12y + (-4y) =$ ________

---

a) $-5x$
b) $8y$

67. Complete each subtraction below by converting it to an addition.

a) $9m - 12m =$ _____ + _____ = _____ b) $15t - 4t =$ _____ + _____ = _____

---

a) $9m + (-12m) = -3m$
b) $15t + (-4t) = 11t$

68. Each subtraction below was performed by converting it to an addition.

$6t - 4t = 6t + (-4t) = 2t$

$5m - 9m = 5m + (-9m) = -4m$

The same subtractions can be performed by simply subtracting the coefficients. That is:

$6t - 4t = (6 - 4)t = 2t$ $5m - 9m = (5 - 9)m = -4m$

Perform each subtraction below by simply subtracting the coefficients.

a) $10y - 3y =$ ________ b) $8x - 5x =$ ________ c) $2R - 9R =$ ________

---

a) $7y$ b) $3x$ c) $-7R$

69. Perform each subtraction below.

a) $4x - 6x =$ ________ b) $12F - 7F =$ ________ c) $8p - 12p =$ ________

---

a) $-2x$
b) $5F$
c) $-4p$

2-9 COMBINING LETTER-TERMS TO SOLVE EQUATIONS

In this section, we will discuss equations that contain two letter-terms on the same side. We will show how they can be solved by combining the letter-terms.

---

70. In the equation below, there are two letter-terms on the left side.

$$2x + 3x = 30$$

When the same letter appears in two terms in an equation, <u>the same number must be substituted for it in both terms</u>.

a) Substitute +4 for both "x's" in the equation above.
   Is +4 the root of the equation? ________

b) Substitute +6 for both "x's" in the equation above.
   Is +6 the root of the equation? ________

---

a) No, since $8 + 12 \neq 30$

b) Yes, since $12 + 18 = 30$

71. To solve an equation with two letter-terms on one side, we:

1) combine the two letter-terms, and
2) then use the shortcut of the multiplication axiom.

An example is shown at the left below. Check "3" as the root of the original equation in the space provided.

$5y + 2y = 21$

$7y = 21$

$y = \frac{21}{7} = 3$

<u>Check</u>: $5y + 2y = 21$

$5(3) + 2(3) = 21$

$15 + 6 = 21$

$21 = 21$

72. By combining the two letter-terms and then using the multiplication axiom, solve each equation below.

a) $8t + 2t = 40$ b) $9x + (-4x) = 25$ c) $18 = (-5m) + 3m$

t = ______ x = ______ m = ______

---

a) $t = 4$ (from $10t = 40$)

b) $x = 5$ (from $5x = 25$)

c) $m = -9$ (from $18 = -2m$)

73. Using the same steps, solve each equation below.

a) $(-6x) + (-3x) = 45$ b) $-48 = (-4V) + (-2V)$

x = ______ V = ______

---

a) $x = -5$ (from $-9x = 45$)

b) $V = 8$ (from $-48 = -6V$)

74. To solve the equation at the left below, we began by performing the subtraction. Use the same steps to solve the other equation.

$10x - 7x = 36$ | $8y - 3y = -20$

$3x = 36$

$x = \frac{36}{3} = 12$ | $y =$ ______

Answer: $y = -4$ (from $5y = -20$)

75. To solve the equation at the left below, we also began by performing the subtraction. Use the same steps to solve the other equation.

$2t - 6t = 40$ | $5d - 7d = -22$

$-4t = 40$

$t = \frac{40}{-4} = -10$ | $d =$ ______

Answer: $d = 11$ (from $-2d = -22$)

76. Solve: a) $-15 = 7x - 4x$ b) $50 = 3y - 8y$

$x =$ ______ $y =$ ______

Answer: a) $x = -5$ (from $-15 = 3x$) b) $y = -10$ (from $50 = -5y$)

77. The root of the equation at the left below is "0". Solve the other two equations.

$3x + 4x = 0$ | a) $8t - 5t = 0$ | b) $2p - 8p = 0$

$7x = 0$

$x = \frac{0}{7} = 0$ | $t =$ ______ | $p =$ ______

Answer: a) $t = 0$ b) $p = 0$

2-10 SUBSTITUTING "1x" FOR "x" IN EQUATIONS

In this section, we will show that any letter without a coefficient really has a coefficient of "1". We will then use that fact as a help in solving equations.

78. Any number can be factored into "1" and itself. For example:

$3 = (1)(3)$ $8 = (1)(8)$

Similarly, any letter without a coefficient can be factored into "1" and itself. For example:

$x = (1)(x)$ $y = (1)(y)$

Therefore, any letter without a coefficient is equal to the same letter with a coefficient of "1". That is:

$x = 1x$ $y = 1y$ a) $t =$ ______ b) $F =$ ______

79. When a letter appears without a coefficient, the coefficient is really "1". Sometimes it is helpful to write the "1" explicitly. For example:

$$7x + x = 7x + 1x = 8x$$

$$y + 4y = 1y + 4y = 5y$$

Write the coefficient "1" explicitly and then complete each addition below.

a) $(-3t) + t = (-3t) +$ ______ $=$ ______　　b) $R + (-7R) =$ ______ $+ (-7R) =$ ______

Answers to 78: a) 1t　b) 1F

---

80. Write the coefficient "1" explicitly and then solve each equation.

a) $6x + x = 28$　　b) $m + 10m = -33$

$x =$ ______　　$m =$ ______

Answers to 79: a) $(-3t) + 1t = -2t$　b) $1R + (-7R) = -6R$

---

81. Solve:　a) $42 = t + (-8t)$　　b) $-20 = (-5d) + d$

$t =$ ______　　$d =$ ______

Answers to 80: a) $x = 4$ (from $7x = 28$)　b) $m = -3$ (from $11m = -33$)

---

82. When a letter without a coefficient follows a subtraction symbol, its coefficient is still "1". That is:

$$3x - x = 3x - 1x$$

The fact above can be used to complete each subtraction below.

a) $6y - y = 6y - 1y =$ ______　　b) $10F - F = 10F - 1F =$ ______

Answers to 81: a) $t = -6$ (from $42 = -7t$)　b) $d = 5$ (from $-20 = -4d$)

---

83. Write the coefficient "1" explicitly and then solve each equation.

a) $9x - x = 40$　　b) $35 = 6d - d$

$x =$ ______　　$d =$ ______

Answers to 82: a) 5y　b) 9F

---

84. Solve:　a) $4y - y = -12$　　b) $-80 = 11R - R$

$y =$ ______　　$R =$ ______

Answers to 83: a) $x = 5$ (from $8x = 40$)　b) $d = 7$ (from $35 = 5d$)

---

85. In each equation below, the coefficient of the letter is "1". Therefore, the root equals the number-term on the right. That is:

$$1x = 5 \qquad 1y = -9$$

$$x = \frac{5}{1} = 5 \qquad y = \frac{-9}{1} = -9$$

(Continued on following page.)

Answers to 84: a) $y = -4$ (from $3y = -12$)　b) $R = -8$ (from $-80 = 10R$)

85. (Continued)

Write the root of each equation below.

a) $1t = 10$ b) $1V = -4$ c) $1p = -12$

$t =$ ______ $V =$ ______ $p =$ ______

a) $t = 10$
b) $V = -4$
c) $p = -12$

---

86. Following the example at the left below, solve the other two equations.

$6x - 5x = 7$ b) $9y - 8y = -3$ b) $2t - t = -9$

$1x = 7$

$x = 7$ $y =$ ______ $t =$ ______

a) $y = -3$ (from $1y = -3$)
b) $t = -9$ (from $1t = -9$)

---

87. In the last frame, we got "-9" as the root of $2t - t = -9$. We checked that root at the left below. Solve the other equation and check its root in the space provided.

$2t - t = -9$ a) $2k - k = 4$ b) Check: $2k - k = 4$

$2(-9) - (-9) = -9$

$(-18) - (-9) = -9$

$(-18) + 9 = -9$

$-9 = -9$ $k =$ ______

a) $k = 4$
b) $2(4) - 4 = 4$
$8 - 4 = 4$
$4 = 4$

---

88. The root of the equation at the left below is "0". Solve the other equations.

$10x - 9x = 0$ a) $5y - 4y = 0$ b) $2k - k = 0$

$1x = 0$

$x = \frac{0}{1} = 0$ $y =$ ______ $k =$ ______

a) $y = 0$ b) $k = 0$

2-11 SUBSTITUTING "-1x" FOR "-x" IN EQUATIONS

In this section, we will show that any letter with only a "-" in front of it really has a coefficient of "-1". We will then use that fact as a help in solving equations.

---

89. Any negative number can be factored into "-1" and its opposite. For example:

$-5 = (-1)(5)$ $-9 = (-1)(9)$

Similarly, any letter with only a "-" in front of it can be factored the same way. For example:

$-x = (-1)(x)$ $-y = (-1)(y)$

Therefore, any letter with only a "-" in front of it is equal to the same letter with a coefficient of "-1". That is:

$-x = -1x$ $-y = -1y$ a) $-p =$ ______ b) $-R =$ ______

90. When a letter appears with only a "-" in front of it, the coefficient is really "-1". Sometimes it is helpful to write the "-1" explicitly. For example:	a) -1p b) -1R

$$5x + (-x) = 5x + (-1x) = 4x$$

$$(-t) + (-8t) = (-1t) + (-8t) = -9t$$

Write the coefficient "-1" explicitly and then complete each addition below.

a) $(-4y) + (-y) = (-4y) +$ ______ $=$ ______ b) $(-R) + 7R =$ ______ $+ 7R =$ ______

---

91. Write the coefficient "-1" explicitly and then solve each equation.

a) $7x + (-x) = 42$ b) $(-q) + (-3q) = -36$

x = ______ q = ______

Answers to 90: a) $(-4y) + (-1y) = -5y$ b) $(-1R) + 7R = 6R$

---

92. By writing the coefficient "-1" explicitly, we solved the equation at the left below. Solve the other equations.

Answers to 91: a) $x = 7$ (from $6x = 42$) b) $q = 9$ (from $-4q = -36$)

$-y = 10$ a) $-x = 7$ b) $3 = -F$

$-1y = 10$

$y = \frac{10}{-1} = -10$ x = ______ F = ______

---

93. By writing the coefficient "-1" explicitly, we solved the equation at the left below. Solve the other equations.

Answers to 92: a) $x = -7$ $\left(\text{from } \frac{7}{-1}\right)$ b) $F = -3$ $\left(\text{from } \frac{3}{-1}\right)$

$-t = -6$ a) $-x = -4$ b) $-1 = -y$

$-1t = -6$

$t = \frac{-6}{-1} = 6$ x = ______ y = ______

---

94. 1x and -1x are a <u>pair of opposites</u> because their sum is "0". That is:

Answers to 93: a) $x = 4$ $\left(\text{from } \frac{-4}{-1}\right)$ b) $y = 1$ $\left(\text{from } \frac{-1}{-1}\right)$

$$1x + (-1x) = [1 + (-1)]x = 0x = 0$$

Since $x = 1x$ and $-x = -1x$, x and -x are also a <u>pair of opposites</u>. Therefore:

a) The opposite of x is ______. b) The opposite of -x is ______.

---

95. Write the opposite of each of the following:

Answers to 94: a) -x b) x

a) 1y ______ b) -1t ______ c) V ______ d) -p ______

Answers to 95: a) -1y b) 1t c) -V d) p

96. Don't confuse equations like $x = 6$ and $y = -4$ with equations like $-x = 6$ and $-y = -4$.

$x = 6$ and $y = -4$ are called "root" equations because the "6" and the "-4" are their roots. That is:

$$x = 6 \qquad\qquad y = -4$$

$$1x = 6 \qquad\qquad 1y = -4$$

$$x = \frac{6}{1} = 6 \qquad\qquad y = \frac{-4}{1} = -4$$

$-x = 6$ and $-y = -4$ are not "root" equations because the "6" and the "-4" are not their roots. That is:

$$-x = 6 \qquad\qquad -y = -4$$

$$-1x = 6 \qquad\qquad -1y = -4$$

$$x = \frac{6}{-1} = -6 \qquad\qquad y = \frac{-4}{-1} = 4$$

Which of the following are "root" equations? ______________

a) $m = -7$ b) $-t = 3$ c) $4 = y$ d) $9 = -b$

Only (a) and (c)

97. Which of the following are not "root" equations? ______________

a) $x = 9$ b) $-y = 7$ c) $-3 = -F$ d) $-8 = D$

Both (b) and (c)

98. We solved the equation at the left below. Use the same steps to solve the other equations.

$4x - 5x = 8$	a) $9y - 10y = -3$	b) $10 = x - 2x$
$-1x = 8$		
$x = \frac{8}{-1} = -8$	$y =$ ________	$x =$ ________

---

a) $y = 3$ (from $-1y = -3$)
b) $x = -10$ (from $10 = -1x$)

99. The root of the equation at the left below is "0". Solve the other equations.

$2y - 3y = 0$	a) $7x - 8x = 0$	b) $0 = t - 2t$
$-1y = 0$		
$y = \frac{0}{-1} = 0$	$x =$ ________	$t =$ ________

---

a) $x = 0$
b) $t = 0$

SELF-TEST 6 (pages 48-59)

Solve each equation.

1. $3x + 6x = 54$	2. $-42 = (-2m) + 8m$	3. $(-y) + (-5y) = 30$
4. $9t - 2t = 14$	5. $20 = 4x - 6x$	6. $p + 8p = -72$
7. $49 = 8R - R$	8. $2d - d = 0$	9. $-F = -8$

ANSWERS:

1. $x = 6$	3. $y = -5$	5. $x = -10$	7. $R = 7$	9. $F = 8$
2. $m = -7$	4. $t = 2$	6. $p = -8$	8. $d = 0$	

2-12 THE ADDITION AXIOM FOR EQUATIONS

To solve equations like $x + 7 = 5$ and $10 = y - 4$, we ordinarily use "the addition axiom for equations". We will discuss the addition axiom for equations in this section.

100. In an equation, each single number and each single letter is a term. We have drawn a box around each term in the equation below.

$$\boxed{10} + \boxed{t} = \boxed{15} \qquad\qquad \boxed{8} = \boxed{x} - \boxed{7}$$

Draw a box around each term in the equations at the right.

a) $y - 1 = 12$

b) $6 = m + 10$

---

a) $\boxed{y} - \boxed{1} = \boxed{12}$

b) $\boxed{6} = \boxed{m} + \boxed{10}$

101. Any equation consists of a left side and a right side connected by an equal sign. The general pattern is:

$$\boxed{\text{Left Side}} = \boxed{\text{Right Side}}$$

Some equations contain a letter and a number on the same side. For example:

In $x + 3 = 7$, there is a letter and a number on the left side.

In $9 = t + 4$, there is a letter and a number on the ________ side.

---

right

102. When an equation contains a letter and a number on the same side, we ordinarily use the addition axiom to solve it. The ADDITION AXIOM FOR EQUATIONS says this:

> If we add the same quantity to both sides of an equation that is true, the new equation is also true.

An example of the addition axiom for equations is shown below.

The equation below is true because both sides equal 5.

$$2 + 3 = 5$$

If we add 4 to both sides of the equation above, we get:

$$2 + 3 + 4 = 5 + 4$$

The new equation is true because both sides equal 9.

9

103. The equation below contains a letter and a number on the left side. You can see that $x = 5$.

$$x + 2 = 7$$

Instead of solving the equation by simply looking at it, we can use the addition axiom to solve it. That is, we can isolate the "x" by "getting rid of" the term "2" on the left side. To do so, we add -2, the opposite of 2, to both sides. We get:

$$x + \underbrace{2 + (-2)}_{0} = \underbrace{7 + (-2)}_{5}$$

$$x + 0 = 5$$

$$x = 5$$

Did we get the same root by the method above? yes

Yes

104. When using the addition axiom to solve an equation like $x + 2 = 7$, we isolate the letter BY ADDING THE OPPOSITE OF THE TERM WE WANT TO "GET RID OF" TO BOTH SIDES. Adding the opposite of 2 in the last frame worked because:

1) The sum of a pair of opposites is "0". That is:

$$2 + (-2) = 0$$

2) If "0" is added to any other term, the sum is the other term. That is:

$$x + 0 = x$$

To solve the equation at the right, we isolated the "y" by "getting rid of" the 10 on the left side. To do so, we added its opposite -10 to both sides.

$$10 + y = 7$$

$$\underbrace{(-10) + 10}_{0} + y = \underbrace{7 + (-10)}_{-3}$$

$$0 + y = -3$$

$$y = -3$$

Is -3 the root of the original equation $10 + y = 7$? yes

Yes, since:

$10 + (-3) = 7$

105. We solved the equation at the left below by adding -1 to both sides. Solve the other equation by adding -4 to both sides.

$8 = t + 1$	$3 = 4 + m$
$[(-1) + 8] = t + [1 + (-1)]$
$7 = t + 0$
$7 = t$ or $t = 7$	m = ______

---

106. The addition axiom is used to "get rid of" the number that is on the same side as the letter. To "get rid of" it, <u>we add its opposite to both sides</u>.

What number would we add to both sides to solve each equation below?

a) $V + 6 = 9$ b) $12 + x = 4$ c) $20 = p + 30$

> $[(-4) + 3] = [(-4) + 4] + m$
>
> $-1 = 0 + m$
>
> $-1 = m$ or $m = -1$

---

107. To solve the equation at the left below, we added the opposite of -3 to both sides. Solve the other equation by adding the opposite of -6 to both sides.

> a) -6
>
> b) -12
>
> c) -30

$t + (-3) = 7$	$(-6) + y = -2$
$t + [(-3) + 3] = 7 + 3$
$t + 0 = 10$
$t = 10$	y = ______

---

108. To solve the equation at the left below, we added the opposite of -5 to both sides. Solve the other equation by adding the opposite of -1 to both sides.

> $y = 4$, from:
>
> $6 + (-6) + y = (-2) + 6$

$9 = (-5) + R$	$-7 = x + (-1)$
$[5 + 9] = [5 + (-5)] + R$
$14 = 0 + R$
$14 = R$ or $R = 14$	x = ______

---

109. What number would we add to both sides to solve each equation below?

> $x = -6$, from:
>
> $1 + (-7) = x + (-1) + 1$

a) $V + (-4) = -5$ b) $(-9) + x = 2$ c) $20 = p + (-10)$

---

110. When an equation contains a subtraction, <u>we convert the subtraction to addition before using the addition axiom</u>. An example is shown at the right.

> a) 4
>
> b) 9
>
> c) 10

$x - 7 = 5$

$x + (-7) = 5$

$x + [(-7) + 7] = [5 + 7]$

$x + 0 = 12$

$x = 12$

Is 12 the root of the original equation $x - 7 = 5$? ______

111. Following the steps below, solve the equation at the right.

$t - 1 = -3$

a) Convert the subtraction to addition.

______ = ______

b) Use the addition axiom to find the root.

t = ______

Yes, since:

$12 - 7 = 5$

---

112. Following the steps below, solve the equation at the right.

$20 = Q - 15$

a) Convert the subtraction to addition.

______ = ______

b) Use the addition axiom to find the root.

Q = ______

a) $t + (-1) = -3$

b) $t = -2$

---

113. We solved $m - 9 = -3$ below. Check "6" as the root by plugging it in for "m" in the original equation.

$m - 9 = -3$

$m + (-9) = -3$

$m + \underbrace{(-9) + 9} = \underbrace{(-3) + 9}$

$m + \quad 0 \quad = \quad 6$

$m = 6$

Check: $m - 9 = -3$

a) $20 = Q + (-15)$

b) $Q = 35$

---

114. There is a "0" on the right side of the equation at the left below. We solved it by adding (-5) to both sides. Solve the other equation by adding -9 to both sides.

$x + 5 = 0$

$x + \underbrace{5 + (-5)} = \underbrace{0 + (-5)}$

$x + \quad 0 \quad = \quad -5$

$x = -5$

$9 + y = 0$

y = ______

$6 - 9 = -3$

$6 + (-9) = -3$

$-3 = -3$

$\underbrace{(-9) + 9} + y = \underbrace{0 + (-9)}$

$0 \quad + y = \quad -9$

$y = -9$

115. We solved the equation at the left below by adding 4 to both sides. Solve the other equation by adding 7 to both sides.

$$t + (-4) = 0$$

$$t + \underbrace{(-4) + 4}_{0} = \underbrace{0 + 4}_{4}$$

$$t + 0 = 4$$

$$t = 4$$

$$(-7) + F = 0$$

F = ________

Answer: $\underbrace{7 + (-7)}_{0} + F = \underbrace{0 + 7}_{7}$; $0 + F = 7$; $F = 7$

---

116. Before using the addition axiom at the left below, we converted the subtraction to addition. Use the same method to solve the other equation.

$$m - 2 = 0$$

$$m + (-2) = 0$$

$$m + \underbrace{(-2) + 2}_{0} = \underbrace{0 + 2}_{2}$$

$$m + 0 = 2$$

$$m = 2$$

$$x - 10 = 0$$

x = ________

Answer: $x = 10$

2-13 USING BOTH AXIOMS TO SOLVE EQUATIONS

To solve equations like $2x + 3 = 11$ or $3y - 5 = 20$, both the addition axiom and the multiplication axiom are used. We will discuss solutions of that type in this section.

---

117. Both axioms are used to solve an equation like $3x + 5 = 26$. That is:

The addition axiom is used to isolate the "3x" by "getting rid of" the 5 on the left side.

The multiplication axiom is then used to "get rid of" the coefficient 3.

The steps used to solve the equation are shown and described below.

1) Addition axiom.
To isolate the "3x", we "get rid of" the 5 on the left side by adding -5 to both sides.

$$3x + 5 = 26$$

$$3x + \underbrace{5 + (-5)}_{0} = \underbrace{26 + (-5)}_{21}$$

$$3x + 0 = 21$$

$$3x = 21$$

2) Multiplication axiom.
To "get rid of" the coefficient 3, we use the shortcut of the multiplication axiom.

$$x = \frac{21}{3}$$

$$x = 7$$

(Continued on following page.)

117. (Continued)

Complete the checking of the root "7" at the right.

$$3x + 5 = 26$$

$$3(7) + 5 = 26$$

$$\underline{\quad\quad} + 5 = 26$$

$$\underline{\quad\quad} = 26$$

$\underline{21} + 5 = 26$
$\underline{26} = 26$

118. Both axioms are used to solve $12 = 37 + 5y$ below. The steps are described.

1) Using the addition axiom to isolate the "5y" by "getting rid of" the 37 on the right side.

$$12 = 37 + 5y$$

$$(-37) + 12 = (-37) + 37 + 5y$$

$$-25 = 0 + 5y$$

$$-25 = 5y$$

2) Using the multiplication axiom to "get rid of" the coefficient 5.

$$y = \frac{-25}{5}$$

$$y = -5$$

Complete the checking of the root "-5" at the right.

$$12 = 37 + 5y$$

$$12 = 37 + 5(-5)$$

$$12 = 37 + \underline{\quad\quad}$$

$$12 = \underline{\quad\quad}$$

$12 = 37 + \underline{(-25)}$
$12 = \underline{12}$

119. When using both axioms to solve an equation, we use the addition axiom first and then the multiplication axiom. Use both axioms to solve each equation below.

a) $24 + 7x = 10$

b) $87 = 9F + 51$

x = ________

F = ________

a) $x = -2$
b) $F = 4$

120. To solve $4d - 13 = 15$ at the left below, we <u>converted the subtraction to addition</u> before using the addition axiom. Use the same steps to solve the other equation.

$$4d - 13 = 15$$
$$4d + (-13) = 15$$
$$4d + \underbrace{(-13) + 13}_{0} = \underbrace{15 + 13}_{28}$$
$$4d + 0 = 28$$
$$4d = 28$$
$$d = \frac{28}{4} = 7$$

$10t - 7 = 83$

t = ______

t = 9

121. To solve $10 - 2x = 54$ at the left below, we again <u>converted the subtraction to addition</u> before using the addition axiom. Use the same steps to solve the other equation.

$$10 - 2x = 54$$
$$10 + (-2x) = 54$$
$$\underbrace{(-10) + 10}_{0} + (-2x) = \underbrace{54 + (-10)}_{44}$$
$$0 + (-2x) = 44$$
$$-2x = 44$$
$$x = \frac{44}{-2} = -22$$

$20 - 5y = 65$

y = ______

y = -9

122. To solve each equation below, convert the subtraction to addition before using the addition axiom.

a) $24 = 3F - 12$

F = ______

b) $95 = 23 - 8p$

p = ______

a) F = 12

b) p = -9

123. In the equation at the left below, the coefficient of "x" is "1". We wrote the coefficient explicitly before converting the subtraction to addition. Use the same steps to solve the other equation.

$$7 - x = 3$$
$$7 - 1x = 3$$
$$7 + (-1x) = 3$$
$$\underbrace{(-7) + 7}_{0} + (-1x) = \underbrace{3 + (-7)}_{-4}$$
$$0 + (-1x) = -4$$
$$-1x = -4$$
$$x = \frac{-4}{-1} = 4$$

$15 = 9 - y$

y = ______

y = -6 (from $6 = -1y$)

124. We used both axioms to solve $3x + 18 = 0$ at the left below. Solve the other equation.

$$3x + 18 = 0$$

$$3x + \underbrace{18 + (-18)} = \underbrace{0 + (-18)}$$

$$3x + 0 = -18$$

$$3x = -18$$

$$x = \frac{-18}{3} = -6$$

$$28 + 7y = 0$$

y = ________

Answer: y = -4 (from $7y = -28$)

---

125. To solve $6p - 30 = 0$ at the left below, we began by converting the subtraction to addition. Solve the other equation.

$$6p - 30 = 0$$

$$6p + (-30) = 0$$

$$6p + \underbrace{(-30) + 30} = \underbrace{0 + 30}$$

$$6p + 0 = 30$$

$$6p = 30$$

$$p = \frac{30}{6} = 5$$

$$20 - 4d = 0$$

d = ________

Answer: d = 5 (from $-4d = -20$)

2-14 EQUATIONS WITH A LETTER-TERM ON BOTH SIDES

To solve equations like $7x = 2x + 20$ or $40 - 3m = 7m$, we also use both axioms. We will solve equations of that type in this section.

---

126. The equation $7x = 2x + 20$ has a letter-term on each side. Both axioms are used to solve it. That is:

The addition axiom is used to get both letter-terms on the same side so that they can be combined.

The multiplication axiom is then used to get rid of the coefficient of the one letter-term.

The steps used to solve the equation are shown and described below.

1) To get both letter-terms on the same side so that they can be combined, we add -2x to both sides.

$$7x = 2x + 20$$

$$\underbrace{7x + (-2x)} = \underbrace{(-2x) + 2x} + 20$$

$$5x = 0 + 20$$

$$5x = 20$$

2) To get rid of the coefficient, we use the shortcut of the multiplication axiom.

$$x = \frac{20}{5} = 4$$

(Continued on following page.)

126. (Continued)

Complete the checking of "4" as the root of the original equation at the right.

$$7x = 2x + 20$$

$$\underbrace{7(4)}_{\downarrow} = \underbrace{2(4)}_{\downarrow} + 20$$

$$\_\_\_\_ = \_\_\_\_ + 20$$

$$\_\_\_\_ = \_\_\_\_$$

Answer: $\underline{28} = \underline{8} + 20$; $\underline{28} = \underline{28}$

127. To solve $5y + 36 = 9y$ at the left below, we added $-5y$ to both sides to get both letter-terms on the same side. Solve the other equation by adding $-3x$ to both sides.

$$5y + 36 = 9y$$

$$\underbrace{(-5y) + 5y}_{\downarrow} + 36 = \underbrace{9y + (-5y)}_{\downarrow}$$

$$0 \quad + 36 = \quad 4y$$

$$36 = 4y$$

$$y = \frac{36}{4} = 9$$

$$7x = 12 + 3x$$

$x =$ ________

Answer: $x = 3$ (from $4x = 12$)

128. When an equation has a letter-term on each side, we use the addition axiom to get both letter-terms on the same side. To do so, <u>we "get rid of" the letter-term on the side that contains two terms</u>. For example:

In $6t = 2t + 28$, we "get rid of" the $2t$ on the right side.

In $10m + 27 = 7m$, we "get rid of" the $10m$ on the left side.

When using the addition axiom, which term would we "get rid of" in each equation below?

a) $8R = 48 + 6R$ b) $56 + 5d = 13d$ c) $8v + 49 = 15v$

Answers: a) $6R$ b) $5d$ c) $8v$

129. Before adding $-9R$ to both sides of the equation at the left below, we <u>converted the subtraction to addition</u>. Solve the other equation.

$$9R - 40 = 4R$$

$$9R + (-40) = 4R$$

$$\underbrace{(-9R) + 9R}_{\downarrow} + (-40) = \underbrace{4R + (-9R)}_{\downarrow}$$

$$0 \quad + (-40) = \quad -5R$$

$$-40 = -5R$$

$$R = \frac{-40}{-5} = 8$$

$$7d = 5d - 60$$

$d =$ ________

Answer: $d = -30$ (from $2d = -60$)

130. To solve $2t = 45 + (-7t)$ at the left below, we added 7t to both sides. Solve the other equation by adding 8m to both sides.

$$2t = 45 + (-7t)$$

$$\underbrace{7t + 2t}_{9t} = 45 + \underbrace{(-7t) + 7t}_{0}$$

$$9t = 45$$

$$t = \frac{45}{9} = 5$$

$36 + (-8m) = 4m$

m = ________

m = 3 (from $36 = 12m$)

131. To solve $7x = 40 - 3x$ at the left below, we converted the subtraction to addition. Solve the other equation.

$$7x = 40 - 3x$$

$$7x = 40 + (-3x)$$

$$\underbrace{3x + 7x}_{10x} = 40 + \underbrace{(-3x) + 3x}_{0}$$

$$10x = 40$$

$$x = \frac{40}{10} = 4$$

$56 - 4y = 4y$

y = ________

y = 7 (from $56 = 8y$)

132. To solve $y = 14 + 3y$ below, we began by writing the coefficient "1" of the "y" explicitly. Solve the other equation.

$$y = 14 + 3y$$

$$1y = 14 + 3y$$

$$\underbrace{(-3y) + 1y}_{-2y} = 14 + \underbrace{3y + (-3y)}_{0}$$

$$-2y = 14$$

$$y = \frac{14}{-2} = -7$$

$2x + 9 = x$

x = ________

x = -9 (from $9 = -1x$)

133. To solve $7t = 18 + t$ below, we began by writing the coefficient "1" of the "t" explicitly. Solve the other equation.

$$7t = 18 + t$$

$$7t = 18 + 1t$$

$$\underbrace{(-1t) + 7t}_{6t} = 18 + \underbrace{1t + (-1t)}_{0}$$

$$6t = 18$$

$$t = \frac{18}{6} = 3$$

$m + 32 = 3m$

m = ________

m = 16 (from $32 = 2m$)

134. To solve $24 - x = 2x$ below, we began by writing the coefficient "1" of the "x" explicitly before converting the subtraction to addition. Solve the other equation.

$24 - x = 2x$	$y = 12 - y$
$24 - 1x = 2x$	
$24 + (-1x) = 2x$	
$24 + \underbrace{(-1x) + 1x}_{0} = \underbrace{2x + 1x}_{3x}$	
$24 + 0 = 3x$	
$24 = 3x$	
$x = \frac{24}{3} = 8$	$y =$ ______

---

y = 6 (from 2y = 12)

SELF-TEST 7 (pages 59-69)

Solve each equation.

1. $x + 9 = 6$	2. $-10 = y - 2$	3. $70 = 8F + 22$
4. $30 - 11t = -3$	5. $12 = 4 - d$	6. $5x + 20 = 0$
7. $4m + 18 = 10m$	8. $6V = V + 35$	9. $20 - x = 9x$

ANSWERS:

1. $x = -3$	3. $F = 6$	5. $d = -8$	7. $m = 3$	9. $x = 2$
2. $y = -8$	4. $t = 3$	6. $x = -4$	8. $V = 7$	

2-15 MULTIPLYING BY THE DISTRIBUTIVE PRINCIPLE OVER ADDITION

To solve equations like $2(x + 7) = 22$ or $5(2y + 3) = 15y$, we have to "multiply by the distributive principle of multiplication over addition". We will discuss multiplications of that type in this section and then use them to solve equations in the next section.

---

135. In each multiplication below, the second factor is an addition. We performed them by the "adding before multiplying" method. The products are 20 and 42.

$$4(3 + 2) \quad \rightarrow \quad 4\ (5) = 20. \qquad\qquad 7(5 + 1) \quad \rightarrow \quad 7\ (6) = 42$$

We can perform the same multiplications by "multiplying before adding". The steps are:

$$4(3 + 2) = 4(3) + 4(2) = 12 + 8 = 20 \qquad\qquad 7(5 + 1) = 7(5) + 7(1) = 35 + 7 = 42$$

Did we get the same results using the second method? ________

Answer: Yes

---

136. In the "multiplying before adding" method, the first step uses the DISTRIBUTIVE PRINCIPLE OF MULTIPLICATION OVER ADDITION. We have diagrammed that principle below. The arrows show that both the "7" and the "4" are multiplied by "5".

$$5(7 + 4) = 5(7) + 5(4)$$

Following the pattern above, fill in the blanks below.

a) $4(2 + 8) =$ ____(2) + ____(8)   b) $9(7 + 1) =$ ____(7) + ____(1)

Answer: a) $\underline{4}(2) + \underline{4}(8)$ b) $\underline{9}(7) + \underline{9}(1)$

---

137. When a letter is involved in the addition-factor, the distributive principle works the same way. For example:

$$3(x + 5) = 3(x) + 3(5) \qquad\qquad 7(4 + y) = 7(4) + 7(y)$$

Following the examples above, fill in the blanks below.

a) $9(t + 6) =$ ____(t) + ____(6)   b) $4(1 + m) =$ ____(1) + ____(m)

Answer: a) $\underline{9}(t) + \underline{9}(6)$ b) $\underline{4}(1) + \underline{4}(m)$

---

138. After using the distributive principle to perform each multiplication below, we simplified the terms in the products.

$$5(x + 3) = 5(x) + 5(3) = 5x + 15 \qquad\qquad 2(7 + d) = 2(7) + 2(d) = 14 + 2d$$

Simplify the terms in each product below.

a) $8(t + 4) = 8(t) + 8(4) =$ ______ + ______

b) $3(5 + y) = 3(5) + 3(y) =$ ______ + ______

Answer: a) $8t + 32$ b) $15 + 3y$

139. In one step, multiply by the distributive principle and write each term of the product in its simplest form.

a) $5(x + 7) =$ _______ + _______ c) $6(10 + y) =$ _______ + _______

b) $4(d + 3) =$ _______ + _______ d) $8(1 + V) =$ _______ + _______

---

a) $5x + 35$
b) $4d + 12$
c) $60 + 6y$
d) $8 + 8V$

140. There is a common error that occurs frequently when multiplying by the distributive principle. Here are two examples:

$$3(x + 8) = 3x + 8$$

$$9(4 + y) = 36 + y$$

The error is forgetting to multiply the "8" by "3" and the "y" by "9". That is:

$3(x + 8)$ really equals $3x + 24$

$9(4 + y)$ really equals $36 + 9y$

Which of the following are performed correctly? _______

a) $5(m + 2) = 5m + 2$ c) $8(2 + V) = 16 + 8V$

b) $10(y + 3) = 10y + 30$ d) $7(7 + b) = 49 + b$

(b) and (c)

141. Which of the following are performed correctly? _______

a) $4(t + 6) = 4t + 24$ c) $10(5 + F) = 50 + F$

b) $7(y + 8) = 7y + 8$ d) $8(9 + x) = 72 + 8x$

(a) and (d)

142. In each multiplication below, the letter has a coefficient. We multiplied by the distributive principle in the usual way.

$$6(3x + 7) = 6(3x) + 6(7)$$

$$2(8y + 5) = 2(8y) + 2(5)$$

To simplify the products above, we must simplify $6(3x)$ and $2(8y)$. Expressions of that type are really three-factor multiplications. That is:

Since $3x$ means $(3)(x)$, $6(3x) = 6(3)(x)$

Since $8y$ means $(8)(y)$, $2(8y) = 2(8)(y)$

To perform such three-factor multiplications, we multiply "two at a time". Therefore:

$6(3x) = \underbrace{6(3)}(x)$ $2(8y) = 2(8)(y)$

$= 18\,(x)$ or $18x$ $=$ ___(y) or _______

$16(y)$ or $16y$

143. Since each expression below is really a three-factor multiplication, we can simplify each by multiplying the two numerical factors. That is:

$5(4x) = 20x$ a) $9(4y) =$ _______ b) $10(2t) =$ _______ c) $7(8F) =$ _______

---

a) $36y$ b) $20t$ c) $56F$

144. Simplify the product for each multiplication below.

a) $3(5x + 7) = 3(5x) + 3(7) =$ ______ + ______

b) $6(4 + 8y) = 6(4) + 6(8y) =$ ______ + ______

---

a) $15x + 21$

b) $24 + 48y$

145. <u>In one step</u>, perform each multiplication and write each term of the product in its simplest form.

a) $10(2t + 7) =$ ______ + ______

b) $8(3b + 4) =$ ______ + ______

c) $5(6 + 8m) =$ ______ + ______

d) $6(1 + 6x) =$ ______ + ______

---

a) $20t + 70$ b) $24b + 32$ c) $30 + 40m$ d) $6 + 36x$

2-16 THE DISTRIBUTIVE PRINCIPLE OVER ADDITION AND EQUATIONS

In this section, we will use the "distributive principle over addition" to solve equations like $2(x + 7) = 22$ and $5(2y + 3) = 15y$.

---

146. To simplify expressions like those below, we multiply by the distributive principle. Therefore, we will call expressions of that type "<u>instances of the distributive principle</u>".

$3(x + 7)$ $5(4 + y)$ $8(2m + 1)$ $7(5 + 4t)$

Draw a circle around each "instance of the distributive principle" in the equations below.

a) $6(x + 3) = 7$ b) $4(2t + 9) = 40$ c) $15 = 2(5 + y)$

There should be a circle around.

a) $6(x + 3)$

b) $4(2t + 9)$

c) $2(5 + y)$

147. The equation at the right contains an "instance of the distributive principle". To solve it, we begin by multiplying by the distributive principle. We get:

$$3(x + 2) = 21$$

$$3x + 6 = 21$$

Now we can complete the solution in the usual way. Do so:

$x =$ ______

$x = 5$ (from $3x = 15$)

148. Whenever an equation contains an "instance of the distributive principle", <u>we begin the solution by multiplying by the distributive principle</u> and then proceed in the usual way. Use that method to solve these.

a) $9(y + 5) = 135$ b) $30 = 5(9 + b)$

$y =$ ______ $b =$ ______

---

a) $y = 10$ b) $b = -3$

149. In the last frame, we got $y = 10$ and $b = -3$ as the solution of the equations below. We checked 10 as the root of the equation at the left. Complete the checking of -3 as the root of the equation at the right.

$$9(y + 5) = 135$$
$$9(10 + 5) = 135$$
$$9(15) = 135$$
$$135 = 135$$

$$30 = 5(9 + b)$$
$$30 = 5[9 + (-3)]$$
$$30 = 5(\quad)$$
$$30 = \_\_\_\_$$

Answer: $30 = 5(6)$; $30 = 30$

---

150. Solve each equation below. Begin by using the distributive principle.

a) $2(7x + 5) = 80$

b) $18 = 6(5 + 2y)$

$x =$ ______ $y =$ ______

Answer: a) $x = 5$ b) $y = -1$

---

151. Solve: a) $5(2y + 3) = 15y$ b) $3x = 4(1 + x)$

$y =$ ______ $x =$ ______

Answer: a) $y = 3$ b) $x = -4$

2-17 THE DISTRIBUTIVE PRINCIPLE OVER SUBTRACTION AND EQUATIONS

To solve equations like $4(t - 1) = 16$ or $7(3m - 4) = 14m$, we can "multiply by the distributive principle of multiplication over subtraction". In this section, we will discuss multiplications of that type and then use them to solve equations like those above.

---

152. In each multiplication below, the second factor is a subtraction. We performed them by the "subtracting before multiplying" method. The products are 6 and 10.

$$3(7 - 5) \rightarrow 3\ (2) = 6$$

$$2(9 - 4) \rightarrow 2\ (5) = 10$$

We can perform the same multiplications by "multiplying before subtracting". The steps are:

$$3(7 - 5) = 3(7) - 3(5) = 21 - 15 = 6$$

$$2(9 - 4) = 2(9) - 2(4) = 18 - 8 = 10$$

Did we get the same results using the second method? ______

Answer: Yes

153. In the "multiplying before subtracting" method, the first step uses the DISTRIBUTIVE PRINCIPLE OF MULTIPLICATION OVER SUBTRACTION. We have diagrammed that principle below. The arrows show that both the "8" and the "5" are multiplied by "4".

$$4(8 - 5) = 4(8) - 4(5)$$

When a letter is involved in the subtraction-factor, the distributive principle works the same way. For example:

$$7(x - 3) = 7(x) - 7(3) \qquad 9(6 - y) = 9(6) - 9(y)$$

Following the examples above, fill in the blanks below.

a) 3(t - 8) = ____(t) - ____(8) b) 6(1 - m) = ____(1) - ____(m)

---

a) 3(t) - 3(8)

b) 6(1) - 6(m)

154. After using the distributive principle to perform each multiplication below, we simplified the terms in the products.

$$9(x - 3) = 9(x) - 9(3) = 9x - 27$$

$$2(7 - d) = 2(7) - 2(d) = 14 - 2d$$

In one step, multiply by the distributive principle and write each term of the product in its simplest form.

a) 6(y - 9) = ______________ c) 3(10 - t) = ______________

b) 8(R - 2) = ______________ d) 5(1 - x) = ______________

---

a) 6y - 54

b) 8R - 16

c) 30 - 3t

d) 5 - 5x

155. When multiplying by the distributive principle over subtraction, don't forget to multiply the second term of the subtraction by the single number. Complete these:

a) 5(F - 1) = ______________ c) 4(9 - y) = ______________

b) 9(x - 7) = ______________ d) 7(8 - V) = ______________

---

a) 5F - 5

b) 9x - 63

c) 36 - 4y

d) 56 - 7V

156. We simplified the terms in each product below.

$$7(2x - 6) = 7(2x) - 7(6) = 14x - 42$$

$$4(6 - 3t) = 4(6) - 4(3t) = 24 - 12t$$

Write the product for each multiplication in its simplest form.

a) 10(3y - 1) = ______________ c) 2(1 - 6x) = ______________

b) 8(4t - 7) = ______________ d) 5(9 - 7F) = ______________

---

a) 30y - 10

b) 32t - 56

c) 2 - 12x

d) 45 - 35F

157. Don't confuse the distributive principle over subtraction with the distributive principle over addition.

In the distributive principle over subtraction, the terms of the product are subtracted. That is:

$$6(x - 5) = 6x \overset{\downarrow}{-} 30$$

In the distributive principle over addition, the terms of the product are added. That is:

$$6(x + 5) = 6x \overset{\downarrow}{+} 30$$

Be careful to write the correct sign in each product below.

a) $4(y - 7) =$ __________ c) $7(3 - 6t) =$ __________

b) $4(y + 7) =$ __________ d) $7(3 + 6t) =$ __________

Answers: a) $4y - 28$ b) $4y + 28$ c) $21 - 42t$ d) $21 + 42t$

158. The equation at the right contains an "instance of the distributive principle over subtraction". To solve it: $4(t - 1) = 16$

We multiply by the distributive principle. $4t - 4 = 16$

We convert the subtraction to addition. $4t + (-4) = 16$

Then we complete the solution in the usual way. Do so.

$t =$ ______

Answer: $t = 5$ (from $4t = 20$)

159. To solve each equation below, multiply by the distributive principle over subtraction, convert the subtraction to addition, and then complete the solution in the usual way.

a) $5(x - 3) = 20$ b) $24 = 8(4 - y)$

$x =$ ______ $y =$ ______

Answers: a) $x = 7$ b) $y = 1$

160. Use the same method to solve these:

a) $9(2b - 4) = 18$ b) $x = 2(9 - 4x)$

$b =$ ______ $x =$ ______

Answers: a) $b = 3$ b) $x = 2$

2-18 EQUATIONS WITH A LETTER-TERM AND A NUMBER-TERM ON BOTH SIDES

To solve equations with a letter-term and a number-term on both sides, the addition axiom is used twice. We will discuss the method in this section.

---

161. The equation below has a letter-term and a number-term on each side.

$$7x + 9 = 4x + 21$$

To solve it, we must get both letter-terms on one side and both number-terms on the other side so that they can be combined. To do so, we must "get rid of" a letter-term from one side and a number-term from the other side. We have a choice of two pairs:

1) getting rid of 4x and 9

or 2) getting rid of 7x and 21

With either choice, we must use the addition axiom twice. Below, we solved the equation by getting rid of "4x" and "9".

To get rid of the "4x" on the right side, we add "-4x" to both sides.

$$7x + 9 = 4x + 21$$

$$\underbrace{(-4x) + 7x}_{3x} + 9 = \underbrace{(-4x) + 4x}_{0} + 21$$

$$3x + 9 = 21$$

To get rid of the "9" on the left side, we add "-9" to both sides.

$$3x + \underbrace{9 + (-9)}_{0} = \underbrace{21 + (-9)}_{12}$$

$$3x = 12$$

Complete the solution. The root is ______.

Answer: 4

162. In the last frame, we obtained "4" as the root of the equation at the right. Check the root.

$$7x + 9 = 4x + 21$$

Answer:

$$7(4) + 9 = 4(4) + 21$$
$$28 + 9 = 16 + 21$$
$$37 = 37$$

163. To solve $5y + 16 = 7y + 10$, we have to get both letter-terms on one side and both number-terms on the other side so that they can be combined. Therefore, we have to get rid of one term from each side. We have a choice of two pairs:

1) getting rid of 5y and 10

or 2) getting rid of 7y and 16

(Continued on following page.)

163. (Continued)

The steps in both possible solutions are shown below.

1) <u>Getting rid of 5y and 10</u>

$$(-5y) + 5y + 16 = (-5y) + 7y + 10$$
$$16 = 2y + 10$$
$$(-10) + 16 = 2y + 10 + (-10)$$
$$6 = 2y$$
$$y = \frac{6}{2} = 3$$

2) <u>Getting rid of 7y and 16</u>

$$(-7y) + 5y + 16 = (-7y) + 7y + 10$$
$$(-2y) + 16 = 10$$
$$(-2y) + 16 + (-16) = 10 + (-16)$$
$$-2y = -6$$
$$y = \frac{-6}{-2} = 3$$

Did we get the same root with both methods? ________

Yes. It is 3.

---

164. For each equation below, identify the two possible pairs of terms that could be eliminated.

a) $7p + 9 = 10p + 18$ ______ and ______ or ______ and ______

b) $6x + 6 = 3x + 21$ ______ and ______ or ______ and ______

---

a) 7p and 18 or 10p and 9

b) 6x and 21 or 3x and 6

---

165. By getting rid of either possible pair of terms, solve each equation below.

a) $3x + 8 = 5x + 2$

x = ______

b) $9y + 1 = 3y + 13$

y = ______

---

a) x = 3

b) y = 2

---

166. To solve each equation below, begin by converting each subtraction to addition.

a) $9a - 4 = 3a + 8$

a = ______

b) $70 - 5d = 30 - 9d$

d = ______

---

a) a = 2

b) d = -10

167. To solve the equation at the right, begin by multiplying by the distributive principle on each side.

$$3(x - 5) = 5(x + 3)$$

$x =$ ______

	$x = -15$

SELF-TEST 8 (pages 70-78)

Solve each equation.

1. $4(x + 9) = 16$
2. $50 = 2(10 + 3y)$
3. $7(m - 6) = m$
4. $7 - 8t = 25 - 2t$
5. $3(x + 2) = 2x + 9$

ANSWERS: 1. $x = -5$ 2. $y = 5$ 3. $m = 7$ 4. $t = -3$ 5. $x = 3$

Unit 3 FRACTIONS AND EQUATIONS: MULTIPLICATION AND DIVISION

In this unit, the procedures for multiplying and dividing numerical fractions are reviewed and extended to multiplications and divisions involving signed fractions. The multiplication axiom is used to solve equations with fractional roots and equations with fractional coefficients and terms.

3-1 THE MEANING OF FRACTIONS

In this section, we will introduce the terminology used to discuss fractions. We will show that any fraction is equal to a division. We will also show that some fractions can be converted to whole numbers by performing the related divisions.

---

1. Any numerical fraction involves two numbers separated by a fraction line. The two numbers are called the "terms" of the fraction. That is:

   In $\frac{3}{4}$, the terms of the fraction are 3 and 4.

   In $\frac{1}{5}$, the terms of the fraction are ____ and ____.

---

1 and 5

2. The term above the fraction line is called the "numerator" of the fraction. The term below the fraction line is called the "denominator" of the fraction. For example:

   In $\frac{5}{7}$: "5" is called the "numerator" of the fraction.
   "7" is called the "denominator" of the fraction.

   In $\frac{1}{8}$: a) The numerator is ____. b) The denominator is ____.

---

a) 1
b) 8

3. Any fraction in which the numerator is smaller than the denominator is called a "proper" fraction. For example:

   $\frac{2}{3}$, $\frac{5}{11}$, and $\frac{1}{4}$ are called "proper" fractions.

   Any fraction in which the numerator is equal to or larger than the denominator is called an "improper" fraction. For example:

   $\frac{7}{4}$, $\frac{15}{5}$, and $\frac{8}{8}$ are called "improper" fractions.

   a) Which fractions at the right are "proper" fractions? ________ $\frac{1}{2}$ $\frac{8}{7}$ $\frac{7}{9}$ $\frac{5}{5}$

   b) Which fractions at the right are "improper" fractions? ________ $\frac{12}{6}$ $\frac{4}{5}$ $\frac{4}{4}$ $\frac{1}{3}$

4. Any fraction is equal to a division in which <u>the numerator is divided by the denominator</u>. That is:

$\frac{8}{2}$ means: 8 divided by 2 $\qquad$ $\frac{3}{4}$ means: 3 divided by 4

Using the above fact, fill in the blanks below.

a) $\frac{5}{5}$ means: __5__ divided by __5__ $\qquad$ b) $\frac{7}{6}$ means: __7__ divided by __6__

a) $\frac{1}{2}$ and $\frac{7}{9}$
b) $\frac{12}{6}$ and $\frac{4}{4}$

---

5. If we divide a first number by a second and the quotient has a "0" remainder, we say that the first number is <u>divisible</u> by the second. For example:

12 <u>is</u> divisible by 3, since 12 divided by 3 = 4

17 <u>is</u> <u>not</u> divisible by 3, since 17 divided by 3 = 5 r2

When the numerator of a fraction is <u>divisible</u> by the denominator, we can convert the fraction to a whole number by performing a division. That is:

$\frac{18}{6} = 18 \div 6 = 3$ $\qquad$ a) $\frac{12}{2}$ = __6__ $\qquad$ b) $\frac{42}{6}$ = __7__ $\qquad$ c) $\frac{24}{12}$ = __2__

a) 5 divided by 5
b) 7 divided by 6

---

6. When the numerator and denominator of a fraction are equal, the fraction equals "1". That is:

$\frac{3}{3} = 3 \div 3 = 1$ $\qquad$ a) $\frac{8}{8}$ = __1__ $\qquad$ b) $\frac{5}{5}$ = __1__ $\qquad$ c) $\frac{17}{17}$ = __1__

a) 6
b) 7
c) 2

---

7. When the denominator of a fraction is "1", the fraction equals its numerator. That is:

$\frac{2}{1} = 2 \div 1 = 2$ $\qquad$ a) $\frac{7}{1}$ = __7__ $\qquad$ b) $\frac{10}{1}$ = __10__ $\qquad$ c) $\frac{1}{1}$ = __1__

a) 1
b) 1
c) 1

---

8. When the numerator of a fraction is "0", the fraction equals "0". That is:

$\frac{0}{5} = 0 \div 5 = 0$ $\qquad$ a) $\frac{0}{3}$ = __0__ $\qquad$ b) $\frac{0}{15}$ = __0__ $\qquad$ c) $\frac{0}{1}$ = __0__

a) 7
b) 10
c) 1

a) 0	b) 0	c) 0

3-2 MULTIPLICATIONS INVOLVING FRACTIONS

In this section, we will discuss multiplications involving two fractions and multiplications involving a whole number and a fraction. Multiplying a fraction by "1" is discussed as a special case.

---

9. To multiply two fractions, we simply <u>multiply the two numerators and the two denominators</u>. That is:

$\left(\frac{2}{3}\right)\left(\frac{5}{7}\right) = \frac{(2)(5)}{(3)(7)} = \frac{10}{21}$ $\qquad$ $\left(\frac{3}{5}\right)\left(\frac{7}{8}\right) = \frac{(3)(7)}{(5)(8)} = \frac{21}{40}$

$\frac{21}{40}$

10. The answer for a multiplication is called the "<u>product</u>". Find each product below.

a) $\left(\frac{1}{2}\right)\left(\frac{3}{4}\right) = \frac{(1)(3)}{(2)(4)} =$ ______ b) $\left(\frac{1}{3}\right)\left(\frac{1}{5}\right) = \frac{(1)(1)}{(3)(5)} =$ ______

---

11. When multiplying fractions, the product can be an improper fraction. For example:

$\left(\frac{5}{6}\right)\left(\frac{7}{3}\right) = \frac{(5)(7)}{(6)(3)} = \frac{35}{18}$ $\left(\frac{11}{2}\right)\left(\frac{1}{3}\right) = \frac{(11)(1)}{(2)(3)} =$ ______

a) $\frac{3}{8}$ b) $\frac{1}{15}$

---

12. Any whole number is equal to a fraction whose denominator is "1". For example:

$2 = \frac{2}{1}$ $5 = \frac{5}{1}$ $9 = \frac{9}{1}$ $12 = \frac{12}{1}$

Therefore, any multiplication of a whole number and a fraction can be converted into a multiplication of two fractions. For example:

$3\left(\frac{4}{7}\right) = \frac{3}{1}\left(\frac{4}{7}\right) = \frac{3(4)}{1(7)} = \frac{12}{7}$ $\frac{1}{6}(13) = \frac{1}{6}\left(\frac{13}{1}\right) = \frac{1(13)}{6(1)} =$ ______

$\frac{11}{6}$

---

13. The number "1" can also be written as a fraction whose denominator is "1". That is:

$$1 = \frac{1}{1}$$

Therefore, any multiplication of "1" and a fraction can be converted into a multiplication of two fractions. That is:

$1\left(\frac{3}{4}\right) = \frac{1}{1}\left(\frac{3}{4}\right) = \frac{1(3)}{1(4)} = \frac{3}{4}$ $1\left(\frac{9}{2}\right) = \frac{1}{1}\left(\frac{9}{2}\right) = \frac{1(9)}{1(2)} =$ ______

$\frac{13}{6}$

---

14. The shorter method below is ordinarily used to multiply a whole number and a fraction. Notice that we got the numerator of each product <u>by simply multiplying the whole number and the original numerator</u>.

$5\left(\frac{2}{3}\right) = \frac{5(2)}{3} = \frac{10}{3}$ $\frac{8}{7}(1) = \frac{8(1)}{7} = \frac{8}{7}$

Using the shorter method above, complete these:

a) $2\left(\frac{7}{5}\right) =$ ______ b) $1\left(\frac{4}{9}\right) =$ ______ c) $\frac{1}{7}(15) =$ ______ d) $\frac{2}{5}(1) =$ ______

$\frac{9}{2}$

---

15. We saw this fact earlier: <u>If any whole number is multiplied by "1", the product is identical to the original whole number</u>. For example:

$1(5) = 5$ $1(17) = 17$

It should be obvious that the same fact is true when a fraction is multiplied by "1". That is:

$1\left(\frac{5}{7}\right) = \frac{5}{7}$ $\frac{8}{3}(1) = \frac{8}{3}$ a) $1\left(\frac{3}{5}\right) =$ ______ b) $\frac{1}{4}(1) =$ ______

a) $\frac{14}{5}$ b) $\frac{4}{9}$ c) $\frac{15}{7}$ d) $\frac{2}{5}$

---

a) $\frac{3}{5}$ b) $\frac{1}{4}$

16. If a fraction is multiplied by "0", the product is "0". For example:

$$0\left(\frac{3}{5}\right) = \frac{0(3)}{5} = \frac{0}{5} = 0 \qquad \frac{7}{4}(0) = \frac{7(0)}{4} = \frac{0}{4} = 0$$

Complete: a) $0\left(\frac{1}{2}\right) =$ ______ b) $\frac{2}{3}(0) =$ ______ c) $\frac{11}{3}(0) =$ ______

a) 0 b) 0 c) 0

3-3 RAISING FRACTIONS TO HIGHER TERMS

In this section, we will define "equivalent" fractions and the meaning of the phase "in higher terms". Then we will show two methods for raising a fraction to an equivalent fraction in higher terms.

---

17. In each pair below, the fractions are equal even though they have different terms.

$$\frac{1}{2} = \frac{3}{6} \qquad \frac{5}{4} = \frac{10}{8}$$

Fractions that are equal even though they have different terms are called "equivalent" fractions. Therefore:

$\frac{1}{2}$ and $\frac{3}{6}$ are a pair of "equivalent" fractions.

$\frac{5}{4}$ and $\frac{10}{8}$ are a pair of "equivalent" fractions.

---

18. Two more pairs of "equivalent" fractions are shown below.

$$\frac{2}{3} = \frac{8}{12} \qquad \frac{6}{5} = \frac{12}{10}$$

In each pair, the terms in the fraction on the right are larger or "higher" than the corresponding terms in the fraction on the left. Therefore, we say that the fraction on the right in each pair is "in higher terms". That is:

$\frac{8}{12}$ is "in higher terms" than $\frac{2}{3}$, since 8 and 12 are larger than 2 and 3.

$\frac{12}{10}$ is "in higher terms" than $\frac{6}{5}$, since 12 and 10 are larger than 6 and 5.

In each pair of equivalent fractions at the right, identify the fraction that is in higher terms. a) $\frac{2}{4} = \frac{10}{20}$ b) $\frac{15}{12} = \frac{5}{4}$

a) $\frac{10}{20}$ b) $\frac{15}{12}$

---

19. All of the fractions at the right are equivalent because each equals "1". Which one is "in highest terms"? ______

$$\frac{2}{2} = \frac{3}{3} = \frac{4}{4} = \frac{5}{5} = \frac{7}{7} = \frac{10}{10}$$

$\frac{10}{10}$

20. Any fraction can be raised to an equivalent fraction in higher terms by multiplying both terms by the same number. For example:

$$\frac{3}{4} = \frac{3(2)}{4(2)} = \frac{6}{8} \qquad \frac{5}{2} = \frac{5(6)}{2(6)} = \frac{30}{12}$$

a) Raise $\frac{1}{2}$ to higher terms by multiplying both terms by 4. $\frac{1}{2} =$ ________

b) Raise $\frac{8}{5}$ to higher terms by multiplying both terms by 3. $\frac{8}{5} =$ ________

---

a) $\frac{4}{8}$

b) $\frac{24}{15}$

21. We raised the fractions below to higher terms by multiplying both terms by 4 at the left and by 2 at the right.

$$\frac{2}{3} = \frac{2(4)}{3(4)} = \frac{8}{12} \qquad \frac{7}{4} = \frac{7(2)}{4(2)} = \frac{14}{8}$$

The method above really involves multiplying each original fraction by a fraction that equals "1". That is:

$$\frac{2}{3} = \frac{2}{3}(1) = \frac{2}{3}\left(\frac{4}{4}\right) = \frac{8}{12} \qquad \frac{7}{4} = \frac{7}{4}(1) = \frac{7}{4}\left(\frac{2}{2}\right) = \frac{14}{8}$$

a) By substituting $\frac{3}{3}$ for the "1" at the right, raise $\frac{1}{4}$ to higher terms. $\frac{1}{4} = \frac{1}{4}(1) = \frac{1}{4}(\quad) =$ ________

b) By substituting $\frac{5}{5}$ for the "1" at the right, raise $\frac{3}{2}$ to higher terms. $\frac{3}{2} = \frac{3}{2}(1) = \frac{3}{2}(\quad) =$ ________

---

a) $\frac{1}{4}\left(\frac{3}{3}\right) = \frac{3}{12}$

b) $\frac{3}{2}\left(\frac{5}{5}\right) = \frac{15}{10}$

22. Any whole number is equal to a fraction whose denominator is "1". For example:

$$7 = \frac{7}{1} \qquad 1 = \frac{1}{1}$$

By converting whole numbers to fractions like those above and then multiplying both terms of that fraction by the same number, we can convert whole numbers to fractions with denominators other than "1". For example:

$$7 = \frac{7}{1} = \frac{7(3)}{1(3)} = \frac{21}{3} \qquad 1 = \frac{1}{1} = \frac{1(6)}{1(6)} = \frac{6}{6}$$

Complete each of the following conversions of a whole number to an improper fraction.

a) $2 = \frac{2}{1} = \frac{2(4)}{1(4)} =$ ________ b) $9 = \frac{9}{1} = \frac{9(2)}{1(2)} =$ ________

---

a) $\frac{8}{4}$

b) $\frac{18}{2}$

23. We converted two whole numbers to improper fractions below.

$$3 = \frac{3}{1} = \frac{3(5)}{1(5)} = \frac{15}{5} \qquad 6 = \frac{6}{1} = \frac{6(3)}{1(3)} = \frac{18}{3}$$

The method above really involves multiplying each whole number by a fraction that equals "1". That is:

$$3 = 3(1) = 3\left(\frac{5}{5}\right) = \frac{15}{5} \qquad 6 = 6(1) = 6\left(\frac{3}{3}\right) = \frac{18}{3}$$

(Continued on following page.)

23. (Continued)

a) By substituting $\frac{2}{2}$ for the "1" at the right, convert "5" to an improper fraction. $5 = 5(1) = 5(\quad) =$ ______

b) By substituting $\frac{4}{4}$ for the "1" at the right, convert "8" to an improper fraction. $8 = 8(1) = 8(\quad) =$ ______

---

a) $5\left(\frac{2}{2}\right) = \frac{10}{2}$ b) $8\left(\frac{4}{4}\right) = \frac{32}{4}$

3-4 REDUCING FRACTIONS TO LOWEST TERMS

In this section, we will define the meaning of the phrases "in lower terms" and "in lowest terms". We will also show two methods for reducing fractions to lowest terms.

---

24. If two fractions are equivalent, we say that the one with the smaller terms is "in lower terms". For example:

$\frac{2}{4} = \frac{4}{8}$, and $\frac{2}{4}$ is in lower terms.

$\frac{10}{25} = \frac{8}{20}$, and ______ is in lower terms.

$\frac{8}{20}$

25. To reduce a fraction to an equivalent fraction in lower terms, we can divide both terms by a common factor. For example:

$$\frac{6}{8} = \frac{6 \div 2}{8 \div 2} = \frac{3}{4} \qquad \frac{15}{6} = \frac{15 \div 3}{6 \div 3} = \frac{5}{2}$$

a) Reduce $\frac{5}{10}$ to lower terms by dividing both terms by 5. $\frac{5}{10} =$ ______

b) Reduce $\frac{16}{12}$ to lower terms by dividing both terms by 4. $\frac{16}{12} =$ ______

---

a) $\frac{1}{2}$ b) $\frac{4}{3}$

26. We have reduced each fraction below to lower terms by dividing both terms by a common factor.

$$\frac{4}{10} = \frac{4 \div 2}{10 \div 2} = \frac{2}{5} \qquad \frac{9}{6} = \frac{9 \div 3}{6 \div 3} = \frac{3}{2}$$

The same reductions can be made by factoring the original terms and then using the common factor to get a fraction that equals "1". That is:

$$\frac{4}{10} = \frac{2(2)}{5(2)} = \left(\frac{2}{5}\right)\left(\frac{2}{2}\right) = \frac{2}{5}(1) = \frac{2}{5} \qquad \frac{9}{6} = \frac{3(3)}{2(3)} = \frac{3}{2}\left(\frac{3}{3}\right) = \frac{3}{2}(1) = \frac{3}{2}$$

Using the second method, complete the reductions below.

a) $\frac{2}{8} = \frac{1(2)}{4(2)} = \left(\frac{1}{4}\right)\left(\frac{2}{2}\right) = \left(\frac{\ }{\ }\right)(\quad) =$ ______

b) $\frac{25}{10} = \frac{5(5)}{2(5)} = \left(\frac{5}{2}\right)\left(\frac{5}{5}\right) = \left(\frac{\ }{\ }\right)(\quad) =$ ______

---

a) $\left(\frac{1}{4}\right)(1) = \frac{1}{4}$ b) $\left(\frac{5}{2}\right)(1) = \frac{5}{2}$

27. Use the same method to complete the reductions below.

a) $\frac{9}{15} = \frac{3(3)}{5(3)} = \left(\frac{\ }{\ }\right)\left(\frac{\ }{\ }\right) = \left(\frac{\ }{\ }\right)(\quad) =$ ______ b) $\frac{24}{20} = \frac{6(4)}{5(4)} = \left(\frac{\ }{\ }\right)\left(\frac{\ }{\ }\right) = \left(\frac{\ }{\ }\right)(\quad) =$ ______

---

a) $\left(\frac{3}{5}\right)\left(\frac{3}{3}\right) = \left(\frac{3}{5}\right)(1) = \frac{3}{5}$

b) $\left(\frac{6}{5}\right)\left(\frac{4}{4}\right) = \left(\frac{6}{5}\right)(1) = \frac{6}{5}$

28. A fraction can be reduced to lower terms <u>only if both terms contain a common factor</u>, because only then can we get a fraction that equals "1". For example:

$\frac{10}{14}$ <u>can</u> be reduced to lower terms because of the common factor "2".

$\frac{6}{25}$ <u>cannot</u> be reduced to lower terms because there is no common factor.

For each fraction below, identify the factor common to both terms <u>if there is one</u>.

a) $\frac{10}{25}$ ______ b) $\frac{4}{6}$ ______ c) $\frac{5}{7}$ ______ d) $\frac{14}{21}$ ______

---

a) 5 c) None

b) 2 d) 7

29. For each fraction below, identify the factor common to both terms <u>if there is one</u>.

a) $\frac{3}{11}$ ______ b) $\frac{22}{55}$ ______ c) $\frac{13}{31}$ ______ d) $\frac{15}{18}$ ______

---

a) None c) None

b) 11 d) 3

30. If the numerator and denominator of a fraction do not contain a common factor, the fraction cannot be reduced to lower terms. We say that a fraction of that type is "<u>in lowest terms</u>".

Which of the fractions below are "<u>in lowest terms</u>"? ______________

a) $\frac{2}{3}$ b) $\frac{14}{16}$ c) $\frac{3}{9}$ d) $\frac{5}{9}$ e) $\frac{12}{16}$

Only (a) and (d)

31. The terms of $\frac{8}{24}$ have three common factors: 2, 4, and 8. We used each of them to reduce the fraction to lower terms at the right.

$$\frac{8}{24} = \left(\frac{4}{12}\right)\left(\frac{2}{2}\right) = \left(\frac{4}{12}\right)(1) = \frac{4}{12}$$

$$\frac{8}{24} = \left(\frac{2}{6}\right)\left(\frac{4}{4}\right) = \left(\frac{2}{6}\right)(1) = \frac{2}{6}$$

$$\frac{8}{24} = \left(\frac{1}{3}\right)\left(\frac{8}{8}\right) = \left(\frac{1}{3}\right)(1) = \frac{1}{3}$$

All three fractions on the right are <u>in lower terms</u>. However, only one is <u>in lowest terms</u>. Which one? ______

$\frac{1}{3}$

32. When the terms of a fraction have <u>more than one common factor</u>:

1) We can reduce <u>to lower terms</u> by factoring out any of the common factors.

2) We can reduce <u>to lowest terms</u> in one step only by factoring out the <u>largest</u> common factor.

Identify the <u>largest</u> common factor for the terms of each fraction below.

a) $\frac{12}{18}$ ______ b) $\frac{24}{36}$ ______ c) $\frac{22}{66}$ ______ d) $\frac{24}{40}$ ______

---

a) 6 b) 12 c) 22 d) 8

33. Since it is sometimes difficult to identify the largest common factor, we sometimes need more than one step to reduce a fraction to lowest terms. An example is discussed below.

We used 8 as the common factor to reduce the fraction at the right.

$$\frac{32}{48} = \left(\frac{4}{6}\right)\left(\frac{8}{8}\right) = \left(\frac{4}{6}\right)(1) = \frac{4}{6}$$

a) Is $\frac{4}{6}$ in lowest terms? ________

b) Therefore, we must reduce further. $\frac{32}{48} = \frac{4}{6} =$ ____

Answers: a) No b) $\frac{2}{3}$

34. Reduce each fraction to lowest terms.

a) $\frac{3}{6} =$ ____ b) $\frac{8}{6} =$ ____ c) $\frac{2}{12} =$ ____

Answers: a) $\frac{1}{2}$ b) $\frac{4}{3}$ c) $\frac{1}{6}$

35. Reduce each fraction to lowest terms.

a) $\frac{14}{21} =$ ____ b) $\frac{25}{10} =$ ____ c) $\frac{4}{20} =$ ____

Answers: a) $\frac{2}{3}$ b) $\frac{5}{2}$ c) $\frac{1}{5}$

36. Reduce each fraction to lowest terms.

a) $\frac{24}{32} =$ ____ b) $\frac{48}{18} =$ ____ c) $\frac{27}{81} =$ ____

Answers: a) $\frac{3}{4}$ b) $\frac{8}{3}$ c) $\frac{1}{3}$

37. We saw earlier that some improper fractions equal whole numbers. They can be converted to whole numbers by reducing to lowest terms. For example:

$$\frac{24}{8} = \left(\frac{3}{1}\right)\left(\frac{8}{8}\right) = \left(\frac{3}{1}\right)(1) = \frac{3}{1} \text{ or } 3 \qquad \frac{30}{6} = \left(\frac{5}{1}\right)\left(\frac{6}{6}\right) = \left(\frac{5}{1}\right)(1) = \frac{5}{1} \text{ or } 5$$

However, it is easier to convert fractions of that type to whole numbers by simply performing a division. That is:

$$\frac{24}{8} = 24 \div 8 = 3 \qquad \frac{30}{6} = 30 \div 6 = \text{\_\_\_\_}$$

Answer: 5

3-5 "CANCELLING" AND MULTIPLICATIONS INVOLVING FRACTIONS

When performing multiplications involving fractions, the products should always be reduced to lowest terms. In this section, we will discuss multiplications that have products that are not in lowest terms. We will show how the "cancelling" process can be used to simplify multiplications of that type.

---

38. We performed two multiplications below. Neither product is in lowest terms.

$$\left(\frac{1}{2}\right)\left(\frac{4}{5}\right) = \frac{4}{10} \qquad \left(\frac{4}{3}\right)\left(\frac{6}{7}\right) = \frac{24}{21}$$

Products should always be reduced to lowest terms. Therefore:

a) $\left(\frac{1}{2}\right)\left(\frac{4}{5}\right) = \frac{4}{10} =$ ____ b) $\left(\frac{4}{3}\right)\left(\frac{6}{7}\right) = \frac{24}{21} =$ ____

39. We reduced the product at the left below to lowest terms. Reduce the other two products to lowest terms.

$5\left(\frac{3}{10}\right) = \frac{15}{10} = \frac{3}{2}$ a) $3\left(\frac{4}{9}\right) =$ ______ b) $2\left(\frac{5}{4}\right) =$ ______

a) $\frac{2}{5}$ b) $\frac{8}{7}$

---

40. Sometimes a product reduces to a whole number. An example is shown below. Complete the other two multiplications.

$8\left(\frac{3}{4}\right) = \frac{24}{4} = 6$ a) $6\left(\frac{2}{3}\right) =$ ______ b) $5\left(\frac{1}{5}\right) =$ ______

a) $\frac{4}{3}$ b) $\frac{5}{2}$

---

41. The product at the right had to be reduced to lowest terms. The need to reduce the product to lowest terms can be avoided by a process called "<u>cancelling</u>". We used "cancelling" for the same multiplication at the right. Notice the steps:

$\left(\frac{4}{5}\right)\left(\frac{3}{8}\right) = \frac{12}{40} = \frac{3}{10}$

$\left(\frac{\overset{1}{\cancel{4}}}{5}\right)\left(\frac{3}{\underset{2}{\cancel{8}}}\right) = \frac{3}{10}$

1) We divided the numerator "4" and the denominator "8" by 4.
2) We got the <u>numerator</u> of the product by multiplying "1" and "3".
3) We got the <u>denominator</u> of the product by multiplying "5" and "2".

Let's use the "cancelling" process for each multiplication below.

a) Divide both the "3" and the "6" by 3 before multiplying. $\left(\frac{3}{4}\right)\left(\frac{5}{6}\right) =$ ______

b) Divide both the "4" and the "6" by 2 before multiplying. $\left(\frac{7}{6}\right)\left(\frac{4}{5}\right) =$ ______

a) 4 b) 1

---

42. "Cancelling" is a process in which we take out common factors <u>before</u> multiplying. We can take out a common factor whenever any numerator and denominator are divisible by the same number. Sometimes two "cancellings" are possible in the same multiplication. An example is shown below. Notice the steps:

1) We divided both 9 and 6 by 3.
2) We divided both 5 and 10 by 5.
3) We got the numerator and denominator of the product by the multiplications 3(1) and 2(2).

$\left(\frac{\overset{3}{\cancel{9}}}{\underset{2}{\cancel{10}}}\right)\left(\frac{\overset{1}{\cancel{5}}}{\underset{2}{\cancel{6}}}\right) = \frac{3}{4}$

Let's use two "cancellings" for each multiplication below.

a) Divide the "4" and "6" by 2 and the "5" and "15" by 5 before multiplying. $\left(\frac{4}{15}\right)\left(\frac{5}{6}\right) =$ ______

b) Divide the "9" and "12" by 3 and the "7" and "14" by 7 before multiplying. $\left(\frac{9}{14}\right)\left(\frac{7}{12}\right) =$ ______

a) $\left(\frac{\overset{1}{\cancel{3}}}{4}\right)\left(\frac{5}{\underset{2}{\cancel{6}}}\right) = \frac{5}{8}$

b) $\left(\frac{7}{\underset{3}{\cancel{6}}}\right)\left(\frac{\overset{2}{\cancel{4}}}{5}\right) = \frac{14}{15}$

43. We used two "cancellings" at the left below. Notice that we got the numerator of the product by multiplying "1" and "1". Complete the other multiplications.

$$\left(\frac{\overset{1}{\cancel{4}}}{\underset{3}{\cancel{21}}}\right)\left(\frac{\overset{1}{\cancel{7}}}{\underset{2}{\cancel{8}}}\right) = \frac{1}{6} \qquad \text{a) } \left(\frac{3}{10}\right)\left(\frac{5}{6}\right) = \_\_\_\_ \qquad \text{b) } \left(\frac{11}{20}\right)\left(\frac{5}{22}\right) = \_\_\_\_$$

a) $\left(\frac{\overset{2}{\cancel{4}}}{\underset{3}{\cancel{15}}}\right)\left(\frac{\overset{1}{\cancel{5}}}{\underset{3}{\cancel{6}}}\right) = \frac{2}{9}$

b) $\left(\frac{\overset{3}{\cancel{9}}}{\underset{2}{\cancel{14}}}\right)\left(\frac{\overset{1}{\cancel{7}}}{\underset{4}{\cancel{12}}}\right) = \frac{3}{8}$

---

44. We used two "cancellings" at the left below. Since we got the denominator of the product by multiplying "1" and "1", the product is a whole number. Complete the other multiplications.

$$\left(\frac{\overset{3}{\cancel{6}}}{\underset{1}{\cancel{5}}}\right)\left(\frac{\overset{1}{\cancel{5}}}{\underset{1}{\cancel{2}}}\right) = \frac{3}{1} = 3 \qquad \text{a) } \left(\frac{8}{7}\right)\left(\frac{7}{4}\right) = \_\_\_\_ \qquad \text{b) } \left(\frac{11}{15}\right)\left(\frac{15}{11}\right) = \_\_\_\_$$

a) $\left(\frac{\overset{1}{\cancel{3}}}{\underset{2}{\cancel{10}}}\right)\left(\frac{\overset{1}{\cancel{5}}}{\underset{2}{\cancel{6}}}\right) = \frac{1}{4}$

b) $\left(\frac{\overset{1}{\cancel{11}}}{\underset{4}{\cancel{20}}}\right)\left(\frac{\overset{1}{\cancel{5}}}{\underset{2}{\cancel{22}}}\right) = \frac{1}{8}$

---

45. We can also "cancel" when multiplying a whole number and a fraction. An example is shown below. Complete the other multiplications.

$$\overset{1}{\cancel{3}}\left(\frac{5}{\underset{3}{\cancel{9}}}\right) = \frac{5}{3} \qquad \text{a) } 5\left(\frac{3}{10}\right) = \_\_\_\_ \qquad \text{b) } \left(\frac{5}{6}\right)(8) = \_\_\_\_$$

a) $\left(\frac{\overset{2}{\cancel{8}}}{\underset{1}{\cancel{7}}}\right)\left(\frac{\overset{1}{\cancel{7}}}{\underset{1}{\cancel{4}}}\right) = \frac{2}{1} = 2$

b) $\left(\frac{\overset{1}{\cancel{11}}}{\underset{1}{\cancel{15}}}\right)\left(\frac{\overset{1}{\cancel{15}}}{\underset{1}{\cancel{11}}}\right) = \frac{1}{1} = 1$

---

46. We used "cancelling" at the left below. Notice that the product is a whole number. Complete the other multiplications.

$$\left(\frac{3}{\underset{1}{\cancel{5}}}\right)(\overset{2}{\cancel{10}}) = \frac{6}{1} = 6 \qquad \text{a) } 8\left(\frac{3}{2}\right) = \_\_\_\_ \qquad \text{b) } \left(\frac{4}{7}\right)(7) = \_\_\_\_$$

a) $\overset{1}{\cancel{5}}\left(\frac{3}{\underset{2}{\cancel{10}}}\right) = \frac{3}{2}$

b) $\left(\frac{5}{\underset{3}{\cancel{6}}}\right)(\overset{4}{\cancel{8}}) = \frac{20}{3}$

---

47. Cancelling is not possible with all multiplications. Some examples are shown below.

$$\left(\frac{3}{7}\right)\left(\frac{4}{5}\right) \qquad 24\left(\frac{2}{5}\right) \qquad \left(\frac{8}{9}\right)(13)$$

And even if you can cancel, be sure to check to see that the product is in lowest terms. Complete each multiplication below.

$$\text{a) } \left(\frac{2}{9}\right)\left(\frac{15}{16}\right) = \_\_\_\_ \qquad \text{b) } 11\left(\frac{4}{5}\right) = \_\_\_\_ \qquad \text{c) } \left(\frac{2}{3}\right)(15) = \_\_\_\_$$

a) $\overset{4}{\cancel{8}}\left(\frac{3}{\underset{1}{\cancel{2}}}\right) = \frac{12}{1} = 12$

b) $\left(\frac{4}{\underset{1}{\cancel{7}}}\right)(\overset{1}{\cancel{7}}) = \frac{4}{1} = 4$

---

a) $\frac{5}{24}$ b) $\frac{44}{5}$ c) 10

SELF-TEST 9 (pages 79-89)

Reduce each fraction to lowest terms.

1. $\frac{8}{32}$ = ____	2. $\frac{20}{12}$ = ____	3. $\frac{30}{45}$ = ____	4. $\frac{30}{24}$ = ____

Complete. Reduce each product to lowest terms.

5. $\left(\frac{4}{3}\right)\left(\frac{2}{5}\right) =$	6. $\left(\frac{1}{3}\right)\left(\frac{1}{4}\right) =$	7. $\left(\frac{3}{2}\right)\left(\frac{10}{9}\right) =$
8. $\left(\frac{25}{16}\right)\left(\frac{4}{5}\right) =$	9. $28\left(\frac{1}{7}\right) =$	10. $\left(\frac{7}{12}\right)(6) =$
11. $\frac{1}{4}(0) =$	12. $5\left(\frac{1}{5}\right) =$	13. $1\left(\frac{6}{5}\right) =$

ANSWERS:

1. $\frac{1}{4}$	3. $\frac{2}{3}$	5. $\frac{8}{15}$	7. $\frac{5}{3}$	9. 4	11. 0	13. $\frac{6}{5}$
2. $\frac{5}{3}$	4. $\frac{5}{4}$	6. $\frac{1}{12}$	8. $\frac{5}{4}$	10. $\frac{7}{2}$	12. 1	

3-6 MULTIPLICATIONS INVOLVING SIGNED FRACTIONS

In this section, we will discuss multiplications involving signed fractions. We will show that the rules for multiplying signed fractions are the same as the rules for multiplying signed whole numbers.

48. When multiplying signed whole numbers, <u>the product is negative when one factor is positive and the other is negative</u>. For example:

$(5)(-2) = -10$ $\qquad$ $(-7)(4) = -28$

The same rule applies when multiplying signed fractions. Therefore:

$\left(\frac{1}{3}\right)\left(-\frac{4}{5}\right) = -\frac{4}{15}$ $\qquad$ $\left(-\frac{5}{2}\right)\left(\frac{3}{7}\right) =$ ____

$-\frac{15}{14}$

49. When multiplying signed whole numbers, <u>the product is positive when both factors are negative</u>. For example:

$(-2)(-7) = 14$ $\qquad$ $(-10)(-5) = 50$

The same rule applies when multiplying signed fractions. Therefore:

$\left(-\frac{7}{8}\right)\left(-\frac{1}{2}\right) = \frac{7}{16}$ $\qquad$ $\left(-\frac{3}{2}\right)\left(-\frac{5}{4}\right) =$ ____

$\frac{15}{8}$

50. Complete each multiplication below.

a) $\left(\frac{1}{2}\right)\left(-\frac{1}{4}\right) =$ ______ b) $\left(-\frac{1}{3}\right)\left(-\frac{1}{5}\right) =$ ______ c) $\left(-\frac{1}{6}\right)\left(\frac{1}{5}\right) =$ ______

---

51. The same rules for the sign of the product apply when one of the factors is a whole number. Therefore:

$2\left(-\frac{3}{5}\right) = -\frac{6}{5}$ $\left(-\frac{3}{7}\right)(-4) = \frac{12}{7}$ $\left(\frac{1}{4}\right)(-7) =$ ______

Answers: a) $-\frac{1}{8}$ b) $\frac{1}{15}$ c) $-\frac{1}{30}$

---

52. When a negative fraction is multiplied by "1", the product is identical to the original fraction. That is:

$(1)\left(-\frac{3}{4}\right) = -\frac{3}{4}$ $\left(-\frac{8}{5}\right)(1) = -\frac{8}{5}$ $(1)\left(-\frac{1}{6}\right) =$ ______

Answer: $-\frac{7}{4}$

---

53. Negative fractions can be reduced to lowest terms. For example:

$-\frac{8}{12} = -\frac{2}{3}$ $-\frac{20}{16} = -\frac{5}{4}$ $-\frac{50}{10} = -5$

Reduce each fraction below to lowest terms.

a) $-\frac{16}{12} =$ ______ b) $-\frac{7}{14} =$ ______ c) $-\frac{9}{9} =$ ______

Answer: $-\frac{1}{6}$

---

54. Cancel where possible in these. <u>Be sure that the product is in lowest terms</u>.

a) $\left(\frac{3}{5}\right)\left(-\frac{7}{6}\right) =$ ______ b) $\left(-\frac{3}{4}\right)\left(-\frac{2}{9}\right) =$ ______ c) $\left(-\frac{10}{7}\right)\left(\frac{14}{5}\right) =$ ______

Answers: a) $-\frac{4}{3}$ b) $-\frac{1}{2}$ c) -1

---

55. Cancel where possible in these. <u>Be sure that the product is in lowest terms</u>.

a) $12\left(-\frac{1}{3}\right) =$ ______ b) $\left(-\frac{3}{8}\right)(-4) =$ ______ c) $\left(-\frac{1}{9}\right)(9) =$ ______

Answers: a) $-\frac{7}{10}$ b) $\frac{1}{6}$ c) -4

---

56. When a negative fraction is multiplied by "0", the product is "0". That is:

$(0)\left(-\frac{3}{7}\right) = 0$ $\left(-\frac{12}{11}\right)(0) = 0$ $(0)\left(-\frac{1}{2}\right) =$ ______

Answers: a) -4 b) $\frac{3}{2}$ c) -1

Answer: 0

3-7 RECIPROCALS

In the last unit, we discussed the reciprocals of signed whole numbers. In this section, we will review that concept and extend it to the reciprocals of signed fractions.

57. When the product of two numbers is "+1", the two numbers are called a "pair of reciprocals".

Since $\left(\frac{1}{3}\right)(3) = +1$, $\frac{1}{3}$ and 3 are called a "pair of reciprocals".

We say: The reciprocal of $\frac{1}{3}$ is 3. The reciprocal of 3 is $\frac{1}{3}$.

Since $\left(-\frac{1}{5}\right)(-5) = +1$, $-\frac{1}{5}$ and -5 are called a "pair of reciprocals".

We say: a) The reciprocal of $-\frac{1}{5}$ is ______. b) The reciprocal of -5 is ______.

---

a) -5
b) $-\frac{1}{5}$

58. The reciprocal of any signed whole number is a fraction with the same sign whose numerator is "1" and whose denominator is the absolute value of the whole number. That is:

The reciprocal of 4 is $\frac{1}{4}$. The reciprocal of -7 is $-\frac{1}{7}$.

Write the reciprocals of the following:

a) 8 ______ b) -6 ______ c) 12 ______ d) -15 ______

---

a) $\frac{1}{8}$ b) $-\frac{1}{6}$ c) $\frac{1}{12}$ d) $-\frac{1}{15}$

59. The reciprocal of any signed fraction whose numerator is "1" is a whole number with the same sign and the same absolute value as the denominator of the fraction. That is:

The reciprocal of $\frac{1}{9}$ is 9. The reciprocal of $-\frac{1}{2}$ is -2.

Write the reciprocals of the following:

a) $\frac{1}{7}$ ______ b) $-\frac{1}{6}$ ______ c) $\frac{1}{11}$ ______ d) $-\frac{1}{20}$ ______

---

a) 7
b) -6
c) 11
d) -20

60. The reciprocal of any signed fraction whose numerator is not "1" is a fraction with the same sign in which the terms of the original fraction are interchanged. For example:

Since $\left(\frac{5}{3}\right)\left(\frac{3}{5}\right) = +1$, $\frac{5}{3}$ and $\frac{3}{5}$ are a pair of reciprocals.

The reciprocal of $\frac{5}{3}$ is $\frac{3}{5}$. The reciprocal of $\frac{3}{5}$ is $\frac{5}{3}$.

Since $\left(-\frac{2}{7}\right)\left(-\frac{7}{2}\right) = +1$, $-\frac{2}{7}$ and $-\frac{7}{2}$ are a pair of reciprocals.

a) The reciprocal of $-\frac{2}{7}$ is ______. b) The reciprocal of $-\frac{7}{2}$ is ______.

---

a) $-\frac{7}{2}$
b) $-\frac{2}{7}$

61. Write the reciprocal of each of the following:

a) $\frac{3}{4}$ ______ b) $-\frac{9}{8}$ ______ c) $\frac{6}{11}$ ______ d) $-\frac{33}{25}$ ______

---

a) $\frac{4}{3}$ b) $-\frac{8}{9}$ c) $\frac{11}{6}$ d) $-\frac{25}{33}$

62. Write the reciprocals of each of these:

a) -10 ______ b) $\frac{9}{4}$ ______ c) $-\frac{2}{3}$ ______ d) $\frac{1}{8}$ ______

---

63. Write the reciprocal of each of these:

a) $-\frac{10}{11}$ ______ b) 15 ______ c) $-\frac{1}{12}$ ______ d) $\frac{19}{15}$ ______

Answers to 62: a) $-\frac{1}{10}$ b) $\frac{4}{9}$ c) $-\frac{3}{2}$ d) 8

---

64. There are two numbers that are their own reciprocals.

a) Since $(+1)(+1) = 1$, the reciprocal of $+1$ is ______.

b) Since $(-1)(-1) = 1$, the reciprocal of -1 is ______.

Answers to 63: a) $-\frac{11}{10}$ b) $\frac{1}{15}$ c) -12 d) $\frac{15}{19}$

---

65. Whenever "0" is a factor in a multiplication, the product is "0". For example:

$(0)(7) = 0$ $(+1)(0) = 0$ $(0)(-1) = 0$

$(0)(0) = 0$ $(0)\left(\frac{2}{5}\right) = 0$ $\left(-\frac{8}{3}\right)(0) = 0$

a) Can we ever get "+1" as a product when one of the factors is "0"? ______

b) Therefore, does "0" have a reciprocal? ______

Answers to 64: a) $+1$ b) -1

---

66. Write the reciprocals of these: a) $+1$ ______ b) -1 ______ c) 0 ______

Answers to 65: a) No b) No. "0" has no reciprocal.

Answers to 66: a) $+1$ b) -1 c) "0" has no reciprocal.

3-8 DIVISION OF SIGNED WHOLE NUMBERS

In this section, we will discuss division of signed whole numbers. We will show that any division can be converted to an equivalent multiplication, and then use conversions of that type to justify the rules for dividing signed whole numbers. We will discuss division by "0" and also show that any negative fraction can be written in three equivalent forms.

---

67. As we saw earlier, any division of whole numbers is written as a fraction in algebra. For example:

$8 \div 2$ is written $\frac{8}{2}$ $21 \div 7$ is written $\frac{21}{7}$

A division is written as a fraction even if the numerator is not divisible by the denominator. That is:

$11 \div 5$ is written $\frac{11}{5}$ $3 \div 10$ is written ______

Answer to 67: $\frac{3}{10}$

68. Divisions involving signed whole numbers are also written as fractions in algebra. For example:

$(-20) \div 4$ is written $\frac{-20}{4}$ $\qquad$ $3 \div (-5)$ is written $\frac{3}{-5}$

Write each division below as a fraction.

a) $24 \div (-6) =$ ______ $\qquad$ b) $(-17) \div 7 =$ ______ $\qquad$ c) $1 \div (-2) =$ ______

---

a) $\frac{24}{-6}$ $\qquad$ b) $\frac{-17}{7}$ $\qquad$ c) $\frac{1}{-2}$

69. Remember that a fraction is simply another way of writing a division. That is:

$\frac{-8}{4}$ means "-8 divided by 4" $\qquad$ $\frac{13}{-3}$ means "13 divided by -3"

a) $\frac{30}{-5}$ means "______ divided by ______" $\qquad$ b) $\frac{-6}{7}$ means "______ divided by ______".

---

a) 30 divided by -5

b) -6 divided by 7

70. Any division (or fraction) can be converted to an equivalent multiplication <u>by multiplying the numerator by the reciprocal of the denominator</u>. For example:

$$\frac{12}{3} = 12(\text{the reciprocal of } 3) = 12\left(\frac{1}{3}\right)$$

$$\frac{5}{6} = 5(\text{the reciprocal of } 6) = 5\left(\frac{1}{6}\right)$$

$$\frac{17}{4} = 17(\text{the reciprocal of } 4) = 17\left(\frac{1}{4}\right)$$

As you can see, each division at the left above is equal to the corresponding multiplication at the right. That is:

$\frac{12}{3} = 12\left(\frac{1}{3}\right)$ $\qquad$ $\frac{5}{6} = 5\left(\frac{1}{6}\right)$ $\qquad$ $\frac{17}{4} = 17\left(\frac{1}{4}\right)$

By multiplying the numerator by the reciprocal of the denominator, convert each division below to a multiplication.

a) $\frac{42}{7} =$ ______ $\qquad$ b) $\frac{19}{8} =$ ______ $\qquad$ c) $\frac{1}{2} =$ ______

---

a) $42\left(\frac{1}{7}\right)$ $\qquad$ b) $19\left(\frac{1}{8}\right)$ $\qquad$ c) $1\left(\frac{1}{2}\right)$

71. Divisions involving signed numbers can also be converted to multiplications <u>by multiplying the numerator by the reciprocal of the denominator</u>. That is:

$$\frac{10}{-5} = 10(\text{the reciprocal of } -5) = 10\left(-\frac{1}{5}\right)$$

a) $\frac{-7}{3} = -7(\text{the reciprocal of } 3) = -7(\quad)$

b) $\frac{2}{-9} = 2(\text{the reciprocal of } -9) = 2(\quad)$

---

a) $-7\left(\frac{1}{3}\right)$

b) $2\left(-\frac{1}{9}\right)$

72. By multiplying the numerator by the reciprocal of the denominator, convert each division below to a multiplication.

a) $\frac{-18}{3}$ = ______ b) $\frac{12}{-7}$ = ______ c) $\frac{5}{-11}$ = ______

---

73. When dividing signed numbers earlier, we saw that <u>the quotient is negative when one number is positive and the other negative</u>. This rule can be justified by converting the division to a multiplication. That is:

$\frac{-16}{8} = -16\left(\frac{1}{8}\right) = -2$ $\frac{35}{-7} = 35\left(-\frac{1}{7}\right) =$ ______

a) $-18\left(\frac{1}{3}\right)$

b) $12\left(-\frac{1}{7}\right)$

c) $5\left(-\frac{1}{11}\right)$

---

74. When dividing signed numbers earlier, we also saw that <u>the quotient is positive when both numbers are negative</u>. This rule can also be justified by converting the division to a multiplication. That is:

$\frac{-8}{-2} = -8\left(-\frac{1}{2}\right) = 4$ $\frac{-28}{-4} = -28\left(-\frac{1}{4}\right) =$ ______

-5

---

75. Complete each division below.

a) $\frac{-27}{3}$ = -9 b) $\frac{-54}{-9}$ = 6 c) $\frac{20}{-2}$ = -10 d) $\frac{-22}{-11}$ = 2

7

---

76. When a signed number is divided by itself, the quotient is "1". That is:

$\frac{3}{3} = 3\left(\frac{1}{3}\right) = 1$ $\frac{-8}{-8} = -8\left(-\frac{1}{8}\right) = 1$ a) $\frac{10}{10}$ = ______ b) $\frac{-12}{-12}$ = ______

a) -9 c) -10
b) 6 d) 2

---

77. When a signed number is divided by "1", the quotient is <u>identical</u> to the signed number. That is:

$\frac{5}{1}$ = 5(the reciprocal of 1) = 5(1) = 5

$\frac{-2}{1}$ = -2(the reciprocal of 1) = -2(1) = -2

a) 1
b) 1

---

78. When a signed number is divided by "-1", the quotient is the <u>opposite</u> of the signed number. That is:

$\frac{7}{-1}$ = 7(the reciprocal of -1) = 7(-1) = -7

$\frac{-4}{-1}$ = -4(the reciprocal of -1) = -4(-1) = -4

-2(1) = -2

---

79. When "0" is divided by a signed number, the quotient is "0". That is:

$\frac{0}{6}$ = 0(the reciprocal of 6) = $0\left(\frac{1}{6}\right) = 0$

$\frac{0}{-9}$ = 0(the reciprocal of -9) = $0\left(-\frac{1}{9}\right)$ = 0

-4(-1) = 4

$0\left(-\frac{1}{9}\right) = 0$

80. When a division of any number by "0" is converted to a multiplication, one of the factors is "the reciprocal of 0". For example:

$$\frac{6}{0} = 6(\text{the reciprocal of } 0) \qquad \frac{-4}{0} = -4(\text{the reciprocal of } 0)$$

Since "0" has no reciprocal, <u>DIVISION BY "0" IS IMPOSSIBLE</u>. Therefore:

Both $\frac{6}{0}$ and $\frac{-4}{0}$ are impossible and cannot be done.

Which of the divisions at the right are impossible to do? ________

a) $\frac{0}{1}$ b) $\frac{1}{0}$ c) $\frac{0}{-3}$ d) $\frac{-3}{0}$

Both (b) and (d)

81. Perform each division below if it can be done.

a) $\frac{0}{8}$ = ______ b) $\frac{8}{0}$ = ______ c) $\frac{-1}{0}$ = ______ d) $\frac{0}{-1}$ = ______

---

a) 0 c) impossible
b) impossible d) 0

82. We converted the division of "0" by itself to a multiplication at the right. Why is it impossible to do the division? ____________

$$\frac{0}{0} = 0(\text{the reciprocal of } 0)$$

Because "0" has no reciprocal.

83. Any fraction with either a <u>negative numerator alone</u> or a <u>negative denominator alone</u> is equal to a negative fraction. For example:

$\frac{-12}{3}$ and $\frac{12}{-3}$ equal $-\frac{12}{3}$ since each equals -4.

$\frac{-10}{2}$ and $\frac{10}{-2}$ equal $-\frac{10}{2}$ since each equals ______.

-5

84. The same fact applies to fractions that do not equal whole numbers. That is:

Both $\frac{-8}{5}$ and $\frac{8}{-5}$ equal $-\frac{8}{5}$. Both $\frac{-1}{7}$ and $\frac{1}{-7}$ equal ______.

$-\frac{1}{7}$

85. a) Write $\frac{-3}{4}$ in two equivalent ways. ______ and ______

b) Write $-\frac{8}{3}$ in two equivalent ways. ______ and ______

---

a) $\frac{3}{-4}$ and $-\frac{3}{4}$ b) $\frac{-8}{3}$ and $\frac{8}{-3}$

3-9 THE MULTIPLICATION AXIOM AND FRACTIONAL ROOTS

Equations like $3x = 7$ and $-5y = 4$ have fractional roots. We will use the multiplication axiom to solve equations of that type in this section.

86. Equations can have fractional roots. For example:

The root of $3x = 7$ is $\frac{7}{3}$, since $3\left(\frac{7}{3}\right) = 7$

The root of $9y = 5$ is $\frac{5}{9}$, since $9\left(\frac{5}{9}\right) = 5$

The multiplication axiom is used to solve equations like those above. For example, we used it to solve $3x = 7$ below. Notice that we multiplied both sides by $\frac{1}{3}$, the reciprocal of 3, in order to "get rid of" the coefficient. Using the same steps, solve $9y = 5$ by multiplying both sides by $\frac{1}{9}$, the reciprocal of 9.

$$3x = 7 \qquad\qquad 9y = 5$$

$$\frac{1}{3}(3x) = \frac{1}{3}(7)$$

$$1x = \frac{7}{3}$$

$$x = \frac{7}{3}$$

$$\frac{1}{9}(9y) = \frac{1}{9}(5)$$

$$1y = \frac{5}{9}$$

$$y = \frac{5}{9}$$

87. To solve $2 = 5t$ below, we multiplied both sides by $\frac{1}{5}$ to "get rid of" the coefficient "5". Solve $7 = 4F$ by multiplying both sides by $\frac{1}{4}$ to "get rid of" the coefficient of "4".

$$2 = 5t \qquad\qquad 7 = 4F$$

$$\frac{1}{5}(2) = \frac{1}{5}(5t)$$

$$\frac{2}{5} = 1t$$

$$t = \frac{2}{5}$$

$$\frac{1}{4}(7) = \frac{1}{4}(4F)$$

$$\frac{7}{4} = 1F$$

$$F = \frac{7}{4}$$

88. To solve $-2p = 3$ below, we multiplied both sides by the reciprocal of -2. Solve $-7 = -10x$ by multiplying both sides by the reciprocal of -10.

$$-2p = 3 \qquad\qquad -7 = -10x$$

$$\left(-\frac{1}{2}\right)(-2p) = \left(-\frac{1}{2}\right)(3)$$

$$1p = -\frac{3}{2}$$

$$p = -\frac{3}{2}$$

$$\left(-\frac{1}{10}\right)(-7) = \left(-\frac{1}{10}\right)(-10x)$$

$$\frac{7}{10} = 1x$$

$$x = \frac{7}{10}$$

89. It should be obvious that using the multiplication axiom is the same as <u>dividing the whole-number term by the coefficient of the letter</u>. Therefore, we can use a shortcut to solve equations like $5x = 8$ and $2y = 11$.

For $5x = 8$, $x = \frac{8}{5}$. For $2y = 11$, $y = \frac{11}{2}$

Use the shortcut to solve each equation below.

a) $10p = 3$ b) $7m = 12$ c) $4d = 1$

p = ______ m = ______ d = ______

Answers: a) $p = \frac{3}{10}$ b) $m = \frac{12}{7}$ c) $d = \frac{1}{4}$

90. When using the shortcut, we always write a negative root as a fraction with a "-" in front of it. For example:

For $2x = -3$, $x = \frac{-3}{2} = -\frac{3}{2}$ For $-5y = 1$, $y = \frac{1}{-5} = -\frac{1}{5}$

Use the shortcut to solve each equation below.

a) $-12t = 5$ b) $8F = -17$ c) $2p = -1$

t = ______ F = ______ p = ______

Answers: a) $t = -\frac{5}{12}$ $\left(\text{from } \frac{5}{-12}\right)$ b) $F = -\frac{17}{8}$ $\left(\text{from } \frac{-17}{8}\right)$ c) $p = -\frac{1}{2}$ $\left(\text{from } \frac{-1}{2}\right)$

91. When both terms of a fractional root are negative, remember that the root is positive. For example:

For $-4x = -5$, $x = \frac{-5}{-4} = \frac{5}{4}$ For $-20t = -13$, $t = \frac{-13}{-20} = \frac{13}{20}$

Use the shortcut to solve each equation below.

a) $-3d = -8$ b) $-12R = -1$ b) $-15q = -7$

d = ______ R = ______ q = ______

Answers: a) $d = \frac{8}{3}$ $\left(\text{from } \frac{-8}{-3}\right)$ b) $R = \frac{1}{12}$ $\left(\text{from } \frac{-1}{-12}\right)$ c) $q = \frac{7}{15}$ $\left(\text{from } \frac{-7}{-15}\right)$

92. When the letter-term is on the right side, <u>be sure to divide the whole-number term by the coefficient of the letter</u>. For example:

For $17 = 5x$, $x = \frac{17}{5}$ For $1 = -3y$, $y = \frac{1}{-3} = -\frac{1}{3}$

Use the shortcut to solve these.

a) $-1 = 7m$ b) $14 = 13S$ c) $-2 = -9b$

m = ______ S = ______ b = ______

Answers: a) $m = -\frac{1}{7}$ b) $S = \frac{14}{13}$ c) $b = \frac{2}{9}$

93. Solve each equation below.

a) $1 = 8x$ b) $-5 = -2t$ c) $10 = -9V$

x = ______ t = ______ V = ______

94. A fractional root should always be reduced to lowest terms. An example is shown below. Solve the other equations.

$10t = 15$ a) $-9p = 12$ b) $-6 = -14V$

$t = \frac{15}{10} = \frac{3}{2}$

p = ______ V = ______

Answers
a) $x = \frac{1}{8}$
b) $t = \frac{5}{2}$
c) $V = -\frac{10}{9}$

---

95. Reduce each root below to lowest terms.

a) $12m = -2$ b) $7 = 28x$ c) $-50 = -15y$

m = ______ x = ______ y = ______

Answers
a) $p = -\frac{4}{3}$
b) $V = \frac{3}{7}$

---

96. To check a root, we simply plug it in for the letter in the original equation to make sure that the equation is then true. An example is shown below. Solve and check the other equation.

$4m = -10$

$m = \frac{-10}{4} = -\frac{5}{2}$

Check

$\overset{2}{\cancel{4}}\left(-\frac{5}{\underset{1}{\cancel{2}}}\right) = -10$

$-10 = -10$

a) $-24 = -18x$ b) Check

Answers
a) $m = -\frac{1}{6}$
b) $x = \frac{1}{4}$
c) $y = \frac{10}{3}$

---

a) $x = \frac{4}{3}$ b) $-24 = -\overset{6}{\cancel{18}}\left(\frac{4}{\underset{1}{\cancel{3}}}\right)$

$-24 = -24$

SELF-TEST 10 (pages 89-99)

Complete. Reduce each product to lowest terms.

1. $\left(-\frac{1}{2}\right)\left(\frac{1}{3}\right) =$	2. $\left(-\frac{8}{15}\right)\left(-\frac{5}{2}\right) =$	3. $\left(-\frac{3}{4}\right)(16) =$	4. $(-1)\left(-\frac{4}{5}\right) =$

Write the reciprocal of each of these.

5. 4 ______	6. $-\frac{1}{7}$ ______	7. $\frac{2}{3}$ ______	8. $-\frac{10}{7}$ ______	9. -1 ______	10. 0 ______

Perform each possible division.

11. $\frac{-3}{-3} =$ ______	12. $\frac{-6}{1} =$ ______	13. $\frac{10}{-1} =$ ______	14. $\frac{0}{-9} =$ ______	15. $\frac{2}{0} =$ ______

(SELF-TEST 10 - Continued on following page.)

SELF-TEST 10 (pages 89-99) - Continued

Solve. Reduce each root to lowest terms.

16. $5x = 6$	17. $-3 = 8y$	18. $-10m = 4$
19. $-30 = -24d$	20. $-2V = 1$	21. $-1 = -7q$

ANSWERS:

1. $-\frac{1}{6}$	4. $\frac{4}{5}$	7. $\frac{3}{2}$	10. None	13. -10	16. $x = \frac{6}{5}$	19. $d = \frac{5}{4}$
2. $\frac{4}{3}$	5. $\frac{1}{4}$	8. $-\frac{7}{10}$	11. 1	14. 0	17. $y = -\frac{3}{8}$	20. $V = -\frac{1}{2}$
3. -12	6. -7	9. -1	12. -6	15. Impossible	18. $m = -\frac{2}{5}$	21. $q = \frac{1}{7}$

3-10 DIVISIONS INVOLVING FRACTIONS

In this section, we will discuss divisions involving fractions. We will show that any division of that type can be performed by converting the division to a multiplication.

97. Any division involving two fractions is written as a "complex" fraction in algebra. Two examples are shown below. In each complex fraction, the longer fraction line is called the "major" fraction line.

$\frac{2}{3} \div \frac{4}{7}$ is written $\dfrac{\frac{2}{3}}{\frac{4}{7}}$ $\qquad$ $\frac{9}{5} \div \frac{1}{8}$ is written $\dfrac{\frac{9}{5}}{\frac{1}{8}}$

The numerator of a complex fraction is above the major fraction line; the denominator of a complex fraction is below the major fraction line. That is:

In $\dfrac{\frac{2}{3}}{\frac{4}{7}}$: The numerator is $\frac{2}{3}$. The denominator is $\frac{4}{7}$.

In $\dfrac{\frac{9}{5}}{\frac{1}{8}}$: a) The numerator is ______. b) The denominator is ______.

a) $\frac{9}{5}$ b) $\frac{1}{8}$

98. Remember that a complex fraction stands for a division in which the numerator is divided by the denominator. That is:

$\dfrac{\frac{3}{5}}{\frac{6}{7}}$ means "divide $\frac{3}{5}$ by $\frac{6}{7}$". $\dfrac{\frac{1}{2}}{\frac{8}{3}}$ means "divide ____ by ____".

---

99. Any division involving a whole number and a fraction is also written as a "complex" fraction in algebra. For example: $\frac{1}{2}$ by $\frac{8}{3}$

$\frac{3}{4} \div 7$ is written $\dfrac{\frac{3}{4}}{7}$ $8 \div \frac{1}{5}$ is written $\dfrac{8}{\frac{1}{5}}$

The numerator and denominator of each complex fraction above are separated by the "major" fraction line. That is:

In $\dfrac{\frac{3}{4}}{7}$: The numerator is $\frac{3}{4}$. The denominator is 7.

In $\dfrac{8}{\frac{1}{5}}$: a) The numerator is ____.   b) The denominator is ____.

---

100. Remember that a complex fraction containing a whole number also stands for a division. That is: a) 8   b) $\frac{1}{5}$

$\dfrac{\frac{2}{3}}{12}$ means "divide $\frac{2}{3}$ by 12". $\dfrac{7}{\frac{1}{2}}$ means "divide ____ by ____".

---

101. Any division in complex-fraction form can be converted to a multiplication <u>by multiplying the numerator by the reciprocal of the denominator</u>. That is: 7 by $\frac{1}{2}$

a) $\dfrac{\frac{3}{7}}{\frac{4}{9}} = \frac{3}{7}\left(\text{the reciprocal of } \frac{4}{9}\right) = \frac{3}{7}\left(\frac{9}{4}\right)$

a) $\dfrac{6}{\frac{5}{3}} = 6\left(\text{the reciprocal of } \frac{5}{3}\right) = 6(\quad)$

b) $\dfrac{\frac{1}{4}}{10} = \frac{1}{4}\left(\text{the reciprocal of } 10\right) = \frac{1}{4}(\quad)$

---

102. Convert each division below to a multiplication.   a) $6\left(\frac{3}{5}\right)$ b) $\frac{1}{4}\left(\frac{1}{10}\right)$

a) $\dfrac{\frac{1}{5}}{\frac{3}{2}} =$ b) $\dfrac{1}{\frac{7}{4}} =$ c) $\dfrac{\frac{3}{8}}{9} =$

103. Having converted a division in complex fraction form to a multiplication, we can find the quotient by performing the multiplication. Find each quotient below.

a) $\dfrac{\frac{3}{4}}{\frac{2}{5}} = (\quad)(\quad) = $ ______ b) $\dfrac{\frac{8}{3}}{\frac{5}{7}} = (\quad)(\quad) = $ ______

a) $\left(\frac{1}{5}\right)\left(\frac{2}{3}\right)$ b) $(1)\left(\frac{4}{7}\right)$ c) $\left(\frac{3}{8}\right)\left(\frac{1}{9}\right)$

---

104. By converting each division to a multiplication, find each quotient below.

a) $\dfrac{3}{\frac{7}{2}} = (\quad)(\quad) = $ ______ b) $\dfrac{\frac{1}{2}}{5} = (\quad)(\quad) = $ ______

a) $\left(\frac{3}{4}\right)\left(\frac{5}{2}\right) = \frac{15}{8}$ b) $\left(\frac{8}{3}\right)\left(\frac{7}{5}\right) = \frac{56}{15}$

---

105. In the division at the left below, the reciprocal of the denominator is the whole number "3". Complete the other division.

$\dfrac{4}{\frac{1}{3}} = (4)(3) = 12$ $\dfrac{\frac{1}{5}}{\frac{1}{4}} = (\quad)(\quad) = $ ______

a) $(3)\left(\frac{2}{7}\right) = \frac{6}{7}$ b) $\left(\frac{1}{2}\right)\left(\frac{1}{5}\right) = \frac{1}{10}$

---

106. Quotients should always be reduced to lowest terms. Complete these.

a) $\dfrac{\frac{9}{8}}{\frac{3}{4}} = $ ______ b) $\dfrac{10}{\frac{5}{3}} = $ ______

$\frac{1}{5}(4) = \frac{4}{5}$

---

a) $\frac{3}{2}$ b) 6

3-11 DIVISIONS INVOLVING SIGNED FRACTIONS

Divisions involving signed fractions are also performed by converting the division to a multiplication. We will discuss divisions of that type in this section.

---

107. When a division involving fractions contains one positive and one negative term, the quotient is negative. For example:

$$\dfrac{-\frac{2}{3}}{\frac{7}{11}} = \left(-\frac{2}{3}\right)\left(\frac{11}{7}\right) = -\frac{22}{21}$$

Complete: a) $\dfrac{2}{-\frac{3}{7}} = (\quad)(\quad) = $ ______ b) $\dfrac{-\frac{1}{4}}{5} = (\quad)(\quad) = $ ______

a) $(2)\left(-\frac{7}{3}\right) = -\frac{14}{3}$ b) $\left(-\frac{1}{4}\right)\left(\frac{1}{5}\right) = -\frac{1}{20}$

108. When a division involving fractions contains two negative terms, the quotient is positive. For example:

$$\frac{-\frac{1}{2}}{-\frac{1}{3}} = \left(-\frac{1}{2}\right)(-3) = \frac{3}{2}$$

Complete: a) $\frac{-1}{-\frac{8}{5}} = (\quad)(\quad) = \underline{\qquad}$ b) $\frac{-\frac{4}{5}}{-3} = (\quad)(\quad) = \underline{\qquad}$

---

a) $(-1)\left(-\frac{5}{8}\right) = \frac{5}{8}$

b) $\left(-\frac{4}{5}\right)\left(-\frac{1}{3}\right) = \frac{4}{15}$

109. Perform each division. Report each quotient in lowest terms.

a) $\frac{-\frac{2}{3}}{-\frac{5}{6}} = \underline{\qquad}$ b) $\frac{-\frac{3}{4}}{9} = \underline{\qquad}$

---

a) $\frac{4}{5}$

b) $-\frac{1}{12}$

110. Report each quotient below in lowest terms.

a) $\frac{\frac{8}{9}}{-\frac{4}{5}} = \underline{\qquad}$ b) $\frac{-7}{-\frac{1}{2}} = \underline{\qquad}$

---

a) $-\frac{10}{9}$

b) 14

111. When a fraction is divided by itself, the quotient is "1". That is:

$$\frac{\frac{3}{5}}{\frac{3}{5}} = \left(\frac{3}{5}\right)\left(\frac{5}{3}\right) = 1 \qquad \frac{-\frac{7}{2}}{-\frac{7}{2}} = \left(-\frac{7}{2}\right)\left(-\frac{2}{7}\right) = \underline{\qquad}$$

1

112. When a fraction is divided by "1", the quotient is identical to the original fraction. For example:

$$\frac{\frac{9}{8}}{1} = \left(\frac{9}{8}\right)(1) = \frac{9}{8} \qquad \frac{-\frac{1}{3}}{1} = \left(-\frac{1}{3}\right)(1) = \underline{\qquad}$$

$-\frac{1}{3}$

113. When a fraction is divided by "-1", the quotient is the opposite of the original fraction. For example:

$$\frac{\frac{4}{5}}{-1} = \left(\frac{4}{5}\right)(-1) = -\frac{4}{5} \qquad \frac{-\frac{8}{3}}{-1} = \left(-\frac{8}{3}\right)(-1) = \underline{\qquad}$$

$\frac{8}{3}$

114. When "0" is divided by a fraction, the quotient is "0". For example:

$$\frac{0}{\frac{1}{2}} = (0)(2) = 0 \qquad \frac{0}{-\frac{6}{5}} = 0\left(-\frac{5}{6}\right) = \underline{\qquad}$$

0

115. We converted two divisions of a fraction by "0" to multiplications below.

$$\frac{\frac{5}{2}}{0} = \frac{5}{2}(\text{the reciprocal of } 0) \qquad \frac{-\frac{6}{7}}{0} = -\frac{6}{7}(\text{the reciprocal of } 0)$$

Why is it impossible to do each division? ____________________

Because "0" has no reciprocal.

116. Complete:

a) $\frac{-\frac{1}{8}}{-\frac{1}{8}} =$ ______ b) $\frac{0}{\frac{12}{13}} =$ ______ c) $\frac{-\frac{5}{3}}{1} =$ ______

a) 1 b) 0 c) $-\frac{5}{3}$

117. Complete:

a) $\frac{\frac{3}{10}}{-1} =$ ______ b) $\frac{-\frac{1}{2}}{0} =$ ______ c) $\frac{\frac{2}{9}}{\frac{2}{9}} =$ ______

a) $-\frac{3}{10}$ b) Impossible c) 1

3-12 EQUATIONS WITH FRACTIONAL COEFFICIENTS AND TERMS

Equations like $\frac{2}{3}x = 5$ and $\frac{3}{4}y = \frac{2}{7}$ contain fractional coefficients or fractional terms or both. We will show how the multiplication axiom is used to solve equations of that type in this section.

118. A letter with a whole-number coefficient indicates a multiplication. That is:

2x means "2 times x" -5y means "-5 times y"

Similarly, a letter with a fractional coefficient indicates a multiplication. That is:

$\frac{2}{3}x$ means "$\frac{2}{3}$ times x" $-\frac{5}{4}y$ means "$-\frac{5}{4}$ times y"

Therefore: a) If x = 15, $\frac{2}{3}x = \frac{2}{3}(15) =$ ______

b) If y = 8, $-\frac{5}{4}y = -\frac{5}{4}(8) =$ ______

a) 10

b) -10

119. We used the multiplication axiom to solve the equation at the left below. Notice that we multiplied both sides by $\frac{3}{2}$, the reciprocal of $\frac{2}{3}$. Solve the other equation by multiplying both sides by $\frac{4}{7}$, the reciprocal of $\frac{7}{4}$.

$$\frac{2}{3}x = 5$$

$$\frac{3}{2}\left(\frac{2}{3}x\right) = \frac{3}{2}(5)$$

$$1x = \frac{15}{2}$$

$$x = \frac{15}{2}$$

$$3 = \frac{7}{4}y$$

$$\frac{4}{7}(3) = \frac{4}{7}\left(\frac{7}{4}y\right)$$

$$\frac{12}{7} = 1y$$

$$y = \frac{12}{7}$$

120. To solve the equation below, we multiplied both sides by $\frac{1}{4}$, the reciprocal of 4. Solve the other equation by multiplying both sides by $\frac{1}{2}$, the reciprocal of 2.

$$4t = \frac{1}{2}$$

$$\frac{1}{4}(4t) = \frac{1}{4}\left(\frac{1}{2}\right)$$

$$1t = \frac{1}{8}$$

$$t = \frac{1}{8}$$

$$-\frac{5}{3} = 2d$$

$$\frac{1}{2}\left(-\frac{5}{3}\right) = \frac{1}{2}(2d)$$

$$-\frac{5}{6} = 1d$$

$$d = -\frac{5}{6}$$

121. To solve the equation below, we multiplied both sides by $\frac{7}{5}$. Solve the other equation by multiplying both sides by $\frac{3}{2}$.

$$\frac{5}{7}p = \frac{3}{4}$$

$$\frac{7}{5}\left(\frac{5}{7}p\right) = \frac{7}{5}\left(\frac{3}{4}\right)$$

$$1p = \frac{21}{20}$$

$$p = \frac{21}{20}$$

$$\frac{1}{5} = \frac{2}{3}y$$

$$\frac{3}{2}\left(\frac{1}{5}\right) = \frac{3}{2}\left(\frac{2}{3}y\right)$$

$$\frac{3}{10} = 1y$$

$$y = \frac{3}{10}$$

122. To solve the equation below, we multiplied both sides by 3. Solve the other equation.

$$\frac{1}{3}m = \frac{4}{5}$$

$$3\left(\frac{1}{3}m\right) = 3\left(\frac{4}{5}\right)$$

$$1m = \frac{12}{5}$$

$$m = \frac{12}{5}$$

$$-7 = \frac{1}{5}p$$

123. When using the multiplication axiom, we multiply both sides by the reciprocal of the coefficient of the letter in order to "get rid of" the coefficient. That is:

To solve $\frac{7}{8}x = 11$, we multiply both sides by $\frac{8}{7}$.

a) To solve $5y = \frac{3}{4}$, we multiply both sides by ______ .

b) To solve $-\frac{1}{2} = \frac{9}{7}d$, we multiply both sides by ______ .

c) To solve $-3 = \frac{1}{8}R$, we multiply both sides by ______ .

$$5(-7) = 5\left(\frac{1}{5}p\right)$$
$$-35 = 1p$$
$$p = -35$$

---

124. To solve the equation below, we multiplied both sides by $-\frac{2}{5}$. Solve the other equation by multiplying both sides by $-\frac{7}{2}$.

$$-\frac{5}{2}x = 3$$
$$-\frac{2}{5}\left(-\frac{5}{2}x\right) = -\frac{2}{5}(3)$$
$$1x = -\frac{6}{5}$$
$$x = -\frac{6}{5}$$

$$-\frac{1}{4} = -\frac{2}{7}y$$

a) $\frac{1}{5}$

b) $\frac{7}{9}$

c) 8

---

125. To solve the equation below, we multiplied both sides by $-\frac{1}{3}$. Solve the other equation by multiplying both sides by -4.

$$-3m = -\frac{7}{8}$$
$$-\frac{1}{3}(-3m) = -\frac{1}{3}\left(-\frac{7}{8}\right)$$
$$1m = \frac{7}{24}$$
$$m = \frac{7}{24}$$

$$\frac{1}{5} = -\frac{1}{4}t$$

$$-\frac{7}{2}\left(-\frac{1}{4}\right) = -\frac{7}{2}\left(-\frac{2}{7}y\right)$$
$$\frac{7}{8} = 1y$$
$$y = \frac{7}{8}$$

---

126. When the coefficient of the letter is negative, its reciprocal is negative. Therefore, we must multiply both sides of the equation by a negative quantity. That is:

To solve $-\frac{5}{9}x = 7$, we multiply both sides by $-\frac{9}{5}$.

a) To solve $-\frac{3}{4} = -9d$, we multiply both sides by ______ .

b) To solve $-\frac{1}{2}y = -4$, we multiply both sides by ______ .

$$-4\left(\frac{1}{5}\right) = -4\left(-\frac{1}{4}t\right)$$
$$-\frac{4}{5} = 1t$$
$$t = -\frac{4}{5}$$

127. Following the example at the left, solve the other two equations.

$$-\frac{1}{6}y = 1$$

a) $\frac{1}{2}x = -1$ b) $-1 = -\frac{1}{7}d$

$$-6\left(-\frac{1}{6}y\right) = -6(1)$$

$$y = -6$$

a) $-\frac{1}{9}$

b) -2

---

128. When solving an equation, <u>the root should be reduced to lowest terms</u>. An example is shown below. Solve the other equation.

$$\frac{4}{3}x = \frac{2}{9}$$

a) $\frac{8}{15} = \frac{2}{5}y$ b) $\frac{2}{3}m = 12$

$$\frac{3}{4}\left(\frac{4}{3}x\right) = \frac{\overset{1}{\cancel{3}}}{\underset{2}{\cancel{4}}}\left(\frac{\overset{1}{\cancel{2}}}{\underset{3}{\cancel{9}}}\right)$$

$$x = \frac{1}{6}$$

a) $x = -2$

b) $d = 7$

---

129. Solve these. Reduce each root to lowest terms.

a) $-\frac{5}{6}d = \frac{7}{12}$ b) $-\frac{4}{5}R = -20$ c) $-3 = \frac{9}{2}x$

a) $y = \frac{4}{3}$

b) $m = 18$

---

130. The root of the equation at the left below is "0". Solve the other equations.

$$\frac{2}{3}x = 0$$

a) $-\frac{7}{5}y = 0$ b) $0 = \frac{1}{4}d$

$$\frac{3}{2}\left(\frac{2}{3}x\right) = \frac{3}{2}(0)$$

$$x = 0$$

a) $d = -\frac{7}{10}$

b) $R = 25$

c) $x = -\frac{2}{3}$

---

131. We solved and checked an equation below. Solve and check the other equation.

$$\frac{1}{3}x = -\frac{2}{7}$$

<u>Check</u>

a) $-2x = \frac{1}{5}$ b) <u>Check</u>

$$3\left(\frac{1}{3}x\right) = 3\left(-\frac{2}{7}\right)$$

$$\frac{1}{\cancel{3}}\left(-\frac{\overset{2}{\cancel{6}}}{7}\right) = -\frac{2}{7}$$

$$x = -\frac{6}{7}$$

$$-\frac{2}{7} = -\frac{2}{7}$$

a) $y = 0$

b) $d = 0$

a) $x = -\frac{1}{10}$

b) Check

$$-\frac{1}{2}\left(-\frac{1}{10}\right) = \frac{1}{5}$$

$$\frac{1}{5} = \frac{1}{5}$$

132. We saw that there is a shortcut for the multiplication axiom when solving equations with whole-number coefficients and terms. That is, we simply divided the whole-number term by the coefficient. For example:

For $4x = 8$, $x = \frac{8}{4} = 2$ For $7y = 5$, $y = \frac{5}{7}$

A similar division could also be used to solve equations with fractional coefficients and terms. For example:

$$\text{For } \frac{2}{5}x = 3, \quad x = \frac{3}{\frac{2}{5}} = 3\left(\frac{5}{2}\right) = \frac{15}{2}$$

$$\text{For } 4x = \frac{1}{6}, \quad x = \frac{\frac{1}{6}}{4} = \frac{1}{6}\left(\frac{1}{4}\right) = \frac{1}{24}$$

$$\text{For } \frac{1}{3}m = \frac{4}{7}, \quad m = \frac{\frac{4}{7}}{\frac{1}{3}} = \frac{4}{7}\left(\frac{3}{1}\right) = \frac{12}{7}$$

However, <u>a division is ordinarily not used to solve equations with fractional roots and terms because using the multiplication axiom in the usual way is easier</u>.

SELF-TEST 11 (pages 99-107)

Complete. Reduce each quotient to lowest terms.

1. $\frac{\frac{5}{6}}{\frac{4}{3}} =$	2. $\frac{1}{\frac{1}{5}} =$	3. $\frac{-\frac{1}{3}}{7} =$	4. $\frac{-\frac{15}{2}}{-\frac{5}{4}} =$
5. $\frac{3}{-\frac{9}{2}} =$	6. $\frac{-\frac{7}{6}}{-\frac{7}{6}} =$	7. $\frac{-\frac{1}{5}}{1} =$	8. $\frac{0}{-\frac{4}{7}} =$

Solve. Reduce each root to lowest terms.

9. $\frac{1}{2}x = 8$	10. $-1 = \frac{7}{5}y$	11. $-4R = -\frac{1}{3}$
12. $\frac{2}{3}p = \frac{3}{4}$	13. $\frac{1}{8} = -\frac{1}{4}d$	14. $-\frac{6}{7}V = 0$

ANSWERS:

1. $\frac{5}{8}$
2. 5
3. $-\frac{1}{21}$
4. 6
5. $-\frac{2}{3}$
6. 1
7. $-\frac{1}{5}$
8. 0
9. $x = 16$
10. $y = -\frac{5}{7}$
11. $R = \frac{1}{12}$
12. $p = \frac{9}{8}$
13. $d = -\frac{1}{2}$
14. $V = 0$

Unit 4 FRACTIONS AND EQUATIONS: ADDITION AND SUBTRACTION

In this unit, the procedure for adding and subtracting numerical fractions are reviewed and extended to additions and subtractions involving signed fractions. Both axioms are used to solve equations with fractional roots and equations with fractional coefficients and terms.

4-1 ADDING FRACTIONS WITH "LIKE" DENOMINATORS

In this section, we will briefly discuss the procedure for adding fractions with "like" denominators.

---

1. To add fractions with identical or "like" denominators, we simply add their numerators and keep the same denominators. That is:

$$\frac{4}{7} + \frac{1}{7} = \frac{4 + 1}{7} = \frac{5}{7} \qquad \frac{2}{9} + \frac{5}{9} = \frac{2 + 5}{9} = \underline{\qquad}$$

$\frac{7}{9}$

2. When adding fractions, the sum can be an improper fraction. For example:

$$\frac{3}{5} + \frac{4}{5} = \frac{3 + 4}{5} = \frac{7}{5} \qquad \frac{7}{3} + \frac{6}{3} = \frac{7 + 6}{3} = \underline{\qquad}$$

$\frac{13}{3}$

3. When adding fractions, <u>we always reduce the sum to lowest terms</u>. For example:

$$\frac{1}{10} + \frac{3}{10} = \frac{4}{10} = \frac{2}{5} \qquad \frac{3}{4} + \frac{3}{4} = \frac{6}{4} = \underline{\qquad}$$

$\frac{3}{2}$

4. When the sum is an improper fraction, it can sometimes be reduced to a whole number. For example:

$$\frac{8}{9} + \frac{1}{9} = \frac{9}{9} = 1 \qquad \frac{2}{3} + \frac{7}{3} = \frac{9}{3} = \underline{\qquad}$$

3

5. Complete. Reduce each sum to lowest terms.

a) $\frac{1}{8} + \frac{1}{8} = \underline{\qquad}$ b) $\frac{5}{6} + \frac{5}{6} = \underline{\qquad}$ c) $\frac{11}{4} + \frac{9}{4} = \underline{\qquad}$

---

a) $\frac{1}{4}$ b) $\frac{5}{3}$ c) 5

4-2 WRITING A FRACTION AS AN EQUIVALENT FRACTION WITH A LARGER DENOMINATOR

To add fractions with "unlike" denominators, we must substitute an equivalent fraction with a larger denominator for one or both fractions. In this section, we will discuss the procedure for writing a fraction as an equivalent fraction with a larger denominator.

---

6. In the problem below, we are asked to convert $\frac{3}{4}$ to an equivalent fraction whose denominator is "8".

$$\frac{3}{4} = \frac{\square}{8}$$

To do so, we multiply $\frac{3}{4}$ by the appropriate fraction that equals "1". That is, we multiply $\frac{3}{4}$ by $\frac{2}{2}$ and get:

$$\frac{3}{4} = \frac{3}{4}\left(\frac{2}{2}\right) = \frac{6}{8}$$

To convert $\frac{3}{4}$ to an equivalent fraction whose denominator is "12" at the right, we must multiply $\frac{3}{4}$ by $\frac{3}{3}$. Do so.

$$\frac{3}{4} = \frac{3}{4}\left(\frac{\quad}{\quad}\right) = \frac{\square}{12}$$

$$\frac{3}{4}\left(\frac{3}{3}\right) = \frac{\boxed{9}}{12}$$

7. In order to decide what fraction to multiply by, we can divide the larger denominator by the original denominator. For example:

Since $10 \div 2 = 5$, we multiplied by $\frac{5}{5}$ at the right.

$$\frac{1}{2} = \frac{1}{2}\left(\frac{5}{5}\right) = \frac{5}{10}$$

Since $12 \div 3 = 4$, we should multiply by $\frac{4}{4}$ at the right. Do so.

$$\frac{5}{3} = \frac{5}{3}\left(\frac{\quad}{\quad}\right) = \frac{\square}{12}$$

$$\frac{5}{3}\left(\frac{4}{4}\right) = \frac{\boxed{20}}{12}$$

8. a) Since $24 \div 4 = 6$, we should multiply by $\frac{6}{6}$ at the right. Do so.

$$\frac{1}{4} = \frac{1}{4}\left(\frac{\quad}{\quad}\right) = \frac{\square}{24}$$

b) Since $14 \div 2 = 7$, we should multiply by $\frac{7}{7}$ at the right. Do so.

$$\frac{5}{2} = \frac{5}{2}\left(\frac{\quad}{\quad}\right) = \frac{\square}{14}$$

---

a) $\frac{1}{4}\left(\frac{6}{6}\right) = \frac{\boxed{6}}{24}$ b) $\frac{5}{2}\left(\frac{7}{7}\right) = \frac{\boxed{35}}{14}$

9. To complete these conversions, multiply by the appropriate fraction that equals "1".

a) $\frac{2}{3} = \frac{2}{3}\left(\frac{\quad}{\quad}\right) = \frac{\square}{15}$ b) $\frac{8}{5} = \frac{8}{5}\left(\frac{\quad}{\quad}\right) = \frac{\square}{10}$

---

a) $\frac{2}{3}\left(\frac{5}{5}\right) = \frac{\boxed{10}}{15}$

b) $\frac{8}{5}\left(\frac{2}{2}\right) = \frac{\boxed{16}}{10}$

10. It should be obvious that the following two-step shortcut can be used to find the new numerator below.

1) Divide 18 by 6 and get 3.
2) Multiply 5 by 3 and get 15.

$$\frac{5}{6} = \frac{\boxed{15}}{18}$$

Using the shortcut, complete these.

a) $\frac{1}{3} = \frac{\square}{6}$ b) $\frac{3}{2} = \frac{\square}{8}$

---

11. Complete: a) $\frac{3}{10} = \frac{\square}{20}$ b) $\frac{5}{4} = \frac{\square}{8}$ c) $\frac{7}{3} = \frac{\square}{9}$

a) $\frac{\boxed{2}}{6}$ b) $\frac{\boxed{12}}{8}$

---

12. Complete: a) $\frac{2}{5} = \frac{\square}{55}$ b) $\frac{11}{4} = \frac{\square}{20}$ c) $\frac{1}{2} = \frac{\square}{100}$

a) $\frac{\boxed{6}}{20}$ b) $\frac{\boxed{10}}{8}$ c) $\frac{\boxed{21}}{9}$

---

a) $\frac{\boxed{22}}{55}$ b) $\frac{\boxed{55}}{20}$ c) $\frac{\boxed{50}}{100}$

4-3 MULTIPLES

In this section, we will discuss the meaning of "multiples". Then in the next section, we will discuss additions of fractions with "unlike" denominators in which the larger denominator is a multiple of the smaller denominator.

---

13. If we divide a number by 3 and the remainder is "0", we say the number is divisible by 3. For example:

12 is divisible by 3, since $12 \div 3 = 4$

17 is not divisible by 3, since $17 \div 3 = 5$ r2

Which of the following are divisible by 3? ____________

a) 6 b) 8 c) 3 d) 5 e) 14 f) 21

(a), (c), and (f)

14. If a number is divisible by 3, we say it is a "multiple" of 3. For example:

27 is a multiple of 3, since $27 \div 3 = 9$

25 is not a multiple of 3, since $25 \div 3 = 8$ r1

Which of the following are multiples of 3? ____________

a) 7 b) 9 c) 15 d) 22 e) 33 f) 37

(b), (c), and (e)

15. A number is a multiple of 4 only if it is divisible by 4. Which of the following are multiples of 4? ____________

a) 8 b) 10 c) 4 d) 15 e) 31 f) 36

(a), (c), and (f)

16. A number is a multiple of 7 only if it is divisible by 7. Which of the following are multiples of 7? ____________

a) 13 b) 21 c) 29 d) 33 e) 42 f) 56

17. In general, one number is a multiple of a second only if it is divisible by the second.

a) Is 14 a multiple of 2? _____ b) Is 23 a multiple of 5? _____ c) Is 48 a multiple of 9? _____

(b), (e), and (f)

---

18. a) Is 34 a multiple of 8? _____ b) Is 80 a multiple of 10? _____

a) Yes b) No c) No

---

a) No b) Yes

4-4 ADDING FRACTIONS WHEN THE LARGER DENOMINATOR IS A MULTIPLE OF THE SMALLER

In this section, we will discuss the procedure for adding fractions with "unlike" denominators when the larger denominator is a multiple of the smaller.

---

19. We can add fractions by simply adding their numerators only when their denominators are "like" or common. To get such an addition when the larger denominator is a multiple of the smaller, we substitute an equivalent fraction for the fraction with the smaller denominator. An example is discussed below.

Two steps were needed to add the fractions at the right. That is:

$$\frac{3}{4} + \frac{1}{8} = \frac{6}{8} + \frac{1}{8} = \frac{7}{8}$$

1) We substitute $\frac{6}{8}$ for $\frac{3}{4}$ to get an equivalent addition of fractions with "like" or common denominators.

2) We then added $\frac{6}{8}$ and $\frac{1}{8}$ in the usual way.

Let's use the same two steps to add $\frac{7}{6}$ and $\frac{2}{3}$ below.

a) Substitute $\frac{4}{6}$ for $\frac{2}{3}$ to get an equivalent addition with "like" denominators. $\frac{7}{6} + \frac{2}{3} = \frac{7}{6} +$ _____

b) Then perform the addition of "like" fractions in the usual way. $\frac{7}{6} + \frac{2}{3} = \frac{7}{6} +$ _____ = _____

---

20. To get an addition of "like" fractions when the larger denominator is a multiple of the smaller, we substitute for the fraction with the smaller denominator. That is:

For $\frac{1}{2} + \frac{5}{8}$, we substitute $\frac{4}{8}$ for $\frac{1}{2}$.

For $\frac{7}{12} + \frac{5}{6}$, we substitute _____ for $\frac{5}{6}$.

a) $\frac{7}{6} + \frac{4}{6}$

b) $\frac{7}{6} + \frac{4}{6} = \frac{11}{6}$

---

21. a) Complete the addition at the right by substituting for $\frac{2}{5}$. $\frac{2}{5} + \frac{3}{10} =$ _____ $+ \frac{3}{10} =$ _____

b) Complete the addition at the right by substituting for $\frac{1}{4}$. $\frac{13}{16} + \frac{1}{4} = \frac{13}{16} +$ _____ = _____

$\frac{10}{12}$

22. By substituting for the fraction with the smaller denominator, complete each addition below.

a) $\frac{1}{9} + \frac{1}{3} = \frac{1}{9} + \_\_\_\_ = \_\_\_\_$ b) $\frac{1}{2} + \frac{11}{12} = \_\_\_\_ + \frac{11}{12} = \_\_\_\_$

Answers: a) $\frac{4}{10} + \frac{3}{10} = \frac{7}{10}$ b) $\frac{13}{16} + \frac{4}{16} = \frac{17}{16}$

---

23. Complete. Reduce each sum to lowest terms.

a) $\frac{5}{12} + \frac{1}{4} = \_\_\_\_\_\_\_\_$

b) $\frac{5}{6} + \frac{2}{3} = \_\_\_\_\_\_\_\_$

Answers: a) $\frac{1}{9} + \frac{3}{9} = \frac{4}{9}$ b) $\frac{6}{12} + \frac{11}{12} = \frac{17}{12}$

---

24. Complete. Reduce each sum to lowest terms.

a) $\frac{1}{2} + \frac{1}{10} = \_\_\_\_\_\_\_\_$

b) $\frac{8}{15} + \frac{4}{5} = \_\_\_\_\_\_\_\_$

Answers: a) $\frac{2}{3}$ $\left(\text{from } \frac{8}{12}\right)$ b) $\frac{3}{2}$ $\left(\text{from } \frac{9}{6}\right)$

Answers: a) $\frac{3}{5}$ $\left(\text{from } \frac{6}{10}\right)$ b) $\frac{4}{3}$ $\left(\text{from } \frac{20}{15}\right)$

4-5 ADDING FRACTIONS WHEN THE LARGER DENOMINATOR IS NOT A MULTIPLE OF THE SMALLER

In this section, we will define what is meant by a "lowest common denominator" and discuss a method for finding lowest common denominators. Then we will use lowest common denominators to add fractions when the larger denominator is not a multiple of the smaller.

---

25. In $\frac{1}{2} + \frac{2}{5}$, the larger denominator is not a multiple of the smaller. To perform the addition, we must substitute for both fractions to get an equivalent addition with "like" or common denominators. Any multiple of both 2 and 5 can be used as the common denominator. For example:

Since 10 is a multiple of both 2 and 5, we used 10 as the common denominator at the right.

$$\frac{1}{2} + \frac{2}{5} = \frac{5}{10} + \frac{4}{10} = \frac{9}{10}$$

Since 20 is a multiple of both 2 and 5, we used 20 as the common denominator at the right.

$$\frac{1}{2} + \frac{2}{5} = \frac{10}{20} + \frac{8}{20} = \frac{18}{20} = \frac{9}{10}$$

Did we get the same sum using both common denominators? ________

Answer: Yes. It is $\frac{9}{10}$.

26. To perform $\frac{3}{4}+\frac{1}{6}$, we can use any multiple of both 4 and 6 as the common denominator. For example:

Since 12 is a multiple of both 4 and 6, we used 12 as the common denominator at the right.

$$\frac{3}{4}+\frac{1}{6}=\frac{9}{12}+\frac{2}{12}=\frac{11}{12}$$

Since 24 is a multiple of both 4 and 6, we used 24 as the common denominator at the right.

$$\frac{3}{4}+\frac{1}{6}=\frac{18}{24}+\frac{4}{24}=\frac{22}{24}=\frac{11}{12}$$

Did we get the same sum using both common denominators? ________

Yes. It is $\frac{11}{12}$.

27. We performed the following additions in the last two frames: $\frac{1}{2}+\frac{2}{5}$ and $\frac{3}{4}+\frac{1}{6}$.

Though more than one common denominator was used for each addition, the smallest possible common denominator is ordinarily used. It is called the "lowest common denominator". That is:

For $\frac{1}{2}+\frac{2}{5}$, the "lowest common denominator" is 10.

For $\frac{3}{4}+\frac{1}{6}$, the "lowest common denominator" is 12.

As you can see, the "lowest common denominator" is the smallest multiple of both denominators. That is:

10 is the smallest multiple of both 2 and 5.

12 is the smallest multiple of both _____ and _____.

---

4 and 6

28. To find the lowest common denominator, we can test each multiple of the larger denominator until we find the smallest one that is also a multiple of the smaller denominator. Two examples are discussed below.

To find the lowest common denominator for $\frac{5}{6}+\frac{3}{8}$, we tested the multiples of 8 below.

Is 8 a multiple of 6? No.
Is 16 a multiple of 6? No.
Is 24 a multiple of 6? Yes.

Therefore, the lowest common denominator for $\frac{5}{6}+\frac{3}{8}$ is 24.

To find the lowest common denominator for $\frac{4}{5}+\frac{1}{6}$, we tested the multiples of 6 below.

Is 6 a multiple of 5? No.
Is 12 a multiple of 5? No.
Is 18 a multiple of 5? No.
Is 24 a multiple of 5? No.
Is 30 a multiple of 5? Yes.

Therefore, the lowest common denominator for $\frac{4}{5}+\frac{1}{6}$ is ________.

30

29. By testing the multiples of the larger denominator, find the lowest common denominator for each addition below.

a) $\frac{1}{5}+\frac{1}{3}$ ________ b) $\frac{11}{12}+\frac{9}{8}$ ________ c) $\frac{1}{6}+\frac{2}{7}$ ________

30. Using the same method, find the lowest common denominator for each addition below.

a) $\frac{11}{15} + \frac{9}{10}$ ______ b) $\frac{5}{6} + \frac{1}{10}$ ______ c) $\frac{7}{8} + \frac{4}{3}$ ______

Answers: a) 15 b) 24 c) 42

---

31. In $\frac{3}{7} + \frac{2}{5}$, 7 is not a multiple of 5. To perform the addition, we use the following steps:

1) Find the lowest common denominator. It is 35.

2) Substitute $\frac{15}{35}$ for $\frac{3}{7}$ and $\frac{14}{35}$ for $\frac{2}{5}$. $\frac{3}{7} + \frac{2}{5} = \frac{15}{35} + \frac{14}{35}$

3) Then add in the usual way. $\frac{3}{7} + \frac{2}{5} = \frac{15}{35} + \frac{14}{35} =$ ______

Answers: a) 30 b) 30 c) 24

---

32. a) The lowest common denominator for the addition below is ______.

b) Following the steps in the last frame, complete the addition. $\frac{1}{4} + \frac{5}{6} =$ ______ + ______ = ______

Answer: $\frac{29}{35}$

---

33. Find the lowest common denominator and then complete each addition below.

a) $\frac{3}{4} + \frac{1}{10} =$ ______ + ______ = ______

b) $\frac{4}{7} + \frac{1}{2} =$ ______ + ______ = ______

Answers: a) 12 b) $\frac{3}{12} + \frac{10}{12} = \frac{13}{12}$

---

34. Complete each addition below.

a) $\frac{5}{8} + \frac{7}{12} =$ ______ + ______ = ______

b) $\frac{1}{9} + \frac{3}{4} =$ ______ + ______ = ______

Answers: a) $\frac{15}{20} + \frac{2}{20} = \frac{17}{20}$ b) $\frac{8}{14} + \frac{7}{14} = \frac{15}{14}$

Answers: a) $\frac{15}{24} + \frac{14}{24} = \frac{29}{24}$ b) $\frac{4}{36} + \frac{27}{36} = \frac{31}{36}$

4-6 ADDING A FRACTION AND A WHOLE NUMBER

In this section, we will show how a whole number can be converted to an equivalent fraction with a specific denominator. Then we will discuss the procedure for adding a fraction and a whole number.

---

35. At the left below, we are asked to convert "3" to an equivalent fraction whose denominator is "4". To do so, we can multiply 3 by $\frac{4}{4}$ as we did at the right below.

$$3 = \frac{\square}{4} \qquad\qquad 3 = 3\left(\frac{4}{4}\right) = \frac{12}{4}$$

To convert "5" to an equivalent fraction whose denominator is "2" at the right, we can multiply 5 by $\frac{2}{2}$. Do so.

$$5 = 5\left(\frac{}{}\right) = \frac{\square}{2}$$

36. To convert a whole number to a fraction, we multiply by a fraction that equals "1". The denominator of the new fraction tells us what fraction to multiply by. That is:

$5\left(\frac{2}{2}\right) = \frac{\boxed{10}}{2}$

a) Since the new denominator is 7, multiply by $\frac{7}{7}$ for this conversion. $6 = 6\left(\frac{}{}\right) = \frac{\square}{7}$

b) Since the new denominator is 10, multiply by $\frac{10}{10}$ for this conversion. $1 = 1\left(\frac{}{}\right) = \frac{\square}{10}$

---

37. It should be obvious that we can make the same conversions <u>by simply multiplying the whole number and the denominator of the fraction</u>. For example:

a) $6\left(\frac{7}{7}\right) = \frac{\boxed{42}}{7}$
b) $1\left(\frac{10}{10}\right) = \frac{\boxed{10}}{10}$

$4 = \frac{\boxed{24}}{6}$ (Since $4 \times 6 = 24$) $8 = \frac{\boxed{16}}{2}$ (Since $8 \times 2 = 16$)

Using the shortcut above, complete each conversion below.

a) $2 = \frac{\square}{5}$ b) $3 = \frac{\square}{8}$ c) $1 = \frac{\square}{9}$ d) $10 = \frac{\square}{7}$

---

38. To add a whole number and a fraction, we simply convert the whole number to a fraction <u>with the same denominator</u>. For example:

a) $\frac{\boxed{10}}{5}$	c) $\frac{\boxed{9}}{9}$
b) $\frac{\boxed{24}}{8}$	d) $\frac{\boxed{70}}{7}$

$3 + \frac{1}{2} = \frac{6}{2} + \frac{1}{2} = \frac{7}{2}$ $\frac{5}{8} + 4 = \frac{5}{8} + \frac{32}{8} = \frac{37}{8}$

Using the same method, complete each addition below.

a) $2 + \frac{4}{5} = \_\_\_\_ + \frac{4}{5} = \_\_\_\_$ b) $\frac{1}{6} + 1 = \frac{1}{6} + \_\_\_\_ = \_\_\_\_$

---

39. Complete each addition below.

a) $\frac{10}{5} + \frac{4}{5} = \frac{14}{5}$
b) $\frac{1}{6} + \frac{6}{6} = \frac{7}{6}$

a) $1 + \frac{7}{8} = \_\_\_\_ + \frac{7}{8} = \_\_\_\_$

b) $\frac{5}{4} + 10 = \frac{5}{4} + \_\_\_\_ = \_\_\_\_$

---

a) $\frac{8}{8} + \frac{7}{8} = \frac{15}{8}$	b) $\frac{5}{4} + \frac{40}{4} = \frac{45}{4}$

4-7 ADDITIONS INVOLVING SIGNED FRACTIONS

In this section, we will discuss the procedure for additions involving signed fractions.

---

40. To add signed fractions with "like" denominators, we simply add their numerators in the usual way. That is:

$\frac{(-1)}{5} + \frac{3}{5} = \frac{(-1) + 3}{5} = \frac{2}{5}$ $\frac{(-6)}{7} + \frac{(-5)}{7} = \frac{(-6) + (-5)}{7} = \_\_\_\_$

41. Complete: a) $\frac{7}{4} + \frac{(-10)}{4} =$ ______ b) $\frac{(-1)}{3} + \frac{(-6)}{3} =$ ______

Answer: $\frac{-11}{7}$ or $-\frac{11}{7}$

---

42. To add signed fractions, <u>all negative signs should appear in the numerator</u>. When a negative sign appears in front of the fraction, we can simply substitute an equivalent form. For example:

In the addition at the right, the negative sign appears in front of the fraction. $\frac{7}{8} + \left(-\frac{2}{8}\right)$

Since $-\frac{2}{8} = \frac{(-2)}{8}$, we can substitute $\frac{(-2)}{8}$ for $-\frac{2}{8}$ to get the negative sign in the numerator. That is:

$$\frac{7}{8} + \left(-\frac{2}{8}\right) = \frac{7}{8} + \frac{(-2)}{8} = \_\_\_\_\_\_$$

Answers: a) $\frac{-3}{4}$ or $-\frac{3}{4}$ b) $\frac{-7}{3}$ or $-\frac{7}{3}$

---

43. Using the same steps, complete each addition below.

a) $\frac{3}{5} + \left(-\frac{7}{5}\right) = \frac{3}{5} + \frac{(\quad)}{5} =$ ______

b) $\left(-\frac{9}{16}\right) + \left(-\frac{2}{16}\right) = \frac{(\quad)}{16} + \frac{(\quad)}{16} =$ ______

Answer: $\frac{5}{8}$

---

44. Complete each addition.

a) $\left(-\frac{6}{7}\right) + \frac{11}{7} =$ ______

b) $\left(-\frac{2}{3}\right) + \left(-\frac{8}{3}\right) =$ ______

Answers: a) $\frac{3}{5} + \frac{(-7)}{5} = \frac{-4}{5}$ or $-\frac{4}{5}$ b) $\frac{(-9)}{16} + \frac{(-2)}{16} = \frac{-11}{16}$ or $-\frac{11}{16}$

---

45. Complete. <u>Reduce each sum to lowest terms</u>.

a) $\left(-\frac{7}{8}\right) + \frac{3}{8} =$ ______ b) $\left(-\frac{3}{2}\right) + \left(-\frac{5}{2}\right) =$ ______

Answers: a) $\frac{5}{7}$ b) $-\frac{10}{3}$

---

46. When the larger denominator is a multiple of the smaller, we must get common denominators first. For example:

$$\left(-\frac{1}{4}\right) + \frac{3}{8} = \left(-\frac{2}{8}\right) + \frac{3}{8} = \frac{(-2)}{8} + \frac{3}{8} = \frac{1}{8}$$

Using the same steps, complete each addition below.

a) $\frac{3}{16} + \left(-\frac{1}{2}\right) =$ ______

b) $\left(-\frac{1}{3}\right) + \left(-\frac{5}{6}\right) =$ ______

Answers: a) $\frac{-1}{2}$ or $-\frac{1}{2}$ b) -4

---

47. Complete. <u>Reduce each sum to lowest terms</u>.

a) $\frac{1}{2} + \left(-\frac{3}{10}\right) =$ ______

b) $\left(-\frac{11}{12}\right) + \left(-\frac{3}{4}\right) =$ ______

Answers: a) $\frac{-5}{16}$ or $-\frac{5}{16}$ b) $\frac{-7}{6}$ or $-\frac{7}{6}$

48. When the larger denominator is not a multiple of the smaller, we must also get common denominators first. For example:

$$\left(-\frac{5}{3}\right) + \frac{1}{2} = \left(-\frac{10}{6}\right) + \frac{3}{6} = \frac{(-10)}{6} + \frac{3}{6} = \frac{-7}{6} \text{ or } -\frac{7}{6}$$

Using the same steps, complete each addition below.

a) $\frac{3}{5} + \left(-\frac{2}{7}\right) =$ ______________

b) $\left(-\frac{5}{6}\right) + \left(-\frac{1}{4}\right) =$ ______________

a) $\frac{1}{5}$

b) $-\frac{5}{3}$

---

49. For additions involving a signed whole number and a signed fraction, we convert the whole number to a fraction. For example:

$$2 + \left(-\frac{4}{5}\right) = \frac{10}{5} + \left(-\frac{4}{5}\right) = \frac{6}{5} \qquad \frac{7}{8} + (-1) = \frac{7}{8} + \left(-\frac{8}{8}\right) = -\frac{1}{8}$$

Using the same steps, complete each addition below.

a) $7 + \left(-\frac{2}{3}\right) =$ ______________

b) $\left(-\frac{1}{2}\right) + (-1) =$ ______________

a) $\frac{11}{35}$

b) $\frac{-13}{12}$ or $-\frac{13}{12}$

---

50. Complete: a) $\left(-\frac{1}{4}\right) + \left(-\frac{1}{3}\right) =$ ______________

b) $1 + \left(-\frac{3}{8}\right) =$ ______________

a) $\frac{19}{3}$

b) $\frac{-3}{2}$ or $-\frac{3}{2}$

---

a) $\frac{-7}{12}$ or $-\frac{7}{12}$ b) $\frac{5}{8}$

4-8 CONVERTING IMPROPER FRACTIONS TO MIXED NUMBERS

In this section, we will discuss the procedure for converting improper fractions to mixed numbers.

---

51. Any addition of a whole number and a fraction can be written as a "mixed number". That is, the whole number and the fraction can be written next to each other with no other symbol. For example:

$1 + \frac{3}{4}$ can be written $1\frac{3}{4}$ $\qquad$ $5 + \frac{1}{3}$ can be written $5\frac{1}{3}$

$1\frac{3}{4}$ and $5\frac{1}{3}$ are called "mixed numbers" because each contains a whole-number part and a fraction part. Remember that any mixed number stands for an addition. That is:

$2\frac{1}{4}$ means $2 + \frac{1}{4}$ $\qquad$ $3\frac{4}{5}$ means _____ + _____

$3 + \frac{4}{5}$

52. Any improper fraction whose numerator is not divisible by its denominator can be converted to a mixed number. An example is shown below. Complete the other conversion.

$$\frac{9}{2} = \frac{8}{2} + \frac{1}{2} \qquad\qquad \frac{18}{7} = \frac{14}{7} + \frac{4}{7}$$

$$= 4 + \frac{1}{2} = 4\frac{1}{2} \qquad\qquad = \_\_\_ + \frac{4}{7} = \_\_\_$$

---

53. The two conversions below were performed in the last frame.

Answer (52): $2 + \frac{4}{7} = 2\frac{4}{7}$

$$\frac{9}{2} = 4\frac{1}{2} \qquad\qquad \frac{18}{7} = 2\frac{4}{7}$$

A shorter method for the same conversions is shown below. The shorter method involves "dividing" the numerator by the denominator.

$$\frac{9}{2} = 2\overline{)9}\ \ (4\text{ r}1) = 4\frac{1}{2} \qquad\qquad \frac{18}{7} = 7\overline{)18}\ \ (2\text{ r}4) = 2\frac{4}{7}$$

Note: 1) The whole-number part of the mixed number is the same as the whole-number part of the quotient.

2) The fraction part of the mixed number is obtained by writing the remainder over the divisor.

Using the shorter "division" method, convert each fraction below to a mixed number.

a) $\frac{8}{3} = 3\overline{)8} = \_\_\_$ b) $\frac{28}{5} = 5\overline{)28} = \_\_\_$ c) $\frac{79}{10} = 10\overline{)79} = \_\_\_$

---

54. When an improper fraction is converted to a mixed number, the fraction part of the mixed number is reduced to lowest terms. That is:

Answers (53):
a) $3\overline{)8}\ \ (2\text{ r}2) = 2\frac{2}{3}$
b) $5\overline{)28}\ \ (5\text{ r}3) = 5\frac{3}{5}$
c) $10\overline{)79}\ \ (7\text{ r}9) = 7\frac{9}{10}$

$$\frac{15}{6} = 2\frac{3}{6} = 2\frac{1}{2} \qquad\qquad \frac{30}{8} = 3\frac{6}{8} = 3\frac{3}{4}$$

Convert each improper fraction to a mixed number.

a) $\frac{6}{4} = \_\_\_$ b) $\frac{26}{10} = \_\_\_$ c) $\frac{40}{12} = \_\_\_$

---

55. A negative improper fraction is equal to a negative mixed number. That is:

Answers (54): a) $1\frac{1}{2}$ b) $2\frac{3}{5}$ c) $3\frac{1}{3}$

$$-\frac{4}{3} = -1\frac{1}{3}$$

a) $-\frac{12}{5} = \_\_\_$ b) $-\frac{42}{8} = \_\_\_$

> Note: In this course, we will ordinarily report answers as improper fractions rather than mixed numbers.

Answers (55): a) $-2\frac{2}{5}$ b) $-5\frac{1}{4}$

SELF-TEST 12 (pages 108-119)

Complete:

1. $\frac{1}{4} = \frac{\square}{12}$

2. $3 = \frac{\square}{8}$

Convert to a mixed number.

3. $\frac{20}{9} =$

4. $-\frac{22}{4} =$

5. $\frac{13}{8} + \frac{3}{8} =$

6. $\frac{1}{3} + \frac{11}{12} =$

7. $\frac{1}{5} + \frac{1}{2} =$

8. $\frac{3}{4} + 2 =$

Perform each addition. Reduce each answer to lowest terms.

9. $\left(-\frac{5}{7}\right) + \frac{2}{7} =$

10. $\frac{1}{2} + \left(-\frac{1}{10}\right) =$

11. $\left(-\frac{5}{4}\right) + \left(-\frac{1}{6}\right) =$

12. $1 + \left(-\frac{4}{5}\right) =$

ANSWERS:

1. $\frac{\boxed{3}}{12}$ 2. $\frac{\boxed{24}}{8}$ 3. $2\frac{2}{9}$ 4. $-5\frac{1}{2}$ 5. 2 6. $\frac{5}{4}$ 7. $\frac{7}{10}$ 8. $\frac{11}{4}$ 9. $-\frac{3}{7}$ 10. $\frac{2}{5}$ 11. $-\frac{17}{12}$ 12. $\frac{1}{5}$

4-9 SUBTRACTIONS INVOLVING FRACTIONS

In this section, we will briefly review subtracting a fraction from a larger fraction and subtracting a fraction from a larger whole number.

56. To subtract a smaller fraction from a larger fraction with the same denominator, we simply subtract their numerators. That is:

$$\frac{4}{5} - \frac{2}{5} = \frac{4-2}{5} = \frac{2}{5} \qquad \frac{8}{3} - \frac{1}{3} = \frac{8-1}{3} = \_\_\_\_$$

57. When subtracting fractions, <u>we always reduce the answer to lowest terms</u>. That is: $\frac{7}{3}$

$$\frac{9}{10} - \frac{7}{10} = \frac{2}{10} = \frac{1}{5} \qquad \frac{11}{4} - \frac{5}{4} = \frac{6}{4} = \_\_\_\_$$

58. When the answer is an improper fraction, it can sometimes be reduced to a whole number. That is: $\frac{3}{2}$

$$\frac{9}{2} - \frac{3}{2} = \frac{6}{2} = 3 \qquad \frac{7}{6} - \frac{1}{6} = \frac{6}{6} = \_\_\_\_$$

1

59. When a fraction is subtracted from itself, the answer is "0". That is:

$$\frac{3}{4} - \frac{3}{4} = \frac{0}{4} = 0 \qquad \frac{11}{5} - \frac{11}{5} = \frac{0}{5} = 0 \qquad \frac{1}{8} - \frac{1}{8} = \frac{0}{8} = \underline{\qquad}$$

Answer: 0

60. When the larger denominator is a multiple of the smaller, we must get common denominators first. For example:

$$\frac{5}{8} - \frac{1}{4} = \frac{5}{8} - \frac{2}{8} = \frac{3}{8}$$

Using the same steps, complete each subtraction below.

a) $\frac{1}{2} - \frac{5}{16} = \underline{\qquad} - \frac{5}{16} = \underline{\qquad}$ b) $\frac{11}{6} - \frac{2}{3} = \frac{11}{6} - \underline{\qquad} = \underline{\qquad}$

Answers: a) $\frac{8}{16} - \frac{5}{16} = \frac{3}{16}$ b) $\frac{11}{6} - \frac{4}{6} = \frac{7}{6}$

61. Complete. Reduce each answer to lowest terms.

a) $\frac{13}{12} - \frac{1}{4} = \underline{\qquad\qquad}$ b) $\frac{7}{3} - \frac{5}{6} = \underline{\qquad\qquad}$

Answers: a) $\frac{5}{6}$ b) $\frac{3}{2}$

62. When the larger denominator is not a multiple of the smaller, we must also get common denominators first. For example:

$$\frac{3}{4} - \frac{2}{3} = \frac{9}{12} - \frac{8}{12} = \frac{1}{12}$$

Using the same steps, complete each subtraction below.

a) $\frac{1}{2} - \frac{1}{5} = \underline{\qquad} - \underline{\qquad} = \underline{\qquad}$ b) $\frac{11}{6} - \frac{3}{8} = \underline{\qquad} - \underline{\qquad} = \underline{\qquad}$

Answers: a) $\frac{5}{10} - \frac{2}{10} = \frac{3}{10}$ b) $\frac{44}{24} - \frac{9}{24} = \frac{35}{24}$

63. To subtract a fraction from a larger whole number, we convert the whole number to a fraction with the same denominator. For example:

$$2 - \frac{5}{8} = \frac{16}{8} - \frac{5}{8} = \frac{11}{8}$$

Using the same steps, complete these.

a) $4 - \frac{1}{2} = \underline{\qquad} - \frac{1}{2} = \underline{\qquad}$ b) $1 - \frac{6}{7} = \underline{\qquad} - \frac{6}{7} = \underline{\qquad}$

Answers: a) $\frac{8}{2} - \frac{1}{2} = \frac{7}{2}$ b) $\frac{7}{7} - \frac{6}{7} = \frac{1}{7}$

4-10 OPPOSITES OF FRACTIONS

In this section, we will discuss the opposites of fractions. Then in the next section, we will use that concept for subtractions involving signed fractions.

---

64. When two fractions with the same terms but different signs are added, their sum is "0". That is:

$$\frac{3}{4} + \left(-\frac{3}{4}\right) = \frac{3}{4} + \frac{(-3)}{4} = \frac{0}{4} = 0 \qquad \frac{5}{2} + \left(-\frac{5}{2}\right) = \frac{5}{2} + \frac{(-5)}{2} = \frac{0}{2} = \underline{\qquad}$$

65. Two fractions are a pair of opposites if their sum is "0". Therefore: | 0

Since $\frac{3}{4} + \left(-\frac{3}{4}\right) = 0$, $\frac{3}{4}$ and $-\frac{3}{4}$ are a pair of opposites.

We say: The opposite of $\frac{3}{4}$ is $-\frac{3}{4}$. The opposite of $-\frac{3}{4}$ is $\frac{3}{4}$.

Since $\frac{5}{2} + \left(-\frac{5}{2}\right) = 0$, $\frac{5}{2}$ and $-\frac{5}{2}$ are a pair of opposites.

We say: a) The opposite of $\frac{5}{2}$ is ______. b) The opposite of $-\frac{5}{2}$ is ______.

Answers: a) $-\frac{5}{2}$ b) $\frac{5}{2}$

66. It should be obvious that two fractions are a pair of opposites if they have the same terms but different signs. That is:

The opposite of $\frac{1}{8}$ is $-\frac{1}{8}$. The opposite of $-\frac{7}{3}$ is $\frac{7}{3}$.

Write the opposite of: a) $\frac{4}{5}$ ______ b) $\frac{9}{4}$ ______ c) $-\frac{1}{2}$ ______

Answers: a) $-\frac{4}{5}$ b) $-\frac{9}{4}$ c) $\frac{1}{2}$

67. Write the opposite of: a) $-\frac{5}{6}$ ______ b) $-\frac{11}{7}$ ______ c) $\frac{1}{3}$ ______

Answers: a) $\frac{5}{6}$ b) $\frac{11}{7}$ c) $-\frac{1}{3}$

68. Don't confuse "opposites" with "reciprocals".

Two fractions are a pair of opposites if their sum is "0".

Two fractions are a pair of reciprocals if their product is "+1".

Therefore: The opposite of $\frac{2}{3}$ is $-\frac{2}{3}$. The reciprocal of $\frac{2}{3}$ is $\frac{3}{2}$.

a) The opposite of $-\frac{7}{4}$ is ______. b) The reciprocal of $-\frac{7}{4}$ is ______.

Answers: a) $\frac{7}{4}$ b) $-\frac{4}{7}$

69. a) The opposite of $\frac{1}{4}$ is ______. c) The opposite of $-\frac{8}{3}$ is ______.

b) The reciprocal of $\frac{1}{4}$ is ______. d) The reciprocal of $-\frac{8}{3}$ is ______.

Answers: a) $-\frac{1}{4}$ b) 4 c) $\frac{8}{3}$ d) $-\frac{3}{8}$

70. a) The opposite of $-\frac{1}{5}$ is ______. c) The opposite of $\frac{10}{9}$ is ______.

b) The reciprocal of $-\frac{1}{5}$ is ______. d) The reciprocal of $\frac{10}{9}$ is ______.

Answers: a) $\frac{1}{5}$ b) -5 c) $-\frac{10}{9}$ d) $\frac{9}{10}$

4-11 SUBTRACTIONS INVOLVING SIGNED FRACTIONS

In this section, we will discuss subtractions in which either the answer or one or both of the terms is a negative fraction. We will show that subtractions of that type can be performed by converting the subtractions to additions.

71. Any subtraction of fractions can be converted to an addition by "adding the opposite of the second term". That is:

$$\frac{2}{5} - \frac{4}{5} = \frac{2}{5} + \left(\text{the opposite of } \frac{4}{5}\right) = \frac{2}{5} + \left(-\frac{4}{5}\right)$$

$$\frac{7}{3} - \left(-\frac{2}{3}\right) = \frac{7}{3} + \left(\text{the opposite of } -\frac{2}{3}\right) = \frac{7}{3} + \frac{2}{3}$$

$$\left(-\frac{5}{9}\right) - \frac{8}{9} = \left(-\frac{5}{9}\right) + \left(\text{the opposite of } \frac{8}{9}\right) = \left(-\frac{5}{9}\right) + \_\_\_\_$$

Answer: $+\left(-\frac{8}{9}\right)$

---

72. Complete each conversion to an addition.

a) $\frac{5}{4} - \frac{7}{4} = \frac{5}{4} + \_\_\_\_$ b) $\frac{7}{8} - \left(-\frac{3}{8}\right) = \frac{7}{8} + \_\_\_\_$ c) $\left(-\frac{1}{2}\right) - \left(-\frac{3}{2}\right) = \left(-\frac{1}{2}\right) + \_\_\_\_$

Answers: a) $+\left(-\frac{7}{4}\right)$ b) $+\frac{3}{8}$ c) $+\frac{3}{2}$

---

73. Complete these by converting to addition and then finding the sum.

a) $\frac{5}{9} - \frac{10}{9} = \_\_\_\_ + \_\_\_\_ = \_\_\_\_$

b) $\left(-\frac{4}{7}\right) - \frac{6}{7} = \_\_\_\_ + \_\_\_\_ = \_\_\_\_$

Answers: a) $\frac{5}{9} + \left(-\frac{10}{9}\right) = -\frac{5}{9}$ b) $\left(-\frac{4}{7}\right) + \left(-\frac{6}{7}\right) = -\frac{10}{7}$

---

74. Complete these by converting to addition and then finding the sum.

a) $\frac{11}{16} - \left(-\frac{6}{16}\right) = \_\_\_\_ + \_\_\_\_ = \_\_\_\_$

b) $\left(-\frac{8}{5}\right) - \left(-\frac{1}{5}\right) = \_\_\_\_ + \_\_\_\_ = \_\_\_\_$

Answers: a) $\frac{11}{16} + \frac{6}{16} = \frac{17}{16}$ b) $\left(-\frac{8}{5}\right) + \frac{1}{5} = -\frac{7}{5}$

---

75. Use the same method for these. Reduce each answer to lowest terms.

a) $\frac{3}{8} - \frac{7}{8} = \_\_\_\_ + \_\_\_\_ = \_\_\_\_$

b) $\left(-\frac{1}{5}\right) - \frac{4}{5} = \_\_\_\_ + \_\_\_\_ = \_\_\_\_$

Answers: a) $\frac{3}{8} + \left(-\frac{7}{8}\right) = -\frac{4}{8} = -\frac{1}{2}$ b) $\left(-\frac{1}{5}\right) + \left(-\frac{4}{5}\right) = -\frac{5}{5} = -1$

---

76. When the larger denominator is a multiple of the smaller, we have to get common denominators before performing the addition. For example:

$$\frac{1}{4} - \frac{3}{2} = \frac{1}{4} + \left(-\frac{3}{2}\right) = \frac{1}{4} + \left(-\frac{6}{4}\right) = -\frac{5}{4}$$

$$\frac{1}{8} - \left(-\frac{3}{4}\right) = \frac{1}{8} + \frac{3}{4} = \_\_\_\_ + \_\_\_\_ = \_\_\_\_$$

Answer: $\frac{1}{8} + \frac{6}{8} = \frac{7}{8}$

---

77. Complete: a) $\frac{5}{12} - \left(-\frac{1}{4}\right) = \_\_\_\_$

b) $\left(-\frac{1}{2}\right) - \left(-\frac{7}{8}\right) = \_\_\_\_$

Answers: a) $\frac{2}{3}$ b) $\frac{3}{8}$

78. When the larger denominator is not a multiple of the smaller, we also have to get common denominators before performing the addition. For example:

$$\frac{1}{3} - \frac{3}{5} = \frac{1}{3} + \left(-\frac{3}{5}\right) = \frac{5}{15} + \left(-\frac{9}{15}\right) = -\frac{4}{15}$$

$$\frac{5}{4} - \left(-\frac{1}{6}\right) = \frac{5}{4} + \frac{1}{6} = \underline{\qquad} + \underline{\qquad} = \underline{\qquad}$$

$$\frac{15}{12} + \frac{2}{12} = \frac{17}{12}$$

79. Complete: a) $\left(-\frac{1}{3}\right) - \frac{1}{2} = \underline{\qquad\qquad}$

b) $\left(-\frac{1}{2}\right) - \left(-\frac{6}{7}\right) = \underline{\qquad\qquad}$

---

a) $-\frac{5}{6}$

b) $\frac{5}{14}$

80. When the subtraction involves a whole number, we have to convert the whole number to a fraction with the same denominator before adding. For example:

$$1 - \frac{7}{4} = 1 + \left(-\frac{7}{4}\right) = \frac{4}{4} + \left(-\frac{7}{4}\right) = -\frac{3}{4}$$

$$\frac{2}{3} - (-2) = \frac{2}{3} + 2 = \underline{\qquad} + \underline{\qquad} = \underline{\qquad}$$

$$\frac{2}{3} + \frac{6}{3} = \frac{8}{3}$$

81. Complete: a) $\left(-\frac{4}{5}\right) - 1 = \underline{\qquad\qquad}$

b) $\left(-\frac{5}{2}\right) - (-3) = \underline{\qquad\qquad}$

---

a) $-\frac{9}{5}$ b) $\frac{1}{2}$

4-12 BOTH AXIOMS AND EQUATIONS WITH FRACTIONAL ROOTS

In this section, we will use both axioms to solve equations with fractional roots. By doing so, we will review some types of equation-solving introduced earlier.

---

82. We used both axioms to solve the equation below. Notice that the root is a fraction. Solve the other equation.

$$4x + 5 = 12$$

$$4x + 5 + (-5) = 12 + (-5)$$

$$4x = 7$$

$$x = \frac{7}{4}$$

$$9 = 8t + 3$$

$$t = \frac{3}{4}$$

83. To solve the equation below, we began by converting the subtraction to addition. Solve the other equation.

$$6x - 1 = 10$$
$$6x + (-1) = 10$$
$$6x + (-1) + 1 = 10 + 1$$
$$6x = 11$$
$$x = \frac{11}{6}$$

$$5 = 3 - 8p$$

$p = -\frac{1}{4}$

84. Solve: a) $6 + 3m = 7$ b) $5 = 5 - 2y$

---

a) $m = \frac{1}{3}$

b) $y = 0$

85. We solved and checked the equation below. Solve and check the other equation.

$$7n + 6 = -2$$
$$7n + 6 + (-6) = (-2) + (-6)$$
$$7n = -8$$
$$n = -\frac{8}{7}$$

Check

$$7\left(-\frac{8}{7}\right) + 6 = -2$$
$$(-8) + 6 = -2$$
$$-2 = -2$$

$$2V - 1 = 4$$

Check

$V = \frac{5}{2}$

Check

$$2\left(\frac{5}{2}\right) - 1 = 4$$
$$5 - 1 = 4$$
$$4 = 4$$

86. To solve the equation below, we began by combining the letter terms. Solve the other equation.

$$3x + 2x = 6$$
$$5x = 6$$
$$x = \frac{6}{5}$$

$$-7 = t + 2t$$

$t = -\frac{7}{3}$

87. Solve: a) $2m - 5m = 2$ b) $5q - q = 2$

88. We solved and checked the equation below. Solve and check the other equation.

a) $m = -\frac{2}{3}$

b) $q = \frac{1}{2}$

$$4x + 5x = 2$$
$$9x = 2$$
$$x = \frac{2}{9}$$

$$4y - y = 1$$

Check

$$4\left(\frac{2}{9}\right) + 5\left(\frac{2}{9}\right) = 2$$
$$\frac{8}{9} + \frac{10}{9} = 2$$
$$\frac{18}{9} = 2$$
$$2 = 2$$

Check

---

89. We used both axioms to solve the equation below. Solve the other equation.

$y = \frac{1}{3}$

Check

$$4\left(\frac{1}{3}\right) - \frac{1}{3} = 1$$
$$\frac{4}{3} - \frac{1}{3} = 1$$
$$\frac{3}{3} = 1$$
$$1 = 1$$

$$9p + 3 = 5p$$
$$(-9p) + 9p + 3 = 5p + (-9p)$$
$$3 = -4p$$
$$p = -\frac{3}{4}$$

$$7x = 2 - x$$

---

90. To solve the equation below, we used the addition axiom twice. Solve the other equation.

$x = \frac{1}{4}$

$$3t - 1 = 6t + 4$$
$$3t + (-1) = 6t + 4$$
$$3t + (-1) + 1 = 6t + 4 + 1$$
$$3t = 6t + 5$$
$$(-6t) + 3t = (-6t) + 6t + 5$$
$$-3t = 5$$
$$t = -\frac{5}{3}$$

$$10 - F = 12 - 7F$$

$F = \frac{1}{3}$

91. We solved and checked the equation below. Solve and check the other equation.

$$4x = 8 - x \qquad\qquad 5x + 7 = x - 2$$

$$4x = 8 + (-1x)$$

$$4x + 1x = 8 + (-1x) + 1x$$

$$5x = 8$$

$$x = \frac{8}{5}$$

Check

$$4\left(\frac{8}{5}\right) = 8 - \frac{8}{5}$$

$$\frac{32}{5} = \frac{40}{5} - \frac{8}{5}$$

$$\frac{32}{5} = \frac{32}{5}$$

Check

$x = -\frac{9}{4}$

Check

$$5\left(-\frac{9}{4}\right) + 7 = \left(-\frac{9}{4}\right) - 2$$

$$\left(-\frac{45}{4}\right) + \frac{28}{4} = \left(-\frac{9}{4}\right) + \left(-\frac{8}{4}\right)$$

$$-\frac{17}{4} = -\frac{17}{4}$$

4-13 THE DISTRIBUTIVE PRINCIPLE AND EQUATIONS WITH FRACTIONAL ROOTS

In this section, we will solve equations with fractional roots that contain instances of the distributive principle. By doing so, we will review the procedure for solving equations of that type.

92. To solve the equation below, we began by multiplying by the distributive principle over addition. Solve the other equation.

$$3(x + 2) = 8 \qquad\qquad 5(1 + y) = 3y$$

$$3x + 6 = 8$$

$$3x + 6 + (-6) = 8 + (-6)$$

$$3x = 2$$

$$x = \frac{2}{3}$$

93. To solve the equation below, we began by multiplying by the distributive principle over subtraction. Solve the other equation.

$y = -\frac{5}{2}$

$$2(m - 1) = 1 \qquad\qquad 4(2 - t) = t$$

$$2m - 2 = 1$$

$$2m + (-2) = 1$$

$$2m + (-2) + 2 = 1 + 2$$

$$2m = 3$$

$$m = \frac{3}{2}$$

$t = \frac{8}{5}$

94. We solved and checked the equation below. Solve and check the other equation.

$$6(F - 1) = F$$
$$6F - 6 = F$$
$$6F + (-6) = F$$
$$(-6F) + 6F + (-6) = F + (-6F)$$
$$-6 = -5F$$
$$F = \frac{6}{5}$$

$$2(x + 3) = 7$$

Check

$$6\left(\frac{6}{5} - 1\right) = \frac{6}{5}$$
$$6\left(\frac{6}{5} - \frac{5}{5}\right) = \frac{6}{5}$$
$$6\left(\frac{1}{5}\right) = \frac{6}{5}$$
$$\frac{6}{5} = \frac{6}{5}$$

Check

Answer: $x = \frac{1}{2}$

Check

$$2\left(\frac{1}{2} + 3\right) = 7$$
$$2\left(\frac{1}{2} + \frac{6}{2}\right) = 7$$
$$2\left(\frac{7}{2}\right) = 7$$
$$7 = 7$$

95. Solve: a) $4(3x + 5) = 30$ b) $5(2 - 3p) = 5p$

Answers: a) $x = \frac{5}{6}$ b) $p = \frac{1}{2}$

96. Solve: a) $2(k - 3) = 5k + 1$ b) $3(1 - V) = 5(1 + V)$

Answers: a) $k = -\frac{7}{3}$ b) $V = -\frac{1}{4}$

SELF-TEST 13 (pages 119-128)

Perform each subtraction. Reduce each answer to lowest terms.

1. $\frac{7}{8} - \frac{5}{8} =$

2. $\frac{13}{10} - \frac{1}{5} =$

3. $\frac{1}{3} - \frac{1}{4} =$

4. $1 - \frac{1}{9} =$

Perform each subtraction. Reduce each answer to lowest terms.

5. $\frac{1}{6} - \frac{5}{6} =$

6. $\frac{7}{8} - \left(-\frac{1}{4}\right) =$

7. $\left(-\frac{5}{12}\right) - \left(-\frac{3}{8}\right) =$

8. $\left(-\frac{1}{2}\right) - (-1) =$

Solve each equation. Reduce each root to lowest terms.

9. $3x + 7 = 8$

10. $5y = 7 - y$

11. $3(t - 1) = 7t + 7$

ANSWERS:

1. $\frac{1}{4}$	3. $\frac{1}{12}$	5. $-\frac{2}{3}$	7. $-\frac{1}{24}$	9. $x = \frac{1}{3}$	11. $t = -\frac{5}{2}$
2. $\frac{11}{10}$	4. $\frac{8}{9}$	6. $\frac{9}{8}$	8. $\frac{1}{2}$	10. $y = \frac{7}{6}$	

4-14 COMBINING LETTER-TERMS WITH FRACTIONAL COEFFICIENTS TO SOLVE EQUATIONS

In this section, we discuss combining letter-terms with fractional coefficients by either adding or subtracting their coefficients. Then we will solve equations in which letter-terms of that type must be combined.

97. To add letter-terms with fractional coefficients, we simply add their coefficients. That is:

$$\frac{2}{5}x + \frac{1}{5}x = \left(\frac{2}{5} + \frac{1}{5}\right)x = \frac{3}{5}x \qquad \frac{1}{9}y + \frac{5}{9}y = \left(\frac{1}{9} + \frac{5}{9}\right)y = \_\_\_\_$$

98. When the fractional coefficients have different denominators, we have to get common denominators first. For example:

$\frac{2}{3}y$

$$\frac{1}{2}t + \frac{3}{4}t = \left(\frac{2}{4} + \frac{3}{4}\right)t = \frac{5}{4}t \qquad \frac{2}{3}m + \frac{1}{5}m = \left(\frac{10}{15} + \frac{3}{15}\right)m = \_\_\_\_$$

$\frac{13}{15}m$

99. When one of the coefficients is a whole number, we convert it to a fraction with the same denominator. For example:

$$p + \frac{1}{3}p = \left(\frac{3}{3} + \frac{1}{3}\right)p = \frac{4}{3}p \qquad \frac{1}{2}R + 2R = \left(\frac{1}{2} + \frac{4}{2}\right)R = \_\_\_\_$$

Answer: $\frac{5}{2}R$

100. One or both coefficients can be negative. For example:

$$x + \left(-\frac{2}{3}x\right) = \left[\frac{3}{3} + \left(-\frac{2}{3}\right)\right]x = \frac{1}{3}x \qquad \left(-\frac{1}{4}y\right) + \left(-\frac{3}{8}y\right) = \left[\left(-\frac{2}{8}\right) + \left(-\frac{3}{8}\right)\right]y = \_\_\_\_$$

Answer: $-\frac{5}{8}y$

101. Sometimes we get a whole number when adding fractional coefficients. For example:

$$\frac{5}{6}t + \frac{1}{6}t = \left(\frac{5}{6} + \frac{1}{6}\right)t = 1t \qquad \left(-\frac{3}{4}F\right) + \left(-\frac{5}{4}F\right) = \left[\left(-\frac{3}{4}\right) + \left(-\frac{5}{4}\right)\right]F = \_\_\_\_$$

Answer: $-2F$

102. To subtract letter-terms with fractional coefficients, we simply subtract their coefficients. That is:

$$\frac{1}{2}k - \frac{1}{3}k = \left(\frac{3}{6} - \frac{2}{6}\right)k = \frac{1}{6}k \qquad V - \frac{3}{5}V = \left(\frac{5}{5} - \frac{3}{5}\right)V = \_\_\_\_$$

Answer: $\frac{2}{5}V$

103. Complete each subtraction of letter-terms below.

a) $\frac{1}{8}y - \frac{3}{4}y = \left(\frac{1}{8} - \frac{6}{8}\right)y = \_\_\_\_$ b) $\frac{1}{2}x - x = \left(\frac{1}{2} - \frac{2}{2}\right)x = \_\_\_\_$

Answer: a) $-\frac{5}{8}y$ b) $-\frac{1}{2}x$

104. Complete these:

a) $\frac{4}{5}d - \frac{4}{5}d = \_\_\_\_$ b) $\frac{2}{3}R + 2R = \_\_\_\_$ c) $m - \frac{3}{2}m = \_\_\_\_$

Answer: a) 0 (from 0d) b) $\frac{8}{3}R$ c) $-\frac{1}{2}m$

105. To solve the equation at the left below, we combined the letter-terms and then used the multiplication axiom. Use the same method to solve the other equation.

$$\frac{1}{4}x + \frac{1}{2}x = 3$$

$$\frac{3}{4}x = 3$$

$$\frac{4}{3}\left(\frac{3}{4}x\right) = \frac{4}{3}(3)$$

$$x = 4$$

$$10 = \frac{2}{3}V - \frac{1}{4}V$$

Answer: $V = 24$

106. Following the example, solve the other equation.

$$t + \frac{3}{4}t = 2 \qquad\qquad 3 = \frac{5}{3}F - F$$

$$\frac{7}{4}t = 2$$

$$\frac{4}{7}\left(\frac{7}{4}t\right) = \frac{4}{7}(2)$$

$$t = \frac{8}{7}$$

---

107. Following the example, solve the other equation.

$F = \frac{9}{2}$

$$\frac{4}{5}y - \frac{3}{5}y = 2 \qquad\qquad m - \frac{5}{6}m = 5$$

$$\frac{1}{5}y = 2$$

$$5\left(\frac{1}{5}y\right) = 5(2)$$

$$y = 10$$

---

108. Following the example, solve the other equation.

$m = 30$

$$2t - \frac{4}{9}t = \frac{7}{3} \qquad\qquad \frac{1}{2}x + \frac{1}{3}x = \frac{1}{5}$$

$$\frac{14}{9}t = \frac{7}{3}$$

$$\frac{9}{14}\left(\frac{14}{9}t\right) = \frac{9}{14}\left(\frac{7}{3}\right)$$

$$t = \frac{3}{2}$$

---

109. Following the example, solve the other equation.

$x = \frac{6}{25}$

$$\left(-\frac{1}{5}d\right) + \left(-\frac{1}{5}d\right) = \frac{4}{3} \qquad\qquad \frac{1}{4}x - x = \frac{5}{8}$$

$$-\frac{2}{5}d = \frac{4}{3}$$

$$-\frac{5}{2}\left(-\frac{2}{5}d\right) = -\frac{5}{2}\left(\frac{4}{3}\right)$$

$$d = -\frac{10}{3}$$

$x = -\frac{5}{6}$

110. We solved and checked the equation below. Solve and check the other equation.

$$\frac{2}{3}x + \frac{1}{6}x = \frac{1}{2}$$

$$\frac{5}{6}x = \frac{1}{2}$$

$$\frac{6}{5}\left(\frac{5}{6}x\right) = \frac{6}{5}\left(\frac{1}{2}\right)$$

$$x = \frac{3}{5}$$

Check

$$\frac{2}{3}\left(\frac{3}{5}\right) + \frac{1}{6}\left(\frac{3}{5}\right) = \frac{1}{2}$$

$$\frac{2}{5} + \frac{1}{10} = \frac{1}{2}$$

$$\frac{4}{10} + \frac{1}{10} = \frac{1}{2}$$

$$\frac{1}{2} = \frac{1}{2}$$

$$m - \frac{3}{4}m = 2$$

Check

$m = 8$

Check

$$8 - \frac{3}{4}(8) = 2$$

$$8 - 6 = 2$$

$$2 = 2$$

4-15 THE ADDITION AXIOM AND EQUATIONS WITH FRACTIONAL TERMS

In this section, we will use the addition axiom to solve equations with fractional terms.

111. To solve the equation below, we used the addition axiom to "get rid of" the fraction on the same side as the letter. Use the same method to solve the other equation.

$$x + \frac{2}{3} = \frac{5}{6}$$

$$x + \frac{2}{3} + \left(-\frac{2}{3}\right) = \frac{5}{6} + \left(-\frac{2}{3}\right)$$

$$x = \frac{5}{6} + \left(-\frac{4}{6}\right)$$

$$x = \frac{1}{6}$$

$$0 = y + \frac{1}{2}$$

$y = -\frac{1}{2}$

112. Following the example, solve the other equation.

$$x + 1 = \frac{2}{5}$$

$$x + 1 + (-1) = \frac{2}{5} + (-1)$$

$$x = -\frac{3}{5}$$

$$2 + t = \frac{11}{8}$$

$t = -\frac{5}{8}$

113. To solve the equation below, we converted the subtraction to addition before using the addition axiom. Solve the other equation.

$$y - \frac{1}{4} = \frac{5}{8}$$

$$y + \left(-\frac{1}{4}\right) = \frac{5}{8}$$

$$y + \left(-\frac{1}{4}\right) + \frac{1}{4} = \frac{5}{8} + \frac{1}{4}$$

$$y = \frac{7}{8}$$

$$\frac{1}{2} = F - 3$$

$F = \frac{7}{2}$

114. We solved and checked the equation below. Solve and check the other equation.

$$x + \frac{1}{2} = \frac{2}{3}$$

$$x + \frac{1}{2} + \left(-\frac{1}{2}\right) = \frac{2}{3} + \left(-\frac{1}{2}\right)$$

$$x = \frac{1}{6}$$

<u>Check</u>

$$\frac{1}{6} + \frac{1}{2} = \frac{2}{3}$$

$$\frac{1}{6} + \frac{3}{6} = \frac{2}{3}$$

$$\frac{2}{3} = \frac{2}{3}$$

$$y - 1 = \frac{4}{5}$$

<u>Check</u>

$y = \frac{9}{5}$

<u>Check</u>

$$\frac{9}{5} - 1 = \frac{4}{5}$$

$$\frac{4}{5} = \frac{4}{5}$$

4-16 BOTH AXIOMS AND EQUATIONS WITH FRACTIONAL COEFFICIENTS AND TERMS

In this section, we will use both axioms to solve equations containing fractional coefficients and terms.

115. To solve the equation below, we used the addition axiom to "get rid of" the $\frac{1}{5}$ and then used the multiplication axiom to "get rid of" the $\frac{2}{5}$. Use the same method to solve the other equation.

$$\frac{2}{5}x + \frac{1}{5} = \frac{4}{5}$$

$$\frac{2}{5}x + \frac{1}{5} + \left(-\frac{1}{5}\right) = \frac{4}{5} + \left(-\frac{1}{5}\right)$$

$$\frac{2}{5}x = \frac{3}{5}$$

$$\frac{5}{2}\left(\frac{2}{5}x\right) = \frac{5}{2}\left(\frac{3}{5}\right)$$

$$x = \frac{3}{2}$$

$$\frac{1}{2}b + \frac{1}{3} = 1$$

$b = \frac{4}{3}$

116. To solve the equation below, we began by converting the subtraction to addition. Then we used both axioms. Use the same method to solve the other equation.

$$\frac{4}{7}p - \frac{1}{7} = \frac{6}{7}$$

$$\frac{4}{7}p + \left(-\frac{1}{7}\right) = \frac{6}{7}$$

$$\frac{4}{7}p + \left(-\frac{1}{7}\right) + \frac{1}{7} = \frac{6}{7} + \frac{1}{7}$$

$$\frac{4}{7}p = 1$$

$$\frac{7}{4}\left(\frac{4}{7}p\right) = \frac{7}{4}(1)$$

$$p = \frac{7}{4}$$

$$2x - \frac{1}{3} = \frac{1}{6}$$

$x = \frac{1}{4}$

117. Following the example, solve the other equation.

$$\frac{3}{4} - \frac{1}{4}x = \frac{1}{4}$$

$$\frac{3}{4} + \left(-\frac{1}{4}x\right) = \frac{1}{4}$$

$$\left(-\frac{3}{4}\right) + \frac{3}{4} + \left(-\frac{1}{4}x\right) = \frac{1}{4} + \left(-\frac{3}{4}\right)$$

$$-\frac{1}{4}x = -\frac{1}{2}$$

$$-4\left(-\frac{1}{4}x\right) = -4\left(-\frac{1}{2}\right)$$

$$x = 2$$

$$1 - \frac{1}{6}x = 0$$

$x = 6$

118. When converting the subtraction to addition below, we wrote "-1x" instead of "-x". Then later, we multiplied both sides by "-1" to "get rid of" the "-1" coefficient. Solve the other equation.

$$\frac{5}{6} - x = \frac{1}{6}$$

$$\frac{5}{6} + (-1x) = \frac{1}{6}$$

$$\left(-\frac{5}{6}\right) + \frac{5}{6} + (-1x) = \frac{1}{6} + \left(-\frac{5}{6}\right)$$

$$-1x = -\frac{2}{3}$$

$$(-1)(-1x) = (-1)\left(-\frac{2}{3}\right)$$

$$x = \frac{2}{3}$$

$$1 - y = \frac{7}{5}$$

$y = -\frac{2}{5}$

119. Following the example, solve the other equation.

$$\frac{4}{9}x - \frac{2}{9} = \frac{5}{9}x$$

$$\frac{4}{9}x + \left(-\frac{2}{9}\right) = \frac{5}{9}x$$

$$\left(-\frac{4}{9}x\right) + \frac{4}{9}x + \left(-\frac{2}{9}\right) = \frac{5}{9}x + \left(-\frac{4}{9}x\right)$$

$$-\frac{2}{9} = \frac{1}{9}x$$

$$9\left(-\frac{2}{9}\right) = 9\left(\frac{1}{9}x\right)$$

$$-2 = x$$

$$\frac{1}{2}y + \frac{1}{4} = \frac{3}{4}y$$

$y = 1$

120. Following the example, solve the other equation.

$$\frac{7}{8}t - \frac{1}{4} = \frac{1}{2}t + \frac{5}{8}$$

$$\frac{7}{8}t + \left(-\frac{1}{4}\right) = \frac{1}{2}t + \frac{5}{8}$$

$$\frac{7}{8}t + \left(-\frac{1}{4}\right) + \frac{1}{4} = \frac{1}{2}t + \frac{5}{8} + \frac{1}{4}$$

$$\frac{7}{8}t = \frac{1}{2}t + \frac{7}{8}$$

$$\left(-\frac{1}{2}t\right) + \frac{7}{8}t = \left(-\frac{1}{2}t\right) + \frac{1}{2}t + \frac{7}{8}$$

$$\frac{3}{8}t = \frac{7}{8}$$

$$\frac{8}{3}\left(\frac{3}{8}t\right) = \frac{8}{3}\left(\frac{7}{8}\right)$$

$$t = \frac{7}{3}$$

$$x + \frac{1}{3} = \frac{2}{3}x + 1$$

$x = 2$

SELF-TEST 14 (pages 128-135)

Solve each equation. Reduce each root to lowest terms.

1. $\frac{2}{3}x + \frac{1}{6}x = 10$	2. $\frac{3}{4}y - y = \frac{1}{8}$	3. $Q - \frac{3}{5}Q = 1$
4. $p + \frac{1}{3} = \frac{1}{4}$	5. $m - \frac{1}{2} = \frac{5}{6}$	6. $\frac{2}{7} = d - 1$
7. $\frac{1}{4}t + \frac{3}{8} = 0$	8. $2 = \frac{7}{5} - b$	9. $V - \frac{1}{8} = 2V + \frac{7}{8}$

ANSWERS:

1. $x = 12$
2. $y = -\frac{1}{2}$
3. $Q = \frac{5}{2}$
4. $p = -\frac{1}{12}$
5. $m = \frac{4}{3}$
6. $d = \frac{9}{7}$
7. $t = -\frac{3}{2}$
8. $b = -\frac{3}{5}$
9. $V = -1$

Unit 5 DECIMAL NUMBERS AND EQUATIONS

In this unit, we will review the four basic operations with decimal numbers and extend them to the same operations with signed decimal numbers. We will solve equations containing decimal coefficients and terms. The rules for rounding decimal numbers and the procedure for converting fractions to decimal numbers are also reviewed.

5-1 DECIMAL NUMBERS

In this section, we will review the names of places to the right of the decimal point and the word names for decimal numbers.

---

1. Any decimal number has a definite number of decimal places. By "decimal places", we mean the number of digits (including 0's) to the right of the decimal point. For example:

   17.28 has two decimal places

   .0407 has four decimal places

   How many decimal places are there in each number below?

   a) 25.9 ________ b) 1.305 ________ c) .000603 ________

---

a) one
b) three
c) six

2. The names of the first six places to the right of the decimal point are shown in the diagram.

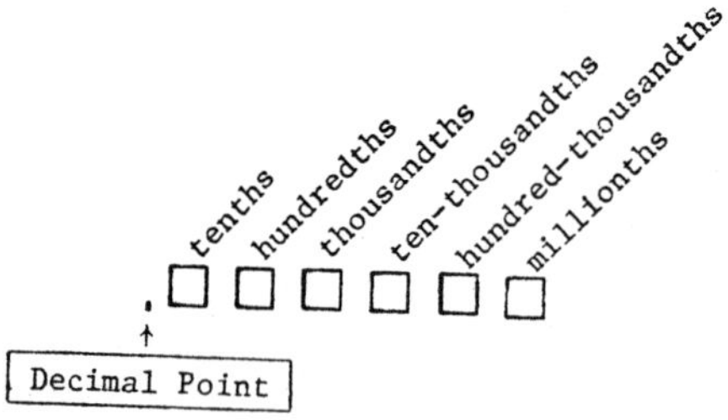

   a) In .27 , the "7" is in the ____________ place.

   b) In .0019 , the "1" is in the ____________ place.

   c) In .005308 , the "3" is in the ____________ place.

---

a) hundredths
b) thousandths
c) ten-thousandths

3. To name a decimal number smaller than "1", we name it as if it were a whole number and then add the name of the place of the last digit. For example:

.3 = three tenths

.55 = fifty-five hundredths

.017 = seventeen thousandths

.000608 = six hundred eight millionths

Write the decimal number corresponding to each word name.

a) eleven hundredths = ________ c) thirty-nine thousandths = ________

b) five millionths = ________ d) one hundred three ten-thousandths = ________

---

a) .11

b) .000005

c) .039

d) .0103

4. When naming decimal numbers larger than "1", the name of the whole-number part and the name of the decimal part are connected by the word "and". For example:

14.5 = fourteen and five tenths

8.92 = eight and ninety-two hundredths

3.007 = three and seven thousandths

Write the decimal number corresponding to each word name.

a) fifty and nine tenths = ________ b) six and twelve thousandths = ________

---

a) 50.9

b) 6.012

5. Sometimes decimal numbers are named by using the word "point" for the decimal point and then naming the digits to the right of the decimal point in order. For example:

7.28 = seven point two eight

.069 = point zero six nine

Write the decimal number corresponding to each word name.

a) twelve point five = ________ b) point five zero seven = ________

---

a) 12.5

b) .507

6. Amounts of money can be written in a decimal-type notation in which two decimal places are used for the "number of cents". For example:

$6.05 means "6 dollars and 5 cents"

$9.60 means "9 dollars and 60 cents"

Amounts of money containing either "0" dollars or "0" cents can also be written in a decimal-type notation. That is:

Instead of $5, we can write $5.00.

Instead of 89¢, we can write $.89.

Write each of these in a decimal-type notation.

a) 10 dollars = ________ b) 35 cents = ________

a) $10.00

b) $.35

5-2 ADDITIONS AND SUBTRACTIONS INVOLVING DECIMAL NUMBERS

In this section, we will discuss additions and subtractions involving decimal numbers.

7. To add decimal numbers, we line up the decimal points so that we are adding the digits in the same places. Two examples are shown below. Complete the other addition.

```
  .184        ²¹
+ .503        25.8        $7.95
  .687        49.2      + $8.50
            + 16.7
              91.7
```

a) $16.45

8. We get an equivalent decimal number <u>if we add one or more final 0's</u> to a decimal number. That is:

.47 = .470 = .4700 = .47000

8.1 = 8.10 = 8.100 = 8.1000

The fact above can be used to add numbers with a different number of decimal places. That is, we can add one or more 0's to the numbers with less decimal places so that all numbers have the same number of decimal places. An example is shown. Complete the other additions.

```
  .418      .418     a)  .008     b)  .3
  .9        .900       + .09          .075
  .32       .320                    + .22
+ .5      + .500
           2.138
```

a) .098

b) .595

9. We get an equivalent number <u>if we add one or more 0's after the decimal point</u> in a whole number. For example:

9 = 9.0 = 9.00 = 9.000

45 = 45.0 = 45.00 = 45.000

The fact above can be used to add whole numbers and decimal numbers. That is, we can again add one or more 0's until all numbers have the same number of decimal places. An example is shown. Complete the other additions.

```
  3.41       3.410    a)  10       b)  25.2
  8          8.000      +  3.85         1.66
+  .275    +  .275                   + 40
            11.685
```

a) 13.85

b) 66.86

10. When an addition is written horizontally, we rewrite the numbers vertically. An example is shown below. Notice that we lined up the decimal points. We also added some 0's before performing the addition. Complete the other additions.

5.03 + 9 + .7

```
  5.03
  9.00
+  .70
 14.73
```

a) 6.5 + 8.1 + 1.4

b) .01 + 3 + .6

---

a) 16.0

b) 3.61

11. To subtract decimal numbers, we also line up the decimal points so that we are subtracting the digits in the same places. Two examples are shown below. Complete the other subtraction.

```
  .947        3 14            $ 2 9.2 5
- .135        4̸ 5̸.11        - $ 1 6.9 9
  .812      - 1 7.6
              2 7.5
```

$12.26

12. To subtract numbers with a different number of decimal places, it is also helpful to add one or more final 0's so that both numbers have the same number of decimal places. An example is shown. Complete the other subtractions.

```
    6          5  9  9         a)  .8 7        b)  1
- .1 4 8       6̸.10̸10̸10         - .2            - .6 6
             - .1 4 8
               5.8 5 2
```

---

a) .67

b) .34

13. When a subtraction is written horizontally, we also rewrite the numbers vertically. An example is shown below. Notice that we lined up the decimal points. We also added some 0's before performing the subtraction. Complete the other subtractions.

$20 - $17.85

```
    1  9  9
  $ 2̸10̸.10̸10
- $ 1 7.8 5
  $   2.1 5
```

a) 1 - .25

b) .648 - .6

---

a) .75

b) .048

5-3 ADDITIONS INVOLVING SIGNED DECIMAL NUMBERS

In this section, we will discuss additions involving signed decimal numbers. Before doing so, however, we will discuss comparisons of decimal numbers and comparisons of the absolute values of decimal numbers.

---

14. To compare decimal numbers with the same number of decimal places, we can use the rules below.

1) If they have different whole-number parts, the one with the larger whole-number part is larger. For example:

   75.3 is larger than 18.7 (since 75 is larger than 18)

2) If they only have decimal parts, we can ignore the decimal points to make the comparison. For example:

   .046 is larger than .033 (since 46 is larger than 33)

3) If they have the same whole-number parts, the one with the larger decimal part is larger. For example:

   5.71 is larger than 5.26 (since .71 is larger than .26)

Which number in each pair is larger?

a) 1.346 or 1.299 b) .04 or .05 c) 19.1 or 18.8

Answers: a) 1.346 b) .05 c) 19.1

15. To compare decimal numbers with a different number of decimal places, we can add as many 0's as necessary until both have the same number of decimal places. For example:

To compare .83 and .5 , we added <u>one</u> "0" to .5 below.

.83	.83	Since .83 is larger than .50 ,
.5	.50	.83 is larger than .5 .

To compare 6.7 and 6.635 , we added <u>two</u> 0's to 6.7 below.

6.7	6.700	Since 6.700 is larger than 6.635 ,
6.635	6.635	6.7 is larger than 6.635 .

Which number in each pair is larger?

a) .21 or .2 b) .0059 or .01 c) 25.8 or 25

Answers: a) .21 b) .01 c) 25.8

16. For practical purposes, the absolute value of a signed decimal number is the number with the sign ignored. For example:

The absolute value of +.075 is .075 .

The absolute value of -8.3 is 8.3 .

Therefore, to compare the absolute values of signed decimal numbers, we can simply ignore their signs. That is:

The absolute value of +5.6 is larger than the absolute value of -1.3 , since 5.6 is larger than 1.3 .

The absolute value of -.075 is larger than the absolute value of +.019 , since .075 is larger than .019 .

Continued on following page.

16. Continued

Which number in each pair has the larger absolute value?

a) +4 or -.5 b) +6.2 or -6.21 c) +.08 or -.079

---

a) +4
b) -6.21
c) +.08

17. The rule for adding two negative decimal numbers is the same as the rule for adding two negative whole numbers. That is:

1) The sign of the sum is negative.
2) The absolute value of the sum is obtained by adding the absolute values of the two terms.

Some examples of the rule above are shown below. Notice that we added the absolute values vertically.

(-2.5) + (-1.3) = -3.8

$$\begin{array}{r} 2.5 \\ +\ 1.3 \\ \hline 3.8 \end{array}$$

(-.6) + (-.275) = -.875

$$\begin{array}{r} .600 \\ +\ .275 \\ \hline .875 \end{array}$$

Complete: a) (-5.6) + (-7) = ________ b) (-.866) + (-.479) = ________

---

a) -12.6
b) -1.345

18. The rule for adding one positive and one negative decimal number is the same as the rule for adding one positive and one negative whole number. That is:

1) The sign of the sum is the same as the sign of the term with the larger absolute value.
2) The absolute value of the sum is obtained by subtracting the smaller absolute value from the larger absolute value.

Some examples of the rule above are shown below. Notice that we subtracted the absolute values vertically.

6.25 + (-4.12) = 2.13

$$\begin{array}{r} 6.25 \\ -\ 4.12 \\ \hline 2.13 \end{array}$$

(-.7) + .55 = -.15

$$\begin{array}{r} .\overset{6}{\not7}{}^{1}0 \\ -\ .5\ 5 \\ \hline .1\ 5 \end{array}$$

Complete: a) 6.4 + (-8.1) = ________ b) (-.02) + .069 = ________

---

a) -1.7 b) .049

19. Complete each addition below.

a) 1 + (-.3) = ________ b) (-.75) + (-2) = ________

---

	a) .7 b) -2.75

5-4 SUBTRACTIONS INVOLVING SIGNED DECIMAL NUMBERS

We will begin this section by discussing the "opposites" of signed decimal numbers. Then we will perform subtractions involving signed decimal numbers by converting the subtraction to addition.

---

20. Two decimal numbers are a pair of opposites if their sum is "0". Therefore:

Since 2.7 + (-2.7) = 0 , 2.7 and -2.7 are a pair of opposites.

Since .61 + (-.61) = 0 , .61 and -.61 are a pair of opposites.

Write the opposite of each decimal number below.

a) 14.2 ________ b) -1.75 ________ c) .08 ________ d) -.105 ________

Answers: a) -14.2 b) 1.75 c) -.08 d) .105

21. Any subtraction involving signed decimal numbers can be converted to an addition <u>by adding the opposite of the second term to the first term</u>. That is:

1.6 - 8.5 = 1.6 + (the opposite of 8.5) = 1.6 + (-8.5)

.04 - (-.02) = .04 + (the opposite of -.02) = .04 + .02

a) (-.7) - .1 = (-.7) + (the opposite of .1) = (-.7) + ________

b) (-9.2) - (-4.7) = (-9.2) + (the opposite of -4.7) = (-9.2) + ________

Answers: a) (-.7) + (-.1) b) (-9.2) + 4.7

22. Any subtraction involving signed numbers can be performed by converting the subtraction to addition and then performing the addition. For example:

3.7 - 9.9 = 3.7 + (-9.9) = -6.2

```
  9.9
- 3.7
  6.2
```

.01 - (-.03) = ______ + ______ = ______

Answer: .01 + .03 = .04

23. Following the example, complete the other subtraction.

(-1.2) - 6.6 = (-1.2) + (-6.6) = -7.8

```
  1.2
+ 6.6
  7.8
```

(-.8) - (-.7) = ______ + ______ = ______

Answer: (-.8) + .7 = -.1

24. Complete each subtraction below.

a) $1 - (-.25) = \_\_\_\_ + \_\_\_\_ = \_\_\_\_$ b) $(-.3) - 1 = \_\_\_\_ + \_\_\_\_ = \_\_\_\_$

---

a) $1 + .25 = 1.25$ b) $(-.3) + (-1) = -1.3$

5-5 THE ADDITION AXIOM AND EQUATIONS WITH DECIMAL TERMS

In this section, we will use the addition axiom to solve equations with decimal terms.

---

25. To get rid of the 1.2 on the left side of the equation below, we added its opposite to both sides. Use the same method to solve the other equation.

$$x + 1.2 = 5.7$$
$$x + 1.2 + (-1.2) = 5.7 + (-1.2)$$
$$x = 4.5$$

$$.08 + y = .03$$

$y = -.05$

26. We solved and checked the equation below. Solve and check the other equation.

$$m + 4.4 = 7.8$$
$$m + 4.4 + (-4.4) = 7.8 + (-4.4)$$
$$m = 3.4$$

$$.9 + F = .7$$

<u>Check</u>

$$3.4 + 4.4 = 7.8$$
$$7.8 = 7.8$$

<u>Check</u>

$F = -.2$

<u>Check</u>

$.9 + (-.2) = .7$

$.7 = .7$

27. To solve the equation below, we began by converting the subtraction to addition. Use the same method to solve the other equation.

$$y - 1.4 = 2.5$$
$$y + (-1.4) = 2.5$$
$$y + (-1.4) + 1.4 = 2.5 + 1.4$$
$$y = 3.9$$

$$x - .38 = .84$$

$x = 1.22$

28. We solved and checked the equation below. Solve and check the other equation.

$$b - .25 = .71$$
$$b + (-.25) = .71$$
$$b + (-.25) + .25 = .71 + .25$$
$$b = .96$$

Check

$.96 - .25 = .71$

$.71 = .71$

$3.4 = d - 8$

Check

$d = 11.4$

Check

$3.4 = 11.4 - 8$

$3.4 = 3.4$

SELF-TEST 15 (pages 136-144)

Perform each addition and subtraction.

1. $8.64 + (-5.91)$	2. $(-.68) + (-.2)$	3. $(-9.7) + 5$
4. $2.1 - 7.3$	5. $(-.42) - .39$	6. $(-.07) - (-1)$

Solve each equation.

7. $p + 2.6 = 8.9$	8. $.7 = 2 + x$	9. $y - .8 = .14$

ANSWERS:

1. 2.73	4. -5.2	7. $p = 6.3$
2. -.88	5. -.81	8. $x = -1.3$
3. -4.7	6. .93	9. $y = .94$

5-6 MULTIPLICATIONS INVOLVING DECIMAL NUMBERS

In this section, we will discuss multiplications involving decimal numbers. The placement of the decimal point in the product is emphasized.

---

29. In the table below, three multiplications involving decimal numbers are shown at the left. The relationship between the number of decimal places in the factors and the number of decimal places in the product is outlined for each multiplication.

Multiplication	Decimal Places First Factor	Decimal Places Second Factor	Decimal Places Product
.3 x .52 = .156	1	2	3
4.1 x 2.3 = 9.43	1	1	2
7 x .0002 = .0014	0	4	4

As you can see from the table: <u>The number of decimal places in a product equals the sum of the number of decimal places in the factors</u>. That is:

.156 has 1 + 2 = 3 decimal places

9.43 has 1 + 1 = 2 decimal places

.0014 has 0 + 4 = 4 decimal places

The correct digits of the product are given for each multiplication below. Place the decimal point correctly in each set of digits.

a) .6 x .8 = "48" ________

c) 5 x .7 = "35" ________

b) 4.11 x 9.6 = "39456" ________

d) 7 x .235 = "1645" ________

---

a) .48

b) 39.456

c) 3.5

d) 1.645

30. We multiplied .71 by .9 below. We got the digits of the product by multiplying 71 by 9. We placed the decimal point in the product by adding the number of decimal places in the factors. Complete the other multiplications.

```
  .71   (2 places)
 x .9   (1 place)
 .639   (3 places)
```

a)
```
   5 7
x .0 3
```

b)
```
 4.6
 x 6
```

---

a) 1.71

b) 27.6

31. To get 4 decimal places in the product below, we had to add <u>two</u> 0's at the <u>left</u> of "18". Complete the other multiplications.

```
  .06    (2 places)
x .03    (2 places)
.0018    (4 places)
 ↑↑
```

a)
```
.0 1
x  7
```

b)
```
.0 2
x .4
```

---

a) .07

b) .008

32. When the digits of the product end with one or more final 0's, the "0" or 0's must be counted when placing the decimal point. For example:

```
 .005          240          800
  x 6        x .005        x .05
 .030         1.200
   ↑            ↑↑
```

40.00

33. Following the example, complete the other multiplication.

```
      2 1 4   (0 places)        .3 2
    x 9.0 1   (2 places)       x .1 4
      2 1 4
    0 0 0
  1 9 2 6
  1,9 2 8.1 4   (2 places)
```

.0448

5-7 MULTIPLICATIONS INVOLVING SIGNED DECIMAL NUMBERS

In this section, we will discuss multiplications involving signed decimal numbers.

---

34. In algebra, multiplications involving decimal numbers are ordinarily written horizontally. To perform horizontal multiplications, we can rewrite the factors vertically as we did with the example below. Complete the other multiplication.

(.3)(.17) = .051 (.4)(.2) = ________

```
  .17
 x .3
 .051
```

.08

35. When one factor is positive and the other negative, the product is negative. An example is shown. Notice how we multiplied vertically to get the absolute value of the product. Complete the other multiplication.

7(-.064) = -.448 (-.09)(.8) = ________

```
 .064
  x 7
 .448
```

-.072

36. When both factors are negative, the product is positive. That is:

(-.9)(-.75) = .675 (-8)(-.6) = ________

```
  .75
 x .9
 .675
```

4.8

37. Following the example, complete the other multiplication.

(2.4)(-1.8) = -4.32 (-3.7)(-55) = ________

```
  1.8
x 2.4
  7 2
3 6
4.3 2
```

203.5

38. When a decimal number is multiplied by "1", the product is identical to the decimal number. That is:

1(.05) = .05 1(-.7) = -.7 a) 1(6.3) = ________ b) 1(-18.5) = ________

a) 6.3 b) -18.5

39. When a decimal number is multiplied by "-1", the product is the opposite of the decimal number. That is:

(-1)(6.6) = -6.6 (-1)(-.4) = .4 a) (-1)(.007) = ______ b) (-1)(-3.9) = ______

a) -.007 b) 3.9

40. When a decimal number is multiplied by "0", the product is "0". That is:

0(12.6) = 0 0(-9.7) = 0 a) 0(.1) = ________ b) 0(-.08) = ________

a) 0 b) 0

5-8 TERMINATING DIVISIONS INVOLVING DECIMAL NUMBERS

In this section, we will discuss terminating divisions involving decimal numbers. Divisions by both whole numbers and decimal numbers are included.

41. When dividing a decimal number by a whole number, we place the decimal point in the quotient directly above the decimal point in the decimal number. (See the arrows in the examples below.) Complete the other division.

```
    4.3          .6 0 2
6)2 5.8       8)4.8 1 6        7)2 1.0 7
 -2 4          -4 8
    1 8          0 1
   -1 8          - 0
                   1 6
                  -1 6
```

3.01

42. In each division below, we got one or more 0's immediately to the right of the decimal point in the quotient (see the arrows). Complete the other division.

```
   ↓              ↓ ↓
  .0 1 7         .0 0 4
5).0 8 5       2).0 0 8        8).0 5 6
  -0             -0
   0 8            0 0
  - 5            -0
     3 5            0 8
    -3 5           -8
```

.007

43. To "terminate" or complete a division, we sometimes have to add one or more 0's to the dividend. For example, we had to add one "0" to .63 to terminate $5\overline{).63}$ below. Complete the other division.

```
   .1 2 6
5) .6 3 0               4)1.5
  -5
   1 3
  -1 0
     3 0
    -3 0
```

.375

44. Following the example, complete the other division. Add as many 0's as necessary to terminate it.

```
      1.6 8
32)5 3.7 6          400)1,7 0 0
  -3 2
    2 1 7
   -1 9 2
      2 5 6
     -2 5 6
```

4.25

45. Any division by a decimal number can be converted to an equivalent division by a whole number. To do so, we shift the decimal point the same number of places to the right in both numbers. That is:

```
 2.6)8.3 2    =   2.6,)8.3,2     =  26)8 3.2

.07).0 0 1 4  =  .07,).0 0,1 4   =  7).1 4

.314).6 2 8   =  .314,).6 2 8,   =  314)6 2 8
```

Usually the decimal-point shifts are made on the original problem. The symbol "^" is used to locate the new place of the decimal point in both numbers. That is:

```
2.6^)8.3^2          .07^).0 0^1 4          .314^).6 2 8^
```

Continued on following page.

45. Continued

Using "‸'s", locate the new places of the decimal points on each original problem below.

a) .7)5.6　　b) 1.24).7 4 4　　c) .003).0 0 0 1 5

Answers
a) .7‸)5.6‸
b) 1.24‸).7 4‸4
c) .003‸).0 0 0‸1 5

46. The three steps needed to perform .3)6.39 are shown and discussed below.

	Step 1	Step 2	Step 3
.3)6.3 9	.3‸)6.3‸9	.3‸)6.3‸9 (decimal point placed above the ‸)	2 1.3 over .3‸)6.3‸9

Step 1: We shifted the decimal points to get the new division 3)63.9 .

Step 2: We placed the decimal point in the quotient <u>directly above the "‸" in the dividend</u>.

Step 3: We divided by "3" and got 21.3 as the quotient.

Since 3)63.9 = 21.3 , .3)6.39 = ________

Answer
21.3

47. When dividing by decimal numbers, all three steps are ordinarily done on the original problem. Two examples are shown below. Complete the other division.

```
       7.              1 1.4
.05‸).3 5‸        .8‸)9.1‸2          .009).0 0 0 7 2
     -3 5             -8
                       1 1
                       -8
                        3 2
                       -3 2
```

Answer
.08

48. Sometimes we have to add one or more 0's to the dividend in order to shift the decimal points. For example, we added <u>two</u> 0's to 5.2 below. Complete the other divisions.

```
         6 5 0.
.008‸)5.2 0 0‸         a) .4)2          b) .002)9.6
     -4 8
        4 0
       -4 0
          0 0
           -0
```

Answers
a) 5
b) 4,800

49. To terminate .04).037 at the right, we had to add <u>two</u> 0's to .037 after shifting the decimal points. Complete the other division.

```
          .9 2 5
.04‸).0 3‸7 0 0          .08)1.4
      -3 6
         1 0
         -8
          2 0
         -2 0
```

50. Following the example, complete the other division.

$$\begin{array}{r} 3\,7\,0. \\ 1.2_{\wedge})\overline{4\,4\,4.0_{\wedge}} \\ -3\,6 \\ \hline 8\,4 \\ -8\,4 \\ \hline 0\,0 \\ -0 \\ \hline \end{array}$$

$$2.4\overline{)\$\,1\,2.6\,0}$$

17.5

$5.25

5-9 TERMINATING DIVISIONS INVOLVING SIGNED DECIMAL NUMBERS

In this section, we will discuss terminating divisions involving signed decimal numbers.

51. In algebra, divisions involving decimal numbers are ordinarily written as fractions. To perform divisions in fraction form, we can rewrite the division in the usual form as we did with the example below. Complete the other division.

$\frac{4.76}{4} = 1.19$

$$\begin{array}{r} 1.1\,9 \\ 4\overline{)4.7\,6} \\ -4 \\ \hline 0\,7 \\ -4 \\ \hline 3\,6 \\ -3\,6 \\ \hline \end{array}$$

$\frac{.028}{7} = $ ________

.004

52. When one number is positive and the other negative, the quotient is negative. An example is shown. Notice how we divided in the usual way to find the absolute value of the quotient. Complete the other division.

$\frac{-9.12}{4} = -2.28$

$$\begin{array}{r} 2.2\,8 \\ 4\overline{)9.1\,2} \\ -8 \\ \hline 1\,1 \\ -8 \\ \hline 3\,2 \\ -3\,2 \\ \hline \end{array}$$

$\frac{-3.42}{.06} = $ ________

-57

53. When both numbers are negative, the quotient is positive. That is:

$\frac{-7.46}{-.2} = 37.3$ $\frac{-13}{-4} =$ ______

$$\begin{array}{r} 3\ 7.3 \\ .2_\wedge\overline{)7.4_\wedge 6} \\ \underline{-6} \\ 1\ 4 \\ \underline{-1\ 4} \\ 0\ 6 \\ \underline{-6} \end{array}$$

3.25

54. Following the example, complete the other division.

$\frac{-4.8}{1.2} = -4$ $\frac{.056}{-.28} =$ ______

$$\begin{array}{r} 4. \\ 1.2_\wedge\overline{)4.8_\wedge} \\ \underline{-4\ 8} \end{array}$$

-.2

55. When a decimal number is divided by itself, the quotient is "1". That is:

$\frac{.74}{.74} = 1$ $\frac{-5.8}{-5.8} = 1$ a) $\frac{.003}{.003} =$ ______ b) $\frac{-1.24}{-1.24} =$ ______

---

a) 1
b) 1

56. When a decimal number is divided by "1", the quotient is identical to the decimal number. That is:

$\frac{.07}{1} = .07$ $\frac{-4.9}{1} = -4.9$ a) $\frac{.025}{1} =$ ______ b) $\frac{-.5}{1} =$ ______

---

a) .025
b) -.5

57. When a decimal number is divided by "-1", the quotient is the opposite of the decimal number. That is:

$\frac{5.7}{-1} = -5.7$ $\frac{-.09}{-1} = .09$ a) $\frac{.078}{-1} =$ ______ b) $\frac{-27.6}{-1} =$ ______

---

a) -.078
b) 27.6

58. When "0" is divided by a decimal number, the quotient is "0". That is:

$\frac{0}{.86} = 0$ $\frac{0}{-2.5} = 0$ a) $\frac{0}{.009} =$ ______ b) $\frac{0}{-.4} =$ ______

---

a) 0 b) 0

5-10 THE MULTIPLICATION AXIOM AND EQUATIONS WITH DECIMAL ROOTS

In this section, we will use the multiplication axiom to solve equations with decimal roots.

59. To solve the equation below, we used the shortcut of the multiplication axiom. That is, we divided the number-term by the coefficient of the letter. Solve the other equation.

$$4t = .56 \qquad\qquad 24.3 = 9R$$

$$t = \frac{.56}{4} = .14$$

R = 2.7

60. When the equation contains negative numbers, we can also solve it by performing a division. An example is shown. Solve the other equation.

$$7m = -2.8 \qquad\qquad -.56 = -8p$$

$$m = \frac{-2.8}{7} = -.4$$

---

p = .07

61. To solve the equation below, we had to divide by a decimal number. Solve the other equation.

$$1.2x = 56.4 \qquad\qquad -9 = -.5d$$

$$x = \frac{56.4}{1.2} = 47$$

```
       4 7.
  1.2)5 6.4↑
     -4 8
        8 4
       -8 4
```

d = 18

SELF-TEST 16 (pages 145-153)

Perform each multiplication and division.

1. $(-5)(4.7) =$	2. $(-1.1)(-3.7) =$	3. $(.06)(-.94) =$
4. $\frac{-6.25}{5} =$	5. $\frac{4.4}{-2.2} =$	6. $\frac{-76.8}{-3.2} =$

(Continued on following page.)

SELF-TEST 16 (pages 145-153) - Continued

Solve each equation.

7. $.15x = .75$	8. $-3y = .21$	9. $-61.6 = -1.1V$

ANSWERS:

1. -23.5
2. 4.07
3. -.0564
4. -1.25
5. -2
6. 24
7. x = 5
8. y = -.07
9. V = 56

5-11 BOTH AXIOMS AND EQUATIONS INVOLVING DECIMAL NUMBERS

In this section, we will use both axioms to solve equations involving decimal coefficients and decimal terms.

62. We used both axioms to solve the equation below. Use both axioms to solve the other equation.

$$.4x + 1.1 = 3.9$$
$$.4x + 1.1 + (-1.1) = 3.9 + (-1.1)$$
$$.4x = 2.8$$
$$x = \frac{2.8}{.4} = 7$$

$$.08 = .06 + .1y$$

y = .2

63. Before using both axioms below, we converted the subtraction to addition. Solve the other equation.

$$1.2t - 4.7 = 2.5$$
$$1.2t + (-4.7) = 2.5$$
$$1.2t + (-4.7) + 4.7 = 2.5 + 4.7$$
$$1.2t = 7.2$$
$$t = \frac{7.2}{1.2} = 6$$

$$.6 - .2x = .8$$

x = -1

64. Following the example, solve the other equation.

$$.7 - m = .1$$
$$.7 + (-1m) = .1$$
$$(-.7) + .7 + (-1m) = .1 + (-.7)$$
$$-1m = -.6$$
$$m = \frac{-.6}{-1} = .6$$

$$5.49 = 2.18 - d$$

d = -3.31

5-12 ROUNDING DECIMAL NUMBERS

In this section, we will review the rules for rounding decimal numbers either to a place to the right of the decimal point or to the nearest whole number.

---

65. The following general rules are used to round to a specific place to the right of the decimal point:

> 1) If the digit in the place immediately to the right of the specific place is a 0, 1, 2, 3, or 4:
>   a) leave the digit in the specific place unchanged, and
>   b) drop all digits to the right of the specific place.
>
> 2) If the digit in the place immediately to the right of the specific place is a 5, 6, 7, 8, or 9:
>   a) add "1" to the digit in the specific place, and
>   b) drop all digits to the right of the specific place.

Some examples of using the rules above are shown below. Notice that we wrote an arrow above the place being rounded to as a help in the rounding process.

Rounding to tenths, 4.827 rounds to 4.8 .
(Because of the "2" in the hundredths place, we left the "8" unchanged.)

Rounding to hundredths, .1354 rounds to .14 .
(Because of the "5" in the thousandths place, we added "1" to the "3" and got "4".)

Rounding to thousandths, .09681 rounds to .097 .
(Because of the "8" in the ten-thousandths place, we added "1" to the "6" and got "7".)

---

66. Round to tenths. a) 2.53 ______ b) 7.061 ______ c) .2539 ______

---

67. Round to hundredths. a) 4.177 ______ b) .0352 ______ c) .2504 ______

Answers to 66: a) 2.5 b) 7.1 c) .3

---

68. Round to thousandths. a) .7055 ______ b) .06914 ______ c) .00862 ______

Answers to 67: a) 4.18 b) .04 c) .25

---

69. To round to the nearest whole number, we examine the digit in the "tenths" place. For example:

15.37 rounds to 15. (Because of the "3" in the tenths place, we left the "5" unchanged.)

27.94 rounds to 28. (Because of the "9" in the tenths place, we added "1" to the "7" and got "8".)

Round to the nearest whole number. a) 231.7 _____ b) 48.5 _____ c) 7.088 _____

Answers to 68: a) .706 b) .069 c) .009

a) 232
b) 49
c) 7

70. When rounding, we can get a "0" in the place rounded to. For example:

Rounding to <u>hundredths</u>, .7025 rounds to .70 .

Rounding to the <u>nearest whole number</u>, 20.46 rounds to 20.

a) Round 1.03 to <u>tenths</u> ________ b) Round .79033 to <u>thousandths</u> ________

---

a) 1.0
b) .790

71. We can also get a "0" in the place rounded to when we add "1" to a "9" in that place. For example:

Rounding to <u>hundredths</u>, 3.195 rounds to 3.20 .

Rounding to the <u>nearest whole number</u>, 49.81 rounds to 50.

a) Round 9.96 to <u>tenths</u> ________ b) Round .05972 to <u>thousandths</u> ________

---

a) 10.0
b) .060

72. When an amount of money is multiplied by a decimal number or divided by a whole number, we can get digits in the place or places to the right of the two "cents" places. Some examples are shown at the right.

```
    $ 3 5.9 5
        x .3 1
      3 5 9 5
  1 0 7 8 5
  $ 1 1.1 4 4 5
```

```
     $   8.3 3 8
  5)$ 4 1.6 9 0
     -4 0
        1 6
       -1 5
          1 9
         -1 5
            4 0
           -4 0
```

In such cases, we round to the nearest "cent" which is the same as rounding to hundredths. That is:

$11.1445 rounds to $11.14 . $8.338 rounds to ________.

$8.34

73. When rounding to the nearest cent, we can get a "0" in the final "cents" place. <u>That final "0" must be kept</u>. For example:

$7.8025 rounds to $7.80 . $5.1966 rounds to $5.20 .

Round to the <u>nearest cent</u>. a) $65.903 ________ b) $14.7975 ________

---

a) $65.90
b) $14.80

74. When rounding to the nearest cent, we can get 0's in both of the "cents" places. <u>Those two 0's can either be kept or dropped</u>. For example:

$8.0045 rounds to $8.00 (or $8). $5.9975 rounds to $6.00 (or $6).

Round to the <u>nearest cent</u>. a) $20.0025 ________ b) $59.99875 ________

---

a) $20.00 (or $20)
b) $60.00 (or $60)

5-13 NON-TERMINATING DIVISIONS

Divisions that do not "end" because no place is reached where the remainder is "0" are called "non-terminating" divisions. In this section, we will discuss divisions of that type and show how they are terminated by rounding to a specific place.

---

75. The division at the right is "non-terminating" because it goes on and on. To show that the division process could continue, we wrote three periods "..." at the right side of the quotient.

```
      2.5 7 1 4 ...
 7)1 8.0 0 0 0
  -1 4
     4 0
    -3 5
       5 0
      -4 9
         1 0
          -7
           3 0
          -2 8
             2
```

To stop a non-terminating division, we ordinarily round the quotient to a specific place. For example:

Rounding to thousandths, 2.5714... rounds to 2.571 .

Rounding to hundredths, 2.5714... rounds to 2.57 .

Rounding to tenths, 2.5714... rounds to ________.

Answer: 2.6

76. When rounding a non-terminating quotient to a definite place, we continue dividing until we get a digit in the place immediately to the right of the definite place and then round back one place. Two examples are shown and discussed.

To round to the nearest whole number below, we divided until we got a digit in the tenths place and then rounded back.

```
    4 8 4.8  or  485
 9)4,3 6 4.0
  -3 6
     7 6
    -7 2
       4 4
      -3 6
         8 0
        -7 2
           8
```

To round to the nearest hundredths below, we divided until we got a digit in the thousandths place and then rounded back.

```
    .1 7 9  or  .18
 3).5 3 8
   -3
    2 3
   -2 1
      2 8
     -2 7
        1
```

Following the method above, round each quotient below to the indicated place.

a) nearest tenth $7\overline{)3\ 5.6}$

b) nearest thousandth $6\overline{)2.5\ 6\ 9}$

77. We rounded the quotient at the right to the nearest tenth. Round the other quotient to the nearest hundredth.

```
      1.8 4 or 1.8
41)7 5.6 0
  -4 1
   3 4 6
  -3 2 8
     1 8 0
    -1 6 4
       1 6
```

```
56)1 0
```

a) 5.1 (from 5.08)

b) .428 (from .4281)

---

78. For non-terminating divisions with decimal divisors, we also round the quotients to a definite place. For example, we rounded to the nearest hundredth at the right. Round the other quotient to the nearest whole number.

```
       5.2 1 4 or 5.21
.7^)3.6^5 0 0
   -3 5
      1 5
     -1 4
        1 0
        -7
          3 0
         -2 8
            2
```

```
.09 )2.3
```

.18 (from .178)

---

79. Divisions involving signed decimal numbers can also be non-terminating. For example, we rounded the quotient below to the nearest tenth. Round the other quotient to the nearest hundredth.

$$\frac{-58.3}{3} = -19.4$$

```
  1 9.4 3  or 19.4
3)5 8.3 0
 -3
  2 8
 -2 7
    1 3
   -1 2
      1 0
       -9
```

$$\frac{-1.06}{-.7} = \_\_\_\_\_\_$$

26 (from 25.5)

---

80. When a non-terminating division is performed on a hand calculator, the quotient shown contains many decimal places. Two examples are given below.

$47\overline{)130} = 2.765957$ $.13\overline{).119} = .915385$

Ordinarily, such quotients are rounded to less places.

a) Round 2.765957 to the nearest tenth. ________

b) Round .915385 to the nearest thousandth. ________

1.51

---

a) 2.8 b) .915

5-14 EQUATIONS WITH NON-TERMINATING ROOTS

In this section, we will discuss equations with non-terminating roots. We will show that non-terminating roots are rounded to a definite place.

---

81. The equation below has a non-terminating root. We rounded it to the nearest hundredth. Solve the other equation and round its root to the nearest tenth.

$3x = 1.7$

$x = \frac{1.7}{3} = .57$

```
   .5 6 6  or  .57
3)1.7 0 0
 -1 5
    2 0
   -1 8
      2 0
     -1 8
```

$-7y = 14.6$

$y = \frac{14.6}{-7} =$ ______

y = -2.1 (from -2.08)

---

82. Following the example, round the other root to the nearest whole number.

$27x = 12,890$

$x = \frac{12,890}{27} = 477$

```
      4 7 7.4  or  477
27)1 2,8 9 0.0
  -1 0 8
     2 0 9
    -1 8 9
       2 0 0
      -1 8 9
         1 1 0
        -1 0 8
```

$1.2y = 223$

$y = \frac{223}{1.2} =$ ______

x = 186 (from 185.8)

5-15 CONVERTING FRACTIONS TO DECIMAL NUMBERS

Though some fractions can be converted to decimal numbers by the "decimal-fraction" method, the "division" method is the general method for such conversions. We will discuss both methods in this section. A table of common fraction-decimal equivalents is shown.

83. A "decimal" fraction is a fraction whose denominator is a number like 10, 100, or 1,000. Any decimal fraction can be converted directly to a decimal number. The decimal number contains the digit or digits of the numerator and <u>as many decimal places as there are 0's in the denominator.</u> For example:

$\frac{7}{10} = .7$ (Since 10 has <u>one</u> "0", .7 has <u>one</u> decimal place.)

$\frac{49}{100} = .49$ (Since 100 has <u>two</u> 0's, .49 has <u>two</u> decimal places.)

Convert each decimal fraction to a decimal number.

$\frac{3}{10} = .3$ $\frac{47}{100} = .47$ a) $\frac{1}{10} =$ ______ b) $\frac{99}{100} =$ ______

---

84. Some fractions can be converted to decimal numbers by converting them to equivalent decimal fractions. That is:

a) .1
b) .99

$\frac{1}{2} = \frac{5}{10} = .5$ $\frac{4}{5} = \frac{8}{10} = .8$ $\frac{3}{4} = \frac{75}{100} = .75$

Complete each conversion of a fraction to a decimal number.

a) $\frac{2}{5} = \frac{4}{10} =$ ______ b) $\frac{1}{4} = \frac{25}{100} =$ ______ c) $\frac{7}{20} = \frac{35}{100} =$ ______

---

85. The "<u>division</u>" method is the general method for converting fractions to decimal numbers. In the division method, <u>we divide the numerator of the fraction by its denominator</u>. Two examples are shown. Complete the other conversion.

a) .4
b) .25
c) .35

```
        .5              .7 5
1/2 = 2)1.0    3/4 = 4)3.0 0    5/8 = 8)5
       -1 0            -2 8
                         2 0
                        -2 0
```

.625

86. When the division method is used, the division is usually non-terminating. An example is shown below. Notice that we rounded <u>to the nearest hundredth.</u> Complete the other conversion, rounding <u>to the nearest hundredth.</u>

```
        .6 6 6  or  .67
2/3 = 3)2.0 0 0                4/7 = 7)4
       -1 8
          2 0
         -1 8
            2 0
           -1 8
              2
```

.57

87. Following the example, convert the other fraction to a decimal number. Round to the nearest thousandth.

$$\frac{9}{13} = 13\overline{)9.0000} = .6923 \text{ or } .692$$

```
      .6 9 2 3  or  .692
 13)9.0 0 0 0
   -7 8
    1 2 0
   -1 1 7
        3 0
       -2 6
          4 0
         -3 9
```

$\frac{41}{70} = 70\overline{)41}$

.586 (from .5857)

88. Some common fraction-decimal equivalents are given in the table below. Because of their usefulness, they should be memorized. Notice that the "thirds" fractions are rounded to the nearest hundredth.

halves	thirds	fourths	fifths	eighths	tenths
$\frac{1}{2}$ = .5	$\frac{1}{3}$ = .33	$\frac{1}{4}$ = .25	$\frac{1}{5}$ = .2	$\frac{1}{8}$ = .125	$\frac{1}{10}$ = .1
	$\frac{2}{3}$ = .67	$\frac{3}{4}$ = .75	$\frac{2}{5}$ = .4	$\frac{3}{8}$ = .375	$\frac{3}{10}$ = .3
			$\frac{3}{5}$ = .6	$\frac{5}{8}$ = .625	$\frac{7}{10}$ = .7
			$\frac{4}{5}$ = .8	$\frac{7}{8}$ = .875	$\frac{9}{10}$ = .9

By memory, write the decimal equivalent for each fraction.

a) $\frac{1}{4}$ = ______ b) $\frac{2}{3}$ = ______ c) $\frac{5}{8}$ = ______ d) $\frac{7}{10}$ = ______

---

a) .25 c) .625
b) .67 d) .7

89. By memory, write the decimal equivalent for each fraction.

a) $\frac{1}{2}$ = ______ b) $\frac{1}{8}$ = ______ c) $\frac{4}{5}$ = ______ d) $\frac{1}{3}$ = ______

---

a) .5
b) .125
c) .8
d) .33

90. When an improper fraction is converted to a decimal number, the decimal number is larger than "1". An example is shown. Complete the other conversion. Round to the nearest hundredth.

```
        2.6 6 6  or  2.67
8/3 = 3)8.0 0 0
       -6
        2 0
       -1 8
          2 0
         -1 8
            2 0
           -1 8
```

$\frac{37}{21} = 21\overline{)37}$

1.76 (from 1.761)

SELF-TEST 17 (pages 153-161)

Solve each equation.

1. $.2x + .8 = 1.4$

2. $.15 - y = .87$

3. Round 17.438 to tenths. ____________

4. Round .06951 to thousandths. ____________

5. Round 487.09 to the nearest whole number. ____________

6. Round \$7.95675 to the nearest cent. ____________

Perform each division. Report each quotient to the indicated place.

7. nearest tenth

$$\frac{51.7}{7}$$

8. nearest whole number

$$\frac{-.4}{.007}$$

9. Solve. Report the root to the nearest hundredth.

$$12.7y = 25.9$$

By memory, convert each fraction to a decimal number.

10. $\frac{3}{4} =$ ______

11. $\frac{1}{3} =$ ______

12. $\frac{3}{8} =$ ______

Convert each fraction to a decimal number to the nearest hundredth.

13. $\frac{5}{6}$

14. $\frac{9}{70}$

ANSWERS:

1. x = 3	3. 17.4	5. 487	7. 7.4	9. 2.04	11. .33	13. .83
2. y = -.72	4. .070	6. \$7.96	8. -57	10. .75	12. .375	14. .13

Unit 6 FRACTIONAL EQUATIONS AND PROPORTIONS

In this unit, we will discuss the methods for solving some fractional equations and proportions. We will define what is meant by the concepts "ratio" and "proportion". Some applied problems involving both concepts and the methods for solving them are discussed.

6-1 ALGEBRAIC FRACTIONS

In this section, we will define what is meant by an algebraic fraction. Some evaluations of algebraic fractions are shown. Two basic fraction principles are extended to algebraic fractions.

---

1. An algebraic fraction is a fraction that contains a letter in either its numerator, its denominator, or both. Some examples are shown below.

$$\frac{x}{3} \qquad \frac{10}{y} \qquad \frac{2d}{5} \qquad \frac{9}{7p} \qquad \frac{m+1}{2} \qquad \frac{t-5}{t}$$

Just as any numerical fraction stands for a division, any algebraic fraction stands for a division. That is:

$\frac{x}{3}$ means: "x" divided by "3". $\qquad$ $\frac{10}{y}$ means: ______ divided by ______.

---

10 divided by y

2. Following the example at the left, complete the statement at the right.

$\frac{2d}{5}$ means: "2d" divided by "5". $\qquad$ $\frac{9}{7p}$ means: ______ divided by ______.

---

9 divided by 7p

3. Following the example at the left, complete the statement at the right.

$\frac{m+1}{2}$ means: "m + 1" divided by "2". $\qquad$ $\frac{t}{t-5}$ means: ______ divided by ____________

---

t divided by t - 5

4. To evaluate fractions like $\frac{x}{3}$ and $\frac{10}{y}$, we simply perform the division. That is:

If $x = 12$, $\frac{x}{3} = \frac{12}{3} = 4$. $\qquad$ If $y = 2$, $\frac{10}{y} = \frac{10}{2} =$ ______.

5

5. To evaluate fractions like $\frac{2d}{5}$ and $\frac{18}{3p}$, we have to perform a multiplication before dividing. That is:

If $d = 10$, $\frac{2d}{5} = \frac{2(10)}{5} = \frac{20}{5} = 4$ If $p = 3$, $\frac{18}{3p} = \frac{18}{3(3)} = \frac{18}{9} = $ ______

Answer: 2

6. To evaluate fractions like $\frac{m+1}{2}$ and $\frac{t}{t-5}$, we have to perform an addition or subtraction before dividing. That is:

If $m = 5$, $\frac{m+1}{2} = \frac{5+1}{2} = \frac{6}{2} = 3$ If $t = 10$, $\frac{t}{t-5} = \frac{10}{10-5} = \frac{10}{5} = $ ______

Answer: 2

7. When the denominator of an algebraic fraction is "1", the fraction equals its numerator. That is:

$\frac{x}{1} = x$ $\frac{4y}{1} = 4y$ $\frac{t+7}{1} = t + 7$

Using the fact above, complete these:

a) $\frac{V}{1} = $ ______ b) $\frac{8m}{1} = $ ______ c) $\frac{d-1}{1} = $ ______

Answers: a) V b) 8m c) d - 1

8. When the numerator and denominator of an algebraic fraction are identical, the fraction equals "1". That is:

$\frac{x}{x} = 1$ $\frac{m+3}{m+3} = 1$ a) $\frac{5b}{5b} = $ ______ b) $\frac{y-9}{y-9} = $ ______

Answers: a) 1 b) 1

6-2 MULTIPLYING AN ALGEBRAIC FRACTION BY ITS DENOMINATOR

When solving fractional equations containing one fraction, the first step involves multiplying the fraction by its denominator. We will discuss multiplications of that type in this section. A shortcut is shown.

---

9. To multiply a fraction by a non-fraction, we can simply multiply the numerator of the fraction by the non-fraction. That is:

$2\left(\frac{x}{5}\right) = \frac{(2)(x)}{5} = \frac{2x}{5}$ $y\left(\frac{3}{4}\right) = \frac{(y)(3)}{4} = $ ______

Answer: $\frac{3y}{4}$

10. In each example below, we multiplied an algebraic fraction by its denominator. Notice that the product in each multiplication is <u>the numerator of the original fraction</u>.

$3\left(\frac{x}{3}\right) = \frac{(3)(x)}{3} = \left(\frac{3}{3}\right)(x) = (1)(x) = x$ $y\left(\frac{5}{y}\right) = \frac{(y)(5)}{y} = \left(\frac{y}{y}\right)(5) = (1)(5) = 5$

A shortcut can be used for the same multiplications. In the shortcut, we simply cancel the non-fraction and the denominator. That is:

$\overset{1}{\cancel{3}}\left(\frac{x}{\underset{1}{\cancel{3}}}\right) = \frac{x}{1} = x$ $\overset{1}{\cancel{y}}\left(\frac{5}{\underset{1}{\cancel{y}}}\right) = \frac{5}{1} = 5$

Continued on following page.

10. Continued

Using the shortcut, complete these: a) $7\left(\frac{d}{7}\right) =$ ______ b) $m\left(\frac{8}{m}\right) =$ ______

---

a) d

b) 8

11. In each example below, we multiplied an algebraic fraction by its denominator. Again, each product is <u>the numerator of the original fraction</u>.

$$5\left(\frac{4y}{5}\right) = \frac{(5)(4y)}{5} = \left(\frac{5}{5}\right)(4y) = (1)(4y) = 4y$$

$$2x\left(\frac{9}{2x}\right) = \frac{(2x)(9)}{2x} = \left(\frac{2x}{2x}\right)(9) = (1)(9) = 9$$

The same type of "cancelling" shortcut can be used for the multiplications above. That is:

$$\overset{1}{\cancel{5}}\left(\frac{4y}{\underset{1}{\cancel{5}}}\right) = \frac{4y}{1} = 4y \qquad \overset{1}{\cancel{2x}}\left(\frac{9}{\underset{1}{\cancel{2x}}}\right) = \frac{9}{1} = 9$$

Using the shortcut, complete these: a) $10\left(\frac{3t}{10}\right) =$ ______ b) $7m\left(\frac{6}{7m}\right) =$ ______

---

a) 3t

b) 6

12. In each example below, we multiplied an algebraic fraction by its denominator. The "cancelling" shortcut was used. Notice again that each product is <u>the numerator of the original fraction.</u>

$$\overset{1}{\cancel{9}}\left(\frac{x+5}{\underset{1}{\cancel{9}}}\right) = \frac{x+5}{1} = x+5 \qquad \overset{1}{\cancel{y}}\left(\frac{y-3}{\underset{1}{\cancel{y}}}\right) = \frac{y-3}{1} = y-3$$

Using the shortcut, complete these: a) $8\left(\frac{m+1}{8}\right) =$ ______ b) $p\left(\frac{7-p}{p}\right) =$ ______

---

a) m + 1

b) 7 - p

13. In each example below, we multiplied an algebraic fraction by its denominator. The "cancelling" shortcut was used. Notice again that each product is <u>the numerator of the original fraction.</u>

$$\overset{1}{\cancel{(m+3)}}\left(\frac{7}{\underset{1}{\cancel{m+3}}}\right) = \frac{7}{1} = 7 \qquad \overset{1}{\cancel{(t-5)}}\left(\frac{t}{\underset{1}{\cancel{t-5}}}\right) = \frac{t}{1} = t$$

Using the shortcut, complete these: a) $(x+8)\left(\frac{4}{x+8}\right) =$ ______ b) $(1-y)\left(\frac{y}{1-y}\right) =$ ______

---

a) 4

b) y

14. Whenever any algebraic fraction is multiplied by its denominator, the product is identical to its numerator. That is:

$$2\left(\frac{m}{2}\right) = m \qquad 7x\left(\frac{1}{7x}\right) = 1 \qquad V\left(\frac{V-7}{V}\right) = V - 7 \qquad (x+4)\left(\frac{5}{x+4}\right) = 5$$

Using the fact above, write the products for these:

a) $y\left(\frac{12}{y}\right) =$ ______ b) $4\left(\frac{3d}{4}\right) =$ ______ c) $6\left(\frac{1+h}{6}\right) =$ ______

---

a) 12
b) 3d
c) $1 + h$

15. Using the same fact, write the products for these.

a) $4d\left(\frac{7}{4d}\right) =$ ______ b) $6\left(\frac{5R}{6}\right) =$ ______ c) $(10 - x)\left(\frac{x}{10 - x}\right) =$ ______

---

a) 7 b) 5R c) x

6-3 SOLVING FRACTIONAL EQUATIONS CONTAINING ONE FRACTION

Any equation that contains an algebraic fraction is called a "fractional" equation. In this section, we will discuss the procedure for solving fractional equations containing one fraction. We will show how the first step in that procedure involves using the multiplication axiom to "clear the fraction" and get a non-fractional equation.

---

16. To "clear the fraction" and solve the equation below, we used the multiplication axiom. That is, we multiplied both sides by "5", the denominator of the fraction. Solve the other equation by multiplying both sides by "8".

$$\frac{x}{5} = 7$$

$$5\left(\frac{x}{5}\right) = 5(7)$$

$$x = 35$$

Check

$$\frac{35}{5} = 7$$

$$7 = 7$$

$$4 = \frac{t}{8}$$

$$8(4) = 8\left(\frac{t}{8}\right)$$

$$32 = t$$

or

$$t = 32$$

17. To clear the fraction below, we multiplied both sides by "y", <u>the denominator of the fraction</u>. Notice that we then had to solve "12 = 3y" to get the root of the original equation. Use the same steps to solve the other equation. Begin by multiplying both sides by "R".

$$\frac{12}{y} = 3$$

$$y\left(\frac{12}{y}\right) = y(3)$$

$$12 = 3y$$

$$y = \frac{12}{3} = 4$$

Check

$$\frac{12}{4} = 3$$

$$3 = 3$$

$$6 = \frac{42}{R}$$

$$R(6) = R\left(\frac{42}{R}\right)$$

$$6R = 42$$

$$R = \frac{42}{6} = 7$$

18. To solve the equation below, we began by multiplying both sides by "3" to clear the fraction. Use the same steps to solve the other equation. Begin by multiplying both sides by "2".

$$\frac{2m}{3} = 6$$

$$3\left(\frac{2m}{3}\right) = 3(6)$$

$$2m = 18$$

$$m = \frac{18}{2} = 9$$

Check

$$\frac{2(9)}{3} = 6$$

$$\frac{18}{3} = 6$$

$$6 = 6$$

$$12 = \frac{3x}{2}$$

$$2(12) = 2\left(\frac{3x}{2}\right)$$

$$24 = 3x$$

$$x = \frac{24}{3} = 8$$

19. To solve the equation below, we began by multiplying both sides by "3t". Use the same steps to solve the other equation. Begin by multiplying both sides by "5p".

$$\frac{24}{3t} = 2$$

$$3t\left(\frac{24}{3t}\right) = 3t(2)$$

$$24 = 6t$$

$$t = \frac{24}{6} = 4$$

Check

$$\frac{24}{3(4)} = 2$$

$$\frac{24}{12} = 2$$

$$2 = 2$$

$$6 = \frac{60}{5p}$$

$$5p(6) = 5p\left(\frac{60}{5p}\right)$$

$$30p = 60$$

$$p = \frac{60}{30} = 2$$

20. Fractional equations can also have fractional roots. An example is shown. Solve the other equations.

$$\frac{5m}{2} = 3$$

$$2\left(\frac{5m}{2}\right) = 2(3)$$

$$5m = 6$$

$$m = \frac{6}{5}$$

a) $4 = \frac{7}{x}$

b) $\frac{6}{5p} = 2$

21. In the last frame, we got $m = \frac{6}{5}$ and $x = \frac{7}{4}$ as the roots for the equations at the right. We checked the first root. Check the second root.

$$\frac{5m}{2} = 3$$

$$\frac{5\left(\frac{6}{5}\right)}{2} = 3$$

$$\frac{6}{2} = 3$$

$$3 = 3$$

$$4 = \frac{7}{x}$$

a) $x = \frac{7}{4}$

b) $p = \frac{3}{5}$

---

22. Following the example, solve the other equations.

$$\frac{3x}{4} = 1$$

$$4\left(\frac{3x}{4}\right) = 4(1)$$

$$3x = 4$$

$$x = \frac{4}{3}$$

a) $\frac{10}{y} = 1$

b) $1 = \frac{5}{9q}$

$$4 = \frac{7}{\frac{7}{4}}$$

$$4 = 7\left(\frac{4}{7}\right)$$

$$4 = 4$$

---

23. To solve the equation below, we began by multiplying both sides by "2". Then we had to use the addition axiom. Solve the other equation.

$$\frac{x + 3}{2} = 5$$

$$2\left(\frac{x + 3}{2}\right) = 2(5)$$

$$x + 3 = 10$$

$$x + 3 + (-3) = 10 + (-3)$$

$$x = 7$$

$$1 = \frac{y - 3}{4}$$

a) $y = 10$

b) $q = \frac{5}{9}$

---

24. To solve the equation below, we began by multiplying both sides by "4". Then we had to use both axioms. Solve the other equation.

$$\frac{2k + 7}{4} = 1$$

$$4\left(\frac{2k + 7}{4}\right) = 4(1)$$

$$2k + 7 = 4$$

$$2k + 7 + (-7) = 4 + (-7)$$

$$2k = -3$$

$$k = -\frac{3}{2}$$

$$2 = \frac{8 - 7V}{3}$$

$y = 7$

$V = \frac{2}{7}$

25. To solve the equation below, we began by multiplying both sides by "x". Solve the other equation.

$$\frac{x + 1}{x} = 3 \qquad\qquad 1 = \frac{3y - 5}{2y}$$

$$x\left(\frac{x + 1}{x}\right) = x(3)$$

$$x + 1 = 3x$$

$$(-1x) + 1x + 1 = 3x + (-1x)$$

$$1 = 2x$$

$$x = \frac{1}{2}$$

---

y = 5

26. To solve the equation below, we began by multiplying both sides by "x + 3". Notice that we then had to multiply by the distributive principle on the right side. Solve the other equation.

$$\frac{7}{x + 3} = 5 \qquad\qquad 4 = \frac{y}{y - 1}$$

$$(x + 3)\left(\frac{7}{x + 3}\right) = 5(x + 3)$$

$$7 = 5x + 15$$

$$(-15) + 7 = 5x + 15 + (-15)$$

$$-8 = 5x$$

$$x = -\frac{8}{5}$$

$y = \frac{4}{3}$

27. Notice how we got "-1x" when converting the subtraction to addition at the right. Solve the other equation.

$$\frac{3 - x}{7} = 2 \qquad\qquad 1 = \frac{5}{10 - d}$$

$$7\left(\frac{3 - x}{7}\right) = 7(2)$$

$$3 - x = 14$$

$$3 + (-1x) = 14$$

$$(-3) + 3 + (-1x) = 14 + (-3)$$

$$-1x = 11$$

$$x = \frac{11}{-1} = -11$$

d = 5

28. Following the example, solve the other equations.

$$\frac{2m}{7} = 0$$

$$7\left(\frac{2m}{7}\right) = 7(0)$$

$$2m = 0$$

$$m = \frac{0}{2} = 0$$

a) $\frac{x}{12} = 0$

b) $\frac{y + 5}{9} = 0$

---

a) $x = 0$ b) $y = -5$

SELF-TEST 18 (pages 162-169)

Solve each equation.

1. $\frac{7}{5x} = 1$

2. $9 = \frac{m}{3}$

3. $2 = \frac{10 - y}{7}$

4. $\frac{t}{t + 1} = 3$

ANSWERS: 1. $x = \frac{7}{5}$ 2. $m = 27$ 3. $y = -4$ 4. $t = -\frac{3}{2}$

6-4 CLEARING THE FRACTIONS IN PROPORTIONS

Each equation below contains only one fraction on each side. Equations of that type are called "proportions".

$$\frac{x}{5} = \frac{3}{7} \qquad \frac{2}{3y} = \frac{5}{8} \qquad \frac{m+4}{m} = \frac{8}{9} \qquad \frac{1}{6} = \frac{5}{d-1}$$

To solve proportions, we begin by clearing the fractions. In this section, we will show how cross-multiplication can be used to do so. The multiplication axiom is used to justify the cross-multiplication process.

---

29. We used the multiplication axiom to clear the fractions in the proportions below. Notice that we multiplied both sides of each proportion by both of its denominators at the same time.

$\frac{x}{5} = \frac{3}{7}$	$\frac{2y}{3} = \frac{7}{4}$
$(5)(7)\left(\frac{x}{5}\right) = (5)(7)\left(\frac{3}{7}\right)$	$(3)(4)\left(\frac{2y}{3}\right) = (3)(4)\left(\frac{7}{4}\right)$
$\frac{(5)(7)(x)}{5} = \frac{(5)(7)(3)}{7}$	$\frac{(3)(4)(2y)}{3} = \frac{(3)(4)(7)}{4}$
$\left(\frac{5}{5}\right)(7)(x) = \left(\frac{7}{7}\right)(5)(3)$	$\left(\frac{3}{3}\right)(4)(2y) = \left(\frac{4}{4}\right)(3)(7)$
$(1)(7x) = (1)(15)$	$(1)(8y) = (1)(21)$
$7x = 15$	$8y = 21$

There is a shortcut called "cross-multiplication" that can be used to clear the fractions in a proportion. The "cross-multiplication" shortcut for each proportion above is shown and discussed below.

$\frac{x}{5} \times \frac{3}{7}$	$\frac{2y}{3} \times \frac{7}{4}$
$(x)(7) = (5)(3)$	$(2y)(4) = (3)(7)$
$7x = 15$	$8y = 21$

Note: 1) To get the left side of each non-fractional equation, we multiplied the the first numerator and the second denominator. That is:

To get "7x", we multiplied "x" and "7".
To get "8y", we multiplied "2y" and "4".

2) To get the right side of each non-fractional equation, we multiplied the first denominator and the second numerator. That is:

To get "15", we multiplied "5" and "3".
To get "21", we multiplied "3" and "7".

Using cross-multiplication, clear the fractions in each proportion at the right.

a) $\frac{9}{8} = \frac{t}{4}$ b) $\frac{10}{7} = \frac{3x}{2}$

30. We used cross-multiplication to clear the fractions and obtain the non-fractional equations below.

$$\frac{2}{3y} = \frac{5}{8}$$

$$(2)(8) = (3y)(5)$$

$$16 = 15y$$

$$\frac{3}{2} = \frac{4}{x}$$

$$(3)(x) = (2)(4)$$

$$3x = 8$$

Use cross-multiplication to clear the fractions in each proportion at the right.

a) $\frac{7}{x} = \frac{10}{9}$ b) $\frac{4}{3} = \frac{11}{6t}$

a) $(9)(4) = (8)(t)$
$36 = 8t$

b) $(10)(2) = (7)(3x)$
$20 = 21x$

---

31. We used cross-multiplication to clear the fractions below. Notice how we multiplied by the distributive principle to get "9m + 36" at the left and "6y - 6" at the right.

$$\frac{m + 4}{m} = \frac{8}{9}$$

$$(m + 4)(9) = (m)(8)$$

$$9(m + 4) = 8m$$

$$9m + 36 = 8m$$

$$\frac{4}{y - 1} = \frac{6}{5}$$

$$(4)(5) = (y - 1)(6)$$

$$20 = 6(y - 1)$$

$$20 = 6y - 6$$

Use cross-multiplication to clear the fractions in the proportions at the right.

a) $\frac{2}{5} = \frac{x - 4}{7}$ b) $\frac{1}{4} = \frac{F}{F + 1}$

a) $(7)(9) = (x)(10)$
$63 = 10x$

b) $(4)(6t) = (3)(11)$
$24t = 33$

---

32. Write the non-fractional equation obtained by cross-multiplying in each proportion below.

a) $\frac{y}{10} = \frac{9}{4}$ b) $\frac{7}{3} = \frac{20}{6x}$ c) $\frac{m + 8}{m} = \frac{3}{4}$

a) $(2)(7) = (5)(x - 4)$
$14 = 5x - 20$

b) $(1)(F + 1) = (4)(F)$
$F + 1 = 4F$

---

a) $4y = 90$

b) $42x = 60$

c) $4m + 32 = 3m$

33. Write the non-fractional equation obtained by cross-multiplying in each proportion below.

a) $\frac{12}{x} = \frac{1}{2}$ b) $\frac{1}{7} = \frac{9}{6t}$ c) $\frac{1}{4} = \frac{5}{d - 7}$

a) $24 = x$
b) $6t = 63$
c) $d - 7 = 20$

6-5 SOLVING PROPORTIONS

To solve a proportion, we clear the fractions and then solve the resulting non-fractional equation. We will discuss the method in this section.

---

34. To solve the proportion below, we cross-multiplied to clear the fractions and then solved "9x = 10". Use the same steps to solve the other proportion.

$$\frac{x}{5} = \frac{2}{9}$$
$$(x)(9) = (5)(2)$$
$$9x = 10$$
$$x = \frac{10}{9}$$

$$\frac{3}{4} = \frac{7}{y}$$

$y = \frac{28}{3}$, from:

$3y = 28$

---

35. To solve the proportion below, we cross-multiplied and then solved "14 = 15y". Use the same steps to solve the other proportion.

$$\frac{2}{5} = \frac{3y}{7}$$
$$(2)(7) = (5)(3y)$$
$$14 = 15y$$
$$y = \frac{14}{15}$$

$$\frac{9}{2d} = \frac{8}{3}$$

$d = \frac{27}{16}$, from:

$27 = 16d$

36. To solve the proportion below, we cross-multiplied and then solved "5t - 15 = 4". Solve the other proportion.

$$\frac{t-3}{4} = \frac{1}{5} \qquad\qquad \frac{3}{10} = \frac{F+1}{7}$$

$$(5)(t-3) = (4)(1)$$
$$5t - 15 = 4$$
$$5t + (-15) = 4$$
$$5t + (-15) + 15 = 4 + 15$$
$$5t = 19$$
$$t = \frac{19}{5}$$

---

37. To solve the proportion below, we cross-multiplied and then solved "3 = 7p + 14". Solve the other proportion.

$F = \frac{11}{10}$, from:

$21 = 10F + 10$

$$\frac{1}{p+2} = \frac{7}{3} \qquad\qquad \frac{1}{3} = \frac{x}{x-5}$$

$$(1)(3) = (p+2)(7)$$
$$3 = 7p + 14$$
$$(-14) + 3 = 7p + 14 + (-14)$$
$$-11 = 7p$$
$$p = -\frac{11}{7}$$

---

38. Solve: a) $\frac{P+2}{6} = \frac{5}{4}$ b) $\frac{7}{3} = \frac{m}{10 - m}$

$x = -\frac{5}{2}$, from:

$x - 5 = 3x$

---

a) $P = \frac{11}{2}$

b) $m = 7$

39. We solved and checked a proportion below. Solve and check the other proportion.

$$\frac{4m}{3} = \frac{5}{6}$$

$$24m = 15$$

$$m = \frac{15}{24} = \frac{5}{8}$$

Check

$$\frac{4\left(\frac{5}{8}\right)}{3} = \frac{5}{6}$$

$$\frac{\frac{5}{2}}{3} = \frac{5}{6}$$

$$\left(\frac{5}{2}\right)\left(\frac{1}{3}\right) = \frac{5}{6}$$

$$\frac{5}{6} = \frac{5}{6}$$

$$\frac{3}{7} = \frac{d}{2}$$

$d = \frac{6}{7}$

Check

$$\frac{3}{7} = \frac{\frac{6}{7}}{2}$$

$$\frac{3}{7} = \left(\frac{6}{7}\right)\left(\frac{1}{2}\right)$$

$$\frac{3}{7} = \frac{3}{7}$$

SELF-TEST 19 (pages 170-174)

Solve each proportion.

1. $\frac{1}{4} = \frac{3}{m}$

2. $\frac{V+1}{3} = \frac{4}{9}$

3. $\frac{3}{5} = \frac{p}{40 - p}$

ANSWERS: 1. $m = 12$ 2. $V = \frac{1}{3}$ 3. $p = 15$

6-6 RATIOS

In this section, we will show how "ratio" language can be used to express comparisons. We will also discuss equivalent ratios.

---

40. Comparisons are sometimes expressed in "ratio" language. For example:

If there are 3 boys and 2 girls in a family, we say that
the ratio of boys to girls is "3 boys to 2 girls" or simply "3 to 2".

If a team wins 4 games and loses 5, we say that
the ratio of wins to losses is "4 wins to 5 losses" or simply "_____ to _____".

---

4 to 5

41. Instead of using the word "to" to indicate a ratio, a colon ":" is sometimes used. For example:

If there are 8 foreign cars and 11 American cars in a block, we say that
the ratio of foreign cars to American cars is "8 to 11" or "8 : 11".

If there are 7 passengers and 20 seats on a bus, we say that
the ratio of passengers to seats is "7 to 20" or ________.

---

7 : 20

42. Instead of using the word "to" or a colon to express a ratio, we ordinarily use a fraction. For example:

If Peggy weighs 101 pounds and her mother weighs 125 pounds, we say that
the ratio of Peggy's weight to her mother's weight is $\frac{101 \text{ pounds}}{125 \text{ pounds}}$ or simply $\frac{101}{125}$.

If 47 females and 35 males voted in an election, we say that
the ratio of female voters to male voters is $\frac{47 \text{ female voters}}{35 \text{ male voters}}$ or simply ______.

$\frac{47}{35}$

43. Ratios can be used to express rates or costs. For example:

If a man drives 158 miles in 3 hours,
the ratio of miles to hours is $\frac{158 \text{ miles}}{3 \text{ hours}}$ or simply $\frac{158}{3}$.

If a woman pays 89 cents for 6 oranges,
the ratio of cost to oranges is $\frac{89 \text{ cents}}{6 \text{ oranges}}$ or simply ______.

$\frac{89}{6}$

44. Ratios can also be used to compare numbers. For example:

The ratio of 7 to 4 is $\frac{7}{4}$. The ratio of 2 to 9 is _____.

$\frac{2}{9}$

45. When ratio language is used, the ratio is ordinarily reduced to lowest terms. For example:

If a student had 10 items correct and 20 items wrong on a test, we ordinarily say that the ratio of items correct to items wrong is $\frac{10}{20}$ or $\frac{1}{2}$ or "1 to 2".

If there are 18 girls and 12 boys in a class, we ordinarily say that the ratio of girls to boys is $\frac{18}{12}$ or $\frac{3}{2}$ or "______ to ______".

---

3 to 2

46. When a ratio is expressed as a fraction that reduces to a whole number, we ordinarily write "1" as the denominator to make it clear that we are comparing two quantities. For example:

If there are 100 students and 4 teachers, we ordinarily say that the ratio of students to teachers is $\frac{100}{4}$ or $\frac{25}{1}$ or "25 to 1".

If an alloy contains 24 grams of tin and 2 grams of silver, we ordinarily say that the ratio of tin to silver is $\frac{24}{2}$ or $\frac{12}{1}$ or "______ to ______".

---

12 to 1

47. Two ratios are "equivalent" if they both reduce to the same lowest-terms ratio. For example, the two ratios below are equivalent because they both reduce to "3 to 4".

15 voters to 20 non-voters 3 to 4 $\left(\text{from } \frac{15}{20} = \frac{3}{4}\right)$

30 voters to 40 non-voters 3 to 4 $\left(\text{from } \frac{30}{40} = \frac{3}{4}\right)$

By reducing to lowest terms, determine whether each pair of ratios below is equivalent or not.

a) 8 females to 2 males
20 females to 5 males

b) 10 hits in 25 times at bat
15 hits in 35 times at bat

---

a) Equivalent, since both equal "4 to 1".

b) Not equivalent.

48. Sometimes we have to subtract in order to find one of the terms of the ratio. Some examples are given below.

Find the ratio of boys to girls if there are 16 boys in a class of 28 students. (Note: To find the number of girls, we must subtract the number of boys from the total number of students.)

The number of girls is $28 - 16 = 12$.

The ratio of boys to girls is $\frac{16}{12}$ or $\frac{4}{3}$ or "4 to 3".

Find the ratio of the number right to the number wrong if a student has 6 wrong on a 30-item test. (Note: To find the number right, we must subtract the number wrong from the total number of items.)

a) The number right is ________.

b) The ratio of number right to number wrong is ________.

---

a) 24 (from $30 - 6$)

b) 4 to 1 $\left(\text{from } \frac{24}{6}\right)$

6-7 APPLIED PROBLEMS INVOLVING PROPORTIONS

In this section, we will discuss some applied problems involving proportions. All of the problems can be solved by solving a proportion.

---

49. A proportion is an equation containing two equivalent ratios. For example:

$$\frac{5 \text{ boys}}{6 \text{ girls}} = \frac{10 \text{ boys}}{12 \text{ girls}}$$ is a proportion, because $\frac{5}{6} = \frac{10}{12}$.

$$\frac{4 \text{ dogs}}{3 \text{ cats}} = \frac{12 \text{ dogs}}{6 \text{ cats}}$$ is not a proportion, because $\frac{4}{3} \neq \frac{12}{6}$.

Which of the following are proportions? ______________

a) $\frac{2 \text{ cars}}{1 \text{ truck}} = \frac{10 \text{ cars}}{5 \text{ trucks}}$ b) $\frac{3 \text{ pets}}{5 \text{ homes}} = \frac{9 \text{ pets}}{15 \text{ homes}}$ c) $\frac{8 \text{ boys}}{3 \text{ men}} = \frac{20 \text{ boys}}{9 \text{ men}}$

Both (a) and (b)

50. To solve the problem below, we set up a proportion with units, solved the same proportion without units, and then stated the solution in terms of the original problem.

A car travels 100 miles in 2 hours. At that rate, how many hours would it take to travel 350 miles?

$$\frac{100 \text{ miles}}{2 \text{ hours}} = \frac{350 \text{ miles}}{x \text{ hours}} \quad \text{or} \quad \frac{100}{2} = \frac{350}{x}$$

$$100x = 700$$

$$x = 7$$

Therefore, it would take ______ hours to travel 350 miles at that rate.

7

51. When solving a problem involving a proportion, the following steps should be used to set up the proportion correctly.

1) Set up the known ratio on the left side.
2) Then make sure that the units in the ratio on the right side correspond to those of the known ratio.

For example, we have set up a proportion correctly below. Set up the other proportion.

A woman saved \$75 in 5 weeks. At that rate, how long would it take her to save \$120?

$$\frac{\$75}{5 \text{ weeks}} = \frac{\$120}{x \text{ weeks}}$$

If you need 3 grams of sodium chloride to make 1 liter of a solution, how many grams would you need to make 3 liters of the same solution?

$$\frac{3 \text{ grams}}{1 \text{ liter}} = \frac{x \text{ grams}}{3 \text{ liters}}$$

52. Solve each problem below.

a) If a car uses 2 gallons of gasoline to travel 30 miles, how many gallons will it need to travel 75 miles?

b) If 12 grams of carbon will produce 44 grams of carbon dioxide, how many grams of carbon dioxide could you get from 9 grams of carbon?

---

a) 5 gallons, from $\frac{2}{30} = \frac{x}{75}$

b) 33 grams, from $\frac{12}{44} = \frac{9}{x}$

---

53. Proportions can be used for conversions of units. Two examples are shown below.

There are 36 inches in 1 yard. How many inches are there in 3 yards?

$$\frac{36 \text{ inches}}{1 \text{ yard}} = \frac{x \text{ inches}}{3 \text{ yards}} \quad \text{or} \quad \frac{36}{1} = \frac{x}{3}$$

$$x = 108$$

Therefore, there are 108 inches in 3 yards.

There are 60 seconds in 1 minute. Therefore, 420 seconds equal how many minutes?

$$\frac{60 \text{ seconds}}{1 \text{ minute}} = \frac{420 \text{ seconds}}{x \text{ minutes}} \quad \text{or} \quad \frac{60}{1} = \frac{420}{x}$$

$$60x = 420$$

$$x = 7$$

Therefore, 420 seconds equal 7 minutes.

Using the same steps, solve these.

a) There are 28 grams in 1 ounce. How many grams are there in 5 ounces?

b) There are 15 milliliters in 1 tablespoon. Therefore, 105 milliliters equal how many tablespoons?

---

a) 140 grams, from $\frac{28}{1} = \frac{x}{5}$ b) 7 tablespoons, from $\frac{15}{1} = \frac{105}{x}$

6-8 APPLIED PROBLEMS INVOLVING RATIO LANGUAGE

In this section, we will discuss some applied problems involving ratio language. All of the problems can be solved by solving a proportion.

54. Proportions can be used to solve problems involving ratio language. An example is shown below.

If the ratio of boys to girls in a class is 4 to 3, how many boys are there if there are 9 girls?

$$\frac{4 \text{ boys}}{3 \text{ girls}} = \frac{x \text{ boys}}{9 \text{ girls}} \quad \text{or} \quad \frac{4}{3} = \frac{x}{9}$$

$$36 = 3x$$

$$x = 12$$

Therefore, there are 12 boys if there are 9 girls.

Using the same steps, solve each problem below.

a) The ratio of non-smokers to smokers at a party is 3 to 1. How many non-smokers are there if there are 6 smokers?

b) In an alloy, the ratio of copper to iron is 3 : 10. How much iron is there if there are 12 pounds of copper?

---

a) 18 non-smokers, from $\frac{3}{1} = \frac{x}{6}$

b) 40 pounds, from $\frac{3}{10} = \frac{12}{x}$

55. We can sometimes set up a ratio even when we are not given a specific number for either term. Some examples are given below.

There are 25 students in a class. Some are male; some are female. Even though we do not know the specific number of males or females, we can set up the ratio of males to females. The steps are:

1) Let "x" equal the number of males.

2) Then "25 - x" is the number of females.

3) Therefore, the ratio of males to females is $\frac{x \text{ males}}{(25 - x) \text{ females}}$

There are 200 families in a neighborhood. Some are homeowners; some are renters. Even though we do not know the specific number of homeowners or renters, we can set up the ratio of homeowners to renters. That is:

Let "x" equal the number of homeowners.

a) Then the number of renters must be ______________.

b) Therefore, the ratio of homeowners to renters is

____

---

a) $200 - x$

b) $\frac{x \text{ homeowners}}{(200 - x) \text{ renters}}$

56. Following the steps in the last frame, we set up the ratio for one problem below. Set up the ratio for the other problem.

There are 80 pounds of an alloy that contain tin and silver. What is the ratio of tin to silver?

$\frac{x \text{ pounds of tin}}{(80 - x) \text{ pounds of silver}}$

A line that is 30 centimeters long is divided into two parts. What is the ratio of the two parts?

$\frac{x \text{ centimeters}}{(30 - x) \text{ centimeters}}$

57. To solve the problem below, we had to set up a ratio with a letter in both terms.

Two cities will send a total of 42 delegates to a convention. The ratio of the population of the two cities is 3 : 4. How many of the delegates should be sent from the smaller city?

$$\frac{3}{4} = \frac{x \text{ delegates}}{(42 - x) \text{ delegates}} \quad \text{or} \quad \frac{3}{4} = \frac{x}{42 - x}$$

$$3(42 - x) = 4x$$

$$126 - 3x = 4x$$

$$126 = 7x$$

$$x = 18$$

Therefore, the smaller city should send 18 delegates.

Continued on following page.

57. Continued

Following the same steps, solve each problem below.

a) The ratio of girls to boys in a graduating class is 3:2. If there are 80 in the class, how many girls are there?

b) The ratio of copper to silver in dimes is 1:9. How much copper is there in 50 pounds of dimes?

	a) 48 girls, from $\frac{3}{2} = \frac{x}{80 - x}$	b) 5 pounds, from $\frac{1}{9} = \frac{x}{50 - x}$

6-9 PROPORTIONS CONTAINING FRACTIONS AND DECIMAL NUMBERS

In this section, we will solve proportions containing fractions and decimal numbers. Some applied problems are included.

58. We used a proportion containing a fraction to solve the problem below.

If $\frac{1}{4}$ inch on a map represents 10 miles, what is the distance between two cities that are 3 inches apart?

$$\frac{\frac{1}{4}\text{ inch}}{10\text{ miles}} = \frac{3\text{ inches}}{x\text{ miles}} \quad\text{or}\quad \frac{\frac{1}{4}}{10} = \frac{3}{x}$$

$$\frac{1}{4}x = 30$$

$$4\left(\frac{1}{4}x\right) = 4(30)$$

$$x = 120$$

Therefore, the two cities are 120 miles apart.

Continued on following page.

58. Continued

Following the same steps, solve each of these.

a) A bicyclist travels 15 miles in $\frac{3}{4}$ hour. At that rate, how far could she travel in 3 hours?

b) Every $\frac{1}{2}$ hour, an automatic bottling machine fills 90 cases of beer. How long would it take to fill 1,080 cases?

a) 60 miles, from: $\frac{15}{\frac{3}{4}} = \frac{x}{3}$

b) 6 hours, from: $\frac{\frac{1}{2}}{90} = \frac{x}{1{,}080}$

59. We need a proportion containing a decimal number to solve the problem below.

1 inch is equal to 2.54 centimeters. How many centimeters are there in 3 inches?

$$\frac{1 \text{ inch}}{2.54 \text{ centimeters}} = \frac{3 \text{ inches}}{x \text{ centimeters}} \quad \text{or} \quad \frac{1}{2.54} = \frac{3}{x}$$

$$x = 3(2.54)$$

$$x = 7.62$$

Therefore, there are 7.62 centimeters in 3 inches.

Following the same steps, solve these:

a) 1 mile is equal to 1.6 kilometers. How many miles are there in 50 kilometers?

b) If 2 pounds of candy cost \$3.50, how much would 5 pounds cost?

a) 31.25 miles, from: $\frac{1}{1.6} = \frac{x}{50}$

b) \$8.75, from: $\frac{2}{\$3.50} = \frac{5}{x}$

SELF-TEST 20 (pages 175-183)

Write each ratio as a fraction in lowest terms.

1. 15 boys to 9 girls
2. 24 eggs to 2 cartons
3. 6:16

4. Is the pair of ratios below equivalent or not?

   20 "Yes" votes to 30 "No" votes

   14 "Yes" votes to 21 "No" votes

5. 1 centimeter equals $\frac{3}{8}$ inch. How many centimeters are there in 60 inches?

6. If 2.5 inches on a map represents 50 miles, what distance does 4 inches represent?

7. For a painting, the ratio of width to height is 3 to 4. If its width is 24 inches, what is its height?

8. If the ratio of women to men in a club is 2:1, how many women are there if the club has 69 members?

ANSWERS:
1. $\frac{5 \text{ boys}}{3 \text{ girls}}$ or $\frac{5}{3}$
2. $\frac{12 \text{ eggs}}{1 \text{ carton}}$ or $\frac{12}{1}$
3. $\frac{3}{8}$
4. equivalent
5. 160 centimeters
6. 80 miles
7. 32 inches
8. 46 women

Unit 7 SQUARES, SQUARE ROOTS, AND EQUATIONS

In this unit, we will review the meaning of squares and square roots. The squaring operation is extended to signed numbers; the square-root operation is extended to include negative square roots. The procedure for solving equations containing a squared letter and the procedure for solving equations containing a square-root radical are discussed.

7-1 SQUARING SIGNED NUMBERS

In this section, we will discuss the meaning of the operation called "squaring". We will square signed whole numbers, signed fractions, and signed decimal numbers.

---

1. "Squaring" a number means "multiplying the number by itself". For example, "3", "7", and "12" are squared below.

   (3)(3) = 9 (7)(7) = 49 (12)(12) = 144

   Complete: a) Square the number "4". ( )( ) = ________

   b) Square the number "10". ( )( ) = ________

---

a) (4)(4) = 16

b) (10)(10) = 100

2. When a number is squared, the new number is called the "square" of the original number. For example:

   36 is called the "square" of 6, since (6)(6) = 36

   81 is called the "square" of 9, since (9)(9) = 81

   Therefore: a) The square of 2 is ________. b) The square of 8 is ________.

---

a) 4

b) 64

3. The "square" of any negative number is a positive number. For example:

   Since (-5)(-5) = 25 , the "square" of -5 is 25.

   Since (-9)(-9) = 81 , the "square" of -9 is 81.

   Therefore: a) The square of -3 is ________. b) The square of -7 is ________.

---

a) 9

b) 49

4. Since the square of any number is positive, any signed number has the same square as its opposite. That is:

   The square of both +2 and -2 is 4.

   The square of both +9 and -9 is ________.

81

5. There is a special notation for the squaring operation. That is:

Instead of writing "square 3", we write 3^2.

Instead of writing "square -5", we write $(-5)^2$.

The small "2" means that "3" and "-5" should be squared. Notice that we wrote parentheses around the negative number. Therefore:

8^2 means (8)(8) a) 2^2 means ( )( ) b) $(-6)^2$ means ( )( )

---

6. Complete:

a) $7^2 = (7)(7) =$ ______ b) $10^2 =$ ______ c) $(-8)^2 =$ ______

a) (2)(2)
b) (-6)(-6)

---

7. The expression 3^2 is called "3 squared". Therefore:

a) 5 squared $= 5^2 =$ ______ b) -4 squared $= (-4)^2 =$ ______

a) 49
b) 100
c) 64

---

8. Complete:

a) $1^2 = (1)(1) =$ ______ b) $(-1)^2 = (-1)(-1) =$ ______ c) $0^2 = (0)(0) =$ ______

a) 25
b) 16

---

9. To square a fraction, we multiply the fraction by itself. That is:

$$\left(\frac{3}{4}\right)^2 = \left(\frac{3}{4}\right)\left(\frac{3}{4}\right) = \frac{9}{16} \qquad \left(-\frac{1}{2}\right)^2 = \left(-\frac{1}{2}\right)\left(-\frac{1}{2}\right) = \frac{1}{4}$$

Complete: a) $\left(\frac{5}{6}\right)^2 =$ ______ b) $\left(-\frac{7}{4}\right)^2 =$ ______ c) $\left(-\frac{1}{3}\right)^2 =$ ______

a) 1
b) 1
c) 0

---

10. When a fraction is squared, the new fraction is called the "square" of the original fraction. That is:

Since $\left(\frac{7}{8}\right)^2 = \frac{49}{64}$, $\frac{49}{64}$ is called the "square" of $\frac{7}{8}$.

Since $\left(-\frac{1}{5}\right)^2 = \frac{1}{25}$, $\frac{1}{25}$ is called the "square" of $-\frac{1}{5}$.

Therefore: a) The square of $\frac{1}{9}$ is ______. b) The square of $-\frac{8}{3}$ is ______.

a) $\frac{25}{36}$
b) $\frac{49}{16}$
c) $\frac{1}{9}$

---

11. Any fraction and its opposite have the same square. That is:

The square of both $\frac{2}{3}$ and $-\frac{2}{3}$ is $\frac{4}{9}$.

The square of both $\frac{1}{4}$ and $-\frac{1}{4}$ is ______.

a) $\frac{1}{81}$
b) $\frac{64}{9}$

---

12. To square a decimal number, we multiply the decimal number by itself. That is:

$(.5)^2 = (.5)(.5) = .25$ $(-1.2)^2 = (-1.2)(-1.2) = 1.44$

Complete: a) $(.03)^2 =$ ______ b) $(-2.5)^2 =$ ______

$\frac{1}{16}$

a) .0009	b) 6.25

13. When a decimal number is squared, the new number is called the square of the original number. That is:

Since $(6.1)^2 = 37.21$, 37.21 is called the "square" of 6.1 .

Since $(-.8)^2 = .64$, .64 is called the "square" of -.8 .

Therefore: a) The square of 2.4 is ______. b) The square of -.01 is ______.

---

a) 5.76

b) .0001

14. Any decimal number and its opposite have the same square. That is:

The square of both 2.1 and -2.1 is 4.41 .

The square of both .09 and -.09 is ______.

.0081

15. The expressions $\left(\frac{2}{5}\right)^2$ and $(.7)^2$ are called "$\frac{2}{5}$ squared" and ".7 squared". Therefore:

a) $\frac{7}{6}$ squared $= \left(\frac{7}{6}\right)^2 =$ ______ b) -.4 squared $= (-.4)^2 =$ ______

---

a) $\frac{49}{36}$ b) .16

7-2 PERFECT SQUARES

In this section, we will define what is meant by a "perfect square". We will show that whole numbers, fractions, and decimal numbers can be perfect squares.

---

16. Any whole number that is the square of a signed whole number is called a "perfect square". That is:

Since $3^2 = 9$, 9 is a perfect square.

Since $(-6)^2 = 36$, 36 is a perfect square.

Since $1^2 = 1$, ___ is a perfect square.

1

17. The first nine perfect-square whole numbers are listed below. Write the next three in the blanks.

1, 4, 9, 16, 25, 36, 49, 64, 81, ______, ______, ______

100, 121, 144

18. Some larger whole numbers are also perfect squares. For example:

Since $(-40)^2 = 1,600$, 1,600 is a perfect square.

Since $125^2 = 15,625$, ______ is a perfect square.

15,625

19. Any fraction that is the square of a signed fraction is called a "perfect square". That is:

Since $\left(\frac{4}{5}\right)^2 = \frac{16}{25}$, $\frac{16}{25}$ is a perfect square.

Since $\left(-\frac{1}{3}\right)^2 = \frac{1}{9}$, ___ is a perfect square.

$\frac{1}{9}$

20. Any decimal number that is the square of a signed decimal number is called a "perfect square". That is:

Since $(1.4)^2 = 1.96$, 1.96 is a perfect square.

Since $(-.8)^2 = .64$, ______ is a perfect square.

.64

7-3 THE SQUARE ROOTS OF POSITIVE NUMBERS

In this section, we will discuss what is meant by the "square root" of a positive number. We will show that any positive number has two square roots. The discussion is limited to the square roots of perfect squares.

---

21. A "square root" of a positive number N is "a number whose square is N". That is:

Since $5^2 = 25$, 5 is a square root of 25.

Since $8^2 = 64$, ____ is a square root of 64.

8

22. Instead of saying "the square root of", we use the symbol $\sqrt{\quad}$. That is:

Since $4^2 = 16$, $\sqrt{16} = 4$ Since $10^2 = 100$, $\sqrt{100}$ = ______

10

23. For expressions like $\sqrt{49}$, the following terminology is used.

1) The entire expression $\sqrt{49}$ is called a square root radical.
2) The $\sqrt{\quad}$ symbol is called a radical sign or simply a radical.
3) The number "49" under the radical sign is called the radicand.

a) In $\sqrt{81}$, the radicand is ______.

b) In $\sqrt{144}$, the $\sqrt{\quad}$ is called the ________________________.

c) Write the square root radical whose radicand is 9. __________

---

a) 81

b) radical sign

c) $\sqrt{9}$

24. A positive number can have a negative square root. That is:

Since $(-6)^2 = 36$, -6 is a square root of 36.

Since $(-9)^2 = 81$, ____ is a square root of 81.

-9

25. Any positive number has two square roots, one positive and one negative. The two square roots are a pair of opposites. For example:

Since $3^2 = 9$, +3 is the positive square root of 9.

Since $(-3)^2 = 9$, -3 is the negative square root of 9.

a) The positive square root of 16 is ______.

b) The negative square root of 100 is ______.

c) The two square roots of 81 are ______ and ______.

a) +4

b) -10

c) +9 and -9

26. The <u>positive</u> square root of a number is ordinarily called its "<u>principal</u>" square root. Therefore:

a) The <u>principal</u> (or positive) square root of 25 is ______.

b) The <u>negative</u> square root of 25 is ______.

---

a) +5
b) -5

27. To indicate a <u>principal</u> square root, we write a "+" sign (or no sign) in front of the radical. To indicate a <u>negative</u> square root, we write a "-" in front of the radical. That is:

$+\sqrt{16} = +4$ and $\sqrt{16} = +4$ $-\sqrt{16} = -4$

Use the correct signed number to complete these.

a) $\sqrt{49} =$ ________ b) $+\sqrt{49} =$ ________ c) $-\sqrt{49} =$ ________

---

a) +7 b) +7 c) -7

28. Use the correct signed number to complete these:

a) $+\sqrt{81} =$ ________ b) $-\sqrt{9} =$ ________ c) $\sqrt{100} =$ ________

---

a) +9
b) -3
c) +10

29. The symbol "±" means "<u>both + and -</u>". When "±" is written in front of a radical, it means that both roots are desired. That is:

$\pm\sqrt{25}$ means both $+\sqrt{25}$ and $-\sqrt{25}$

$\pm\sqrt{81}$ means both $+\sqrt{81}$ and $-\sqrt{81}$

Therefore: $\pm\sqrt{25} = +5$ and -5 $\pm\sqrt{81} =$ ______ and ______

+9 and -9

30. When "±" is written in front of a number, it means both signed values of that number. That is:

±3 means both +3 and -3 ±8 means both +8 and -8

Therefore: Instead of $\pm\sqrt{9} = +3$ and -3, we can write $\pm\sqrt{9} = \pm 3$.

Instead of $\pm\sqrt{64} = +8$ and -8, we can write $\pm\sqrt{64} =$ ______ .

±8

31. a) Since $0^2 = 0$, $\sqrt{0} =$ ________

b) Since both $(+1)^2$ and $(-1)^2 = 1$, $\pm\sqrt{1} =$ ________

---

a) 0
b) ±1

32. Any whole number that is a perfect square has two whole numbers as its square roots. That is:

Since both 30^2 and $(-30)^2 = 900$, $\pm\sqrt{900} = \pm 30$.

Since both 256^2 and $(-256)^2 = 65,536$, $\pm\sqrt{65,536} =$ __________

±256

33. Any decimal number that is a perfect square has two decimal numbers as its square roots. That is:

Since both $(1.2)^2$ and $(-1.2)^2 = 1.44$, $\pm\sqrt{1.44} = \pm 1.2$

Since both $(.04)^2$ and $(-.04)^2 = .0016$, $\pm\sqrt{.0016} =$ __________

±.04

34. Any fraction that is a perfect square has two fractions as its square roots. That is:

Since both $\left(\frac{5}{8}\right)^2$ and $\left(-\frac{5}{8}\right)^2 = \frac{25}{64}$, $\pm\sqrt{\frac{25}{64}} = \pm\frac{5}{8}$

Since both $\left(\frac{1}{2}\right)^2$ and $\left(-\frac{1}{2}\right)^2 = \frac{1}{4}$, $\pm\sqrt{\frac{1}{4}} =$ ______

---

35. A fraction is a perfect square only if both of its terms are perfect squares. That is: | $\pm\frac{1}{2}$

$\frac{1}{36}$ is a perfect square, since both " 1 " and "36" are perfect squares.

$\frac{81}{64}$ is a perfect square, since both "81" and "64" are perfect squares.

Therefore: a) $\pm\sqrt{\frac{1}{36}} =$ ______ b) $\pm\sqrt{\frac{81}{64}} =$ ______ c) $\pm\sqrt{\frac{49}{100}} =$ ______

---

36. Since the square of any number (positive or negative) is positive, there is no ordinary number that is the square root of a negative number. For example:

$\sqrt{-16} \neq +4$, since $(+4)^2 = +16$ $\sqrt{-16} \neq -4$, since $(-4)^2 = +16$

Though mathematicians have developed a method for handling the square roots of negative numbers, we will not discuss that method in this book.

a) $\pm\frac{1}{6}$

b) $\pm\frac{9}{8}$

c) $\pm\frac{7}{10}$

7-4 A TABLE OF SQUARE ROOTS FOR WHOLE NUMBERS FROM 1 TO 200

In this section, we will give and discuss a table of square roots for whole numbers from 1 to 200. We will show how the table can be used to approximate the square roots of decimal numbers from 1 to 200.

---

37. A table of square roots for whole numbers from 1 to 200 is given on page 190. All of the square roots in the table are given to three decimal places, including the square roots of perfect squares. For example, the table has these entries:

Number	Square Root
9	3.000
81	9.000
169	13.000

Since the three numbers are perfect squares, the three 0's after the decimal point should be ignored. That is:

$\sqrt{9} = 3$ $\sqrt{81} = 9$ $\sqrt{169} =$ ______

13

SQUARE ROOTS OF WHOLE NUMBERS FROM 1 TO 200

Number	Square root	Number	Square root	Number	Square root	Number	Square root
1	1.000	51	7.141	101	10.050	151	12.288
2	1.414	52	7.211	102	10.100	152	12.329
3	1.732	53	7.280	103	10.149	153	12.369
4	2.000	54	7.348	104	10.198	154	12.410
5	2.236	55	7.416	105	10.247	155	12.450
6	2.449	56	7.483	106	10.296	156	12.490
7	2.646	57	7.550	107	10.344	157	12.530
8	2.828	58	7.616	108	10.392	158	12.570
9	3.000	59	7.681	109	10.440	159	12.610
10	3.162	60	7.746	110	10.488	160	12.649
11	3.317	61	7.810	111	10.536	161	12.689
12	3.464	62	7.874	112	10.583	162	12.728
13	3.606	63	7.937	113	10.630	163	12.767
14	3.742	64	8.000	114	10.677	164	12.806
15	3.873	65	8.062	115	10.724	165	12.845
16	4.000	66	8.124	116	10.770	166	12.884
17	4.123	67	8.185	117	10.817	167	12.923
18	4.243	68	8.246	118	10.863	168	12.961
19	4.359	69	8.307	119	10.909	169	13.000
20	4.472	70	8.367	120	10.954	170	13.038
21	4.583	71	8.426	121	11.000	171	13.077
22	4.690	72	8.485	122	11.045	172	13.115
23	4.796	73	8.544	123	11.091	173	13.153
24	4.899	74	8.602	124	11.136	174	13.191
25	5.000	75	8.660	125	11.180	175	13.229
26	5.099	76	8.718	126	11.225	176	13.266
27	5.196	77	8.775	127	11.269	177	13.304
28	5.292	78	8.832	128	11.314	178	13.342
29	5.385	79	8.888	129	11.358	179	13.379
30	5.477	80	8.944	130	11.402	180	13.416
31	5.568	81	9.000	131	11.446	181	13.454
32	5.657	82	9.055	132	11.489	182	13.491
33	5.745	83	9.110	133	11.533	183	13.528
34	5.831	84	9.165	134	11.576	184	13.565
35	5.916	85	9.220	135	11.619	185	13.601
36	6.000	86	9.274	136	11.662	186	13.638
37	6.083	87	9.327	137	11.705	187	13.675
38	6.164	88	9.381	138	11.747	188	13.711
39	6.245	89	9.434	139	11.790	189	13.748
40	6.325	90	9.487	140	11.832	190	13.784
41	6.403	91	9.539	141	11.874	191	13.820
42	6.481	92	9.592	142	11.916	192	13.856
43	6.557	93	9.644	143	11.958	193	13.892
44	6.633	94	9.695	144	12.000	194	13.928
45	6.708	95	9.747	145	12.042	195	13.964
46	6.782	96	9.798	146	12.083	196	14.000
47	6.856	97	9.849	147	12.124	197	14.036
48	6.928	98	9.899	148	12.166	198	14.071
49	7.000	99	9.950	149	12.207	199	14.107
50	7.071	100	10.000	150	12.247	200	14.142

38. Except for the perfect squares, the square roots of all other whole numbers from 1 to 200 are non-terminating decimal numbers. In the table, they are rounded to the nearest thousandth. For example, the table has these entries:

Number	Square Root	
19	4.359	(Rounded from 4.358898...)
185	13.601	(Rounded from 13.601470...)

Though the entries above are not exact because they are rounded, we will treat them as if they were exact. That is:

We will say: $\sqrt{19} = 4.359$ $\sqrt{185} =$ ________

13.601

39. When the square-root entry has a "0" in the thousandths place, that "0" can be ignored. For example:

$\sqrt{53} = 7.280$ or 7.28 $\sqrt{116} = 10.770$ or ________

10.77

40. Though only principal (or positive) square roots are given in the table, it can be used to get both square roots. For example:

$\pm\sqrt{196} = \pm 14$ $\pm\sqrt{47} = \pm 6.856$ $\pm\sqrt{83} =$ ________

± 9.110 or ± 9.11

41. When using the table, we can round the square-root entries even further. For example, we rounded the entry below to the nearest hundredth. Round the other entries to the nearest hundredth.

$\sqrt{62} = 7.87$ (from 7.874) a) $\sqrt{103} =$ ________ b) $\sqrt{178} =$ ________

a) 10.15
b) 13.34

42. We rounded the entry below to the nearest tenth. Round the other entries to the nearest tenth.

$\sqrt{129} = 11.4$ (from 11.358) a) $\sqrt{199} =$ ______ b) $\sqrt{98} =$ ______

a) 14.1
b) 9.9

43. The table can also be used to find a close approximation to the square roots of decimal numbers from 1 to 200. To do so, we simply round the radicand to the nearest whole number and then use the whole-number's entry. For example:

$\sqrt{7.14}$ is very close to $\sqrt{7}$ or 2.646
$\sqrt{58.7}$ is very close to $\sqrt{59}$ or 7.681
$\sqrt{131.5}$ is very close to $\sqrt{}$ or ________

$\sqrt{132}$ or 11.489

44. When rounding a decimal number to a whole number, we occasionally get a perfect-square whole number. In such cases, we get a whole number as the approximation of the square root of a decimal number. For example:

$\sqrt{36.29}$ is very close to $\sqrt{36}$ or 6
$\sqrt{120.5}$ is very close to $\sqrt{}$ or ______

$\sqrt{121}$ or 11

45. When using the table to approximate the square roots of decimal numbers, we can also round the entries further. For example:

a) $\sqrt{13.45}$ is approximately $\sqrt{13}$ or 3.606 .

Rounding to hundredths, we get ________.

b) $\sqrt{180.75}$ is approximately $\sqrt{181}$ or 13.454 .

Rounding to tenths, we get ________.

a) 3.61 b) 13.5

7-5 CALCULATORS AND SQUARE ROOTS

Various methods can be used to find square roots. More detailed tables than the one presented in this book are available. And other methods like the averaging method, the square-root algorithm, and logarithms can be used. However, now that square roots can quickly be found on many pocket calculators, using a pocket calculator is the most efficient. We will briefly discuss the use of calculators to find square roots in this section.

---

46. When finding a square root of a perfect-square whole number, a calculator might report some 0's after the decimal point. Those 0's should be ignored. For example:

If the calculator gives $\sqrt{529}$ = 23.00 , it should be read 23.

If the calculator gives $\sqrt{209,764}$ = 458.000000 , it should be read ________.

458

47. When finding a square root of a perfect-square decimal number, a calculator might report one or more final 0's after the decimal point. Those 0's can be ignored. For example:

If the calculator gives $\sqrt{5.76}$ = 2.40 , it can be read 2.4 .

If the calculator gives $\sqrt{.0289}$ = .170000 , it can be read ________.

.17

48. When finding a non-terminating square root, the calculator reports a rounded answer. For example, the square roots below were reported to two decimal places (or hundredths).

$\sqrt{59,677}$ = 244.29 $\sqrt{641}$ = 25.32 $\sqrt{.89}$ = .94

Sometimes it makes sense to round even further. For example:

Rounding to the nearest whole number, $\sqrt{59,677}$ = 244.29 or 244.

Rounding to the nearest tenth, $\sqrt{641}$ = 25.32 or ________.

25.3

49. Some calculators report non-terminating square roots to as many as six decimal places or more. For example:

$\sqrt{8,125}$ = 90.138782 $\sqrt{.497}$ = .704982269

In such cases, it almost always makes sense to round further.

a) Rounding to the nearest tenth, $\sqrt{8,125}$ = 90.138782 or ________.

b) Rounding to the nearest thousandth, $\sqrt{.497}$ = .704982269 or ________.

a) 90.1 b) .705

SELF-TEST 21 (pages 184-193)

Find each square.

1. $(-5)^2$ = ______	2. $\left(\frac{1}{3}\right)^2$ = ______	3. $(-1.2)^2$ = ______	4. $(30)^2$ = ______

Find each square root.

5. $\sqrt{36}$ = ______ 6. $-\sqrt{81}$ = ______

Using the table, find each square root.

7. $\sqrt{94}$ = ______ 8. $\pm\sqrt{137}$ = ______

Using the table: 9. Find $\sqrt{87}$ to the nearest hundredth. $\sqrt{87}$ = ______

10. Find $-\sqrt{146}$ to the nearest tenth. $-\sqrt{146}$ = ______

By rounding to the nearest whole number and then using the table, find a close approximation to each square root below. Round each answer to the nearest tenth.

11. $\sqrt{71.44}$ = ______ 12. $-\sqrt{183.6}$ = ______

Round each "calculator" square root to the indicated place.

13. nearest whole number $\sqrt{87,595} = 295.964525$ ______

14. nearest hundredth $\sqrt{3.44} = 1.854723699$ ______

ANSWERS:	1. 25	3. 1.44	5. 6	7. 9.695	9. 9.33	11. 8.4	13. 296
	2. $\frac{1}{9}$	4. 900	6. -9	8. ±11.705	10. -12.1	12. -13.6	14. 1.85

7-6 THE MEANING OF EXPRESSIONS CONTAINING A SQUARED LETTER

In this section, we will show the meaning of expressions containing a squared letter. To do so, we will evaluate some expressions of that type.

50. To square a letter, we simply multiply the letter by itself. That is:

$x^2 = (x)(x)$ $y^2 = (y)(y)$ $V^2 = (\ \ )(\ \ )$

(V)(V)

51. To evaluate a squared letter, we simply substitute the number and then square it. For example:

If $x = 5$, $x^2 = 5^2 = 25$ If $F = -3$, $F^2 = (-3)^2 = 9$

By substituting and then squaring, complete these:

a) If $y = 10$, y^2 = ______ b) If $t = -12$, t^2 = ______

a) 100

b) 144

52. Following the example, complete the other evaluation.

If $m = 2$, $m^2 + 3 = (2)^2 + 3 = 4 + 3 = 7$

If $p = -10$, $p^2 - 75 = (-10)^2 - 75$ = ______ - ______ = ______

100 - 75 = 25

53. Following the example, complete the other evaluation.

If $t = 4$, $5t^2 = 5(4)^2 = 5(16) = 80$

If $s = -8$, $\frac{1}{2}s^2 = \frac{1}{2}(-8)^2 = (\quad)(\quad) =$ ______

$\frac{1}{2}(64) = 32$

54. Following the example, complete the other evaluation.

If $V = 6$, $\frac{V^2}{3} = \frac{6^2}{3} = \frac{36}{3} = 12$

If $F = -2$, $\frac{40}{F^2} = \frac{40}{(-2)^2} = \text{———} =$ ______

$\frac{40}{4} = 10$

55. Following the example, complete the other evaluation.

If $x = 3$, $\frac{4x^2}{9} = \frac{4(3)^2}{9} = \frac{4(9)}{9} = \frac{36}{9} = 4$

If $y = -2$, $\frac{24}{3y^2} = \frac{24}{3(-2)^2} = \frac{24}{3(4)} = \text{———} =$ ______

$\frac{24}{12} = 2$

7-7 SOLVING EQUATIONS CONTAINING A SQUARED LETTER

In this section, we will discuss the procedure for solving some equations containing a squared letter.

---

56. The equation $x^2 = 25$ contains a squared letter. You can see that both "+5" and "-5" are roots of the equation since:

$(+5)(+5) = 25$ and $(-5)(-5) = 25$

Similarly: a) The two roots of $y^2 = 49$ are ______ and ______

b) The two roots of $p^2 = 16$ are ______ and ______

---

a) +7 and -7

b) +4 and -4

57. When an equation contains a squared letter on one side and a number on the other side, it has two roots. <u>The two roots are the two square roots of the number</u>. That is:

For $y^2 = 36$, $y = \pm\sqrt{36} = +6$ and -6.

a) For $t^2 = 81$, $t = \pm\sqrt{81} =$ ______ and ______.

b) For $d^2 = 1$, $d = \pm\sqrt{1} =$ ______ and ______.

---

a) +9 and -9

b) +1 and -1

58. Using the table of square roots on page 190, complete these:

For $s^2 = 175$, $s = \pm\sqrt{175} = +13.229$ and -13.229

a) For $m^2 = 33$, $m = \pm\sqrt{33} =$ ________ and ________

b) For $r^2 = 101$, $r = \pm\sqrt{101} =$ ________ and ________

---

a) +5.745 and -5.745

b) +10.05 and -10.05

59. By rounding to the nearest whole number and then using the square-root table, complete these:

a) For $x^2 = 17.8$, $x = \pm\sqrt{17.8}$ or $\pm\sqrt{18} =$ ________ and ________

b) For $y^2 = 87.26$, $y = \pm\sqrt{87.26}$ or $\pm\sqrt{87} =$ ________ and ________

---

a) +4.243 and -4.243

b) +9.327 and -9.327

60. To solve the equation below, we used the addition axiom to isolate "x^2" and then found both square roots of 9.

$$x^2 + 4 = 13$$
$$x^2 + 4 + (-4) = 13 + (-4)$$
$$x^2 = 9$$
$$x = \pm\sqrt{9} = +3 \text{ and } -3$$

We checked "+3" as one root of the equation at the right. Check "-3" as the other root in the space provided.

Checking +3	Checking -3
$x^2 + 4 = 13$	$x^2 + 4 = 13$
$(3)^2 + 4 = 13$	$(-3) + 4 = 13$
$9 + 4 = 13$	$9 + 4 = 13$
$13 = 13$	$13 = 13$

$(-3)^2 + 4 = 13$

$9 + 4 = 13$

$13 = 13$

61. To solve the equation below, we used the addition axiom to isolate "y^2" and then found both square roots of 64.

$$y^2 - 64 = 0$$
$$y^2 + (-64) = 0$$
$$y^2 + (-64) + 64 = 0 + 64$$
$$y^2 = 64$$
$$y = \pm\sqrt{64} = +8 \text{ and } -8$$

At the right, we checked +8 as one root of the above equation. Check -8 as the other root in the space provided.

Checking +8	Checking -8
$y^2 - 64 = 0$	$y^2 - 64 = 0$
$8^2 - 64 = 0$	
$64 - 64 = 0$	
$0 = 0$	

$(-8)^2 - 64 = 0$

$64 - 64 = 0$

$0 = 0$

62. Use the addition axiom to isolate the squared letter and then find both roots of each equation below.

a) $135 = 35 + r^2$ b) $m^2 - 1 = 0$

---

a) $r = +10$ and -10

b) $m = +1$ and -1

63. Solve each equation. Use the square-root table.

a) $t^2 - 8 = 40$ b) $F^2 - 12.6 = 0$

---

a) $t = \pm 6.928$

b) $F = \pm\sqrt{12.6}$ or $\sqrt{13} = \pm 3.606$

64. When the number is a perfect-square fraction, the two roots of the equation are the two square roots of the fraction. That is:

For $x^2 = \frac{4}{9}$, $x = \pm\sqrt{\frac{4}{9}} = \frac{2}{3}$ and $-\frac{2}{3}$

a) For $y^2 = \frac{81}{49}$, $y = \pm\sqrt{\frac{81}{49}} =$ ____ and ____

b) For $t^2 = \frac{1}{16}$, $t = \pm\sqrt{\frac{1}{16}} =$ ____ and ____

---

a) $\frac{9}{7}$ and $-\frac{9}{7}$

b) $\frac{1}{4}$ and $-\frac{1}{4}$

65. To solve the equation below, we used the multiplication axiom to isolate "x^2" and then found both square roots of $\frac{9}{16}$. Solve the other equation.

$$16x^2 = 9$$

$$x^2 = \frac{9}{16}$$

$$x = \frac{3}{4} \text{ and } -\frac{3}{4}$$

$$25y^2 = 64$$

$y = \frac{8}{5}$ and $-\frac{8}{5}$

66. To solve the equation below, we used both axioms to isolate "x^2" and then wrote the two roots. Use the same method to solve the other equation.

$$16x^2 + 10 = 91$$
$$16x^2 + 10 + (-10) = 91 + (-10)$$
$$16x^2 = 81$$
$$x^2 = \frac{81}{16}$$
$$x = +\frac{9}{4} \text{ and } -\frac{9}{4}$$

$$4m^2 - 1 = 0$$

$m = \frac{1}{2}$ and $-\frac{1}{2}$

67. To solve the equation below, we used the multiplication axiom to isolate "y^2". Solve the other equation.

$$\frac{1}{2}y^2 = 18$$
$$2\left(\frac{1}{2}y^2\right) = 2(18)$$
$$y^2 = 36$$
$$y = +6 \text{ and } -6$$

$$27 = \frac{1}{3}d^2$$

---

a) $d = \pm 9$

68. In solving the equation below, we obtained a fraction that is not a perfect square. Therefore, we performed a division and then used the square-root table. Solve the other equation.

$$5x^2 = 120$$
$$x^2 = \frac{120}{5}$$
$$x^2 = 24$$
$$x = \pm 4.899$$

$$2y^2 = 350$$

$y = \pm\sqrt{175} = \pm 13.229$

---

69. To solve the equation below, we performed a division. Notice that we divided until we got a tenths digit, then rounded to a whole number and used the square-root table. Solve the other equation.

$$7m^2 = 125$$
$$m^2 = \frac{125}{7}$$
$$m^2 = 17.8 \text{ or } 18$$
$$m = \pm 4.243$$

$$1.1r^2 = 78.9$$

$r = \pm\sqrt{71.7}$ or $\pm\sqrt{72} = \pm 8.485$

7-8 SOLVING FRACTIONAL EQUATIONS AND PROPORTIONS CONTAINING A SQUARED LETTER

In this section, we will discuss the procedure for solving some fractional equations and proportions containing a squared letter.

---

70. To clear the fraction in the equation at the right, we multiplied both sides by 3. Solve the other equation.

$$\frac{x^2}{3} = 12 \qquad\qquad 25 = \frac{y^2}{4}$$

$$3\left(\frac{x^2}{3}\right) = 3(12)$$

$$x^2 = 36$$

$$x = \pm 6$$

---

y = ±10

71. Following the example, solve the other equation.

$$\frac{4x^2}{5} = 5 \qquad\qquad 16 = \frac{9y^2}{4}$$

$$5\left(\frac{4x^2}{5}\right) = 5(5)$$

$$4x^2 = 25$$

$$x^2 = \frac{25}{4}$$

$$x = \pm\frac{5}{2}$$

$y = \pm\frac{8}{3}$

72. When a squared-letter term is divided by itself, the quotient is "1". That is:

$$\frac{x^2}{x^2} = 1 \qquad \frac{2y^2}{2y^2} = 1 \qquad \frac{3V^2}{3V^2} = 1$$

Therefore, when a fraction with a squared letter in its denominator is multiplied by its denominator, <u>the product equals its numerator</u>. For example:

$$x^2\left(\frac{5}{x^2}\right) = \frac{(x^2)(5)}{x^2} = \left(\frac{x^2}{x^2}\right)(5) = (1)(5) = 5$$

$$2y^2\left(\frac{1}{2y^2}\right) = \frac{(2y^2)(1)}{2y^2} = \left(\frac{2y^2}{2y^2}\right)(1) = (\quad)(\quad) = \underline{\qquad}$$

---

(1)(1) = 1

73. Following the example, solve the other equations.

$$\frac{25}{4d^2} = 9 \qquad\qquad \text{a) } 81 = \frac{49}{t^2} \qquad\qquad \text{b) } \frac{4}{25V^2} = 1$$

$$4d^2\left(\frac{25}{4d^2}\right) = (4d^2)(9)$$

$$25 = 36d^2$$

$$d^2 = \frac{25}{36}$$

$$d = \pm\frac{5}{6}$$

a) $t = \pm\frac{7}{9}$

b) $V = \pm\frac{2}{5}$

74. We performed a division before using the square-root table to solve the equation at the right. Solve the other equation.

$$\frac{2x^2}{7} = 10$$

$$7\left(\frac{2x^2}{7}\right) = 7(10)$$

$$2x^2 = 70$$

$$x^2 = \frac{70}{2} = 35$$

$$x = \pm 5.916$$

$$\frac{180}{3t^2} = 2$$

$t = \pm 5.477$

---

75. To solve the proportion at the right, we began by cross-multiplying. Solve the other proportion.

$$\frac{3}{4} = \frac{y^2}{3}$$

$$(3)(3) = (4)(y^2)$$

$$9 = 4y^2$$

$$y^2 = \frac{9}{4}$$

$$y = \pm\frac{3}{2}$$

$$\frac{25}{18} = \frac{2}{d^2}$$

$d = \pm\frac{6}{5}$

SELF-TEST 22 (pages 193-200)

Solve each equation.

1. $x^2 - 25 = 0$	2. $y^2 = \frac{1}{9}$	3. $36t^2 + 20 = 45$
4. $8 = \frac{1}{2}s^2$	5. $\frac{2x^2}{5} = 40$	6. $\frac{49}{64} = \frac{1}{m^2}$

Continued on following page.

SELF-TEST 22 (pages 193-200) - Continued

Using the table, solve each equation. Round decimal numbers to whole numbers before using the table.

7. $3b^2 = 339$	8. $1 = \frac{75}{F^2}$	9. $15.5 = 47.8 - p^2$

ANSWERS:

1. $x = \pm 5$
2. $y = \pm\frac{1}{3}$
3. $t = \pm\frac{5}{6}$
4. $s = \pm 4$
5. $x = \pm 10$
6. $m = \pm\frac{8}{7}$
7. $b = \pm 10.63$
8. $F = \pm 8.66$
9. $p = \pm 5.657$

7-9 THE MEANING OF SQUARE-ROOT RADICALS CONTAINING A LETTER

In this section, we will show the meaning of square-root radicals containing a letter. To do so, we will evaluate some radicals of that type.

Note: In this section and the following sections, the discussion will be limited to operations and equations involving only principal square roots.

76. To evaluate a square-root radical containing a letter, we substitute the number for the letter and then find the square root of the number. For example:

If $x = 25$, $\sqrt{x} = \sqrt{25} = 5$ If $y = 81$, $\sqrt{y} = \sqrt{81} =$ ____

77. To evaluate when the letter is only part of an algebraic expression, we substitute, simplify, and then find the square root of the resulting number. For example: 9

If $t = 8$, $\sqrt{t+1} = \sqrt{8+1} = \sqrt{9} = 3$

If $r = 12$, $\sqrt{3r} = \sqrt{3(12)} = \sqrt{36} =$ ____

78. Following the example, complete the other evaluation. 6

If $d = 32$, $\sqrt{\frac{d}{8}} = \sqrt{\frac{32}{8}} = \sqrt{4} = 2$ If $p = 2$, $\sqrt{\frac{128}{p}} = \sqrt{\frac{128}{2}} = \sqrt{\quad} =$ ____

$\sqrt{64} = 8$

79. Following the example, complete the other evaluation.

$$\text{If } F = 99, \quad \sqrt{\frac{F+1}{4}} = \sqrt{\frac{99+1}{4}} = \sqrt{\frac{100}{4}} = \sqrt{25} = 5$$

$$\text{If } b = 5, \quad \sqrt{\frac{8}{b+3}} = \sqrt{\frac{8}{5+3}} = \sqrt{\frac{8}{8}} = \sqrt{\quad} = \_\_\_\_$$

$\sqrt{1} = 1$

80. Following the example, complete the other evaluation.

$$\text{If } x = 6, \quad \sqrt{\frac{2x}{3}} = \sqrt{\frac{2(6)}{3}} = \sqrt{\frac{12}{3}} = \sqrt{4} = 2$$

$$\text{If } y = 2, \quad \sqrt{\frac{54}{3y}} = \sqrt{\frac{54}{3(2)}} = \sqrt{\frac{54}{6}} = \sqrt{\quad} = \_\_\_\_$$

$\sqrt{9} = 3$

7-10 SQUARING SQUARE-ROOT RADICALS CONTAINING A LETTER

In order to solve equations containing a square-root radical, we must be able to square the square-root radical. We will discuss that squaring process in this section. We will show that the square of any square-root radical is equal to the original radicand.

81. To indicate a multiplication of two radicals, we ordinarily use a dot as the multiplication symbol. That is:

$\sqrt{3} \cdot \sqrt{5}$ means "multiply $\sqrt{3}$ and $\sqrt{5}$".

$\sqrt{7} \cdot \sqrt{11}$ means "multiply ________ and ________".

$\sqrt{7}$ and $\sqrt{11}$

82. When multiplying two square-root radicals, we get <u>a single radical in which the two original radicands are multiplied</u>. That is:

$$\sqrt{3} \cdot \sqrt{5} = \sqrt{(3)(5)} \qquad \sqrt{7} \cdot \sqrt{x} = \sqrt{(7)(x)} \qquad \sqrt{y} \cdot \sqrt{y} = \sqrt{(\ \ )(\ \ )}$$

$\sqrt{(y)(y)}$

83. To square a square-root radical, we simply multiply the radical times itself. For example:

$$\left(\sqrt{8}\right)^2 = \sqrt{8} \cdot \sqrt{8} = \sqrt{(8)(8)} \qquad \left(\sqrt{x}\right)^2 = \sqrt{x} \cdot \sqrt{x} = \sqrt{(\ \ )(\ \ )}$$

$\sqrt{(x)(x)}$

84. When squaring a square-root radical, the square can be written as a single radical whose radicand is the square of the original radicand. That is:

$$\left(\sqrt{7}\right)^2 = \sqrt{7} \cdot \sqrt{7} = \sqrt{(7)(7)} = \sqrt{7^2} \qquad \left(\sqrt{t}\right)^2 = \sqrt{t} \cdot \sqrt{t} = \sqrt{(t)(t)} = \sqrt{\quad}$$

$\sqrt{t^2}$

85. The square root of a quantity squared is the quantity itself. That is:

$$\sqrt{5^2} = 5\,, \quad \text{since } \sqrt{5^2} = \sqrt{25} = 5 \qquad\qquad \sqrt{8^2} = 8\,, \quad \text{since } \sqrt{8^2} = \sqrt{64} = 8$$

The fact above also applies to the square root of a letter squared. That is:

$$\sqrt{x^2} = x \qquad \sqrt{y^2} = y \qquad \text{a) } \sqrt{m^2} = \_\_\_\_ \qquad \text{b) } \sqrt{V^2} = \_\_\_\_$$

Answers: a) m b) V

86. Since the square root of a quantity squared is the quantity itself, the square of a square-root radical is the original radicand. That is:

$$\left(\sqrt{3}\right)^2 = \sqrt{3}\cdot\sqrt{3} = \sqrt{(3)(3)} = \sqrt{3^2} = 3 \qquad \left(\sqrt{x}\right)^2 = \sqrt{x}\cdot\sqrt{x} = \sqrt{(x)(x)} = \sqrt{x^2} = x$$

The fact above can be used as a shortcut for squaring a radical. That is, the "square" is simply the radicand. For example:

$$\left(\sqrt{7}\right)^2 = 7 \qquad \left(\sqrt{y}\right)^2 = y \qquad \text{a) } \left(\sqrt{10}\right)^2 = \_\_\_\_ \qquad \text{b) } \left(\sqrt{r}\right)^2 = \_\_\_\_$$

Answers: a) 10 b) r

87. The same shortcut applies when the original radicand is a letter term. That is:

$$\left(\sqrt{2t}\right)^2 = 2t \qquad \left(\sqrt{5d}\right)^2 = 5d \qquad \text{a) } \left(\sqrt{8y}\right)^2 = \_\_\_\_ \qquad \text{b) } \left(\sqrt{11t}\right)^2 = \_\_\_\_$$

Answers: a) 8y b) 11t

88. The same shortcut applies when the original radicand is a fraction. That is:

$$\left(\sqrt{\frac{3}{4}}\right)^2 = \frac{3}{4} \qquad \left(\sqrt{\frac{x}{5}}\right)^2 = \frac{x}{5} \qquad \text{a) } \left(\sqrt{\frac{9}{t}}\right)^2 = \_\_\_\_ \qquad \text{b) } \left(\sqrt{\frac{7R}{3}}\right)^2 = \_\_\_\_$$

Answers: a) $\frac{9}{t}$ b) $\frac{7R}{3}$

89. The same shortcut applies when the original radicand is an addition or subtraction. That is:

$$\left(\sqrt{x+1}\right)^2 = x + 1 \qquad \left(\sqrt{y-5}\right)^2 = y - 5 \qquad \left(\sqrt{4b+3}\right)^2 = \_\_\_\_$$

Answer: 4b + 3

90. The square of any square-root radical is equal to the original radicand. That is:

$$\left(\sqrt{\frac{m+1}{3}}\right)^2 = \frac{m+1}{3} \qquad \left(\sqrt{\frac{100}{v+1}}\right)^2 = \frac{100}{v+1} \qquad \left(\sqrt{\frac{2x-1}{5}}\right)^2 = \_\_\_\_$$

Answer: $\frac{2x - 1}{5}$

7-11 SOLVING EQUATIONS CONTAINING A SQUARE-ROOT RADICAL

In this section, we will discuss the procedure for solving equations containing a square-root radical when the radical is isolated on one side of the equation.

91. When a square-root radical is isolated on one side of an equation, we can eliminate the radical and obtain a non-radical equation by squaring both sides of the original equation. For example:

$$\sqrt{2x} = 4 \qquad\qquad \sqrt{y+3} = 5$$

$$\left(\sqrt{2x}\right)^2 = (4)^2 \qquad\qquad \left(\sqrt{y+3}\right)^2 = (5)^2$$

$$2x = 16 \qquad\qquad y + 3 = 25$$

Square both sides of each equation below and write the resulting non-radical equation.

a) $\sqrt{m} = 3$ b) $\sqrt{4t - 1} = 2$ c) $\sqrt{\frac{5t}{9}} = 7$

_____ = _____ _____ = _____ _____ = _____

---

a) $m = 9$

b) $4t - 1 = 4$

c) $\frac{5t}{9} = 49$

92. When squaring to eliminate a radical in an equation, be sure to square both sides of the equation. What non-radical equation is obtained by squaring both sides of these?

a) $\sqrt{\frac{x}{3}} = 5$ b) $\frac{1}{4} = \sqrt{2d}$ c) $\sqrt{\frac{4}{y}} = \frac{4}{3}$

_____ = _____ _____ = _____ _____ = _____

---

a) $\frac{x}{3} = 25$

b) $\frac{1}{16} = 2d$

c) $\frac{4}{y} = \frac{16}{9}$

93. To solve a radical equation when the square-root radical is isolated on one side of the equation, we use the following two steps:

1) Eliminate the radical by squaring both sides of the equation.

2) Then solve the resulting non-radical equation.

An example solution is shown at the right. Check "12" as the root of $\sqrt{3x} = 6$ in the space provided.

$$\sqrt{3x} = 6$$

$$3x = 36$$

$$x = \frac{36}{3}$$

$$x = 12$$

Check

$\sqrt{3(12)} = 6$

$\sqrt{36} = 6$

$6 = 6$

94. Another example solution is shown at the right. Check "11" as the root of $\sqrt{m - 2} = 3$ in the space provided.

$$\sqrt{m-2} = 3$$

$$\left(\sqrt{m-2}\right)^2 = (3)^2$$

$$m - 2 = 9$$

$$m = 11$$

Check

$\sqrt{11 - 2} = 3$

$\sqrt{9} = 3$

$3 = 3$

95. Following the example, solve the other equations.

$$\sqrt{p} = \frac{1}{3}$$

$$p = \frac{1}{9}$$

a) $\sqrt{r} = 5$

b) $\frac{6}{5} = \sqrt{t}$

---

a) $r = 25$

b) $t = \frac{36}{25}$

96. Following the example, solve the other equation.

$$\sqrt{2x} = 3$$

$$2x = 9$$

$$x = \frac{9}{2}$$

$4 = \sqrt{11y}$

$y = \frac{16}{11}$

97. Following the example, solve the other equation.

$$\sqrt{3p} = \frac{1}{2}$$

$$3p = \frac{1}{4}$$

$$\frac{1}{3}(3p) = \frac{1}{3}\left(\frac{1}{4}\right)$$

$$p = \frac{1}{12}$$

$\frac{4}{3} = \sqrt{2d}$

$d = \frac{8}{9}$

98. Following the example, solve the other equation.

$$\sqrt{x + 2} = 3$$

$$x + 2 = 9$$

$$x + 2 + (-2) = 9 + (-2)$$

$$x = 7$$

$1 = \sqrt{y - 4}$

$y = 5$

99. Following the example, solve the other equations.

$$\sqrt{\frac{x}{2}} = \frac{1}{3}$$

$$\frac{x}{2} = \frac{1}{9}$$

$$(x)(9) = (2)(1)$$

$$x = \frac{2}{9}$$

a) $\sqrt{\frac{9}{5t}} = 2$

b) $\frac{3}{4} = \sqrt{\frac{1}{y}}$

a) $t = \frac{9}{20}$

b) $y = \frac{16}{9}$

100. Following the example, solve the other equation.

$$3 = \sqrt{\frac{x - 1}{2}}$$

$$9 = \frac{x - 1}{2}$$

$$2(9) = 2\left(\frac{x - 1}{2}\right)$$

$$18 = x - 1$$

$$18 = x + (-1)$$

$$18 + 1 = x + (-1) + 1$$

$$x = 19$$

$$\sqrt{\frac{30}{V + 5}} = 1$$

Answer: $V = 25$

7-12 SOLVING EQUATIONS WITH A RADICAL CONTAINING A SQUARED LETTER

In this section, we will discuss the procedure for solving equations with a square-root radical containing a squared letter.

101. When the square-root radical contains a squared letter, the equation has two roots. An example is shown. Solve the other equations.

$$\sqrt{x^2 - 9} = 0$$

$$x^2 - 9 = 0$$

$$x^2 = 9$$

$$x = \pm 3$$

$$0 = \sqrt{16y^2 - 25}$$

Answer: $y = \pm\frac{5}{4}$

102. Following the example, solve the other equation.

$$\sqrt{4p^2 - 9} = 4$$

$$4p^2 - 9 = 16$$

$$4p^2 = 25$$

$$p^2 = \frac{25}{4}$$

$$p = \pm\frac{5}{2}$$

$$5 = \sqrt{2V^2 - 7}$$

Answer: $V = \pm 4$

103. Following the example, solve the other equation.

$$\sqrt{10 - x^2} = 1$$

$$10 - x^2 = 1$$

$$10 + (-1x^2) = 1$$

$$(-10) + 10 + (-1x^2) = 1 + (-10)$$

$$-1x^2 = -9$$

$$x^2 = \frac{-9}{-1} = 9$$

$$x = \pm 3$$

$$6 = \sqrt{100 - p^2}$$

Answer: $p = \pm 8$

104. We used the square-root table to find the two roots at the right. Use the table to solve the other problem.

$$\sqrt{y^2 + 10} = 13$$
$$y^2 + 10 = 169$$
$$y^2 = 159$$
$$y = \pm 12.61$$

$$\sqrt{4x^2 - 360} = 0$$

$x = \pm 9.487$

SELF-TEST 23 (pages 200-206)

Write the non-radical equation obtained by squaring both sides of each equation below.

1. $\frac{1}{3} = \sqrt{x}$

____ = ____

2. $\sqrt{2y - 7} = 5$

____ = ____

3. $6 = \sqrt{\frac{200}{V - 1}}$

____ = ____

Solve each equation. Use the square-root table when necessary.

4. $\sqrt{t} = 9$

5. $1 = \sqrt{3d + 13}$

6. $\sqrt{\frac{x - 5}{3}} = 4$

7. $\frac{1}{2} = \sqrt{\frac{7}{p}}$

8. $\sqrt{m^2 - 1} = 0$

9. $3 = \sqrt{35 - y^2}$

ANSWERS:

1. $\frac{1}{9} = x$
2. $2y - 7 = 25$
3. $36 = \frac{200}{V - 1}$
4. $t = 81$
5. $d = -4$
6. $x = 53$
7. $p = 28$
8. $m = \pm 1$
9. $y = \pm 5.099$

Unit 8 FORMULA EVALUATION

Formulas are frequently used in science, business, statistics, geometry, and other parts of mathematics to state relationships and solve applied problems. In this unit, we will discuss the meaning of formulas and formula evaluation. Major emphasis is given to the numerical part of formula evaluation. Some evaluations that require equation-solving are included. In order to emphasize the formula-evaluation process, the numbers used are generally quite simple.

8-1 THE MEANING OF FORMULAS AND FORMULA EVALUATION

In this section, we will discuss the meaning of formulas and formula evaluation. Some typical formulas from physical science and technology, business, statistics, geometry, and other parts of mathematics are shown. The general characteristics of formulas are discussed.

1. A formula is a shorthand way of stating a relationship that can also be stated in words. An example is given below.

   There is a relationship between the retail selling price of an item, its cost to the store, and the amount of markup. That is:

   The retail selling price of an item can be found by adding its cost to the amount of markup.

   The relationship above can be written as a formula containing words. That is:

Retail selling price = Cost + Markup

   However, it is usually written with letters that are abbreviations for the words. That is:

   $$R = C + M$$

   where: R = Retail selling price
   C = Cost
   M = ______________

Markup

2. By substituting numbers in place of the letters, we can use formulas to solve applied problems. This process is called "formula evaluation". Some examples are given below.

Using the formula $\boxed{R = C + M}$, we can find the retail selling price of an item if we know its cost to the store and the amount of markup that the store uses for that item. For example:

If a store paid \$20 for a pocket calculator and marked it up \$10, the retail selling price is \$30 since:

$$R = C + M = \$20 + \$10 = \$30$$

If a car dealer paid \$3,500 for a car and used a \$450 markup, the retail selling price is \$3,950 since:

$$R = C + M = \$3,500 + \$450 = \_\_\_\_\_\_$$

\$3,950

3. Some typical formulas from (1) physical science and technology, (2) business, (3) statistics (used in social science, physical science, and business), (4) geometry, and (5) other parts of mathematics are shown below.

(1) (Uniform Velocity) $\boxed{v = \frac{s}{t}}$ (Engine Horsepower) $\boxed{H = \frac{D^2N}{2.5}}$

(2) (Retail Selling Price) $\boxed{R = C + M}$ (Net Price) $\boxed{N = L - D}$

(3) (Standard Deviation) $\boxed{s = \sqrt{V}}$ (Standard Score) $\boxed{z = \frac{X - M}{s}}$

(4) (Area of a Circle) $\boxed{A = \pi r^2}$ (Hypotenuse of a Right Triangle) $\boxed{c = \sqrt{a^2 + b^2}}$

(5) (Slope-Intercept Form of a Straight Line) $\boxed{y = mx + b}$ (Quadratic Formula) $\boxed{x = \frac{-b \pm \sqrt{b^2 - 4ac}}{2a}}$

By examining the formulas above, you can see these facts:

1) Formulas can contain either capital or small letters. Some also contain numbers.

2) Formulas can involve addition, subtraction, multiplication, division, squaring, finding a square root, or various combinations of those operations.

> Since the purpose of this unit is to introduce the numerical part of formula evaluation rather than a method for dealing with the "units" in formulas, "units" are generally avoided in the remaining sections.

8-2 FORMULAS INVOLVING ADDITION AND SUBTRACTION

In this section, we will discuss formulas that involve only addition and subtraction.

---

4. In some formulas, the algebraic expression on the right side is simply an addition or subtraction. Some examples are shown below.

$$R = C + M \qquad D = R + S + T \qquad C = K - 273$$

To perform evaluations with formulas of that type, we simply substitute and then perform the addition or subtraction. An example is shown below. Complete the other evaluation.

In $\boxed{R = C + M}$, if $C = \$12.50$ and $M = \$3.45$, $R = \$15.95$ since:

$$R = C + M = \$12.50 + \$3.45 = \$15.95$$

In the formula below, find the value of D if $R = 40$, $S = 30$, and $T = 100$.

$D = R + S + T =$ ______ + ______ + ______ = ______

---

5. In $\boxed{C = A - L}$, if $A = 30$ and $L = 10$, $C = 20$ since:

$$C = A - L = 30 - 10 = 20$$

In the formula below, find the value of N if $L = \$11.95$ and $D = \$2.39$.

$N = L - D =$ ______ - ______ = ______

> D = 170 , from:
> 40 + 30 + 100

---

6. In $\boxed{M = V + 100}$, if $V = 65$, $M = 165$ since:

$$M = V + 100 = 65 + 100 = 165$$

In the formula below, find the value of C if $K = 350$.

$C = K - 273 =$ ________ - ________ = ________

> N = $9.56 , from:
> $11.95 - $2.39

---

7. On the right side of each equation below, the second term of the subtraction is an addition enclosed in parentheses.

$$A = 180 - (B + C) \qquad A = 360 - (B + C + D)$$

The parentheses mean that the addition should be performed before subtracting. An example is shown below. Complete the other evaluation.

In $\boxed{A = 180 - (B + C)}$, if $B = 45$ and $C = 55$, $A = 80$ since:

$$A = 180 - (B + C) = 180 - (45 + 55) = 180 - 100 = 80$$

In the formula below, find A if $B = 70$, $C = 110$, and $D = 90$.

$A = 360 - (B + C + D) =$ ____________________

> C = 77 , from:
> 350 - 273

> A = 90 , from:
> 360 - 270

8-3 FORMULAS INVOLVING MULTIPLICATIONS

In this section, we will discuss formulas involving multiplication. Some also involve addition or subtraction.

---

8. In each formula below, the algebraic expression on the right side is either a number and a letter or two letters written next to each other.

$$v = 5t \qquad y = mx \qquad E = IR$$

Expressions of that type stand for multiplication. That is:

5t means: multiply "5" and "t".

mx means: multiply "m" and "x".

IR means: multiply "I" and "R".

Therefore, to perform evaluations with formulas of that type, we simply substitute and then perform the multiplication. An example is shown below. Complete the other evaluation.

In $\boxed{P = 4s}$, if $s = 5$, $P = 20$ since:

$$P = 4s = 4(5) = 20$$

In the formula below, find R if $N = 10$ and $P = \$50$.

$$R = NP = (\quad)(\quad) = \underline{\qquad}$$

---

9. In each formula below, the algebraic expression on the right side is either three letters or a number and two letters written next to each other.

R = \$500 , from: (10)(\$50)

$$V = LWH \qquad A = \frac{1}{2}bh$$

Expressions of that type stand for three-factor multiplications. That is:

LWH means: (L)(W)(H)

$\frac{1}{2}bh$ means: $\left(\frac{1}{2}\right)(b)(h)$

Therefore, to perform evaluations with formulas of that type, we simply substitute and then perform the three-factor multiplication. An example is shown below. Complete the other evaluation.

In $\boxed{V = LWH}$, if $L = 10$, $W = 2$, and $H = 6$, $V = 120$ since:

$$V = LWH = (10)(2)(6)$$
$$= (20)\ (6) = 120$$

In the formula below, find A if $b = 20$ and $h = 8$.

$$A = \frac{1}{2}bh = \underline{\qquad\qquad}$$

10. In each formula below, one of the terms in the addition on the right side stands for a multiplication.	A = 80, from: $\frac{1}{2}(20)(8)$ → 10 (8)

$$d = f + at \qquad y = mx + b$$

To perform evaluations with formulas of that type, we perform the multiplication before adding. An example is shown below. Complete the other evaluation.

In $\boxed{d = f + at}$, if f = 30, a = 4, and t = 5, d = 50 since:

$$d = f + at = 30 + (4)(5) = 30 + 20 = 50$$

In the formula below, find "y" when $m = \frac{1}{2}$, x = 10, and b = 2.

$$y = mx + b = \left(\frac{1}{2}\right)(10) + 2$$

$$= \_\_\_\_ + \_\_\_\_ = \_\_\_\_$$

11. In each formula below, both terms in the addition on the right side stand for multiplications.	y = 7, from: 5 + 2

$$P = 2L + 2W \qquad h = ab + cd$$

To perform evaluations with formulas of that type, we perform both multiplications before adding. An example is shown. Complete the other evaluation.

In $\boxed{P = 2L + 2W}$, if L = 15 and W = 10, P = 50 since:

$$P = 2L + 2W = 2(15) + 2(10) = 30 + 20 = 50$$

In the formula below, find "h" when a = 2, b = 3, c = 4, and d = 5.

$$h = ab + cd = (2)(3) + (4)(5)$$

$$= \_\_\_\_ + \_\_\_\_ = \_\_\_\_$$

12. In the formulas below, the right side is a multiplication in which one factor is an addition or subtraction enclosed in parentheses.	h = 26, from: 6 + 20

$$D = P(Q + R) \qquad C = \frac{5}{9}(F - 32)$$

To perform evaluations with formulas of that type, we perform the operation in parentheses before multiplying. An example is shown. Complete the other evaluation.

In $\boxed{D = P(Q + R)}$, if P = 10, Q = 3, and R = 4, D = 70 since:

$$D = P(Q + R) = 10(3 + 4) = 10(7) = 70$$

In the formula below, find C when F = 50.

$$C = \frac{5}{9}(F - 32) = \_\_\_\_\_\_\_\_$$

C = 10, from: $\frac{5}{9}(18)$

8-4 FORMULAS INVOLVING DIVISION

In this section, we will discuss formulas involving division. Some also involve addition, subtraction, multiplication, or some combination of those operations.

---

13. In each formula below, the right side is a fraction.

$$I = \frac{E}{R} \qquad\qquad D = \frac{B}{N}$$

Since a fraction stands for a division, we perform evaluations with formulas like those above by dividing. An example is shown. Complete the other evaluation.

In $\boxed{I = \frac{E}{R}}$, if $E = 40$ and $R = 5$, $I = 8$ since:

$$I = \frac{E}{R} = \frac{40}{5} = 8$$

In the formula below, find D when $B = \$675$ and $N = 20$.

$$D = \frac{B}{N} = \frac{\$675}{20} = \_\_\_\_\_\_\_\_$$

D = \$33.75

14. In the formulas below, either the numerator or denominator of the fraction is a multiplication.

$$V = \frac{4st}{a} \qquad\qquad b = \frac{c}{dt}$$

To perform evaluations with formulas of that type, we perform the multiplication before dividing. An example is shown. Complete the other evaluation.

In $\boxed{V = \frac{4st}{a}}$, if $s = 5$, $t = 3$, and $a = 10$, $V = 6$ since:

$$V = \frac{4st}{a} = \frac{4(5)(3)}{10} = \frac{60}{10} = 6$$

In the formula below, find "b" when $c = 200$, $d = 5$, and $t = 10$.

$$b = \frac{c}{dt} = \frac{200}{(5)(10)} = \_\_\_\_\_\_ = \_\_\_\_\_\_$$

$b = 4$, from $\frac{200}{50}$

15. In each formula below, the numerator of the fraction is "1". Following the example, complete the other evaluation.

In $\boxed{t = \frac{1}{f}}$, if $f = 100$, $t = .01$ since: $t = \frac{1}{f} = \frac{1}{100} = .01$

In $\boxed{m = \frac{1}{2cd}}$, if $c = 2$ and $d = 5$, $m = \frac{1}{2(2)(5)} = \_\_\_\_\_\_\_\_$

16. In the fraction below, either the numerator or denominator or both is an addition or subtraction.

m = .05, from $\frac{1}{20}$

$$R = \frac{p + q}{2} \qquad t = \frac{a - b}{c - d}$$

To perform evaluations with formulas of that type, we perform all additions and subtractions before dividing. An example is shown. Complete the other evaluation.

In $\boxed{R = \frac{p + q}{2}}$, if p = 20, and q = 30, R = 25 since:

$$R = \frac{p + q}{2} = \frac{20 + 30}{2} = \frac{50}{2} = 25$$

In the formula below, find "t" when a = 30, b = 10, c = 10, and d = 6.

$$t = \frac{a - b}{c - d} = \frac{30 - 10}{10 - 6} = \underline{\hspace{4cm}}$$

---

17. When evaluating with a formula in which the right side is a complicated fraction, we always simplify the numerator and denominator before dividing. An example is shown. Complete the other evaluation.

t = 5, from $\frac{20}{4}$

In $\boxed{S = \frac{PT}{P + T}}$, if P = 4 and T = 6, S = 2.4 since:

$$S = \frac{PT}{P + T} = \frac{(4)(6)}{4 + 6} = \frac{24}{10} = 2.4$$

In the formula below, find C when F = 50.

$$C = \frac{5F - 160}{9} = \frac{5(50) - 160}{9} = \underline{\hspace{4cm}}$$

---

18. In the addition on the right side of the formula below, one term is a fraction or division. Notice that we perform that division before adding. Complete the other evaluation.

C = 10, from $\frac{90}{9}$

In $\boxed{F = \frac{9C}{5} + 32}$, if C = 10, F = 50 since:

$$F = \frac{9C}{5} + 32 = \frac{9(10)}{5} + 32 = \frac{90}{5} + 32 = 18 + 32 = 50$$

In the same formula, find F when C = 15.

$$F = \frac{9C}{5} + 32 = \underline{\hspace{4cm}}$$

F = 59, from 27 + 32

8-5 FORMULAS INVOLVING SQUARES

In this section, we will discuss formulas involving squares. Most of the formulas also require a multiplication or division or both.

---

19. In $\boxed{A = s^2}$, the right side is a squared letter. We can find A by simply squaring the value of "s". Therefore:

If $s = 8$, $A = s^2 = 8^2 = 64$ If $s = 30$, $A = s^2 = 30^2 =$ ________

900

20. In each formula below, the right side is a multiplication in which one of the factors is a squared letter.

$$P = I^2R \qquad s = \frac{1}{2}at^2 \qquad A = 3.14r^2$$

To perform evaluations with formulas of that type, we square that factor before multiplying. An example is shown. Complete the other evaluation.

In $\boxed{P = I^2R}$, if $I = 3$ and $R = 10$, $P = 90$ since:

$$P = I^2R = (3^2)(10) = 9(10) = 90$$

In the formula below, find A when $r = 2$.

$$A = 3.14r^2 = \text{\_\_\_\_\_\_\_\_}$$

A = 12.56, from 3.14(4)

21. In $\boxed{s = \frac{1}{2}at^2}$, if $a = 12$ and $t = 4$, $s = 96$ since:

$$s = \frac{1}{2}at^2 = \frac{1}{2}(12)(4^2) = \frac{1}{2}(12)(16) = 6(16) = 96$$

In the formula below, find E when $m = 8$ and $v = 10$.

$$E = \frac{1}{2}mv^2 = \text{\_\_\_\_\_\_\_\_}$$

E = 400, from $\frac{1}{2}(8)(100)$

22. In each formula below, the right side is a fraction that contains a squared letter in either its numerator or denominator.

$$P = \frac{E^2}{R} \qquad H = \frac{D^2N}{2.5} \qquad V = \frac{Fcd}{r^2}$$

To perform evaluations with formulas of that type, we simplify the numerator or denominator or both and then divide. An example is shown. Complete the other evaluation.

In $\boxed{P = \frac{E^2}{R}}$, if $E = 9$ and $R = 10$, $P = 8.1$ since: $P = \frac{E^2}{R} = \frac{9^2}{10} = \frac{81}{10} = 8.1$

In $\boxed{h = \frac{a}{s^2}}$, if $a = 300$ and $s = 10$, $h = \frac{a}{s^2} = \frac{300}{10^2} =$ ________

23. In $\boxed{V = \frac{Fcd}{r^2}}$, if F = 5 , c = 3 , d = 6 , and r = 2 , V = 22.5 since:

h = 3, from $\frac{300}{100}$

$$V = \frac{Fcd}{r^2} = \frac{(5)(3)(6)}{2^2} = \frac{90}{4} = 22.5$$

In the formula below, find "a" when p = 3 , q = 5 , r = 10 , and t = 5 .

$$a = \frac{pqr}{t^2} = \_\_\_\_\_\_\_\_\_\_$$

a = 6, from $\frac{150}{25}$

8-6 FORMULAS INVOLVING SQUARE ROOTS

In this section, we will discuss formulas involving square roots. Most of the formulas also involve one or more operations.

> Note: In this section, use the square-root table on page 190 when necessary.

24. In $s = \sqrt{A}$, we can find "s" by simply finding the square root of A. Therefore:

If A = 64 , $s = \sqrt{A} = \sqrt{64} = 8$ If A = 196 , $s = \sqrt{A} = \sqrt{196} = \_\_\_\_\_$

14

25. In each formula below, the radicand is an expression involving at least one other operation.

$$v = \sqrt{2as} \qquad I = \sqrt{\frac{P}{R}} \qquad c = \sqrt{a^2 + b^2}$$

To perform evaluations with formulas of that type, <u>we simplify the expression under the radical</u> and then find its square root. An example is shown. Complete the other evaluation.

In $\boxed{v = \sqrt{2as}}$, if a = 3 and s = 6 , v = 6 since:

$$v = \sqrt{2as} = \sqrt{2(3)(6)} = \sqrt{36} = 6$$

In the formula below, find E when P = 4 and R = 16 .

$$E = \sqrt{PR} = \_\_\_\_\_\_\_\_\_\_$$

E = 8, from $\sqrt{64}$

26. In $\boxed{I = \sqrt{\frac{P}{R}}}$, if P = 400 and R = 4 , I = 10 since:

$$I = \sqrt{\frac{P}{R}} = \sqrt{\frac{400}{4}} = \sqrt{100} = 10$$

In the formula below, find "t" when s = 64 and g = 32 .

$$t = \sqrt{\frac{2s}{g}} = \_\_\_\_\_\_\_\_\_\_$$

t = 2, from $\sqrt{4}$

27. In $\boxed{r = \sqrt{\frac{A}{3.14}}}$, if A = 6.28 , r = 1.414 since:

$$r = \sqrt{\frac{A}{3.14}} = \sqrt{\frac{6.28}{3.14}} = \sqrt{2} = 1.414$$

In the formula below, find "d" when A = 8.635 .

$$d = \sqrt{\frac{A}{.785}} = \underline{\hspace{6cm}}$$

---

d = 3.317, from $\sqrt{11}$

28. In $\boxed{c = \sqrt{a^2 + b^2}}$, if a = 3 and b = 4 , c = 5 since:

$$c = \sqrt{a^2 + b^2} = \sqrt{3^2 + 4^2} = \sqrt{9 + 16} = \sqrt{25} = 5$$

In the formula below, find R when Z = 10 and X = 7 .

$$R = \sqrt{Z^2 - X^2} = \underline{\hspace{6cm}}$$

R = 7.141, from $\sqrt{51}$

29. In $\boxed{d = \sqrt{b^2 - 4ac}}$, if b = 10 , a = 1 , and c = 6 , d = 8.718 since:

$$d = \sqrt{b^2 - 4ac} = \sqrt{10^2 - 4(1)(6)} = \sqrt{100 - 24} = \sqrt{76} = 8.718$$

In the formula below, find "a" when d = 12 , b = 2 , and c = 8 .

$$a = \sqrt{d^2 - 2bc} = \underline{\hspace{6cm}}$$

---

a = 10.583, from $\sqrt{112}$

30. In each formula below, the square-root radical is a factor in a multiplication.

$$r = a\sqrt{w} \qquad N = K\sqrt{\frac{V}{H}}$$

To perform evaluations with formulas of that type, <u>we simplify the radical before multiplying</u>. An example is shown. Complete the other evaluation.

In $\boxed{r = a\sqrt{w}}$, if a = 3 and w = 16 , r = 12 since:

$$r = a\sqrt{w} = 3\sqrt{16} = 3(4) = 12$$

In the formula below, find N when K = 7 , V = 18 , and H = 2 .

$$N = K\sqrt{\frac{V}{H}} = \underline{\hspace{6cm}}$$

N = 21, from $7\sqrt{9}$

SELF-TEST 24 (pages 207 - 217)

(pages 207 - 217)

1. Find A if B = 45 and C = 60 .

$$A = 180 - (B + C)$$

2. Find F if k = .5 and s = 12 .

$$F = ks$$

3. Find C if F = 77 .

$$C = \frac{5}{9}(F - 32)$$

4. Find I if E = 12.5 and R = 2.5 .

$$I = \frac{E}{R}$$

5. Find A if B = 9 .

$$A = \frac{B}{B + 1}$$

6. Find A if r = 10 .

$$A = 3.14r^2$$

7. Find s if a = 10 and t = 4 .

$$s = \frac{1}{2}at^2$$

8. Find H if D = 5 and N = 2 .

$$H = \frac{D^2N}{2.5}$$

9. Find v if a = 5 and s = 10 .

$$v = \sqrt{2as}$$

10. Find I if P = 100 and R = 25 .

$$I = \sqrt{\frac{P}{R}}$$

11. Find a if c = 9 and b = 7 .

$$a = \sqrt{c^2 - b^2}$$

ANSWERS:

1. A = 75
2. F = 6
3. C = 25
4. I = 5
5. A = .9
6. A = 314
7. s = 80
8. H = 20
9. v = 10
10. I = 2
11. a = 5.657

8-7 SUBSCRIPTS IN FORMULAS

Some formulas contain literal or numerical subscripts. In this section, we will discuss the meaning of subscripts and do some evaluations with formulas that contain subscripts.

---

31. When it makes sense to use the same letter more than once in a formula, subscripts are used. Letters can be used as the subscripts. An example is given below.

There is a relationship between average velocity (or speed), original velocity, and final velocity when acceleration is constant. That is:

Average velocity is found by adding the original velocity and final velocity and dividing by 2.

The relationship above can be written as a formula containing words. That is:

$$\text{average velocity} = \frac{\text{original velocity} + \text{final velocity}}{2}$$

However, it is usually written as a formula containing letters. Since it makes sense to use the letter "v" for the three different velocities, the subscripts "av", "o", and "f" are used to distinguish them. That is:

$$v_{av} = \frac{v_o + v_f}{2}$$

where: v_{av} = average velocity
v_o = original velocity
v_f = ________ velocity

---

final velocity

32. Instead of using letters for subscripts, numbers can be used. An example is given below.

There is a relationship between the total force applied in a situation and all of the separate forces applied. That is:

The total force applied is the sum of the separate forces applied.

The relationship above can be written as a formula containing words. That is, if there are three separate forces:

total force = first force + second force + third force

However, it is usually written as a formula containing letters. Since it makes sense to use the letter "F" for the four different forces, the subscripts "T", "$_1$", "$_2$", "$_3$" are used to distinguish them. That is:

$$F_T = F_1 + F_2 + F_3$$

where: F_T = total force
F_1 = first force
F_2 = second force
F_3 = ________ force

third force

33. The abbreviation "sub" is used when naming a letter with a subscript. For example:

V_2 is called "V sub 2"

V_t is called "V sub t"

Write a letter with a subscript for each of these:

a) F sub 1 = ______ b) v sub f = ______ c) P sub 2 = ______

---

a) F_1

b) v_f

c) P_2

34. In evaluations, a letter with a subscript is treated like a letter without a subscript. An example is shown. Complete the other evaluation.

In $\boxed{v_{av} = \frac{v_o + v_f}{2}}$, if $v_o = 20$ and $v_f = 40$, $v_{av} = 30$ since:

$$v_{av} = \frac{v_o + v_f}{2} = \frac{20 + 40}{2} = \frac{60}{2} = 30$$

In the formula below, find "a" when $v_2 = 24.6$, $v_1 = 8.4$, and $t = 2$.

$$a = \frac{v_2 - v_1}{t} = \rule{5cm}{0.4pt}$$

$a = 8.1$, from $\frac{16.2}{2}$

35. In $\boxed{E = \frac{P_o}{P_i}}$, if $P_o = 60$ and $P_i = 12$, $E = 5$ since:

$$E = \frac{P_o}{P_i} = \frac{60}{12} = 5$$

In the formula below, find V_1 when $P_2 = 5$, $V_2 = 8$, and $P_1 = 10$.

$$V_1 = \frac{P_2V_2}{P_1} = \rule{5cm}{0.4pt}$$

$V_1 = 4$, from $\frac{40}{10}$

36. In $\boxed{v = \frac{s_2 - s_1}{t_2 - t_1}}$, if $s_2 = 50$, $s_1 = 20$, $t_2 = 10$ and $t_1 = 4$, $v = 5$ since:

$$v = \frac{s_2 - s_1}{t_2 - t_1} = \frac{50 - 20}{10 - 4} = \frac{30}{6} = 5$$

In the formula below, find "m" when $y_2 = 20$, $y_1 = 12$, $x_2 = 10$, and $x_1 = 8$.

$$m = \frac{y_2 - y_1}{x_2 - x_1} = \rule{5cm}{0.4pt}$$

$m = 4$, from $\frac{8}{2}$

37. In $\boxed{R = \frac{R_1R_2}{R_1 + R_2}}$, if $R_1 = 7$ and $R_2 = 3$, $R = 2.1$ since:

$$R = \frac{R_1R_2}{R_1 + R_2} = \frac{(7)(3)}{7 + 3} = \frac{21}{10} = 2.1$$

In the formula below, find C_2 when $C_1 = .8$ and $C_t = .6$.

$$C_2 = \frac{C_1C_t}{C_1 - C_t} = \underline{\hspace{6cm}}$$

$C_2 = 2.4$, from $\frac{.48}{.2}$

38. In $\boxed{H = ms(t_2 - t_1)}$, if $m = 2$, $s = 4$, $t_2 = 10$, and $t_1 = 5$, $H = 40$ since:

$$H = ms(t_2 - t_1) = (2)(4)(10 - 5) = 8(5) = 40$$

In the formula below, find A when $h = 10$, $b_1 = 8$ and $b_2 = 6$.

$$A = \frac{1}{2}h(b_1 + b_2) = \underline{\hspace{6cm}}$$

$A = 70$, from $\left(\frac{1}{2}\right)(10)(14)$

---

39. Don't confuse the subscript "2" with the "2" that is the symbol for squaring.

The subscript "2" is written slightly below the letter. It usually means "second".

v_2 means "the second velocity".

The symbol for squaring is written slightly above the letter. That is:

v^2 means "square the value of v".

a) Does t_2 or t^2 mean "square the value of t"? ______

b) Does V_2 or V^2 mean "the second volume"? ______

---

a) t^2

b) V_2

40. In $F = \frac{m_1m_2}{rd^2}$: a) Does m_2 mean "the second mass" or "the mass squared"? ______________________

b) Does d^2 mean "the second distance" or "the distance squared"? ______________________

---

a) the second mass

b) the distance squared

41. a) In the formula below, find F when $m_1 = 8$, $m_2 = 10$, $r = 5$ and $d = 2$.

$$F = \frac{m_1m_2}{rd^2} = \underline{\hspace{6cm}}$$

b) In the formula below, find "v_o" when $v_f = 7$, $g = 3$, and $s = 4$.

$$v_o = \sqrt{v_f^2 - 2gs} = \underline{\hspace{6cm}}$$

a) $F = 4$, from $\frac{80}{5(4)}$

b) $v_o = 5$, from $\sqrt{49 - 24}$

8-8 EVALUATIONS REQUIRING EQUATION-SOLVING

In this section, we will discuss some formula evaluations that require solving an equation.

---

42. In many formulas, there is one letter alone on the left side. That letter is called the "solved for" letter. For example:

In $N = L - D$, the "solved for" letter is N.

In $A = LW$, the "solved for" letter is ______.

Answer: A

---

43. When evaluating to find the value of the "solved for" letter, we simply substitute the numbers given for the other letters and then perform some arithmetic operations. However, when evaluating to find the value of one of the "non-solved-for" letters, we have to solve an equation after substituting the given numbers. An example is given below. Complete the other evaluation.

In the formula below, find C when R = \$80 and M = \$20.

$$R = C + M$$
$$\$80 = C + \$20$$
$$(-\$20) + \$80 = C + \$20 + (-\$20)$$
$$C = \$60$$

In the formula below, find K when C = 27.

$$C = K - 273$$

Answer: K = 300, from: $27 = K - 273$

---

44. In this frame and many of the following frames, we will give an example that is solved and ask you to complete an evaluation.

In the formula below, find "s" when P = 15.

$$P = 3s$$
$$15 = 3s$$
$$s = \frac{15}{3} = 5$$

In the formula below, find "t" when v = 70.

$$v = 10t$$

Answer: t = 7, from: $70 = 10t$

---

45. In the formula below, find L when V = 80, W = 2, and H = 4.

$$V = LWH$$
$$80 = L(2)(4)$$
$$80 = 8L$$
$$L = \frac{80}{8} = 10$$

In the formula below, find "b" when A = 40 and h = 4.

$$A = \frac{1}{2}bh$$

Answer: A = 20, from: $40 = 2b$

46. In the formula below, find E when $I = 10$ and $R = 3$.

$$I = \frac{E}{R}$$

$$10 = \frac{E}{3}$$

$$3(10) = 3\left(\frac{E}{3}\right)$$

$$E = 30$$

In the formula below, find V_2 when $V_1 = 8$, $P_2 = 2$, $P_1 = 3$.

$$V_1 = \frac{P_2V_2}{P_1}$$

---

47. In the formula below, find "p" when $R = 12$ and $q = 10$.

$$R = \frac{p + q}{2}$$

$$12 = \frac{p + 10}{2}$$

$$2(12) = 2\left(\frac{p + 10}{2}\right)$$

$$24 = p + 10$$

$$p = 14$$

$V_2 = 12$, from:

$$8 = \frac{2V_2}{3}$$

In the formula below, find "v_1" when $a = 20$, $v_2 = 100$, and $t = 3$.

$$a = \frac{v_2 - v_1}{t}$$

$v_1 = 40$, from:

$$20 = \frac{100 - v_1}{3}$$

8-9 EVALUATIONS AND SOLVING EQUATIONS CONTAINING A SQUARED LETTER OR A RADICAL

In this section, we will discuss some evaluations that require solving an equation containing a squared letter or a square-root radical.

---

48. In the last unit, we saw that equations like $x^2 = 25$ have two roots, one positive and one negative. However, when the squared letter appears in a formula, the negative root does not ordinarily make sense. For example:

In $s^2 = 100$, where "s" stands for "distance", -10 does not ordinarily make sense as a value for distance.

In $r^2 = 49$, where "r" stands for the radius of a circle, -7 does not ordinarily make sense as a value for the radius.

Therefore, when solving equations like those above, <u>we report only the positive root</u>. That is:

For $s^2 = 100$, we report $s = 10$.

For $r^2 = 49$, we report $r =$ ________

$r = 7$

49. When solving for a squared letter in a formula, we only report the positive root because ordinarily only the positive root makes sense. An example is given below. Complete the other evaluation.

In the formula below, find "s" when A = 25 .

$$A = s^2$$

$$25 = s^2$$

$$s = \sqrt{25} = 5$$

In the formula below, find I when P = 72 and R = 2 .

$$P = I^2R$$

I = +6 (not -6), from $72 = 2I^2$

50. Following the example, complete the other evaluation. Use the square-root table. Remember to report only the positive root.

In the formula below, find "t" when s = 100 and a = 10 .

$$s = \frac{1}{2}at^2$$

$$100 = \frac{1}{2}(10)(t^2)$$

$$100 = 5t^2$$

$$t^2 = \frac{100}{5} = 20$$

$$t = \sqrt{20} = 4.472$$

In the formula below, find "v" when E = 600 and m = 20 .

$$E = \frac{1}{2}mv^2$$

E = 7.746 (from $\sqrt{60}$)

51. In the formula below, find "t" when s = 100 and g = 5 .

$$s = \frac{gt^2}{2}$$

$$100 = \frac{5t^2}{2}$$

$$2(100) = 2\left(\frac{5t^2}{2}\right)$$

$$200 = 5t^2$$

$$t^2 = \frac{200}{5} = 40$$

$$t = \sqrt{40} = 6.325$$

In the formula below, find "d" when F = 3 , $m_1 = 6$, $m_2 = 10$, and r = 2 .

$$F = \frac{m_1m_2}{rd^2}$$

F = 3.162 (from $\sqrt{10}$)

52. To perform the evaluation below, we had to solve an equation containing a square-root radical. Complete the other evaluation.

In the formula below, find V when s = 10 .

$$s = \sqrt{V}$$

$$10 = \sqrt{V}$$

$$V = 100$$

In the formula below, find "s" when v = 8 and a = 2 .

$$v = \sqrt{2as}$$

---

s = 16, from:

$$8 = \sqrt{4s}$$

$$64 = 4s$$

53. In the formula below, find R when I = 6 and P = 72 .

$$I = \sqrt{\frac{P}{R}}$$

$$6 = \sqrt{\frac{72}{R}}$$

$$36 = \frac{72}{R}$$

$$R(36) = R\left(\frac{72}{R}\right)$$

$$36R = 72$$

$$R = 2$$

In the formula below, find "s" when t = 10 and g = 3 .

$$t = \sqrt{\frac{2s}{g}}$$

---

s = 150, from:

$$100 = \frac{2s}{3}$$

54. a) In the formula below, find I_c when d = 5 , $I_a = 100$, and m = 2 .

$$d = \sqrt{\frac{I_a - I_c}{m}}$$

b) In the formula below, find "a" when c = 10 and b = 6 .

$$c = \sqrt{a^2 + b^2}$$

---

a) $I_c = 50$, from:

$$25 = \frac{100 - I_c}{2}$$

b) a = 8, from:

$$100 = a^2 + 36$$

8-10 FORMULAS IN PROPORTION FORM

In this section, we will discuss some formulas that are written in proportion form.

55. The formulas below are written in proportion form because they have only one fraction on each side.

$$\frac{F_1}{F_2} = \frac{d_1}{d_2} \qquad \frac{P_1V_1}{T_1} = \frac{P_2V_2}{T_2} \qquad \frac{I_1}{I_2} = \frac{(d_2)^2}{(d_1)^2}$$

To perform evaluations with formulas of that type, we simply substitute and then solve the proportion. An example is shown. Complete the other evaluation.

In the formula below, find "d_1" when $F_1 = 12$, $F_2 = 20$, and $d_2 = 40$.

$$\frac{F_1}{F_2} = \frac{d_1}{d_2}$$

$$\frac{12}{20} = \frac{d_1}{40}$$

$$(12)(40) = (20)(d_1)$$

$$480 = 20d_1$$

$$d_1 = \frac{480}{20} = 24$$

In the formula below, find P_2 when $P_1 = 8$, $V_2 = 12$, and $V_1 = 18$.

$$\frac{P_1}{P_2} = \frac{V_2}{V_1}$$

$P_2 = 12$, from:

$144 = 12P_2$

56. By substituting and then solving the proportion, complete each evaluation below.

a) In the formula below, find V_1 when $P_1 = 2$, $T_1 = 5$, $P_2 = 3$, $V_2 = 4$, and $T_2 = 6$.

$$\frac{P_1V_1}{T_1} = \frac{P_2V_2}{T_2}$$

b) In the formula below, find "d_2" when $I_1 = 1$, $I_2 = 4$, and $d_1 = 10$.

$$\frac{I_1}{I_2} = \frac{(d_2)^2}{(d_1)^2}$$

a) $V_1 = 5$, from:

$$\frac{2V_1}{5} = \frac{12}{6}$$

b) $d_2 = 5$, from:

$$\frac{1}{4} = \frac{(d_2)^2}{100}$$

SELF-TEST 25 (pages 218-226)

1. Find C_2 if $C_1 = 3$ and $C_T = 2.4$.

$$C_2 = \frac{C_1 C_T}{C_1 - C_T}$$

2. Find A if $h = 20$, $b_1 = 3$ and $b_2 = 2$.

$$A = \frac{1}{2}h(b_1 + b_2)$$

3. Find "v_o" if $v_f = 10$, $g = 32$ and $s = 1$.

$$v_o = \sqrt{v_f^2 - 2gs}$$

4. Find L if N = \$64 and C = \$16 .

$$N = L - C$$

5. Find I if $E = 20$ and $R = 4$.

$$E = IR$$

6. Find "v_2" if $a = 30$, $v_1 = 20$ and $t = 2$.

$$a = \frac{v_2 - v_1}{t}$$

7. Find "v" if $E = 100$ and $m = 10$.

$$E = \frac{1}{2}mv^2$$

8. Find "s" if $t = 10$ and $g = 32$.

$$t = \sqrt{\frac{2s}{g}}$$

9. Find V_1 if $P_1 = 10$, $P_2 = 3$, and $V_2 = 4.5$.

$$\frac{P_1}{P_2} = \frac{V_2}{V_1}$$

ANSWERS:

1. $C_2 = 12$
2. $A = 50$
3. $v_o = 6$
4. $L = \$80$
5. $I = 5$
6. $v_2 = 80$
7. $v = 4.472$
8. $s = 1{,}600$
9. $V_1 = 1.35$

Unit 9 FORMULA REARRANGEMENT

In order to make evaluations easier, formulas are frequently rearranged to solve for a non-solved-for letter. We will discuss formula rearrangements in this unit. We will show that the same algebraic principles used to solve equations are used to rearrange formulas.

9-1 THE PURPOSE OF FORMULA REARRANGEMENT

In this section, we will show that formula rearrangement is used to avoid equation-solving in formula evaluations.

---

1. To perform the evaluation below, we substituted the numbers in the equation as it stands. Since N is non-solved-for, we had to solve an equation to find its value.

Find N when $D = 10$ and $B = 70$.

$$D = \frac{B}{N}$$

$$10 = \frac{70}{N}$$

$$N(10) = N\left(\frac{70}{N}\right)$$

$$10N = 70$$

$$N = \frac{70}{10} = 7$$

However, we can use algebraic principles to rearrange the formula above to solve for N. That is:

Rearranging $D = \frac{B}{N}$, we get $N = \frac{B}{D}$

Using the rearranged formula in which N is solved-for, we can perform the same evaluation without solving an equation. We did so below.

Find N when $D = 10$ and $B = 70$.

$$N = \frac{B}{D} = \frac{70}{10} = 7$$

Did we get the same value for N with both methods? ___________

Yes. It is 7.

2. To perform the evaluation below, we also substituted the numbers in the equation as it stands. Since "v_1" is non-solved-for, we had to solve an equation to find its value.

Find "v_1" when $a = 10$, $v_2 = 60$, and $t = 2$.

$$a = \frac{v_2 - v_1}{t}$$

$$10 = \frac{60 - v_1}{2}$$

$$2(10) = 2\left(\frac{60 - v_1}{2}\right)$$

$$20 = 60 - v_1$$

$$-40 = -v_1$$

$$v_1 = 40$$

Using algebraic principles, we can rearrange the above formula to solve for "v_1". That is:

Rearranging $a = \dfrac{v_2 - v_1}{t}$, we get $v_1 = v_2 - at$

Using the rearranged formula in which "v_1" is solved-for, we can perform the same evaluation without solving an equation. We did so below.

Find "v_1" when $a = 10$, $v_2 = 60$, and $t = 2$.

$$v_1 = v_2 - at = 60 - 10(2) = 60 - 20 = 40$$

Did we get the same value for "v_1" with both methods? ________

Yes. It is 40.

> The main purpose of formula rearrangement is to avoid the need for equation-solving when doing a formula evaluation. A rearrangement is especially useful when it avoids an equation-solving that is not simple.

9-2 MULTIPLICATIONS INVOLVING LITERAL FRACTIONS

In this section, we will discuss the procedure for multiplying two literal fractions or a literal fraction and a non-fraction.

3. To multiply two numerical fractions, we simply multiply their numerators and multiply their denominators. The same procedure is used to multiply two literal fractions. That is:

$$\left(\frac{a}{b}\right)\left(\frac{c}{d}\right) = \frac{(a)(c)}{(b)(d)} = \frac{ac}{bd} \qquad \left(\frac{P}{Q}\right)\left(\frac{T}{S}\right) = \frac{(P)(T)}{(Q)(S)} = \_\_\_\_\_\_$$

$\dfrac{PT}{QS}$

4. When multiplying literal fractions, there is an accepted order for writing the factors in the numerator and denominator of the product. That is:

1. Number-factors are always written in front of letter-factors.
2. Letter-factors are usually written in alphabetical order.

The product at the left below is written in the accepted order. Complete the other multiplication.

$$\left(\frac{T}{v^2}\right)\left(\frac{s}{5}\right) = \frac{(T)(s)}{(v^2)(5)} = \frac{sT}{5v^2} \qquad \left(\frac{m^2}{t}\right)\left(\frac{2p}{q}\right) = \frac{(m^2)(2p)}{(t)(q)} = \underline{\qquad}$$

$\frac{2m^2p}{qt}$

5. When any literal expression is divided by "1", the quotient is identical to the literal expression. For example:

$$\frac{t}{1} = t \qquad \frac{a+b}{1} = a+b \qquad \frac{2at^2}{1} = 2at^2 \qquad \frac{d(x-y)}{1} = d(x-y)$$

Therefore, any literal expression can be written as a fraction whose denominator is "1". That is:

$$a = \frac{a}{1} \qquad b + c = \frac{b+c}{1} \qquad \text{a) } 3.14r^2 = \underline{\qquad} \qquad \text{b) } ms(t_2 - t_1) = \underline{\qquad}$$

---

a) $\frac{3.14r^2}{1}$

b) $\frac{ms(t_2 - t_1)}{1}$

6. Since any literal expression can be written as a fraction whose denominator is "1", any multiplication of a fraction and a non-fraction can be converted into a multiplication of two fractions. That is:

$$\left(\frac{V}{R}\right)(a) = \left(\frac{V}{R}\right)\left(\frac{a}{1}\right) = \frac{aV}{R} \qquad \left(\frac{1}{d^2}\right)(P^2T) = \left(\frac{1}{d^2}\right)\left(\frac{P^2T}{1}\right) = \underline{\qquad}$$

$\frac{P^2T}{d^2}$

7. Following the example, complete the other multiplication.

$$m^2\left(\frac{a}{b}\right) = \left(\frac{m^2}{1}\right)\left(\frac{a}{b}\right) = \frac{am^2}{b} \qquad 2as\left(\frac{1}{v^2}\right) = \left(\frac{2as}{1}\right)\left(\frac{1}{v^2}\right) = \underline{\qquad}$$

$\frac{2as}{v^2}$

8. Following the example, complete the other multiplication.

$$(p+q)\left(\frac{c}{d}\right) = \left(\frac{p+q}{1}\right)\left(\frac{c}{d}\right) = \frac{c(p+q)}{d} \qquad \left(\frac{1}{V}\right)(m)(t_2 - t_1) = \frac{1}{V}\left[\frac{(m)(t_2 - t_1)}{1}\right] = \underline{\qquad}$$

$\frac{m(t_2 - t_1)}{V}$

9. Do these multiplications.

$$\text{a) } 5a^2\left(\frac{M^2}{R^2}\right) = \underline{\qquad} \qquad \text{b) } (c - d)\left(\frac{h}{V}\right) = \underline{\qquad}$$

a) $\frac{5a^2M^2}{R^2}$

b) $\frac{h(c - d)}{V}$

10. Do these multiplications.

a) $\frac{1}{P_1}(P_2V_2) =$ ______ b) $\frac{1}{t^2}(V_2 - V_1) =$ ______

Answer: a) $\frac{P_2V_2}{P_1}$ b) $\frac{V_2 - V_1}{t^2}$

9-3 RECIPROCALS OF LITERAL EXPRESSIONS

In this section, we will discuss the reciprocals of literal expressions.

11. When any literal expression is multiplied by "1", the product is identical to the literal expression. That is:

$(1)(v^2) = v^2$ $(1)(a - c) = a - c$ a) $(1)(P_2V_2) =$ ______ b) $(1)(F_o + F_i) =$ ______

Answer: a) P_2V_2 b) $F_o + F_i$

12. When any literal expression is divided by itself, the quotient is "1". That is:

$\frac{T}{T} = 1$ $\frac{p + q}{p + q} = 1$ a) $\frac{2ab^2}{2ab^2} =$ ______ b) $\frac{m^2(t_2 - t_1)}{m^2(t_2 - t_1)} =$ ______

Answer: a) 1 b) 1

13. To complete each multiplication below, we substituted "1" for the fraction.

$\left(\frac{v^2}{v^2}\right)(t) = (1)(t) = t$ $\left(\frac{x + y}{x + y}\right)(2ab) = (1)(2ab) = 2ab$

Use the same method to complete these:

a) $\left(\frac{3t^2}{3t^2}\right)(pqr) = (\quad)(pqr) =$ ______ b) $\left(\frac{t_2 - t_1}{t_2 - t_1}\right)(s_2 - s_1) = (\quad)(s_2 - s_1) =$ ______

Answer: a) $(1)(pqr) = pqr$ b) $(1)(s_2 - s_1) = s_2 - s_1$

14. Two quantities are a pair of reciprocals if their product is "1".

Since $\frac{1}{LW}(LW) = \frac{LW}{LW} = 1$, the reciprocal of LW is $\frac{1}{LW}$.

Since $\frac{1}{c + d}(c + d) = \frac{c + d}{c + d} = 1$, the reciprocal of $c + d$ is ______.

Answer: $\frac{1}{c + d}$

15. It should be obvious that the reciprocal of a literal expression is a fraction whose numerator is "1" and whose denominator is the literal expression. For example:

The reciprocal of P_1 is $\frac{1}{P_1}$. The reciprocal of $2r^2$ is $\frac{1}{2r^2}$.

Write the reciprocal of each expression.

a) c ______ b) $(d_1)^2$ ______ c) at^2 ______ d) $4b^2d$ ______

16. The reciprocals of two more literal expressions are given below.

The reciprocal of $v_1 - v_2$ is $\frac{1}{v_1 - v_2}$. The reciprocal of $a(b + c)$ is $\frac{1}{a(b + c)}$.

Write the reciprocal of each expression.

a) $x - y$ ______ b) $4(p + q)$ ______ c) $AK(t_2 - t_1)$ = ______

Answers: a) $\frac{1}{c}$ b) $\frac{1}{(d_1)^2}$ c) $\frac{1}{at^2}$ d) $\frac{1}{4b^2d}$

Answers: a) $\frac{1}{x - y}$ b) $\frac{1}{4(p + q)}$ c) $\frac{1}{AK(t_2 - t_1)}$

9-4 IDENTIFYING TERMS IN FORMULAS

Since a formula is an equation, each side of a formula contains one or more terms. In this section, we will discuss the types of terms in formulas and identify terms in formulas.

---

17. In a formula, terms are separated by addition or subtraction symbols. Any single letter or squared letter (with or without a subscript) is a term. For example, we have drawn boxes around each term in the formulas below.

$$\boxed{C} = \boxed{A} - \boxed{L} \qquad \boxed{F_T} = \boxed{F_1} + \boxed{F_2} + \boxed{F_3} \qquad \boxed{A} = \boxed{s^2}$$

Draw a box around each term in these formulas.

a) $D = R + S + T$ b) $I_E = I_C + I_B$ c) $c^2 = a^2 + b^2$

---

18. Any multiplication (no matter how many factors it contains) is one term. For example, we have drawn boxes around each term in the formulas below.

$$\boxed{A} = \boxed{LW} \qquad \boxed{V_1T_2} = \boxed{T_1V_2} \qquad \boxed{I_A} = \boxed{I_C} + \boxed{md^2}$$

Draw a box around each term in these formulas.

a) $y = mx + b$ b) $E = IR_1 + IR_2$ c) $E = \frac{1}{2}mv^2$

Answers:
a) $\boxed{D} = \boxed{R} + \boxed{S} + \boxed{T}$
b) $\boxed{I_E} = \boxed{I_C} + \boxed{I_B}$
c) $\boxed{c^2} = \boxed{a^2} + \boxed{b^2}$

---

19. A multiplication is one term even if one of the factors is an addition or a subtraction. For example, we have drawn a box around each term in the formulas below.

$$\boxed{D} = \boxed{P(Q + R)} \qquad \boxed{A} = \boxed{\frac{1}{2}h(b_1 + b_2)}$$

Draw a box around each term in these formulas.

a) $C = \frac{5}{9}(F - 32)$ b) $H = ms(t_2 - t_1)$ c) $W_S = P(V_1 - V_2)$

Answers:
a) $\boxed{y} = \boxed{mx} + \boxed{b}$
b) $\boxed{E} = \boxed{IR_1} + \boxed{IR_2}$
c) $\boxed{E} = \boxed{\frac{1}{2}mv^2}$

20. Any single fraction (no matter how complicated its numerator or denominator) is one term. For example, we have drawn a box around each term in the formulas below.

$$\boxed{I} = \boxed{\frac{E}{R}} \qquad \boxed{H} = \boxed{\frac{D^2N}{2.5}} \qquad \boxed{w} = \boxed{\frac{F_o + F_i}{t}}$$

Draw a box around each term in these formulas.

a) $\frac{P_1}{P_2} = \frac{V_2}{V_1}$ b) $s = \frac{gt^2}{2}$ c) $m = \frac{y_2 - y_1}{x_2 - x_1}$

a) $\boxed{C} = \boxed{\frac{5}{9}(F - 32)}$

b) $\boxed{H} = \boxed{ms(t_2 - t_1)}$

c) $\boxed{W_S} = \boxed{P(V_1 - V_2)}$

---

21. Draw a box around each term in these formulas.

a) $P = I^2R$ c) $v_{av} = \frac{v_o + v_f}{2}$

b) $P = 2L + 2W$

a) $\boxed{\frac{P_1}{P_2}} = \boxed{\frac{V_2}{V_1}}$

b) $\boxed{s} = \boxed{\frac{gt^2}{2}}$

c) $\boxed{m} = \boxed{\frac{y_2 - y_1}{x_2 - x_1}}$

---

a) $\boxed{P} = \boxed{I^2R}$ b) $\boxed{P} = \boxed{2L} + \boxed{2W}$ c) $\boxed{v_{av}} = \boxed{\frac{v_o + v_f}{2}}$

9-5 COEFFICIENTS OF LETTERS IN TERMS

In this section, we will define what is meant by the coefficient of a letter in a term.

---

22. In any two-factor term containing a number and a letter, the number is the coefficient of the letter. For example:

In 4s , the coefficient of "s" is 4.

In 1.5t , the coefficient of "t" is ________.

1.5

23. In any two-factor multiplication containing two letters, the coefficient of each factor is the other factor. For example:

In LW , the coefficient of W is L.

In IR , the coefficient of I is ________.

R

24. In any multiplication with more than two factors, the coefficient of each factor is the other factors. For example:

In LWH , the coefficient of H is LW.

In $.24I^2Rt$, the coefficient of R is ________.

$.24I^2t$

25. In any term, the coefficient of any one factor is the remaining factor or factors. For example:

In $P(Q + R)$: the coefficient of $(Q + R)$ is P.

the coefficient of P is $(Q + R)$.

In $ms(t_2 - t_1)$: a) the coefficient of $(t_2 - t_1)$ is ________.

b) the coefficient of "s" is ______________.

c) the coefficient of "m" is ______________.

Answers: a) ms b) $m(t_2 - t_1)$ c) $s(t_2 - t_1)$

26. a) In Ftg, the coefficient of "t" is ________.

b) In $2a(b + c)$, the coefficient of "a" is ______________.

c) In $AKT(t_2 - t_1)$, the coefficient of K is ______________.

Answers: a) Fg b) $2(b + c)$ c) $AT(t_2 - t_1)$

9-6 THE MULTIPLICATION AXIOM AND FORMULA REARRANGEMENT

When a formula has only one non-fractional term on each side, we can use the multiplication axiom to solve for a single letter. We will discuss the method in this section. A shortcut for the multiplication axiom is shown.

---

27. When a formula has only one non-fractional term on each side, we use the multiplication axiom to solve for a single letter. To do so, <u>we multiply both sides by the reciprocal of the coefficient of that letter</u>. An example is shown below. Following the instructions, complete the other rearrangement.

To solve for "h" in the formula at the right, we multiplied both sides by $\frac{1}{.433}$, <u>the reciprocal of the coefficient of "h"</u>. Solve for "t" in the other formula by multiplying both sides by $\frac{1}{10}$.

$$P = .433h$$

$$\frac{1}{.433}(P) = \frac{1}{.433}(.433h)$$

$$\frac{P}{.433} = 1h$$

$$h = \frac{P}{.433}$$

$$v = 10t$$

Answer: $t = \frac{v}{10}$

28. To solve for R in the formula at the right, we multiplied both sides by $\frac{1}{I^2}$, <u>the reciprocal of the coefficient of R</u>. Solve for V_1 in the other formula by multiplying both sides by $\frac{1}{P_1}$.

$$P = I^2R$$

$$\frac{1}{I^2}(P) = \frac{1}{I^2}(I^2R)$$

$$\frac{P}{I^2} = 1R$$

$$R = \frac{P}{I^2}$$

$$P_1V_1 = P_2V_2$$

Answer: $V_1 = \frac{P_2V_2}{P_1}$

29. To solve for "s" at the right, we multiplied both sides by $\frac{1}{2a}$. Solve for L in the other formula by multiplying both sides by $\frac{1}{WH}$.

$$v^2 = 2as$$
$$\frac{1}{2a}(v^2) = \frac{1}{2a}(2as)$$
$$\frac{v^2}{2a} = 1s$$
$$s = \frac{v^2}{2a}$$

$$V = LWH$$

---

30. To solve for P below, we multiplied both sides by $\frac{1}{Q + R}$. Solve for "b" in the other formula by multiplying both sides by $\frac{1}{c - d}$.

$L = \frac{V}{WH}$

$$D = P(Q + R)$$
$$\frac{1}{Q + R}(D) = P(Q + R)\left(\frac{1}{Q + R}\right)$$
$$\frac{D}{Q + R} = P \cdot 1$$
$$P = \frac{D}{Q + R}$$

$$a = b(c - d)$$

---

31. Some of the solutions from the preceding frames are shown below.

$b = \frac{a}{c - d}$

If $P = .433h$, $h = \frac{P}{.433}$

If $V = LWH$, $L = \frac{V}{WH}$

If $D = P(Q + R)$, $P = \frac{D}{Q + R}$

As you can see from the examples above, there is a shortcut for the multiplication axiom. That is, to solve for a letter, we can simply divide the other side of the formula by the coefficient of that letter. For example:

In $E = IR$: The coefficient of I is R.

Therefore, $I = \frac{E}{R}$

In $v^2 = 2as$: a) The coefficient of "a" is ________.

b) Therefore, a = ________

---

32. Two more examples of the shortcut are shown below.

a) 2s

b) $\frac{v^2}{2s}$

In $2RF = W(R - r)$: The coefficient of W is (R - r).

Therefore, $W = \frac{2RF}{R - r}$

In $H = ms(t_2 - t_1)$: a) The coefficient of "s" is ________________.

b) Therefore, s = ________________

33. When rearranging a formula, a capital letter should not be changed to a small letter and a small letter should not be changed to a capital letter. That is:

Don't change "A" to "a".

Don't change "e" to "E".

Use the shortcut of the multiplication axiom to solve for the indicated letter in this frame and the next.

a) Solve for W: $A = LW$ $W =$ ______

b) Solve for F: $d^2Fr = m_1m_2$ $F =$ ______

c) Solve for R: $H = .24I^2Rt$ $R =$ ______

Answers: a) $m(t_2 - t_1)$ b) $s = \frac{H}{m(t_2 - t_1)}$

---

34. a) Solve for "b". $2ab = c(d - f)$ $b =$ ______

b) Solve for "m". $H = ms(t_2 - t_1)$ $m =$ ______

Answers: a) $W = \frac{A}{L}$ b) $F = \frac{m_1m_2}{d^2r}$ c) $R = \frac{H}{.24I^2t}$

Answers: a) $b = \frac{c(d - f)}{2a}$ b) $m = \frac{H}{s(t_2 - t_1)}$

SELF-TEST 26 (pages 227-235)

1. Solve for "h": $P = 62.4h$

2. Solve for "L": $A = LW$

3. Solve for "d_1": $F_1d_1 = F_2d_2$

4. Solve for "t": $H = .24I^2Rt$

5. Solve for "m": $2mt = d(x - y)$

6. Solve for "t": $K = mt(x - r)$

ANSWERS:

1. $h = \frac{P}{62.4}$ 2. $L = \frac{A}{W}$ 3. $d_1 = \frac{F_2d_2}{F_1}$ 4. $t = \frac{H}{.24I^2R}$ 5. $m = \frac{d(x - y)}{2t}$ 6. $t = \frac{K}{m(x - r)}$

9-7 CLEARING THE FRACTION IN FORMULAS CONTAINING ONLY ONE FRACTION

When a formula contains only one fraction, we can "clear the fraction" by multiplying both sides by its denominator. We will discuss that use of the multiplication axiom in this section.

---

35. In each example below, we used the "cancelling" shortcut to multiply a literal fraction by its denominator. Notice that each product is the numerator of the original fraction.

$$\overset{1}{\cancel{R}}\left(\frac{E}{\underset{1}{\cancel{R}}}\right) = \frac{E}{1} = E \qquad \overset{1}{\cancel{dt^2}}\left(\frac{ac}{\underset{1}{\cancel{dt^2}}}\right) = \frac{ac}{1} = ac$$

Using the fact that the product is the numerator of the original fraction, complete these.

a) $v^2\left(\frac{4st}{v^2}\right) =$ ______ b) $P_1\left(\frac{P_2V_2}{P_1}\right) =$ ______ c) $2\left(\frac{gt^2}{2}\right) =$ ______

---

36. In each example below, we used the "cancelling" shortcut to multiply a literal fraction by its denominator. Notice again that each product is the numerator of the original fraction.

$$\overset{1}{\cancel{t}}\left(\frac{v_2 - v_1}{\underset{1}{\cancel{t}}}\right) = \frac{v_2 - v_1}{1} = v_2 - v_1$$

$$\overset{1}{\cancel{(t_2 - t_1)}}\left(\frac{s_2 - s_1}{\underset{1}{\cancel{t_2 - t_1}}}\right) = \frac{s_2 - s_1}{1} = s_2 - s_1$$

Using the fact that the product is the numerator of the original fraction, complete these.

a) $2\left(\frac{v_o + v_f}{2}\right) =$ ______ b) $(c - d)\left(\frac{4x}{c - d}\right) =$ ______

Answers to 35: a) 4st b) P_2V_2 c) gt^2

---

37. When multiplying a fraction whose numerator is "1" by its denominator, the product is "1". That is:

$$2cd\left(\frac{1}{2cd}\right) = \frac{2cd}{2cd} = 1 \qquad \frac{1}{f}(f) = \frac{f}{f} = \text{\_\_\_\_\_}$$

Answers to 36: a) $v_o + v_f$ b) 4x

---

38. When a formula contains one fraction, we can "clear the fraction" by multiplying both sides by its denominator. An example is shown. Clear the fractions in the other formulas.

$$v = \frac{s}{t} \qquad \text{a) } Ft = \frac{mv}{g} \qquad \text{b) } H = \frac{D^2N}{2.5}$$

$$t(v) = t\left(\frac{s}{t}\right)$$

$$tv = s$$

Answer to 37: 1

Answers to 38: a) $gFt = mv$ b) $2.5H = D^2N$

39. Following the example, clear the fractions in the other formulas.

$$v_{av} = \frac{v_o + v_f}{2}$$

$$2(v_{av}) = 2\left(\frac{v_o + v_f}{2}\right)$$

$$2v_{av} = v_o + v_f$$

a) $a = \dfrac{v_2 - v_1}{t}$

b) $F = \dfrac{m_1 m_2}{rd^2}$

Answers: a) $ta = v_2 - v_1$ b) $rd^2F = m_1 m_2$

40. Following the example, clear the fraction in the other formula.

$$v = \frac{s_2 - s_1}{t_2 - t_1}$$

$$(t_2 - t_1)v = (t_2 - t_1)\left(\frac{s_2 - s_1}{t_2 - t_1}\right)$$

$$v(t_2 - t_1) = s_2 - s_1$$

$$m = \frac{y_2 - y_1}{x_2 - x_1}$$

Answer: $m(x_2 - x_1) = y_2 - y_1$

9-8 REARRANGING FORMULAS CONTAINING ONLY ONE FRACTION

In this section, we will discuss the method for rearranging formulas that contain only one fraction. The rearrangements are limited to those that can be done by the multiplication axiom alone.

41. To solve for B at the right, we simply cleared the fraction. Complete the other rearrangement.

Solve for B.

$$D = \frac{B}{N}$$

$$N(D) = N\left(\frac{B}{N}\right)$$

$$DN = B$$

$$B = DN$$

Solve for "a".

$$h = \frac{a}{s^2}$$

Answer: $a = hs^2$

42. To solve for "s" at the right, we cleared the fraction and then used the shortcut of the multiplication axiom. Complete the other rearrangement.

Solve for "s".

$$P = \frac{Fs}{t}$$

$$t(P) = t\left(\frac{Fs}{t}\right)$$

$$Pt = Fs$$

$$s = \frac{Pt}{F}$$

Solve for m_2.

$$F = \frac{m_1 m_2}{rd^2}$$

Answer: $m_2 = \dfrac{Frd^2}{m_1}$

43. Following the example, complete the other rearrangements.

Solve for R.

$$I = \frac{E}{R}$$

$$R(I) = R\left(\frac{E}{R}\right)$$

$$IR = E$$

$$R = \frac{E}{I}$$

a) Solve for P_1.

$$V_1 = \frac{P_2V_2}{P_1}$$

b) Solve for N.

$$H = \frac{D^2N}{2.5}$$

---

a) $P_1 = \frac{P_2V_2}{V_1}$

b) $N = \frac{2.5H}{D^2}$

44. Following the example, complete the other rearrangement.

Solve for "t".

$$a = \frac{v_2 - v_1}{t}$$

$$t(a) = t\left(\frac{v_2 - v_1}{t}\right)$$

$$at = v_2 - v_1$$

$$t = \frac{v_2 - v_1}{a}$$

Solve for "r".

$$m = \frac{c - d}{2r}$$

$r = \frac{c - d}{2m}$

45. Following the example, complete the other rearrangement.

Solve for A.

$$H = \frac{AKT(t_2 - t_1)}{L}$$

$$L(H) = L\left[\frac{AKT(t_2 - t_1)}{L}\right]$$

$$HL = AKT(t_2 - t_1)$$

$$A = \frac{HL}{KT(t_2 - t_1)}$$

Solve for R.

$$H = \frac{\pi dR(F_1 - F_2)}{33,000}$$

$R = \frac{33,000H}{\pi d(F_1 - F_2)}$

46. Since $\frac{1}{2}bh = \frac{1}{2}(bh) = \frac{bh}{2}$, we can write the formula below in two equivalent forms.

$$\boxed{A = \frac{1}{2}bh} \qquad \boxed{A = \frac{bh}{2}}$$

It is easier to clear the fraction in the form on the right above. Therefore, before clearing the fraction in formulas of that type, we should write them in that form. That is:

$s = \frac{1}{2}at^2$ should be written $s = \frac{at^2}{2}$

$A = \frac{1}{2}h(b_1 + b_2)$ should be written $A =$ ________

$\frac{h(b_1 + b_2)}{2}$

47. Following the example, complete the other rearrangement.

Solve for "h".

$$A = \frac{1}{2}bh$$

$$A = \frac{bh}{2}$$

$$2(A) = 2\left(\frac{bh}{2}\right)$$

$$2A = bh$$

$$h = \frac{2A}{b}$$

Solve for "a".

$$s = \frac{1}{2}at^2$$

$$a = \frac{2s}{t^2}$$

48. Following the example, complete the other rearrangement.

Solve for "h".

$$A = \frac{1}{2}h(b_1 + b_2)$$

$$A = \frac{h(b_1 + b_2)}{2}$$

$$2(A) = 2\left[\frac{h(b_1 + b_2)}{2}\right]$$

$$2A = h(b_1 + b_2)$$

$$h = \frac{2A}{b_1 + b_2}$$

Solve for v^2.

$$E = \frac{1}{2}(m_1 + m_2)v^2$$

$$v^2 = \frac{2E}{m_1 + m_2}$$

9-9 REARRANGING PROPORTION FORMULAS

In this section, we will discuss the method for rearranging some proportion formulas. Before doing so, we will show how cross-multiplication can be used to clear the fractions in formulas of that type.

49. To clear the fractions in a proportion formula, we can use the cross-multiplication shortcut. An example is shown. Clear the fractions in the other formulas.

$$\frac{P_1}{P_2} = \frac{V_2}{V_1}$$

$$P_1V_1 = P_2V_2$$

a) $\frac{F}{G} = \frac{A}{B}$ ______ = ______

b) $\frac{V_1}{V_2} = \frac{T_1}{T_2}$ ______ = ______

---

a) $FB = GA$

b) $V_1T_2 = V_2T_1$

50. Following the example, cross-multiply to clear the fractions in the other formulas.

$$\frac{I_1}{I_2} = \frac{(d_2)^2}{(d_1)^2}$$

$$I_1(d_1)^2 = I_2(d_2)^2$$

a) $\frac{P_1V_1}{T_1} = \frac{P_2V_2}{T_2}$ ______ = ______

b) $\frac{a}{b} = \frac{c^2}{d^2}$ ______ = ______

---

a) $P_1V_1T_2 = T_1P_2V_2$

b) $ad^2 = bc^2$

51. To solve for P_1 below, we cross-multiplied to clear the fractions and then used the multiplication axiom. Use the same method for the other rearrangements.

Solve for P_1.

$$\frac{P_1}{P_2} = \frac{V_2}{V_1}$$

$$P_1V_1 = P_2V_2$$

$$P_1 = \frac{P_2V_2}{V_1}$$

a) Solve for T_1.

$$\frac{V_1}{V_2} = \frac{T_1}{T_2}$$

b) Solve for d.

$$\frac{D}{d} = \frac{F}{f}$$

---

a) $T_1 = \dfrac{V_1T_2}{V_2}$

b) $d = \dfrac{Df}{F}$

52. Following the example, solve for the indicated letter in the other formulas.

Solve for I_1.

$$\frac{I_1}{I_2} = \frac{(d_2)^2}{(d_1)^2}$$

$$I_1(d_1)^2 = I_2(d_2)^2$$

$$I_1 = \frac{I_2(d_2)^2}{(d_1)^2}$$

a) Solve for T_2.

$$\frac{P_1V_1}{T_1} = \frac{P_2V_2}{T_2}$$

b) Solve for I_2.

$$\frac{I_1}{I_2} = \frac{(d_2)^2}{(d_1)^2}$$

---

a) $T_2 = \dfrac{T_1P_2V_2}{P_1V_1}$

b) $I_2 = \dfrac{I_1(d_1)^2}{(d_2)^2}$

9-10 SQUARED LETTERS AND FORMULA REARRANGEMENT

In this section, we will discuss the procedure for solving for a letter that is squared in a formula.

---

53. Equations like $x^2 = 16$ have two roots, one positive and one negative. However, as we saw in the last unit, the negative root does not ordinarily make sense in a formula evaluation. For example:

If $s^2 = 100$ when "s" stands for "distance", "-10" does not ordinarily make sense as a value for distance.

If $v^2 = 36$ when "v" stands for "velocity", ordinarily only one of the two roots (+6 and -6) makes sense as a value for "velocity". Which one? ________

+6

54. When rearranging formulas to solve for a letter that is squared, negative square roots also do not ordinarily make sense. Therefore, to solve for "s" in the formula below, <u>we simply take the principle (or positive) square root of the other side</u>. That is:

$$s^2 = A$$

$$s = \sqrt{A}$$

In the formula above, "s" stands for the length of a side of a square. Would the solution $s = -\sqrt{A}$ make sense? ________

No, because a length is not ordinarily negative.

55. To solve for E below, we simply took the principle square root of the other side. Complete the other solutions.

If $E^2 = PR$, a) If $v^2 = 2as$, b) If $s^2 = \frac{a}{h}$,

$E = \sqrt{PR}$ $v =$ ________ $s =$ ________

Answers: a) $v = \sqrt{2as}$ b) $s = \sqrt{\frac{a}{h}}$

56. When solving for a letter that is squared, <u>be sure to take the square root of all of the other side</u>. For example:

If $a^2 = \frac{b + c}{d}$, If $a^2 = b^2 + c^2$,

$a = \sqrt{\frac{b + c}{d}}$ $a = \sqrt{b^2 + c^2}$

a) If $d^2 = \frac{I_A - I_C}{m}$, b) If $v_o^2 = v_f^2 - 2gs$,

$d =$ ________ $v_o =$ ________

Answers: a) $d = \sqrt{\frac{I_A - I_C}{m}}$ b) $v_o = \sqrt{v_f^2 - 2gs}$

57. To solve for I at the right, we had to isolate I^2 first. Complete the other rearrangement.

Solve for I.

$P = I^2R$

$I^2 = \frac{P}{R}$

$I = \sqrt{\frac{P}{R}}$

Solve for "d".

$A = .785d^2$

Answer: $d = \sqrt{\frac{A}{.785}}$

58. To isolate "v^2" at the left below, we began by clearing the fraction. Complete the other rearrangement.

Solve for "v".

$F = \frac{mv^2}{r}$

$r(F) = r\left(\frac{mv^2}{r}\right)$

$Fr = mv^2$

$v^2 = \frac{Fr}{m}$

$v = \sqrt{\frac{Fr}{m}}$

Solve for "d".

$R = \frac{pL}{d^2}$

Answer: $d = \sqrt{\frac{pL}{R}}$

59. a) Solve for "D". $H = \frac{D^2N}{2.5}$ b) Solve for "t". $s = \frac{gt^2}{2}$

---

60. To solve for "t" below, we began by writing $\frac{1}{2}at^2$ as $\frac{at^2}{2}$.

Complete the other rearrangement.

a) $D = \sqrt{\frac{2.5H}{N}}$

b) $t = \sqrt{\frac{2s}{g}}$

Solve for "t".

$$s = \frac{1}{2}at^2$$

$$s = \frac{at^2}{2}$$

$$2s = at^2$$

$$t^2 = \frac{2s}{a}$$

$$t = \sqrt{\frac{2s}{a}}$$

Solve for "v".

$$E = \frac{1}{2}mv^2$$

$v = \sqrt{\frac{2E}{m}}$

SELF-TEST 27 (pages 236-243)

1. Solve for "E". $W = \frac{EI}{S}$	2. Solve for "R". $m = \frac{t_2 - t_1}{R}$	3. Solve for "d". $H = \frac{\pi dR(F_1 - F_2)}{33,000}$
4. Solve for "d_1". $\frac{F_1}{F_2} = \frac{d_1}{d_2}$	5. Solve for "T_2". $\frac{P_1V_1}{T_1} = \frac{P_2V_2}{T_2}$	6. Solve for "I". $I^2 = bv$
7. Solve for "r". $A = \pi r^2$	8. Solve for "Z". $Z^2 = X^2 + R^2$	9. Solve for "c". $a = \frac{1}{2}bc^2$

ANSWERS:

1. $E = \frac{SW}{I}$
2. $R = \frac{t_2 - t_1}{m}$
3. $d = \frac{33,000H}{\pi R(F_1 - F_2)}$
4. $d_1 = \frac{F_1d_2}{F_2}$
5. $T_2 = \frac{P_2V_2T_1}{P_1V_1}$
6. $I = \sqrt{bv}$
7. $r = \sqrt{\frac{A}{\pi}}$
8. $Z = \sqrt{X^2 + R^2}$
9. $c = \sqrt{\frac{2a}{b}}$

9-11 THE ADDITION AXIOM AND FORMULA REARRANGEMENT

When a formula contains more than one term on one side, we need the addition axiom to solve for a letter on that side. We will discuss that use of the addition axiom in this section. Before doing so, however, we will extend the concept of "opposites" to literal terms.

---

61. Two literal terms are a pair of opposites if their sum is "0". Therefore:

Since $4P + (-4P) = 0$, "4P" and "-4P" are a pair of opposites.

Since $2ab + (-2ab) = 0$, "2ab" and "-2ab" are a pair of opposites.

Write the opposite of each term.

a) 10t ________ b) -3s ________ c) 7VT ________ d) -5pq ________

---

62. To show that the following terms are a pair of opposites, we wrote "1" and "-1" explicitly.

Since $m + (-m) = 1m + (-1m) = 0$, "m" and "-m" are a pair of opposites.

Since $AB + (-AB) = 1AB + (-1AB) = 0$, "AB" and "-AB" are a pair of opposites.

Write the opposite of each term.

a) t ________ b) $-I_a$ ________ c) F_1r_1 ________ d) -pq ________

Answers to 61: a) -10t b) 3s c) -7VT d) 5pq

---

63. To convert the subtraction below to addition, we added the opposite of the second term.

$$a - b = a + (\text{the opposite of } b) = a + (-b)$$

Convert each of these subtractions to addition.

a) $A - L =$ ________ b) $t_2 - t_1 =$ ________ c) $F_1r_1 - F_2r_2 =$ ________

Answers to 62: a) -t b) I_a c) $-F_1r_1$ d) pq

---

64. To solve for C in the formula below, we added -M to both sides to isolate the "C" term. Solve for V in the other formula by adding -100 to both sides.

$$R = C + M$$

$$R + (-M) = C + \underbrace{M + (-M)}_{0}$$

$$R + (-M) = C + 0$$

$$C = R + (-M)$$

$$M = V + 100$$

Answers to 63: a) $A + (-L)$ b) $t_2 + (-t_1)$ c) $F_1r_1 + (-F_2r_2)$

Answer to 64: $V = M + (-100)$

65. Here are the two solutions obtained in the last frame.

$C = R + (-M)$ $V = M + (-100)$

The right side of each is an addition in which <u>the second term is negative</u>. Ordinarily, an addition of that type is converted to subtraction in order to eliminate the negative term. To do so, we simply <u>subtract the opposite of the negative term</u>. That is:

$C = R + (-M)$ $V = M + (-100)$

$C = R -$ (the opposite of $-M$) $V = M -$ (the opposite of -100)

$C = R - M$ $V =$ _____ $-$ _____

Answer: $M - 100$

66. Convert each of these additions to subtraction form.

a) $a + (-b) =$ __________ c) $c + (-de) =$ __________

b) $V_2 + (-V_1) =$ __________ d) $F_1r_1 + (-F_2r_2) =$ __________

Answers: a) $a - b$ b) $V_2 - V_1$ c) $c - de$ d) $F_1r_1 - F_2r_2$

67. In the addition below, both the second and third terms are negative. We converted both to subtraction form.

$p + (-q) + (-r) = p - q - r$

Convert all additions of negative terms to subtraction form.

a) $D + (-S) + (-T) =$ __________ b) $R_t + (-R_1) + (-R_2) =$ __________

Answers: a) $D - S - T$ b) $R_t - R_1 - R_2$

68. When solving for a letter in a formula, the solution should not be left in a form that contains an addition of a negative term. Additions of that type should be converted to subtractions. That is:

Instead of $m = p + (-q)$, we write $m = p - q$.

Instead of $T_3 = T_T + (-T_1) + (-T_2)$, we write $T_3 =$ __________

Answer: $T_T - T_1 - T_2$

69. Solve for the indicated letter. Write each solution in subtraction form.

a) Solve for E: $H = E + PV$ b) Solve for R: $D = R + S + T$

70. To solve for A in the formula below, we converted the subtraction to addition and then added L to both sides. Solve for L in the other formula by converting to subtraction and then adding D to both sides.

a) $E = H - PV$

b) $R = D - S - T$

$$C = A - L \qquad\qquad N = L - D$$

$$C = A + (-L)$$

$$L + C = A + \underbrace{(-L) + L}_{0}$$

$$L + C = A + 0$$

$$A = L + C$$

---

71. Using the same steps, solve for the indicated letter in each formula below.

$L = N + D$

a) Solve for K: $C = K - 273$

b) Solve for E: $e = E - Ir$

---

72. To solve for P in the formula below, we used the addition axiom to isolate PV. Then we used the multiplication axiom to "get rid of" V. Use the same method to solve for "m" in the other formula.

a) $K = C + 273$

b) $E = e + Ir$

$$H = E + PV \qquad\qquad I_a = I_c + md^2$$

$$H + (-E) = (-E) + E + PV$$

$$H - E = PV$$

$$P = \frac{H - E}{V}$$

---

73. After clearing the fraction below, we had to use the addition axiom to solve for v_2. Solve for F_i in the other formula.

$$m = \frac{I_a - I_c}{d^2}$$

$$a = \frac{v_2 - v_1}{t} \qquad\qquad w = \frac{F_o + F_i}{t}$$

$$at = v_2 - v_1$$

$$at = v_2 + (-v_1)$$

$$at + v_1 = v_2 + (-v_1) + v_1$$

$$v_2 = at + v_1$$

$F_i = wt - F_o$

74. To show that the following terms are a pair of opposites, we wrote "1" and "-1" explicitly.

Since $b^2 + (-b^2) = 1b^2 + (-1b^2) = 0$, "$b^2$" and "$-b^2$" are a pair of opposites.

Therefore: a) The opposite of c^2 is ________. b) The opposite of $-v_o^2$ is ________.

---

a) $-c^2$

b) v_o^2

75. To solve for X below, we had to isolate X^2 first. Solve for "b" in the other formula.

$$Z^2 = X^2 + R^2 \qquad\qquad c^2 = a^2 + b^2$$

$$Z^2 + (-R^2) = X^2 + R^2 + (-R^2)$$

$$Z^2 - R^2 = X^2$$

$$X = \sqrt{Z^2 - R^2}$$

$b = \sqrt{c^2 - a^2}$

76. To solve for "v_f" below, we had to isolate v_f^2 first. Solve for "c" in the other formula.

$$v_o^2 = v_f^2 - 2gs \qquad\qquad a^2 = c^2 - b^2$$

$$v_o^2 = v_f^2 + (-2gs)$$

$$v_o^2 + 2gs = v_f^2 + (-2gs) + 2gs$$

$$v_o^2 + 2gs = v_f^2$$

$$v_f = \sqrt{v_o^2 + 2gs}$$

$c = \sqrt{a^2 + b^2}$

9-12 THE OPPOSITING PRINCIPLE AND FORMULA REARRANGEMENTS

When using the addition axiom to isolate a term in a formula, we occasionally get results like those below.

$$-D = N - L \qquad\qquad -v_1 = at - v_2$$

Though "D" and "v_1" have been isolated, they are not solved for because there is a "-" in front of each. In this section, we will show how to complete the solution for "D" and "v_1" by replacing each side of the formula by its opposite. Before doing so, however, we will discuss the opposites of subtractions and the oppositing principle for equations.

---

77. We want to show that we can obtain the opposite of a subtraction by simply interchanging the two terms. That is:

The opposite of $(5 - 3)$ is $(3 - 5)$.

The opposite of $(2x - 1)$ is $(1 - 2x)$.

The opposite of $(a - b)$ is (____ - ____) .

$b - a$

78. When two numbers are involved, it is easy to show that we can obtain the opposite of a subtraction by interchanging the two terms. For example:

The opposite of $(5 - 3)$ is $(3 - 5)$ since:

$$5 - 3 = +2$$

$$3 - 5 = -2 \qquad [\text{and } +2 + (-2) = 0]$$

$(5 - 3)$ and $(3 - 5)$ are a pair of opposites because their sum is ______.

Answer: 0

79. To show that $(2x - 1)$ and $(1 - 2x)$ are a pair of opposites, we added them below to show that their sum is 0.

$$\begin{aligned}(2x - 1) + (1 - 2x) &= 2x - 1 + 1 - 2x \\ &= 2x + (-1) + 1 + (-2x) \\ &= \underline{2x + (-2x)} + \underline{1 + (-1)} \\ &= \quad 0 \quad + \quad 0 \quad = 0\end{aligned}$$

Show that $(a - b)$ and $(b - a)$ are a pair of opposites by completing their addition below.

$$\begin{aligned}(a - b) + (b - a) &= a - b + b - a \\ &= a + (-b) + b + (-a) \\ &= \underline{a + (-a)} + \underline{b + (-b)} \\ &= \_\_\_\_ + \_\_\_\_ = \_\_\_\_\end{aligned}$$

Answer: $0 + 0 = 0$

80. Write the opposite of each subtraction by simply interchanging the two terms.

a) $10 - 5$ ______ b) $4y - 1$ ______ c) $p - q$ ______

Answers: a) $5 - 10$ b) $1 - 4y$ c) $q - p$

81. Write the opposite of each subtraction.

a) $at - v_2$ ______ b) $a^2 - c^2$ ______ c) $v_f^2 - 2gs$ ______

Answers: a) $v_2 - at$ b) $c^2 - a^2$ c) $2gs - v_f^2$

82. If we replace each side of an equation by its opposite, the new equation is equivalent to the original one. For example:

$-5x = 15$ and $5x = -15$ are equivalent because the root of each is "-3".

$x - 4 = 2$ and $4 - x = -2$ are equivalent because the root of each is ______.

83. Replacing each side of an equation by its opposite is called the "oppositing principle for equations". That principle can be used to solve for "y" in the formula below.

$$-y = a - 2b$$

$$y = 2b - a$$

Solve for "t" in each formula below by simply replacing each side by its opposite.

a) $-t = p - q$ b) $-t = cd - V$ c) $-t = m_1 - 4F$

$t =$ ________ $t =$ ________ $t =$ ________

Answers: a) $t = q - p$ b) $t = V - cd$ c) $t = 4F - m_1$

---

84. a) Solve for I: $-I = E_1 - E_2$ b) Solve for "v_1": $-v_1 = at - v_2$

Answers: a) $I = E_2 - E_1$ b) $v_1 = v_2 - at$

---

85. In solving for L below, we had to use the oppositing principle for the final step. Solve for D in the other formula.

$$C = A - L$$

$$C = A + (-L)$$

$$C + (-A) = (-A) + A + (-L)$$

$$C - A = -L$$

$$L = A - C$$

$$N = L - D$$

Answer: $D = L - N$

---

86. We solved for "v_1" in the formula below. Use the same steps to solve for "d" in the other formula.

$$a = \frac{v_2 - v_1}{t}$$

$$at = v_2 - v_1$$

$$at = v_2 + (-v_1)$$

$$at + (-v_2) = (-v_2) + v_2 + (-v_1)$$

$$at - v_2 = -v_1$$

$$v_1 = v_2 - at$$

$$m = \frac{b - d}{r}$$

Answer: $d = b - mr$

87. We solved for "I" below. Use the same steps to solve for "s" in the other formula.

$$e = E - Ir \qquad\qquad v_o^2 = v_f^2 - 2gs$$

$$e = E + (-Ir)$$

$$e + (-E) = (-E) + E + (-Ir)$$

$$e - E = -Ir$$

$$Ir = E - e$$

$$I = \frac{E - e}{r}$$

$$s = \frac{v_f^2 - v_o^2}{2g}$$

88. We solved for "b" below. Use the same steps to solve for "q" in the other formula.

$$a^2 = c^2 - b^2 \qquad\qquad p^2 - q^2 = r^2$$

$$a^2 = c^2 + (-b^2)$$

$$a^2 + (-c^2) = (-c^2) + c^2 + (-b^2)$$

$$a^2 - c^2 = -b^2$$

$$b^2 = c^2 - a^2$$

$$b = \sqrt{c^2 - a^2}$$

$$q = \sqrt{p^2 - r^2}$$

9-13 CONTRASTING COEFFICIENTS AND TERMS

Confusing a subtraction of "two terms" with a "negative factor and its coefficient" is a common error. We will contrast those two different expressions in this section.

89. "$F_1 - F_2$" and "$F_1(-F_2)$" are two different algebraic expressions.

In $F_1 - F_2$, the "-" is the symbol for subtraction. Therefore, F_1 and F_2 are <u>two different terms</u>.

In $F_1(-F_2)$, the parentheses indicate a multiplication. The "-" is the sign of F_2. Therefore, the whole expression is <u>only one term</u>.

Which of these indicate a multiplication? ________ a) $X_L - X_C$ b) $X_L(-X_C)$

b) $X_L(-X_C)$

90. Which of these is a subtraction? ________ a) $E_1 - E_o$ b) $E_1(-E_o)$

a) $E_1 - E_o$

91. a) In $V_1 - V_2$, is "V_1" a term or the coefficient of "$-V_2$"? ________________

b) In $c(-d)$, is "c" a term or the coefficient of "-d"? ________________

a) a term

b) the coefficient of "-d"

92. In which of the following is V_0 a coefficient? ____________

a) V_0V_1 b) $V_0 - V_1$ c) $V_0(-V_1)$

Both (a) and (c)

93. To solve for X_C below, we had to eliminate X_L. Since X_L is a term in a subtraction, we used the addition axiom, not the multiplication axiom. Solve for E_0 in the other formula.

$$X = X_L - X_C$$

$$X = X_L + (-X_C)$$

$$X + (-X_L) = (-X_L) + X_L + (-X_C)$$

$$X - X_L = -X_C$$

$$X_C = X_L - X$$

$$I = E_1 - E_0$$

$E_0 = E_1 - I$

SELF-TEST 28 (pages 244-252)

1. Solve for T: $D = H + T + S$	2. Solve for F: $F - R = N$	3. Solve for "b": $c = d + ab$
4. Solve for "t": $v^2 = t^2 + m^2$	5. Solve for E_i: $I_a = \frac{E_i - E_o}{R_a}$	6. Solve for "p": $-p = q - r$

Continued on following page.

SELF-TEST 28 (pages 244-252) - Continued

7. Solve for D:	8. Solve for "b":	9. Solve for "s":
$M = H - D$	$c = \dfrac{a - b}{d}$	$v^2 = h^2 - s^2$

ANSWERS:

1. $T = D - H - S$
2. $F = N + R$
3. $b = \dfrac{c - d}{a}$
4. $t = \sqrt{v^2 - m^2}$
5. $E_i = I_aR_a + E_o$
6. $p = r - q$
7. $D = H - M$
8. $b = a - cd$
9. $s = \sqrt{h^2 - v^2}$

9-14 THE DISTRIBUTIVE PRINCIPLE AND FORMULA REARRANGEMENT

In this section, we will discuss formula rearrangements that involve multiplying by the distributive principle or factoring by the distributive principle.

94. When letters are involved, we multiply by the distributive principle in the usual way. That is:

$$P(Q + R) = PQ + PR \qquad C(T_1 - T_2) = CT_1 - CT_2$$

Complete each multiplication below.

a) $h(b_1 + b_2) =$ __________ b) $3.14(d_2 - d_1) =$ __________

95. In $D = P(Q + R)$, the letter "Q" appears within the parentheses. To solve for Q, we can begin by multiplying by the distributive principle as we have done below. Use the same method to solve for "t" in the other formula.

a) $hb_1 + hb_2$

b) $3.14d_2 - 3.14d_1$

$$D = P(Q + R)$$
$$D = PQ + PR$$
$$D + (-PR) = PQ + PR + (-PR)$$
$$D - PR = PQ$$
$$Q = \frac{D - PR}{P}$$

$$m = b(d + t)$$

$t = \dfrac{m - bd}{b}$

96. In the formula below, T_1 is within the parentheses. Therefore, to solve for it, we began by multiplying by the distributive principle. Solve for L in the other formula.

$$Q = C(T_1 - T_2)$$

$$Q = CT_1 - CT_2$$

$$Q = CT_1 + (-CT_2)$$

$$Q + CT_2 = CT_1 + (-CT_2) + CT_2$$

$$Q + CT_2 = CT_1$$

$$T_1 = \frac{Q + CT_2}{C}$$

$$M = P(L - X)$$

$$L = \frac{M + PX}{P}$$

97. To solve for V_2 in the formula below, we began by multiplying by the distributive principle. Notice that we had to use the oppositing principle for equations. Solve for "t_1" in the other formula.

$$W_S = P(V_1 - V_2)$$

$$W_S = PV_1 - PV_2$$

$$W_S = PV_1 + (-PV_2)$$

$$W_S + (-PV_1) = (-PV_1) + PV_1 + (-PV_2)$$

$$W_S - PV_1 = -PV_2$$

$$PV_1 - W_S = PV_2$$

$$V_2 = \frac{PV_1 - W_S}{P}$$

$$H = ms(t_2 - t_1)$$

$$t_1 = \frac{mst_2 - H}{ms}$$

98. To solve for either "b_1" or "b_2" below, we have to get them out of the parentheses. Notice the steps.

$$A = \frac{1}{2}h(b_1 + b_2)$$

$$A = \frac{h(b_1 + b_2)}{2}$$

$$2A = h(b_1 + b_2)$$

$$2A = hb_1 + hb_2$$

Beginning with the last formula, solve for both "b_1" and "b_2".

a) Solve for "b_1".

b) Solve for "b_2".

99. When the same letter appears in both terms of an addition, or subtraction, we can factor by the distributive principle. That is:

$$ab + ac = a(b + c) \qquad CV - CT = C(V - T)$$

Factor out the common letter in each expression below.

a) $IR_1 + IR_2 =$ ________ b) $mt_1 - mt_2 =$ ________

Answers: a) $b_1 = \frac{2A - hb_2}{h}$ b) $b_2 = \frac{2A - hb_1}{h}$

---

100. In $E = IR_1 + IR_2$, the letter "I" appears in both terms on the right side. To solve for "I", we must begin by factoring by the distributive principle as we have done below. Use the same method to solve for S in the other formula.

$$E = IR_1 + IR_2 \qquad MS + QS = R$$

$$E = I(R_1 + R_2)$$

$$I = \frac{E}{R_1 + R_2}$$

Answers: a) $I(R_1 + R_2)$ b) $m(t_1 - t_2)$

Answer: $S = \frac{R}{M + Q}$

9-15 SQUARING LITERAL TERMS

In order to eliminate the radical in radical formulas, we have to square literal terms. We will discuss squaring terms in this section.

---

101. To indicate the square of a letter, we simply use a small "2". That is:

$$(x)^2 = (x)(x) = x^2 \qquad (y)^2 = (y)(y) = y^2 \qquad (V_1)^2 = (V_1)(V_1) = \text{\_\_\_\_\_\_}$$

---

102. To square a letter-term, we simply multiply the letter-term by itself. For example:

$$(2x)^2 = (2x)(2x) = (2)(2)(x)(x) = 4x^2$$

$$(ab)^2 = (ab)(ab) = (a)(a)(b)(b) = a^2b^2$$

Complete each of the following squarings.

a) $(5y)^2 = (5y)(5y) =$ ________

b) $(mt)^2 = (mt)(mt) =$ ________

Answer: V_1^2

---

103. It should be obvious that squaring a letter-term is equivalent to squaring each factor in the letter-term. For example:

$$(3y)^2 = 3^2y^2 = 9y^2$$

$$(4cd)^2 = 4^2c^2d^2 = 16c^2d^2$$

Complete each of the following squarings.

a) $(6R)^2 =$ ________ b) $(V_1V_2)^2 =$ ________ c) $(2pq)^2 =$ ________

Answers: a) $25y^2$ b) m^2t^2

a) $36R^2$
b) $V_1^2V_2^2$
c) $4p^2q^2$

104. To square a fraction, we simply multiply the fraction by itself. For example:

$$\left(\frac{a}{b}\right)^2 = \left(\frac{a}{b}\right)\left(\frac{a}{b}\right) = \frac{(a)(a)}{(b)(b)} = \frac{a^2}{b^2}$$

a) $\left(\frac{2x}{3}\right)^2 = \left(\frac{2x}{3}\right)\left(\frac{2x}{3}\right) = \frac{(2x)(2x)}{(3)(3)} = $ ________

b) $\left(\frac{xy}{d}\right)^2 = \left(\frac{xy}{d}\right)\left(\frac{xy}{d}\right) = \frac{(xy)(xy)}{(d)(d)} = $ ________

---

a) $\frac{4x^2}{9}$
b) $\frac{x^2y^2}{d^2}$

105. It should be obvious that squaring a fraction is equivalent to squaring both its numerator and its denominator. That is:

$$\left(\frac{p}{q}\right)^2 = \frac{p^2}{q^2} \qquad \left(\frac{3m}{2}\right)^2 = \frac{(3m)^2}{(2)^2} = \frac{9m^2}{4} \qquad \left(\frac{1}{cd}\right)^2 = \frac{(1)^2}{(cd)^2} = \text{\_\_\_\_\_\_}$$

$\frac{1}{c^2d^2}$

106. Complete.

a) $(2w)^2 = $ ________ b) $\left(\frac{M}{K}\right)^2 = $ ________ c) $\left(\frac{1}{3xy}\right)^2 = $ ________

---

a) $4w^2$ b) $\frac{M^2}{K^2}$ c) $\frac{1}{9x^2y^2}$

9-16 ELIMINATING THE RADICAL IN A RADICAL FORMULA

In order to solve for a letter under a radical in a radical formula, we have to get rid of the radical first. We will discuss the procedure for doing so in this section.

---

107. As we saw earlier, the square of a square-root radical is its radicand. That is:

$$\left(\sqrt{A}\right)^2 = A \qquad \left(\sqrt{2gs}\right)^2 = 2gs \qquad \text{a) } \left(\sqrt{R}\right)^2 = \text{\_\_\_\_\_\_} \qquad \text{b) } \left(\sqrt{L_1L_2}\right)^2 = \text{\_\_\_\_\_\_}$$

---

a) R
b) L_1L_2

108. Following the example, write the squares of the other radicals.

$$\left(\sqrt{\frac{P}{R}}\right)^2 = \frac{P}{R} \qquad \text{a) } \left(\sqrt{\frac{pL}{R}}\right)^2 = \text{\_\_\_\_\_\_} \qquad \text{b) } \left(\sqrt{\frac{A}{.785}}\right)^2 = \text{\_\_\_\_\_\_}$$

---

a) $\frac{pL}{R}$
b) $\frac{A}{.785}$

109. Following the example, write the squares of the other radicals.

$$\left(\sqrt{b + c}\right)^2 = b + c \qquad \text{a) } \left(\sqrt{h + 2r}\right)^2 = \text{\_\_\_\_\_\_} \qquad \text{b) } \left(\sqrt{X^2 + Y^2}\right)^2 = \text{\_\_\_\_\_\_}$$

---

a) $h + 2r$
b) $X^2 + Y^2$

110. To get rid of the radical in a formula, we simply square both sides of the formula. For example:

$$s = \sqrt{A} \qquad \text{a)}\quad I = \sqrt{\frac{P}{R}} \qquad \text{b)}\quad v = \sqrt{2gs}$$

$$(s)^2 = \left(\sqrt{A}\right)^2 \qquad (I)^2 = \left(\sqrt{\frac{P}{R}}\right)^2 \qquad (v)^2 = \left(\sqrt{2gs}\right)^2$$

$$s^2 = A \qquad \underline{\quad} = \underline{\quad} \qquad \underline{\quad} = \underline{\quad}$$

Answers: a) $I^2 = \frac{P}{R}$ b) $v^2 = 2gs$

111. Following the example, get rid of the radicals in the other formulas.

$$\sqrt{\frac{P}{K}} = mt \qquad \text{a)}\quad \sqrt{h + 2r} = sd \qquad \text{b)}\quad \sqrt{L_1L_2} = \frac{M}{K}$$

$$\left(\sqrt{\frac{P}{K}}\right)^2 = (mt)^2 \qquad \left(\sqrt{h + 2r}\right)^2 = (sd)^2 \qquad \left(\sqrt{L_1L_2}\right)^2 = \left(\frac{M}{K}\right)^2$$

$$\frac{P}{K} = m^2t^2 \qquad \underline{\quad} = \underline{\quad} \qquad \underline{\quad} = \underline{\quad}$$

Answers: a) $h + 2r = s^2d^2$ b) $L_1L_2 = \frac{M^2}{K^2}$

112. When squaring to get rid of a radical in a formula, be sure to square both sides. Complete these.

$$\text{a)}\quad b = \sqrt{c^2 - a^2} \qquad \text{b)}\quad \sqrt{d} = 2w \qquad \text{c)}\quad \sqrt{a} = \frac{r}{t}$$

$$\underline{\quad} = \underline{\quad\quad} \qquad \underline{\quad} = \underline{\quad\quad} \qquad \underline{\quad} = \underline{\quad\quad}$$

Answers: a) $b^2 = c^2 - a^2$ b) $d = 4w^2$ c) $a = \frac{r^2}{t^2}$

9-17 REARRANGING RADICAL FORMULAS

In this section, we will discuss the procedure for solving for a letter under the radical in a radical formula.

---

113. To solve for a letter under the radical in a radical formula, we square both sides to get a non-radical formula. Then we complete the solution in the usual way. An example is shown below. Solve for the indicated letter in the other formulas.

Solve for A: a) Solve for R: b) Solve for "s":

$$r = \sqrt{\frac{A}{\pi}} \qquad I = \sqrt{\frac{P}{R}} \qquad v = \sqrt{2gs}$$

$$r^2 = \frac{A}{\pi}$$

$$\pi(r^2) = \pi\left(\frac{A}{\pi}\right)$$

$$A = \pi r^2$$

114. Square both sides and then solve for the indicated letters.

a) Solve for A:

$$s = \sqrt{A}$$

b) Solve for "b":

$$a = \sqrt{b + c}$$

c) Solve for "g":

$$t = \sqrt{\frac{2s}{g}}$$

Answers:

a) $R = \frac{P}{I^2}$

b) $s = \frac{v^2}{2g}$

---

115. Following the example, solve for the indicated letter in the other formula.

Solve for P:

$$\sqrt{\frac{P}{K}} = mt$$

$$\frac{P}{K} = m^2t^2$$

$$K\left(\frac{P}{K}\right) = K(m^2t^2)$$

$$P = Km^2t^2$$

Solve for "h":

$$\sqrt{h + 2r} = sd$$

Answers:

a) $A = s^2$

b) $b = a^2 - c$

c) $g = \frac{2s}{t^2}$

---

116. Following the example, solve for the indicated letter in the other formula.

Solve for L_1:

$$\sqrt{L_1L_2} = \frac{M}{K}$$

$$L_1L_2 = \frac{M^2}{K^2}$$

$$K^2(L_1L_2) = K^2\left(\frac{M^2}{K^2}\right)$$

$$K^2L_1L_2 = M^2$$

$$L_1 = \frac{M^2}{K^2L_2}$$

Solve for "x":

$$\sqrt{ax} = \frac{b}{c}$$

Answer: $h = s^2d^2 - 2r$

Answer: $x = \frac{b^2}{ac^2}$

117. Following the example, solve for the indicated letters.

Solve for "a":

$$c = \sqrt{a^2 + b^2}$$

$$c^2 = a^2 + b^2$$

$$a^2 = c^2 - b^2$$

$$a = \sqrt{c^2 - b^2}$$

a) Solve for "Z":

$$R = \sqrt{Z^2 - X^2}$$

b) Solve for "p":

$$m = \sqrt{h^2 - p^2}$$

a) $Z = \sqrt{R^2 + X^2}$

b) $p = \sqrt{h^2 - m^2}$

SELF-TEST 29 (pages 252-258)

1. Solve for "R":

$$D = P(Q + R)$$

2. Solve for "X":

$$M = P(L - X)$$

3. Solve for "P_2":

$$I = P_1P_2 + RP_2$$

4. Solve for "P":

$$E = \sqrt{PR}$$

5. Solve for "d":

$$m = \sqrt{\frac{v}{d}}$$

6. Solve for "a":

$$\sqrt{a - c} = pq$$

7. Solve for "R":

$$\sqrt{MR} = \frac{1}{t}$$

8. Solve for "p":

$$d = \sqrt{p^2 - q^2}$$

ANSWERS:

1. $R = \dfrac{D - PQ}{P}$
2. $X = \dfrac{PL - M}{P}$
3. $P_2 = \dfrac{I}{P_1 + R}$
4. $P = \dfrac{E^2}{R}$
5. $d = \dfrac{v}{m^2}$
6. $a = p^2q^2 + c$
7. $R = \dfrac{1}{Mt^2}$
8. $p = \sqrt{d^2 + q^2}$

Unit 10 GEOMETRIC FORMULAS

In this unit, we will review some formulas for the following geometric figures: rectangles, squares, parallelograms, triangles, right triangles, circles, rectangular solids, cubes, and cylinders. The formulas are used to solve some applied problems.

> Note: The formulas reviewed in this section are summarized on page 280. They should be memorized.

10-1 RECTANGLES

In this section, we will review the formulas for the perimeter and area of a rectangle. The formulas are used to solve some applied problems.

---

1. In a rectangle, there are two pairs of equal sides and four right angles (90° angles).

   The longer sides are called the "length" (or L).

   The shorter sides are called the "width" (or W).

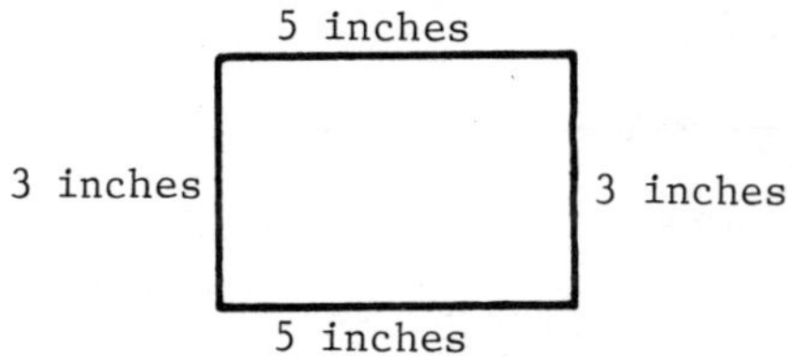

The perimeter (P) of a rectangle is the distance around it. We can find the perimeter by adding the lengths of its four sides. Or we can double its length and double its width and then add. That is:

$$P = 2L + 2W$$

When using the formula above, we treat the units as if they were letters. That is, for the rectangle above:

$$\begin{aligned} P &= 2(5 \text{ inches}) + 2(3 \text{ inches}) \\ &= 10 \text{ inches} + 6 \text{ inches} \\ &= (10 + 6) \text{ inches} \\ &= \_\_\_\_\_\_ \text{ inches} \end{aligned}$$

---

16 inches

2. Lengths can be measured with units from either the English system or the metric system. Some common units from each system and their abbreviations are shown in the table below. Notice that periods are not used in the abbreviations.

	Unit	Abbrev.
English System	inches	in
	feet	ft
	yards	yd
Metric System	centimeters	cm
	meters	m

As you can see from the table:

The abbreviation for "inches" is "in".

a) The abbreviation for "feet" is ________.

b) The abbreviation for "meter" is ________.

a) ft

b) m

---

3. The area of a geometric figure is the number of "unit squares" needed to fill it. Some "unit squares" are shown below.

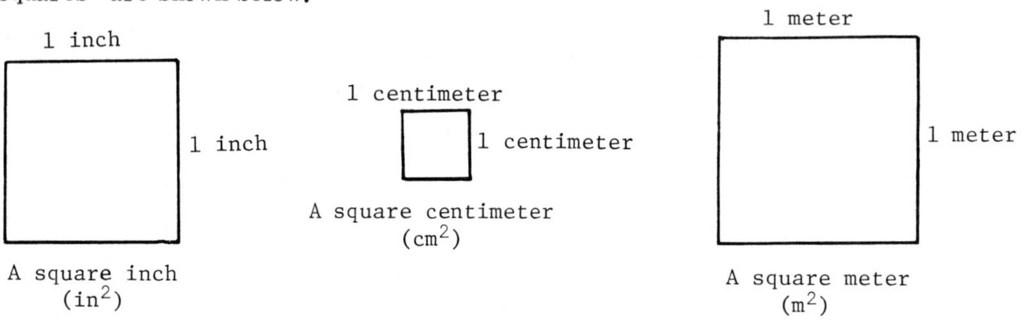

Notice that we use a small "2" in the abbreviation for "unit squares". Therefore:

The abbreviation for "a square foot" is "ft^2".

The abbreviation for "a square yard" is ________.

yd^2

---

4. Two word names are used for square units. For example:

$10\ ft^2$ is called "10 feet squared" or "10 square feet".

$47\ m^2$ is called "47 meters squared" or "47 square meters".

Write the abbreviation for each word name.

a) 25 inches squared = ________ b) 84 square centimeters = ________

a) $25\ in^2$

b) $84\ cm^2$

---

5. It takes 6 square centimeters to fill up the rectangle at the right. Therefore, its area is $6\ cm^2$. Notice that we can find its area by multiplying its length and width. That is:

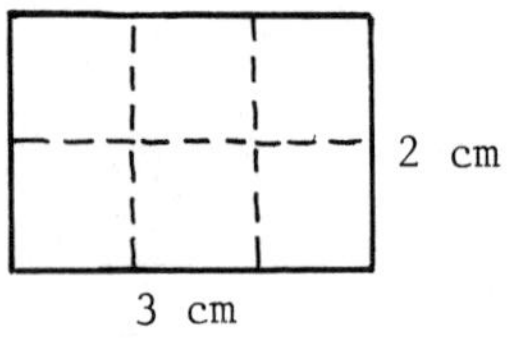

$$\boxed{A = LW}$$

When using the formula above, we also treat the units as if they were letters. That is, for the rectangle above:

$$A = (3\ cm)(2\ cm)$$

$$= (3)(2)(cm)(cm)$$

$$= \_\_\_\_\_\ cm^2$$

$6\ cm^2$

6. Use the formulas to find the perimeter and area of the rectangle at the right.

a) $P = 2L + 2W$ b) $A = LW$

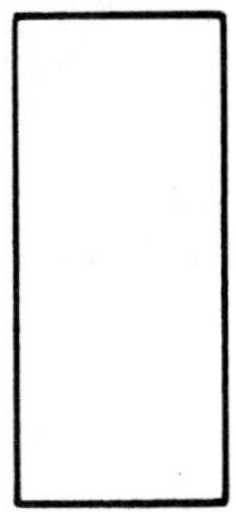

---

a) $P = 12$ m
b) $A = 8\ m^2$

7. To find the width of a rectangle if its area is 40 ft^2 and its length is 10 ft, we can use either the equation-solving method or the rearrangement method. For example:

Equation-Solving Method

$$A = LW$$
$$40 = 10W$$
$$W = \frac{40}{10} = 4 \text{ ft}$$

Rearrangement Method

$$A = LW$$
$$W = \frac{A}{L}$$
$$W = \frac{40}{10} = 4 \text{ ft}$$

Using either method, solve the problems below.

a) Find the width of a rectangle if its area is 54 in^2 and its length is 9 in.

b) Find the length of a rectangle if its area is 20 cm^2 and its width is 2 cm.

---

a) $W = 6$ in
b) $L = 10$ cm

8. To find the length of a rectangle if its perimeter is 16 cm and its width is 3 cm, we can also use either the equation-solving method or the rearrangement method. For example:

Equation-Solving Method

$$P = 2L + 2W$$
$$16 = 2L + 2(3)$$
$$16 = 2L + 6$$
$$16 + (-6) = 2L + 6 + (-6)$$
$$10 = 2L$$
$$L = \frac{10}{2} = 5 \text{ cm}$$

Rearrangement Method

$$P = 2L + 2W$$
$$P + (-2W) = 2L + 2W + (-2W)$$
$$P - 2W = 2L$$
$$L = \frac{P - 2W}{2}$$
$$L = \frac{16 - 2(3)}{2} = \frac{16 - 6}{2} = \frac{10}{2} = 5 \text{ cm}$$

Continued on the following page.

8. Continued

Using either method, solve the problems below.

a) Find the <u>length</u> of a rectangle if its perimeter is 30 yd and its width is 5 yd.

b) Find the <u>width</u> of a rectangle if its perimeter is 18 m and its length is 7 m.

a) L = 10 yd   b) W = 2 m

10-2 SQUARES

In this section, we will review the formulas for the perimeter and area of a square. The formulas are used to solve some applied problems.

9. A square is a special type of rectangle in which all four sides (s) are equal. Therefore:

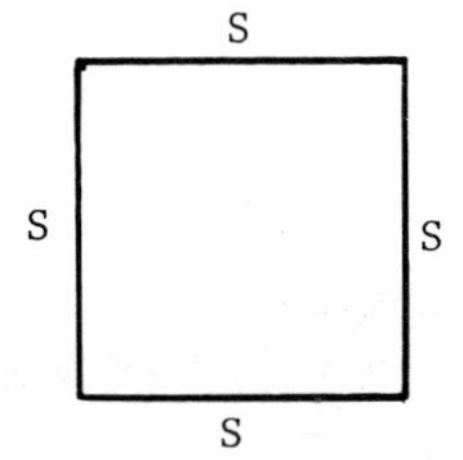

The <u>perimeter</u> of a square is s + s + s + s. That is:

$$P = 4s$$

The <u>area</u> of a square is (s)(s). That is:

$$A = s^2$$

Using the formulas, find the perimeter and area of a square if its side is 10 cm.

a) $P = 4s$

b) $A = s^2$

a) P = 40 cm

b) A = 100 cm^2

10. To find the side of a square if its perimeter is 12 in, we can use either method below.

<u>Equation-Solving Method</u>

$$P = 4s$$

$$12 = 4s$$

$$s = \frac{12}{4} = 3 \text{ in}$$

<u>Rearrangement Method</u>

$$P = 4s$$

$$s = \frac{P}{4}$$

$$s = \frac{12}{4} = 3 \text{ in}$$

Using either method, solve each problem.

a) Find the <u>side</u> of a square if its perimeter is 100 m.

b) Find the <u>side</u> of a square if if its perimeter is 48 ft.

---

a) s = 25 m

b) s = 12 ft

11. To find the side of a square if its area is 36 cm^2, we can use either method below.

<u>Equation-Solving Method</u>

$$A = s^2$$

$$36 = s^2$$

$$s = \sqrt{36} = 6 \text{ cm}$$

<u>Rearrangement Method</u>

$$A = s^2$$

$$s = \sqrt{A}$$

$$s = \sqrt{36} = 6 \text{ cm}$$

Using either method, solve each problem.

a) Find the <u>side</u> of a square if its area is 81 in^2.

b) Find the <u>side</u> of a square if its area is 4 m^2.

a) s = 9 in   b) s = 2m

10-3 PARALLELOGRAMS

In this section, we will review the formula for the area of a parallelogram and use it to solve some applied problems.

---

12. A parallelogram is a four-sided figure with two pairs of parallel sides. To show that the area of a parallelogram is equal to its base (b) times its height (h), we can cut off part of the parallelogram and use it to form a rectangle as we have done below.

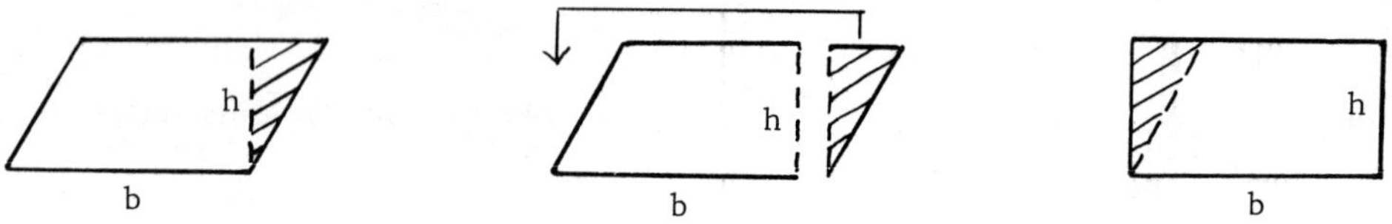

Since the areas of the parallelogram and rectangle are equal, we can use the following formula for the area of a parallelogram.

$$\boxed{A = bh}$$

Using the formula, find the area of each parallelogram at the right.

a) 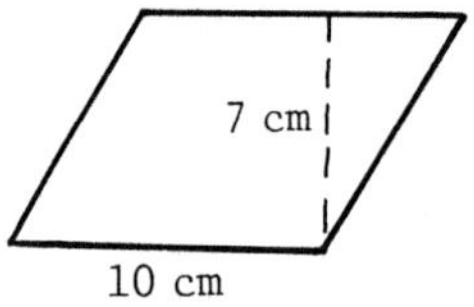

b) 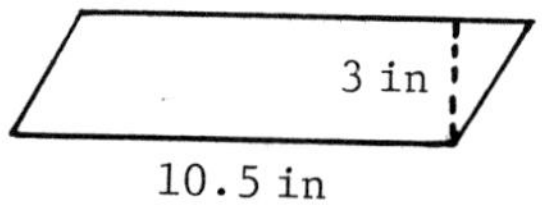

---

a) $A = 70\ cm^2$
b) $A = 31.5\ in^2$

13. To find the base (or height) of a parallelogram when its area and height (or base) are given, we can use either the equation-solving method or the rearrangement method. Use either method to solve the problems below.

a) Find the base of a parallelogram if its area is $30\ m^2$ and its height is 6 m.

b) Find the height of a parallelogram if its area is $100\ ft^2$ and its base is 25 ft.

a) $b = 5$ m b) $h = 4$ ft

10-4 TRIANGLES

In this section, we will review the formulas for the perimeter and area of a triangle and the angle-sum formula for triangles. The formulas are used to solve some applied problems.

14. The perimeter of a triangle can be found by adding its three sides. If the three sides are "a", "b", and "c", we can state that fact with the formula below.

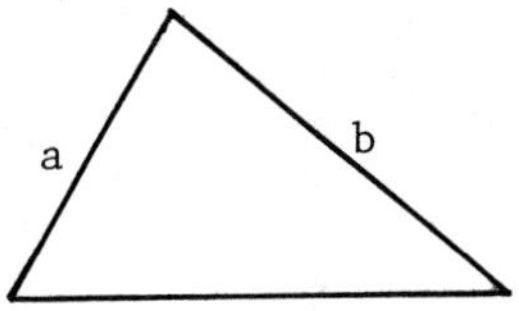

$$P = a + b + c$$

Use the formula above to find the perimeter of each triangle at the right.

a) 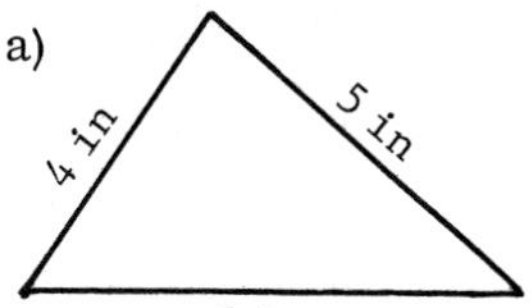

b) 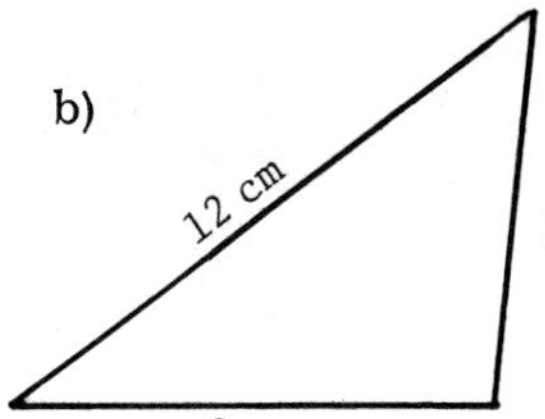

---

a) P = 15 in

b) P = 28 cm

15. To find side "a" when the perimeter of a triangle is 30 ft, side "b" is 12 ft, and side "c" is 8 ft, we can use either method below.

Equation-Solving Method

$$P = a + b + c$$
$$30 = a + 12 + 8$$
$$30 = a + 20$$
$$30 + (-20) = a + 20 + (-20)$$
$$a = 10 \text{ ft}$$

Rearrangement Method

$$P = a + b + c$$
$$a = P - b - c$$
$$a = \underbrace{30 - 12} - 8$$
$$a = 18 - 8 = 10 \text{ ft}$$

Use either method to solve these problems.

a) Find "b" if the perimeter of a triangle is 50 cm, a = 10 cm, and c = 25 cm.

b) Find "c" if the perimeter of a triangle is 100 m, a = 35 m, and b = 40 m.

---

a) b = 15 cm

b) c = 25 m

16. In any triangle, the sum of the three angles is 180°. If the three angles are A, B, and C, we can state the angle-sum fact with the formula below.

$$\boxed{A + B + C = 180°}$$

To find the size of angle A in the triangle at the right, we can use either method below.

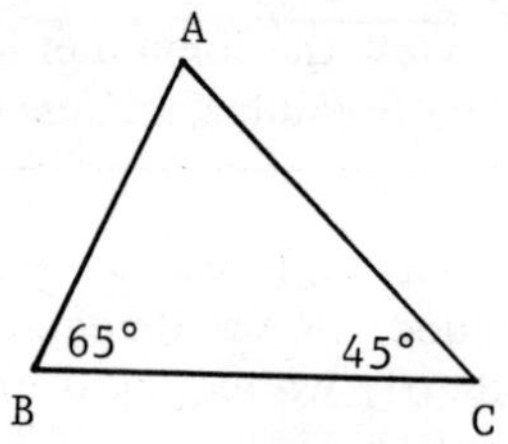

<u>Equation-Solving Method</u>

$$A + B + C = 180°$$
$$A + 65° + 45° = 180°$$
$$A + 110° = 180°$$
$$A + 110° + (-110°) = 180° + (-110°)$$
$$A = 70°$$

<u>Rearrangement Method</u>

$$A + B + C = 180°$$
$$A = 180° - B - C$$
$$A = \underbrace{180° - 65°}_{\downarrow} - 45°$$
$$A \quad 115° \quad - 45° = 70°$$

Using either method, find the size of the missing angle in each triangle at the right.

a)

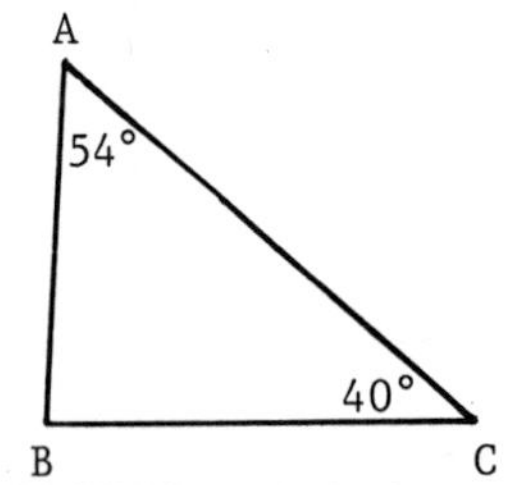

b)

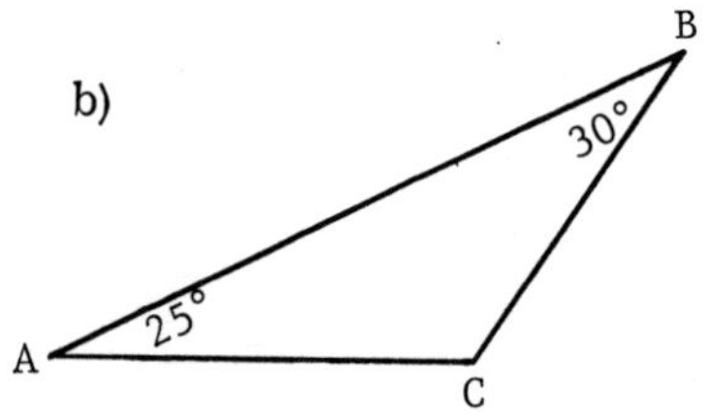

---

a) B = 86°

b) C = 125°

17. As you can see from the figure at the right, a triangle is simply half a parallelogram. Since the area of the parallelogram is "bh", the area of the triangle is half of "bh". That is:

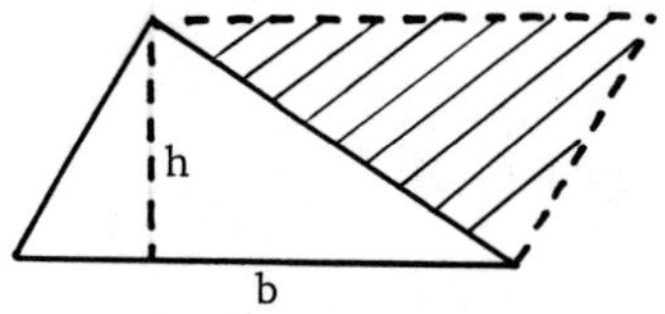

$$\boxed{A = \frac{1}{2}bh}$$

Using the formula above, find the area of each triangle at the right.

a)

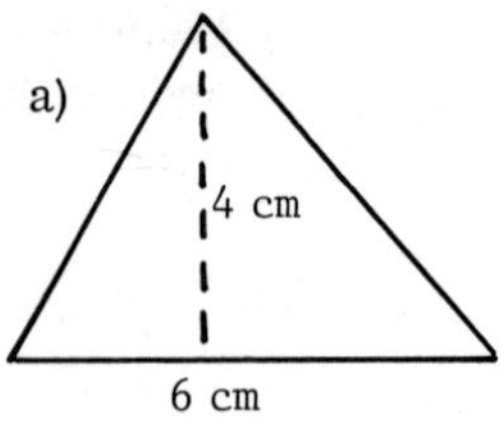

b)

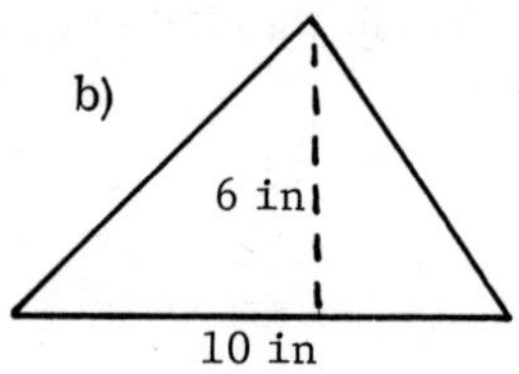

18. To find the height of a triangle if its area is 36 m^2 and its base is 8 m, we can use either method below.

a) $A = 12\ cm^2$
b) $A = 30\ in^2$

<u>Equation-Solving Method</u>

$$A = \frac{1}{2}bh$$

$$36 = \frac{1}{2}(8)h$$

$$36 = 4h$$

$$h = \frac{36}{4} = 9\ m$$

<u>Rearrangement Method</u>

$$A = \frac{1}{2}bh$$

$$2A = bh$$

$$h = \frac{2A}{b}$$

$$h = \frac{2(36)}{8} = \frac{72}{8} = 9\ m$$

Using either method, solve the problems below.

a) Find the <u>height</u> of a triangle if its area is 80 cm^2 and its base is 10 cm.

b) Find the <u>base</u> of a triangle if its area is 100 in^2 and its height is 20 in.

a) $h = 16$ cm
b) $b = 10$ in

10-5 RIGHT TRIANGLES

In this section, we will discuss some formulas related to right triangles and use them to solve some applied problems.

---

19. A right triangle is a triangle that contains a right angle (90°). The right angle is marked with a small box. In a right triangle:

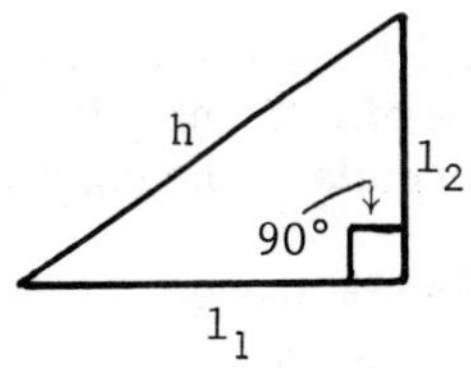

1) The longest side (the one opposite the right angle) is called the "hypotenuse" (h).

2) The two shorter sides are called the "legs" (l_1 and l_2).

The perimeter of a right triangle is simply the sum of the hypotenuse and the legs. That is:

$$P = h + l_1 + l_2$$

Using the formula above, find the perimeter of the right triangle at the right.

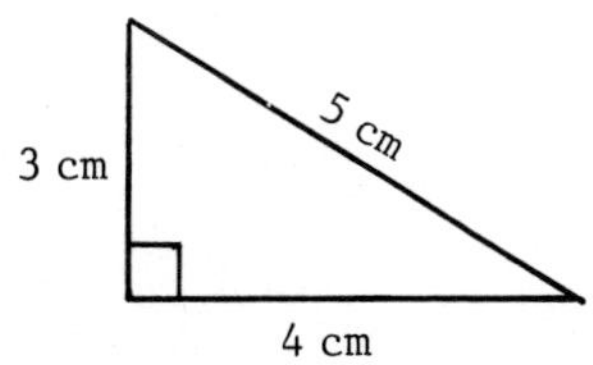

20. Using either the equation-solving method or the rearrangement method, solve each problem below.

a) For a right triangle, find l_1 if $P = 30$ in, $h = 13$ in, and $l_2 = 12$ in.

b) For a right triangle, find l_2 if $P = 120$ ft, $h = 50$ ft, and $l_1 = 30$ ft.

$P = 12$ cm

---

21. Find the missing angle in each right triangle. (Note: Each right angle is 90°.)

a) 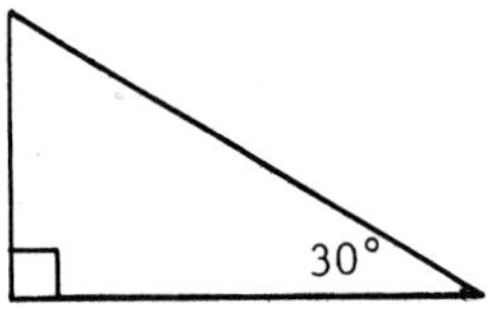

b) 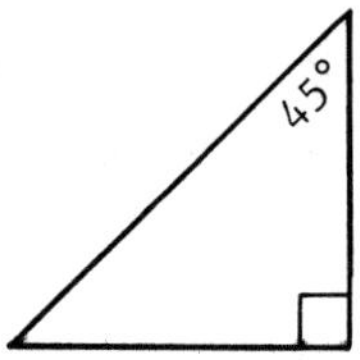

a) $l_1 = 5$ in

b) $l_2 = 40$ ft

---

22. In the right triangle at the right:

1) leg 1 (or l_1) is the "base".

2) leg 2 (or l_2) is the "height".

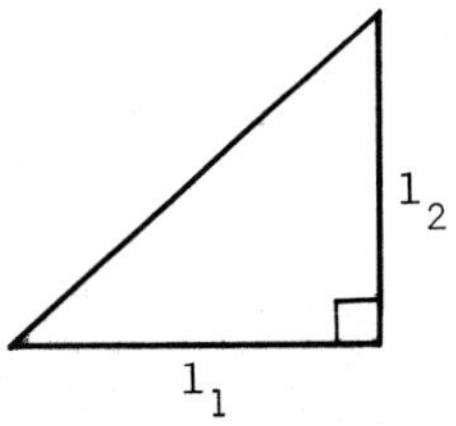

Since the area of a triangle is half the base times the height, we use the formula at the right to find the area of a right triangle.

$$A = \frac{1}{2}l_1l_2$$

Using the formula above, find the area of each right triangle.

a) 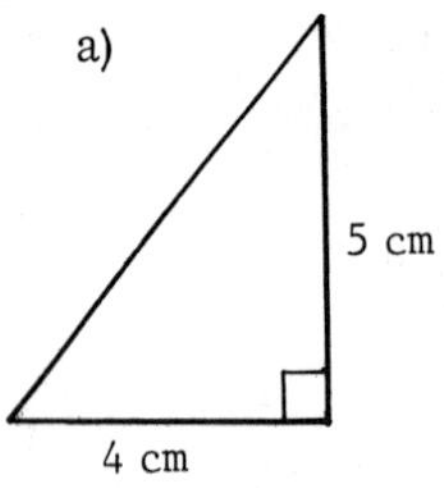

b) 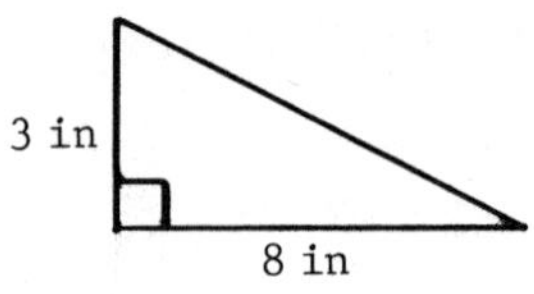

a) 60°

b) 45°

23. Using either the equation-solving method or the rearrangement method, solve each problem.

a) In a right triangle, find l_2 if $A = 50\ m^2$ and $l_1 = 20$ m.

b) In a right triangle, find l_2 if $A = 48\ ft^2$ and $l_1 = 8$ ft.

a) $A = 10\ cm^2$
b) $A = 12\ in^2$

a) $l_2 = 5$ m
b) $l_1 = 12$ ft

SELF-TEST 30 (pages 259-270)

Find the perimeter of each figure.

1.

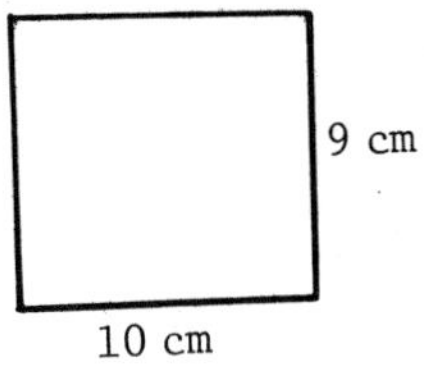

2.

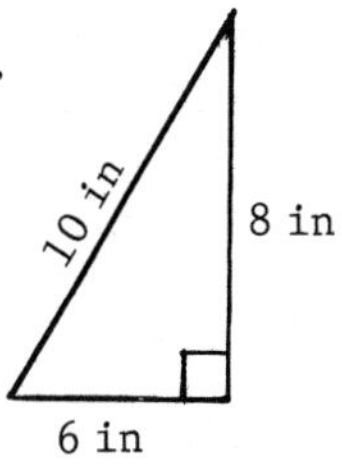

Find the area of each figure.

3.

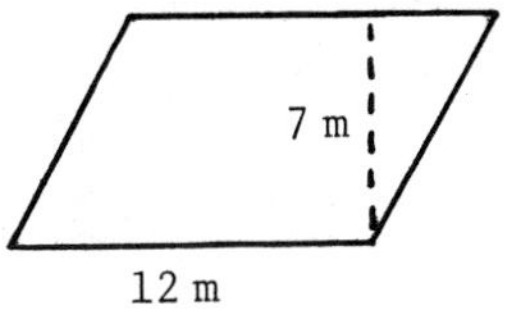

4.

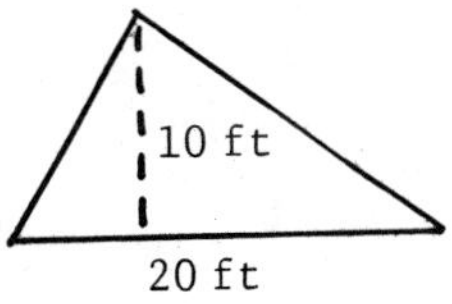

Continued on the following page.

SELF-TEST 30 (pages 259-270) - Continued

5. Find the length of a rectangle if its area is 48 cm^2 and its width is 6 cm.	6. Find the side of a square if its perimeter is 44 in.
7. Find angle B. 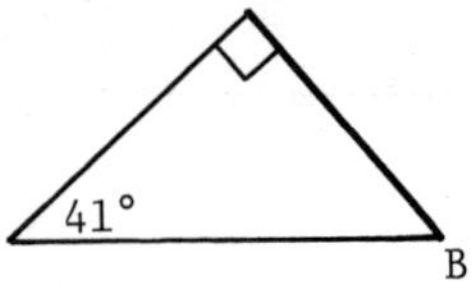	8. Find the height of a triangle if its area is 50 m^2 and its base is 10 m.

ANSWERS:

1. P = 38 cm	3. A = 84 m^2	5. L = 8 cm	7. B = 49°
2. P = 24 in	4. A = 100 ft^2	6. s = 11 in	8. h = 10 m

10-6 THE PYTHAGOREAN THEOREM

The Pythagorean Theorem is a formula relating the hypotenuse and legs of a right triangle. We will discuss that formula in this section and use it to solve some applied problems.

24. To label the sides of a right triangle, we frequently use the letters "a", "b", and "c".

"c" is the hypotenuse

"a" and "b" are the legs

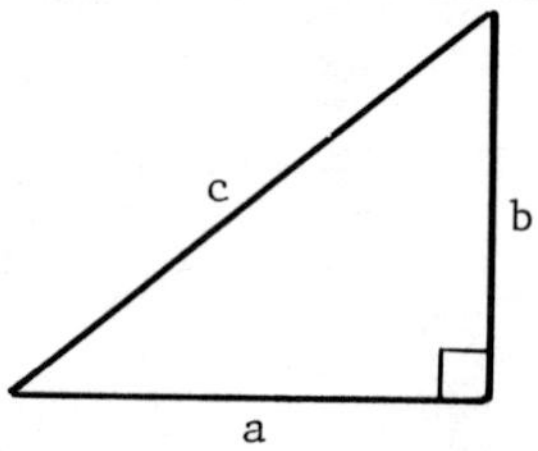

The Pythagorean Theorem says this: In any right triangle, the square of the hypotenuse is equal to the sum of the squares of the two legs. That is:

$$c^2 = a^2 + b^2$$

Continued on the following page.

24. Continued

We used the formula on the preceding page to find the length of the hypotenuse of the right triangle below. Use it to find the length of the hypotenuse of the other triangle. (Note: Use the square-root table on page 190).

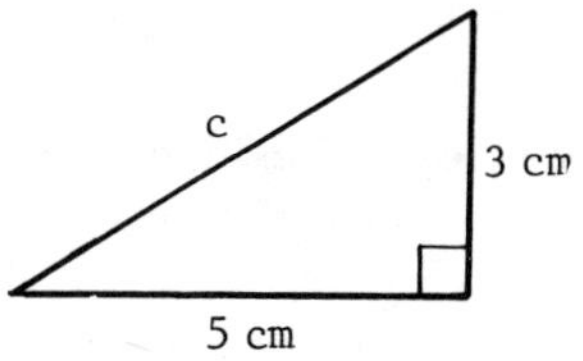

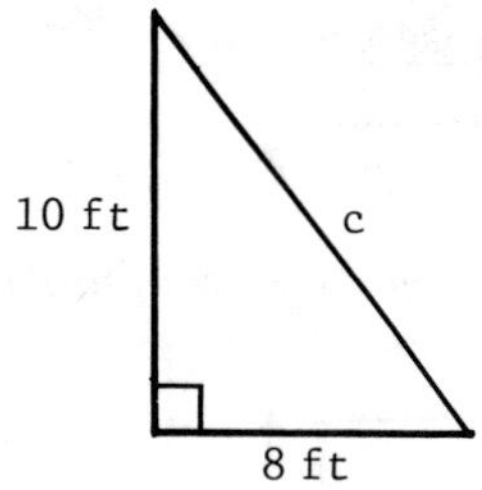

$c^2 = a^2 + b^2$

$c^2 = 5^2 + 3^2$

$c^2 = 25 + 9$

$c^2 = 34$

$c = \sqrt{34} = 5.831$ cm

$c = \sqrt{164} = 12.806$ ft

25. The same formula can be used to find leg "a" in the right triangle at the right. Either of the two methods below can be used.

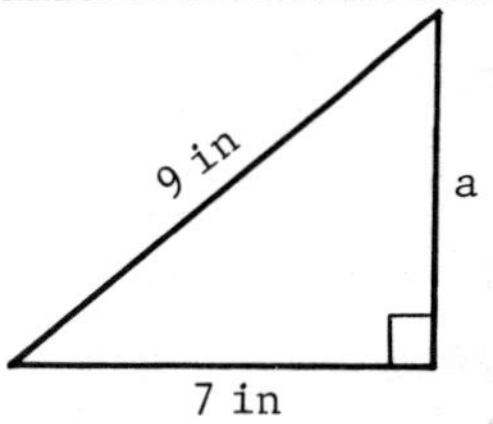

Equation-Solving Method	Rearrangement Method
$c^2 = a^2 + b^2$	$c^2 = a^2 + b^2$
$9^2 = a^2 + 7^2$	$a^2 = c^2 - b^2$
$81 = a^2 + 49$	$a = \sqrt{c^2 - b^2}$
$a^2 = 32$	$a = \sqrt{9^2 - 7^2}$
$a = \sqrt{32} = 5.657$ in	$a = \sqrt{81 - 49} = \sqrt{32} = 5.657$ in

Using either method above, solve each problem below.

a) Find leg "a".

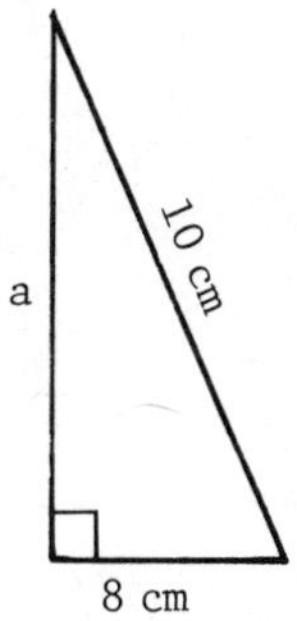

b) Find leg "b".

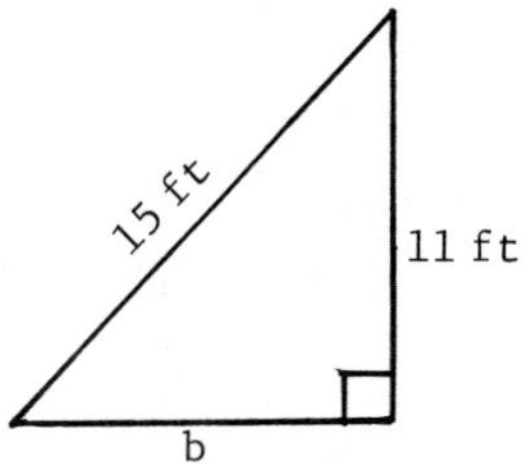

a) $a = \sqrt{36} = 6$ cm

b) $b = \sqrt{104} = 10.198$ ft

26. The distance between the opposite corners of a rectangle or square is called the "diagonal" of the rectangle or square. In each figure, the diagonal is the hypotenuse of a right triangle.

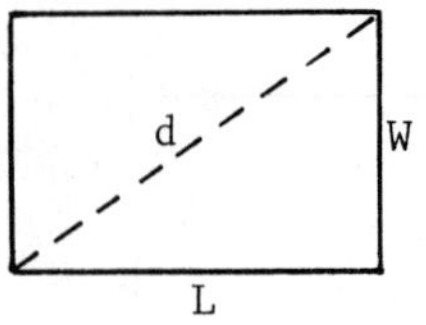

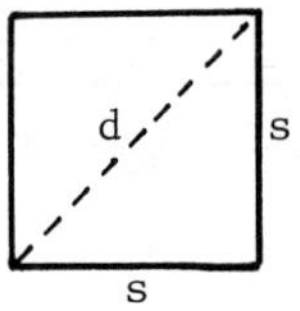

We used the Pythagorean Theorem to find the diagonal of the rectangle below. Find the diagonal of the square.

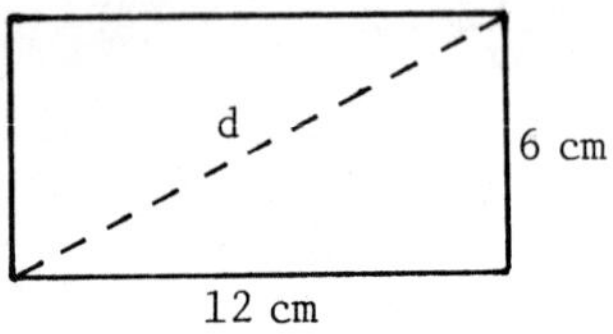

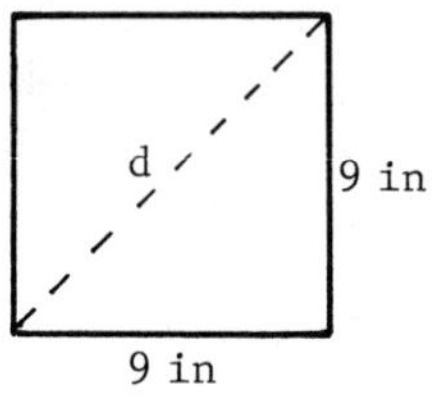

$$d^2 = 12^2 + 6^2$$

$$d^2 = 144 + 36$$

$$d^2 = 180$$

$$d = \sqrt{180} = 13.416 \text{ cm}$$

$d = \sqrt{162} = 12.728$ in

27. Use the Pythagorean Theorem to solve each problem below. (Hint: Make a sketch and label the parts.)

a) Find the length of a rectangle if its diagonal is 12 cm and its width is 4 cm.

b) Find the width of a rectangle if its diagonal is 10 ft and its length is 9 ft.

---

a) $L = \sqrt{128} = 11.314$ cm

b) $W = \sqrt{19} = 4.359$ ft

10-7 CIRCLES

In this section, we will review the formulas for the circumference and area of a circle and use them to solve some applied problems.

---

28. Point 0 is the center of the circle at the right.

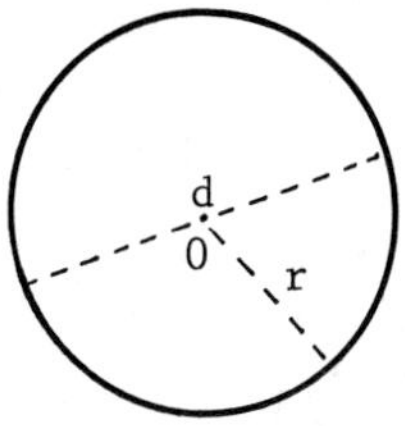

The line "d" is a <u>diameter</u> of the circle.

The line "r" is a <u>radius</u> of the circle.

As you can see, a diameter is twice as long as a radius. That is:

$$d = 2r$$

Using the formula above, solve each problem below.

a) Find the diameter of a circle if its radius is 10 cm.

b) Find the radius of a circle if its diameter is 12 in.

---

a) d = 20 cm

b) r = 6 in

29. The distance around a circle is called its "circumference". If we divide the circumference of any circle by its diameter, we always get the same number. The Greek letter "π" (pronounced "pie") is used for that number. Though π stands for the unending number 3.1415925636..., we usually use 3.14 for π. Therefore:

$$\frac{C}{d} = \pi \quad \text{or} \quad \frac{C}{d} = 3.14$$

Solving for C, we get the following "circumference" formula.

$$C = \pi d \quad \text{or} \quad C = 3.14d$$

Using the above formula, solve each problem below.

a) Find the circumference of a circle if its diameter is 10 cm.

b) Find the diameter of a circle if its circumference is 9.42 in.

30. Since $d = 2r$, we can also get a circumference formula in terms of the radius. That is:

$$C = \pi d$$

$$C = \pi(2r)$$

or

$$\boxed{C = 2\pi r}$$

Using the formula above, solve each problem below.

a) Find the <u>circumference</u> of a circle if its radius is 4 in.

b) Find the <u>radius</u> of a circle if its circumference is 12.56 cm.

a) $C = 31.4$ cm
b) $d = 3$ in

---

31. To find the area of a circle, we can use the formula below.

$$\boxed{A = \pi r^2}$$

Using the formula above, solve each problem.

a) Find the <u>area</u> of a circle if its radius is 2 in.

b) Find the <u>area</u> of a circle if its radius is 10 cm.

a) $C = 25.12$ in
b) $r = 2$ cm

---

32. To find the radius of a circle whose area is 28.26 in^2, we can use either method below.

<u>Equation-Solving Method</u>

$$A = \pi r^2$$

$$28.26 = 3.14r^2$$

$$r^2 = \frac{28.26}{3.14}$$

$$r^2 = 9$$

$$r = \sqrt{9} = 3 \text{ in}$$

<u>Rearrangement Method</u>

$$A = \pi r^2$$

$$r = \sqrt{\frac{A}{\pi}}$$

$$r = \sqrt{\frac{28.26}{3.14}}$$

$$r = \sqrt{9} = 3 \text{ in}$$

a) $A = 12.56$ in^2
b) $A = 314$ cm^2

Continued on the following page.

32. Continued

Using either method from the preceding page, solve this problem.

Find the radius of a circle if its area is 3.14 m^2.

r = 1 m

10-8 VOLUMES

In this section, we will review the formulas for the volume of a rectangular solid, a cube, and a cylinder. The formulas are then used to solve some applied problems.

---

33. The volume of a solid figure is the number of "unit cubes" needed to fill it up. Some "unit cubes" are shown below.

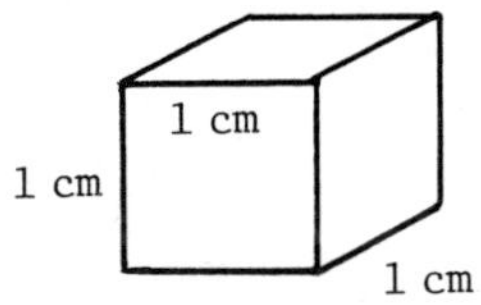

A cubic centimeter (cm^3)

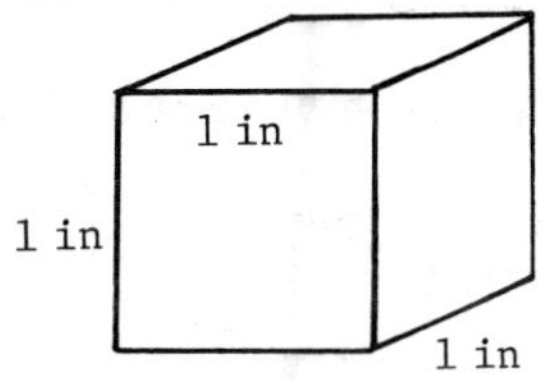

A cubic inch (in^3)

Notice that we use a small "3" in the abbreviations for "cubic units". Therefore:

The abbreviation for "a cubic meter" is "m^3".

The abbreviation for "a cubic foot" is "ft^3".

The abbreviation for "a cubic yard" is ________?

yd^3

---

34. A small "3" means that the quantity should be cubed. That is, it should be used as a factor three times. That is:

$2^3 = (2)(2)(2)$ $x^3 = (x)(x)(x)$ $cm^3 = (cm)(cm)(cm)$

Similarly, any three-factor multiplication with identical factors can be written with a small "3".

$(5)(5)(5) = 5^3$ $(y)(y)(y) = y^3$ $(in)(in)(in) =$ ________

in^3

35. Two word names are used for cubic units. For example:

$25\ cm^3$ is called "25 centimeters cubed" or "25 cubic centimeters".

$80\ ft^3$ is called "80 feet cubed" or "80 cubic feet".

Write the abbreviation for each word name.

a) 37 inches cubed = ________ b) 100 cubic meters = ________

---

a) $37\ in^3$

b) $100\ m^3$

36. It takes 24 cubic centimeters to fill up the rectangular solid at the right. Therefore, its volume is $24\ cm^3$. Notice that we can find its volume by multiplying its length, width, and height. That is:

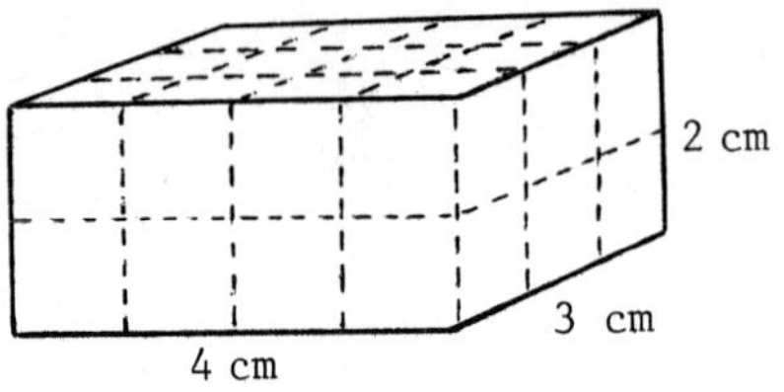

$$V = LWH$$

Using the formula for the rectangular solid above, we get:

$$V = (4\ cm)(3\ cm)(2\ cm)$$
$$= (4)(3)(2)(cm)(cm)(cm)$$
$$= \_\_\_\_\_\_\_\_\ cm^3$$

$24\ cm^3$

37. Use the formula to find the volume of each solid.

a)

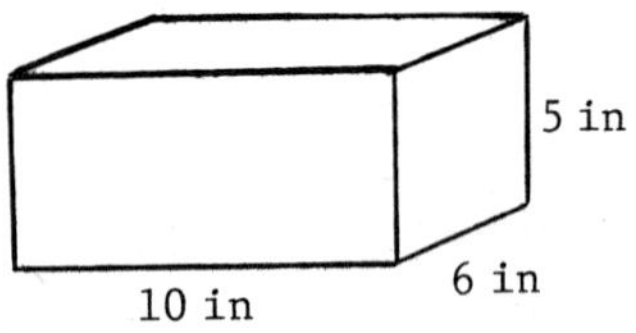

b)

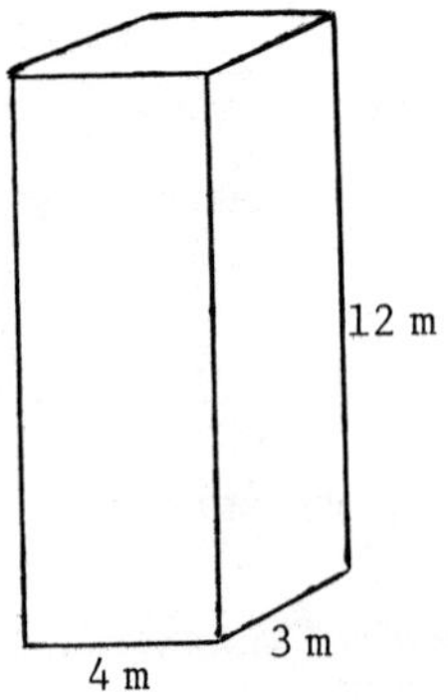

a) $V = 300\ in^3$

b) $V = 144\ m^3$

38. To find the width of a rectangular solid when its volume is 60 in^3, its length is 5 in, and its height is 6 in, we can use either method below.

Equation-Solving Method

$$V = LWH$$
$$60 = (5)(W)(6)$$
$$60 = 30W$$
$$W = \frac{60}{30} = 2 \text{ in}$$

Rearrangement Method

$$V = LWH$$
$$W = \frac{V}{LH}$$
$$W = \frac{60}{(5)(6)} = \frac{60}{30} = 2 \text{ in}$$

Using either method, solve each problem.

a) Find the height of a rectangular solid if $V = 120\ m^3$, $L = 4$ m, and $W = 3$ m.

b) Find the length of a rectangular solid if $V = 24\ yd^3$, $W = 2$ yd, and $H = 2$ yd.

---

a) $H = 10$ m

b)) $L = 6$ yd

39. A cube is a special type of rectangular solid in which all the sides are equal. Therefore, the volume of a cube equals (s)(s)(s). That is:

$$\boxed{V = s^3}$$

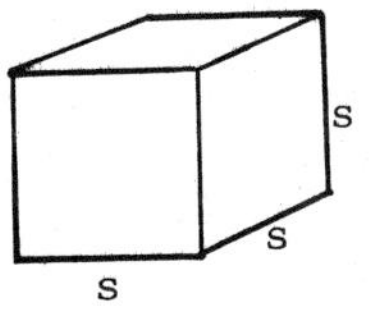

Using the formula above, we found the volume of one cube below. Find the volume of the other cube.

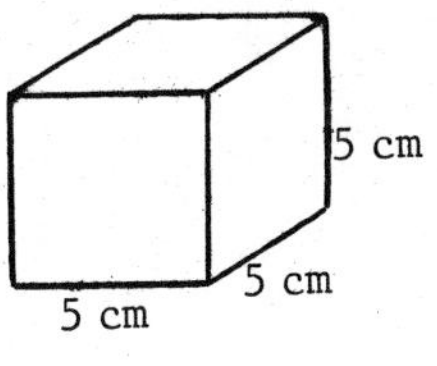

$$V = s^3$$
$$V = (5\ cm)^3$$
$$V = (5^3)(cm)^3$$
$$V = 125\ cm^3$$

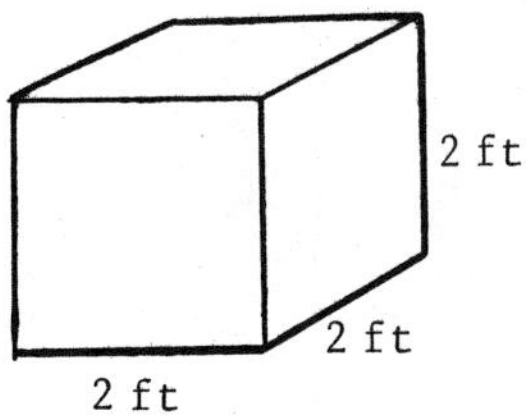

$V = 8\ ft^3$

40. The formula for the volume of a rectangular solid is the same as multiplying the area of the base by the height. That is:

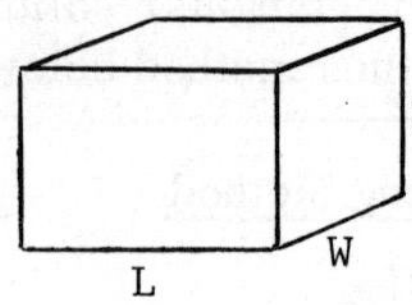

Area of the rectangular base = LW

Height = H

Volume = (LW)(H) or LWH

To find the volume of a cylinder, we also multiply the area of the base by the height. That is:

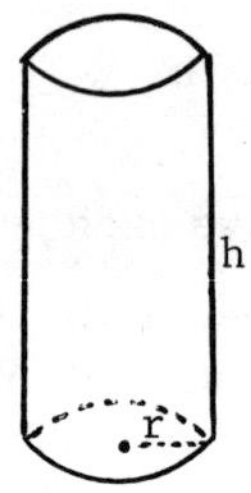

Area of the circular base $= \pi r^2$

Height = h

Volume $= (\pi r^2)(h) = \pi r^2 h$

or

$$V = \pi r^2 h$$

We used the formula above to find the volume of one cylinder below. Find the volume of the other cylinder.

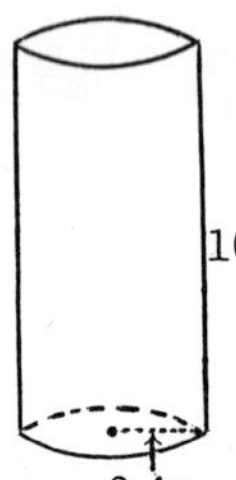

$V = \pi r^2 h$

$V = (3.14)(2^2)(10)$

$V = (3.14)(4)(10)$

$V = (12.56)(10)$

$V = 125.6 \text{ in}^3$

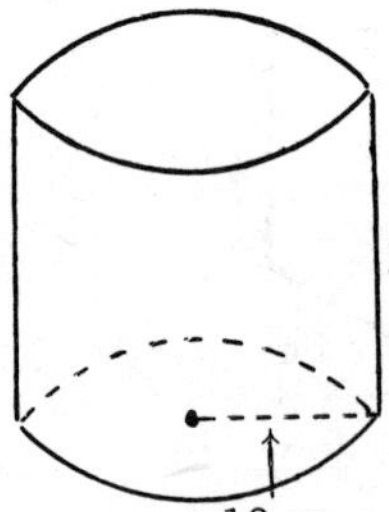

$V = 3,768 \text{ cm}^3$

SELF-TEST 31 (pages 270-279)

1. Find the hypotenuse "c".

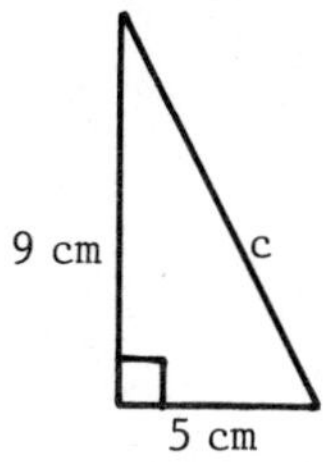

2. Find the width W.

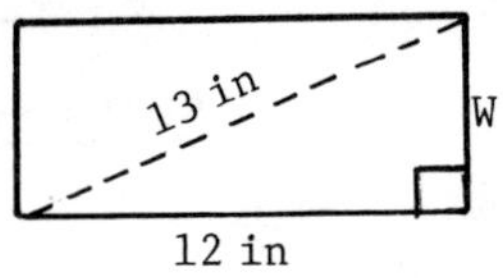

3. Find the circumference.

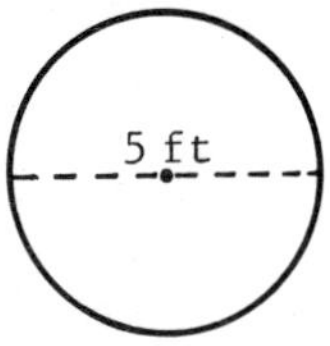

4. Find the area.

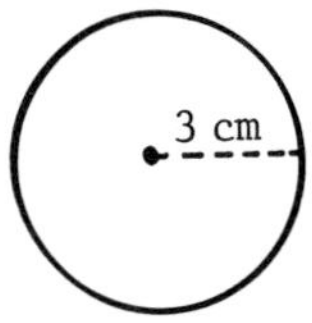

5. Find the volume.

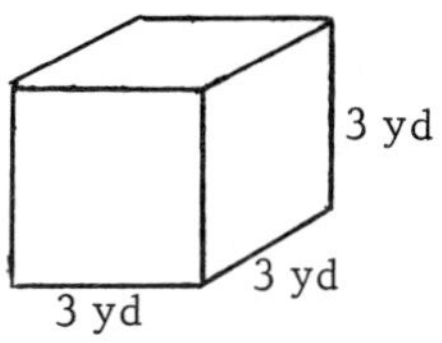

6. Find the volume.

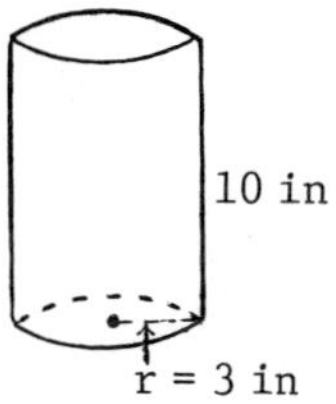

7. Find the radius of a circle if its circumference is 37.68 m.

8. Find the height of a rectangular solid if its volume is 30 cm^2, its length is 3 cm, and its width is 2 cm.

ANSWERS:

1. c = 10.296 cm
2. W = 5 in
3. C = 15.7 ft
4. A = 28.26 cm^2
5. V = 27 yd^3
6. V = 282.6 in^3
7. r = 6 m
8. H = 5 cm

SUMMARY OF FORMULAS

Rectangle:

$P = 2L + 2W$

$A = LW$

Square:

$P = 4s$

$A = s^2$

Parallelogram:

$A = bh$

Triangle:

$P = a + b + c$

$A + B + C = 180°$

$A = \frac{1}{2}bh$

Right Triangle:

$P = h + l_1 + l_2$

$A = \frac{1}{2}l_1 l_2$

$c^2 = a^2 + b^2$ (Pythagorean Theorem)

Circle:

$d = 2r$

$C = \pi d$ or $C = 2\pi r$

$A = \pi r^2$

Rectangular Solid:

$V = LWH$

Cube:

$V = s^3$

Cylinder:

$V = \pi r^2 h$

Unit 11 SCIENTIFIC NOTATION

In this unit, we will define the meaning of powers of ten and scientific notation. Methods for converting numbers in scientific notation to ordinary numbers and for converting ordinary numbers to scientific notation are discussed. The use of scientific notation in science, in calculator answers, and in some applied problems is shown.

11-1 POWERS OF TEN WITH POSITIVE EXPONENTS

In this section, we will discuss the power-of-ten form of numbers like 10, 100 , 1,000 and so on. In the power-of-ten form of these numbers, the exponents are positive whole numbers.

---

1. Expressions like 10^1, 10^2, 10^3, 10^4, 10^5, and 10^6 are called "powers of ten". In each one, 10 is called the base; the small number is called the exponent. That is:

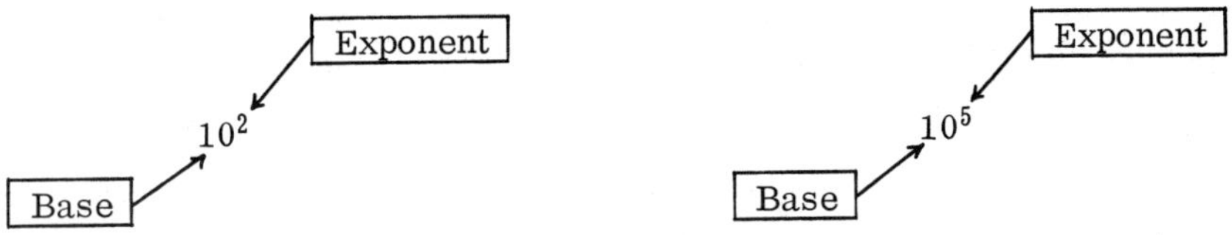

In 10^4: a) The base is ______. b) The exponent is ______.

---

a) 10
b) 4

2. The following pattern is used to name powers of ten.

10^2 is called "10 to the second power".

10^3 is called "10 to the third power".

10^5 is called "10 to the __________ power".

---

fifth

3. Following the example, write each of these in base-exponent form.

10 to the first power = 10^1

a) 10 to the fourth power = ______

b) 10 to the seventh power = ______

a) 10^4
b) 10^7

4. When naming a power of ten, we sometimes omit the word "power". For example:

Instead of "10 to the <u>third</u> power", we simply say "10 to the <u>third</u>".

Instead of "10 to the <u>sixth</u> power", we simply say "10 to the <u>sixth</u>".

Therefore: 10 to the <u>second</u> = 10^2 10 to the <u>eighth</u> = ______

Answer: 10^8

5. We frequently use the words "<u>squared</u>" and "<u>cubed</u>" for the second and third power of ten. That is:

We say "10 <u>squared</u>" instead of "10 to the <u>second</u> power".

We say "10 <u>cubed</u>" instead of "10 to the <u>third</u> power".

Therefore: 10 squared = 10^2 10 cubed = ______

Answer: 10^3

6. Any power of ten is a short way of writing a multiplication of 10's. The <u>exponent</u> tells us how many 10's to use. For example:

$10^2 = 10 \cdot 10$ (The exponent tells us to multiply <u>two</u> 10's.)

$10^5 = 10 \cdot 10 \cdot 10 \cdot 10 \cdot 10$ (The exponent tells us to multiply <u>five</u> 10's.)

Write each power of ten as a multiplication of 10's.

a) 10^3 = ____________ b) 10^4 = ______________ c) 10^6 = ________________

Answers:
a) $10 \cdot 10 \cdot 10$
b) $10 \cdot 10 \cdot 10 \cdot 10$
c) $10 \cdot 10 \cdot 10 \cdot 10 \cdot 10 \cdot 10$

7. By counting the number of 10's, we can determine each exponent below.

$10 \cdot 10 \cdot 10 = 10^3$ (Since there are <u>three</u> 10's, the exponent is 3.)

$10 \cdot 10 \cdot 10 \cdot 10 \cdot 10 \cdot 10 = 10^6$ (Since there are <u>six</u> 10's, the exponent is 6.)

Write the corresponding power of ten.

a) $10 \cdot 10$ = ______ b) $10 \cdot 10 \cdot 10 \cdot 10$ = ______ c) $10 \cdot 10 \cdot 10 \cdot 10 \cdot 10$ = ______

Answers:
a) 10^2
b) 10^4
c) 10^5

8. Powers of ten equal ordinary numbers like 10, 100, 1000 and so on. For example:

$10^1 = 10$ $10^2 = 10 \cdot 10 = 100$ $10^3 = 10 \cdot 10 \cdot 10$ = __________

Answer: 1,000

9. By examining the table at the right, you can see this fact:

<u>The number of 0's in the ordinary number equals the exponent of the power of ten.</u>

10^6	=	1,000,000
10^5	=	100,000
10^4	=	10,000
10^3	=	1,000
10^2	=	100
10^1	=	10

Using the fact above, convert each of these to an ordinary number.

a) 10^7 = ____________________

b) 10^8 = ____________________

Answers: a) 10,000,000 b) 100,000,000

10. Using the exponent to determine the number of 0's, convert each of these to an ordinary number.

a) 10^2 = ______ b) 10^4 = ______ c) 10^6 = ______

---

a) 100
b) 10,000
c) 1,000,000

11. By examining the table at the right, you can see this fact:

<u>The exponent of the power of ten equals the number of 0's in the ordinary number</u>.

1,000,000	=	10^6
100,000	=	10^5
10,000	=	10^4
1,000	=	10^3
100	=	10^2
10	=	10^1

Using the fact above, convert each of these to a power of ten.

a) 10,000,000 = ______

b) 10,000,000,000 = ______

---

a) 10^7
b) 10^{10}

12. By counting the number of 0's to find the exponent, convert each of these to a power of ten.

a) 1,000 = ______ b) 10 = ______ c) 100,000,000 = ______

a) 10^3 b) 10^1 c) 10^8

11-2 MULTIPLYING AND DIVIDING POWERS OF TEN WITH POSITIVE EXPONENTS

In this section, we will introduce the laws of exponents for multiplying and dividing powers of ten.

---

13. To perform the multiplications of powers of ten below, we converted each factor to an ordinary number and then converted the products back to powers of ten.

$$\begin{array}{ccccc} 10^1 & \times & 10^3 & = & 10^4 \\ \downarrow & & \downarrow & & \uparrow \\ 10 & \times & 1,000 & = & 10,000 \end{array} \qquad \begin{array}{ccccc} 10^4 & \times & 10^2 & = & 10^6 \\ \downarrow & & \downarrow & & \uparrow \\ 10,000 & \times & 100 & = & 1,000,000 \end{array}$$

From the examples, you can see that <u>the exponent of each product can be obtained by adding the exponents of the factors</u>. That is:

$10^1 \times 10^3 = 10^{1+3} = 10^4$ $\qquad$ $10^4 \times 10^2 = 10^{4+2} = 10^{\square}$

10^6

14. To multiply powers of ten, <u>we simply add their exponents</u>. That is:

$$\boxed{10^a \times 10^b = 10^{a+b}}$$

Using the above law of exponents, complete these:

a) $10^4 \times 10^1 = 10^{4+1}$ = ______ b) $10^2 \times 10^5 = 10^{2+5}$ = ______

---

a) 10^5
b) 10^7

15. Complete these by simply adding the exponents of the factors.

a) $10^2 \times 10^2$ = ______ b) $10^3 \times 10^7$ = ______ c) $10^8 \times 10^1$ = ______

a) 10^4 b) 10^{10} c) 10^9

16. To perform the divisions of powers of ten below, we converted each term to an ordinary number and then converted the quotients back to powers of ten.

$$\frac{10^3}{10^1} = 10^2 \qquad \frac{10^5}{10^2} = 10^3$$

$$\frac{1,000}{10} = 100 \qquad \frac{100,000}{100} = 1,000$$

From the examples, you can see that <u>the exponent of each quotient can be obtained by subtracting the exponent of the denominator from the exponent of the numerator</u>. That is:

$$\frac{10^3}{10^1} = 10^{3-1} = 10^2 \qquad \frac{10^5}{10^2} = 10^{5-2} = 10^{\square}$$

Answer: 10^3

17. To divide powers of ten, <u>we simply subtract the exponent of the denominator from the exponent of the numerator</u>. That is:

$$\boxed{\frac{10^a}{10^b} = 10^{a-b}}$$

Using the above law of exponents, complete these:

a) $\frac{10^6}{10^2} = 10^{6-2} =$ ______ b) $\frac{10^9}{10^4} = 10^{9-4} =$ ______

Answers: a) 10^4 b) 10^5

18. Complete these by subtracting the exponents of the terms.

a) $\frac{10^7}{10^1} =$ ______ b) $\frac{10^8}{10^7} =$ ______ c) $\frac{10^{12}}{10^3} =$ ______

Answers: a) 10^6 b) 10^1 c) 10^9

19. Remember: To <u>multiply</u> powers of ten, we <u>add</u> the exponents.
To <u>divide</u> powers of ten, we <u>subtract</u> the exponents.

Complete: a) $10^4 \times 10^5 =$ ______ c) $10^{12} \times 10^6 =$ ______

b) $\frac{10^9}{10^8} =$ ______ d) $\frac{10^{10}}{10^1} =$ ______

Answers: a) 10^9 b) 10^1 c) 10^{18} d) 10^9

11-3 THE MEANING OF 10^0

In this section, we will define the meaning of 10^0 and use 10^0 in multiplications and divisions.

---

20. The pattern in the table at the right suggests the following definition for 10^0.

$$\boxed{10^0 = 1}$$

10^3	= 1,000
10^2	= 100
10^1	= 10
10^0	= 1

Continued on following page.

20. Continued

If that definition makes sense, it should fit with the laws of exponents. That is:

$$10^0 \times 10^2 = 10^{0+2} = 10^2 \qquad \frac{10^3}{10^3} = 10^{3-3} = 10^0$$

To show that it fits with the laws of exponents, we have converted each problem above to ordinary numbers.

$$\begin{array}{ccccc} 10^0 & \times & 10^2 & = & 10^2 \\ \downarrow & & \downarrow & & \uparrow \\ 1 & \times & 100 & = & 100 \end{array} \qquad \begin{array}{ccc} \frac{10^3}{10^3} & = & 10^0 \\ \downarrow & & \uparrow \\ \frac{1,000}{1,000} & = & 1 \end{array}$$

Therefore 10^0 equals what ordinary number? ______

1

21. Write each product as a power of ten and as an ordinary number.

a) $10^1 \times 10^0$ = ______ = ______ b) $10^0 \times 10^0$ = ______ = ______

---

a) $10^1 = 10$

b) $10^0 = 1$

22. Write each quotient as a power of ten and as an ordinary number.

a) $\frac{10^1}{10^0}$ = ______ = ______ b) $\frac{10^0}{10^0}$ = ______ = ______

---

a) $10^1 = 10$

b) $10^0 = 1$

23. Use the laws of exponents to complete these.

a) $\frac{10^8}{10^8}$ = ______ b) $10^0 \times 10^5$ = ______ c) $\frac{10^6}{10^0}$ = ______

---

a) 10^0 (or 1) b) 10^5 c) 10^6

11-4 POWERS OF TEN WITH NEGATIVE EXPONENTS

In this section, we will discuss the power-of-ten form of numbers like .1 , .01 , .001 and so on. In the power-of-ten form of those numbers, the exponents are negative whole numbers.

24. The following pattern is used to name powers of ten with negative exponents.

10^{-1} is called "10 to the negative-one power".

10^{-2} is called "10 to the negative-two power".

10^{-4} is called "10 to the ____________ power".

---

negative-four

25. We sometimes use the word "minus" instead of the word "negative". For example:

10^{-1} is called "10 to the minus-one power".

10^{-3} is called "10 to the minus-three power".

10^{-6} is called "10 to the ____________ power".

---

minus-six

26. Sometimes the word "power" is omitted in the names. For example:

10^{-2} is called "10 to the negative-two" or "10 to the minus-two".

10^{-5} is called "10 to the ________" or "10 to the ________".

---

negative-five ... minus-five

27. Write the corresponding power of ten.

a) 10 to the negative-one power = ______ c) 10 to the negative-seven = ______

b) 10 to the minus-four power = ______ d) 10 to the minus-nine = ______

---

a) 10^{-1}

b) 10^{-4}

c) 10^{-7}

d) 10^{-9}

28. The following definition is used for powers of ten with negative exponents.

$$\boxed{10^{-a} = \frac{1}{10^a}}$$

Therefore: $10^{-1} = \frac{1}{10^1}$ $10^{-3} = \frac{1}{10^3}$ $10^{-5} =$ ______

$\frac{1}{10^5}$

29. Because of the definition, any power of ten with a negative exponent equals a decimal fraction whose numerator is "1". That is:

$10^{-1} = \frac{1}{10^1} = \frac{1}{10}$ $10^{-2} = \frac{1}{10^2} = \frac{1}{100}$ $10^{-4} = \frac{1}{10^4} =$ ______

$\frac{1}{10,000}$

30. Any power of ten with a negative exponent can be converted to a decimal number. That is:

$10^{-1} = \frac{1}{10^1} = \frac{1}{10} = .1$ $10^{-3} = \frac{1}{10^3} = \frac{1}{1,000} = .001$

Write the correct decimal number in each blank.

a) $10^{-2} = \frac{1}{10^2} = \frac{1}{100} =$ ______ b) $10^{-5} = \frac{1}{10^5} = \frac{1}{100,000} =$ ______

---

a) .01

b) .00001

31. By examining the table at the right, you can see this fact:

<u>The number of decimal places in each ordinary number equals the absolute value of the exponent</u>.

Using the fact above, convert each of these to an ordinary number.

a) $10^{-7} =$ ______

b) $10^{-8} =$ ______

10^{-1}	= .1
10^{-2}	= .01
10^{-3}	= .001
10^{-4}	= .0001
10^{-5}	= .00001
10^{-6}	= .000001

---

a) .0000001

b) .00000001

32. Using the exponent to determine the number of decimal places, convert each of these to an ordinary number.

a) $10^{-1} =$ ____ b) $10^{-4} =$ ______ c) $10^{-9} =$ ______

---

a) .1 b) .0001 c) .000000001

33. By examining the table at the right, you can see this fact:

.1	=	10^{-1}
.01	=	10^{-2}
.001	=	10^{-3}
.0001	=	10^{-4}
.00001	=	10^{-5}
.000001	=	10^{-6}

The absolute value of the exponent equals the number of decimal places in the ordinary number.

Using the fact above, convert each of these to a power of ten.

a) .0000001 = ________

b) .000000001 = ________

---

a) 10^{-7}

b) 10^{-9}

34. By counting the number of decimal places to find the exponent, convert each of these to a power of ten.

a) .001 = ________ b) .1 = ________ c) .00000001 = ________

---

a) 10^{-3} b) 10^{-1} c) 10^{-8}

11-5 MULTIPLYING AND DIVIDING POWERS OF TEN WITH NEGATIVE EXPONENTS

In this section, we will show that the same laws of exponents can be used for multiplications and divisions involving powers of ten with negative exponents.

35. The multiplications below contain powers of ten with negative exponents. To perform them, we converted the factors to ordinary numbers and then converted the products back to powers of ten.

$$\begin{array}{ccccc} 10^{-1} & \times & 10^{2} & = & 10^{1} \\ \downarrow & & \downarrow & & \uparrow \\ .1 & \times & 100 & = & 10 \end{array} \qquad \begin{array}{ccccc} 10^{-3} & \times & 10^{-2} & = & 10^{-5} \\ \downarrow & & \downarrow & & \uparrow \\ .001 & \times & .01 & = & .00001 \end{array}$$

Notice that the law of exponents for multiplying powers of ten can be used to get the same products. That is:

$$10^{-1} \times 10^{2} = 10^{(-1)+2} = 10^{1} \qquad 10^{-3} \times 10^{-2} = 10^{(-3)+(-2)} = 10^{\square}$$

10^{-5}

36. The law of exponents for multiplying powers of ten is shown again below. It is used with positive, negative, and "0" exponents.

$$10^{a} \times 10^{b} = 10^{a+b}$$

Use the above law to complete these:

a) $10^{5} \times 10^{-3}$ = ________ b) $10^{-1} \times 10^{-8}$ = ________ c) $10^{0} \times 10^{-4}$ = ________

---

a) 10^{2}

b) 10^{-9}

c) 10^{-4}

37. To perform each division below, we converted to ordinary numbers. Each quotient is a power of ten with a negative exponent.

$$\frac{10^2}{10^3} = 10^{-1} \qquad \frac{10^1}{10^4} = 10^{-3}$$

$$\frac{100}{1,000} = .1 \qquad \frac{10}{10,000} = .001$$

Notice that the law of exponents for dividing powers of ten can be used to get the same quotients. That is:

$$\frac{10^2}{10^3} = 10^{2-3} = 10^{-1} \qquad \frac{10^1}{10^4} = 10^{1-4} = 10^{\square}$$

Answer: 10^{-3}

---

38. The divisions below contain powers of ten with negative exponents. To perform them, we converted to ordinary numbers.

$$\frac{10^{-3}}{10^1} = 10^{-4} \qquad \frac{10^{-2}}{10^{-4}} = 10^2$$

$$\frac{.001}{10} = .0001 \qquad \frac{.01}{.0001} = 100$$

Notice again that the law of exponents can be used to get the same quotients. That is:

$$\frac{10^{-3}}{10^1} = 10^{(-3)-1} = 10^{(-3)+(-1)} = 10^{-4}$$

$$\frac{10^{-2}}{10^{-4}} = 10^{(-2)-(-4)} = 10^{(-2)+4} = 10^{\square}$$

Answer: 10^2

---

39. The law of exponents for dividing powers of ten is shown again below. It is used with positive, negative, and "0" exponents.

$$\boxed{\frac{10^a}{10^b} = 10^{a-b}}$$

Use the above law to complete these:

a) $\frac{10^0}{10^3} =$ ________ c) $\frac{10^5}{10^{-1}} =$ ________

b) $\frac{10^{-2}}{10^4} =$ ________ d) $\frac{10^{-5}}{10^{-4}} =$ ________

Answers: a) 10^{-3} b) 10^{-6} c) 10^6 d) 10^{-1}

---

40. Complete: a) $10^5 \times 10^{-5} =$ ________ c) $10^{-5} \times 10^{-5} =$ ________

b) $\frac{10^{-3}}{10^{-3}} =$ ________ d) $\frac{10^3}{10^{-3}} =$ ________

Answers: a) 10^0 (or 1) b) 10^0 (or 1) c) 10^{-10} d) 10^6

SELF-TEST 32 (pages 281-289)

Convert each power of ten to an ordinary number.

1. 10^5 = ______	2. 10^{-1} = ______	3. 10^0 = ______	4. 10^{-4} = ______

Convert each ordinary number to a power of ten.

5. 1,000,000 = ______	6. .01 = ______	7. 10 = ______	8. .000001 = ______

Complete each multiplication and division.

9. $10^5 \times 10^6$ = ______	10. $10^{-8} \times 10^4$ = ______	11. $10^{-4} \times 10^{-7}$ = ______
12. $\frac{10^2}{10^7}$ = ______	13. $\frac{10^{-5}}{10^3}$ = ______	14. $\frac{10^{-7}}{10^{-8}}$ = ______

ANSWERS:

1. 100,000	3. 1	5. 10^6	7. 10^1	9. 10^{11}	12. 10^{-5}
2. .1	4. .0001	6. 10^{-2}	8. 10^{-6}	10. 10^{-4}	13. 10^{-8}
				11. 10^{-11}	14. 10^1

11-6 MULTIPLYING BY POWERS OF TEN WITH POSITIVE OR "0" EXPONENTS

In this section, we will discuss the decimal-point-shift method for multiplying by powers of ten with positive exponents.

41. To multiply by numbers like 10, 100, and 1,000 , we can simply shift the decimal point to the right. The number of 0's in the multiplier tells us how many places to shift it. For example:

10 x 8.25 = 8.25 = 82.5 (Shifted one place. 10 has one "0".)

100 x 9.16 = 9.16 = 916 (Shifted two places. 100 has two 0's.)

1,000 x 43 = 43.000 = 43,000 (Shifted three places. 1,000 has three 0's.)

Use the decimal-point-shift method for these:

a) 10 x 350 = ______ b) 100 x .661 = ______ c) 1,000 x 8.9 = ______

Answers: a) 3,500 b) 66.1 c) 8,900

42. Since $10^1 = 10$, $10^2 = 100$, and $10^3 = 1,000$, we can use the decimal-point-shift method to multiply by those powers of ten. The exponent tells us how many places to shift the decimal point to the right. For example:

10^1 x .56 = .56 = 5.6 (Shifted one place. Exponent is "1".)

10^2 x .39 = .39 = 39 (Shifted two places. Exponent is "2".)

10^3 x 9.8 = 9.800 = 9,800 (Shifted three places. Exponent is "3".)

Use the decimal-point-shift method for these:

a) 10^1 x 470 = ______ b) 10^2 x 6.08 = ______ c) 10^3 x 6.5 = ______

43. The decimal-point-shift method can also be used to multiply by powers of ten with larger exponents. <u>The number of places shifted depends on the exponent</u>. For example:

$10^4 \times 2.56 = 2.5600 = 25,600$ (Shifted <u>four</u> places.)

$10^7 \times .0029 = .0029000 = 29,000$ (Shifted <u>seven</u> places.)

Use the decimal-point-shift method for these:

a) $10^6 \times .00098$ = ________ b) $10^8 \times .000125$ = ________

Answers: a) 4,700 b) 608 c) 6,500

---

44. The decimal-point-shift method can also be used when the power of ten is the second factor. For example:

$775 \times 10^3 = 775.000 = 775,000$

$.094 \times 10^5 = .09400 = 9,400$

Use the decimal-point-shift method for these:

a) 1.5×10^1 = ______ b) $.27 \times 10^4$ = ________ c) $.00007 \times 10^6$ = ________

Answers: a) 980 b) 12,500

---

45. Since $10^0 = 1$, when multiplying by 10^0, the product is <u>identical</u> to the other factor. That is:

$10^0 \times 55 = 55$ $1.39 \times 10^0 = 1.39$

Therefore, when multiplying by 10^0, we do not shift the decimal point. Complete these:

a) $10^0 \times .009$ = ________ b) 27.5×10^0 = ________

Answers: a) 15 b) 2,700 c) 70

Answers: a) .009 b) 27.5

11-7 MULTIPLYING BY POWERS OF TEN WITH NEGATIVE EXPONENTS

In this section, we will discuss the decimal-point-shift method for multiplying by powers of ten with negative exponents.

---

46. To multiply by numbers like .1, .01, and .001, we can simply shift the decimal point to the left. <u>The number of decimal places in the multiplier tells us how many places to shift it</u>. For example:

$.1 \times 56 = 5.6. = 5.6$ (Shifted <u>one</u> place; .1 has <u>one</u> decimal place.)

$.01 \times 2,900 = 2,9.00. = 29$ (Shifted <u>two</u> places; .01 has <u>two</u> decimal places.)

$.001 \times 3.4 = .003.4 = .0034$ (Shifted <u>three</u> places; .001 has <u>three</u> decimal places.)

Use the decimal-point-shift method for these.

a) $.1 \times 350$ = ________ b) $.01 \times 2.9$ = ________ c) $.001 \times .03$ = ________

Answers: a) 35 b) .029 c) .00003

47. Since $10^{-1} = .1$, $10^{-2} = .01$, and $10^{-3} = .001$, we can use the decimal-point-shift method to multiply by those powers of ten. The absolute value of the exponent tells us how many places to shift the decimal point to the left. For example:

$10^{-1} \times 3.8 = 3.8 = .38$ (Shifted one place. Exponent is "-1".)

$10^{-2} \times 7,500 = 7,500. = 75$ (Shifted two places. Exponent is "-2".)

$10^{-3} \times .91 = 000.91 = .00091$ (Shifted three places. Exponent is "-3".)

Use the decimal-point-shift method for these:

a) $10^{-1} \times 250 =$ ______ b) $10^{-2} \times 487 =$ ______ c) $10^{-3} \times 6.4 =$ ______

---

a) 25
b) 4.87
c) .0064

48. The decimal-point-shift method can also be used to multiply by powers of ten with larger negative exponents. The number of places shifted depends on the absolute value of the exponent. For example:

$10^{-4} \times 36 = 0036. = .0036$ (Shifted four places.)

$10^{-7} \times 90,000,000 = 90,000,000. = 9$ (Shifted seven places.)

Use the decimal-point-shift method for these:

a) $10^{-6} \times 391 =$ ______ b) $10^{-8} \times 45,000,000 =$ ______

---

a) .000391
b) .45

49. The same method can be used when the power of ten is the second factor. For example:

$510 \times 10^{-3} = 510. = .51$

$1,200 \times 10^{-5} = 01,200. = .012$

Use the decimal-point-shift method for these:

a) $58 \times 10^{-1} =$ ______ b) $700 \times 10^{-4} =$ ______ c) $2.4 \times 10^{-6} =$ ______

---

a) 5.8
b) .07
c) .0000024

50. Don't confuse the direction of the decimal-point-shift when multiplying by powers of ten.

If the exponent is positive, we shift to the right since we are multiplying by 10, 100, 1000, and so on. That is:

$39 \times 10^1 = 39.0 = 390$ $5.3 \times 10^3 = 5.300 = 5,300$

If the exponent is negative, we shift to the left since we are multiplying by .1, .01, .001, and so on. That is:

$39 \times 10^{-1} = 3.9. = 3.9$ $5.3 \times 10^{-3} = 005.3 = .0053$

Use the decimal-point-shift method for these:

a) $72 \times 10^2 =$ ______ c) $.48 \times 10^4 =$ ______

b) $72 \times 10^{-2} =$ ______ d) $.48 \times 10^{-4} =$ ______

a) 7,200
b) .72
c) 4,800
d) .000048

11-8 SCIENTIFIC NOTATION

Any number can be written in a special form called "scientific notation". Though it is called "scientific" notation because it is an especially useful way of writing the very large and very small numbers that occur in science, it is a useful way of writing numbers of any size. We will discuss "scientific notation" in this section.

51. In the table at the right, each whole number is written in scientific notation. Notice these two points about the multiplication called "scientific notation":

1) The first factor is a number between 1 and 10.

2) The second factor is a power of ten.

WHOLE NUMBER	SCIENTIFIC NOTATION
520,000	5.2×10^5
37,000	3.7×10^4
1,600	1.6×10^3
480	4.8×10^2
95	9.5×10^1
↑ A number between 1 and 10	↑ A power of ten

Which of the following are written in scientific notation? That is, which are multiplications of a number between 1 and 10 and a power of ten? ______________

a) 6.7×10^2 b) 25×10^8 c) $.41 \times 10^5$ d) 9.3×10^6

Only (a) and (d)

52. In the table at the right, each decimal number is written in scientific notation. Notice the same points about "scientific notation":

1) The first factor is a number between 1 and 10.

2) The second factor is a power of ten.

DECIMAL NUMBER	SCIENTIFIC NOTATION
.47	4.7×10^{-1}
.018	1.8×10^{-2}
.0092	9.2×10^{-3}
.00053	5.3×10^{-4}
.000066	6.6×10^{-5}
↑ A number between 1 and 10	↑ A power of ten

Which of the following are written in scientific notation? That is, which are multiplications of a number between 1 and 10 and a power of ten? ______________

a) $.34 \times 10^{-4}$ b) 9.6×10^{-1} c) 44×10^{-7} d) 2.8×10^{-5}

Only (b) and (d)

53. To convert a number written in scientific notation to its ordinary form, we simply perform the multiplication by the decimal-point-shift method. For example:

$$2.7 \times 10^3 = 2.700. = 2,700$$

$$9.6 \times 10^{-5} = .00009.6 = .000096$$

Convert each of these to ordinary numbers.

a) 1.3×10^6 = ______________ b) 5.8×10^{-1} = ______________

a) 1,300,000
b) .58

54. Each number below is in scientific notation because the first factor is a number between 1 and 10. Convert each to an ordinary number.

a) 3×10^4 = ____________ c) 2.54×10^2 = ____________

b) 3×10^{-4} = ____________ d) 2.54×10^{-2} = ____________

a) 30,000
b) .0003
c) 254
d) .0254

55. A number written in scientific notation can stand for a very large or very small number. For example:

$$5.1 \times 10^7 = 5.1000000_\wedge = 51{,}000{,}000$$

$$3.9 \times 10^{-9} = {}_\wedge000000003.9 = .0000000039$$

Convert each of these to an ordinary number.

a) 7.13×10^9 = ________ b) 5×10^{-7} = ________

a) 7,130,000,000	b) .0000005

11-9 CONVERTING NUMBERS LARGER THAN "1" TO SCIENTIFIC NOTATION

In this section, we will discuss a method for converting numbers larger than "1" to scientific notation.

---

56. To convert a number larger than "1" to scientific notation, we must find both the first factor and the exponent for the power of ten. Since the first factor must be a number between 1 and 10, we can find it by writing a caret (‸) after the first digit on the left. That is:

For 7‸3,000 , the first factor is 7.3 .

For 9‸00,000 , the first factor is 9 .

For 1‸8.4 , the first factor is 1.84 .

For each number below, write the first factor of its scientific-notation form.

a) 2,400 ________ b) 6,000,000 ________ c) 85.08 ________ d) 195 ________

a) 2.4
b) 6
c) 8.508
d) 1.95

57. Having found the first factor, we can find the exponent of the power of ten by counting the number of places from the caret to the decimal point. That is:

7‸3,000. = 7.3×10^4 (Four places to the decimal point)

8‸00,000. = 8×10^5 (Five places to the decimal point)

9‸1.7 = 9.17×10^1 (One place to the decimal point)

By counting the number of places from the caret to the decimal point, write the correct exponent in each box below.

a) 2‸80. = $2.8 \times 10^{\square}$ c) 3‸00,000. = $3 \times 10^{\square}$

b) 5‸250. = $5.25 \times 10^{\square}$ d) 6‸8.7 = $6.87 \times 10^{\square}$

a) 2.8×10^2
b) 5.25×10^3
c) 3×10^5
d) 6.87×10^1

58. Write each number in scientific notation.

a) 47,000 = ________ x ________ c) 88.55 = ________ x ________

b) 200 = ________ x ________ d) 9,160,000 = ________ x ________

a) 4.7×10^4	b) 2×10^2	c) 8.855×10^1	d) 9.16×10^6

59. When the ordinary number is quite large, the exponent is quite large. For example:

$$7_{\wedge}1{,}000{,}000{,}000 = 7.1 \times 10^{10}$$

Write each number in scientific notation.

a) 3,000,000,000 = ______ x ______ b) 259,000,000,000 = ______ x ______

a) 3×10^9	b) 2.59×10^{11}

11-10 CONVERTING NUMBERS SMALLER THAN "1" TO SCIENTIFIC NOTATION

In this section, we will discuss a method for converting numbers smaller than "1" to scientific notation.

---

60. To convert a number smaller than "1" to scientific notation, we must again find both the first factor and the exponent for the power of ten. Since the first factor must be <u>a number between 1 and 10</u>, we can find it by writing a caret <u>after the first non-zero digit in the number</u>. That is:

For .06‸8 , the first factor is 6.8 .

For .1‸95 , the first factor is 1.95 .

For .00007‸ , the first factor is 7 .

For each number below, write the first factor of its scientific-notation form.

a) .00034 ______ b) .008 ______ c) .0175 ______ d) .000001 ______

---

61. Having found the first factor, we can find the exponent of the power of ten <u>by counting the number of places from the caret to the decimal point</u>. Since we count <u>to the left</u>, the exponent is <u>negative</u>. That is:

.06‸8 = 6.8×10^{-2} (<u>Two</u> places to the decimal point)

.1‸95 = 1.95×10^{-1} (<u>One</u> place to the decimal point)

.00007‸ = 7×10^{-5} (<u>Five</u> places to the decimal point)

By counting the number of places from the caret to the decimal point, write the correct exponent in each box below.

a) .003‸5 = $3.5 \times 10^{\square}$ c) .000002‸ = $2 \times 10^{\square}$

b) .7‸08 = $7.08 \times 10^{\square}$ d) .0006‸49 = $6.49 \times 10^{\square}$

Answers (60)
a) 3.4
b) 8
c) 1.75
d) 1

---

62. Write each number in scientific notation.

a) .00025 = ______ x ______ c) .06718 = ______ x ______

b) .4 = ______ x ______ d) .0000019 = ______ x ______

Answers (61)
a) 3.5×10^{-3}
b) 7.08×10^{-1}
c) 2×10^{-6}
d) 6.49×10^{-4}

---

a) 2.5×10^{-4}	c) 6.718×10^{-2}
b) 4×10^{-1}	d) 1.9×10^{-6}

63. When the original number is quite small, the absolute value of the exponent is quite large. For example:

$$.000000009_{\wedge}47 = 9.47 \times 10^{-9}$$

Write each number in scientific notation.

a) .000000044 = ________ x ________ b) .0000000006 = ________ x ________

a) 4.4×10^{-8}

b) 6×10^{-10}

64. In scientific notation, the exponent can be either positive or negative.

If we count "to the right" from the caret to the decimal point, the exponent is positive. For example:

$$9_{\wedge}1,300 = 9.13 \times 10^{4}$$

If we count "to the left" from the caret to the decimal point, the exponent is negative. For example:

$$.009_{\wedge}13 = 9.13 \times 10^{-3}$$

Write each number in scientific notation.

a) 5,600 = ________ x ________ c) 300,000,000 = ________ x ________

b) .056 = ________ x ________ d) .000000003 = ________ x ________

a) 5.6×10^{3} b) 5.6×10^{-2} c) 3×10^{8} d) 3×10^{-9}

11-11 THE USE OF SCIENTIFIC NOTATION IN SCIENCE

Scientific notation is frequently used to express the very large and very small numbers that occur in science. We will give some examples in this section.

65. a) The average distance from the earth to the sun is 9.3×10^{7} miles. Express the distance as an ordinary number.

________________ miles

b) The size of a virus is 6×10^{-5} millimeter. Express the size as an ordinary number.

________________ millimeter

a) 93,000,000 miles

b) .00006 millimeter

66. a) The speed of light is 299,800,000 meters per second. Express the speed in scientific notation.

________ x ________ meters per second

b) The radius of an atom is .00000000209 inch. Express the radius in scientific notation.

________ x ________ inch

a) 2.998×10^{8} meters per second b) 2.09×10^{-9} inch

67. a) The charge on an electron is 4.8×10^{-10} electrostatic units. Express the charge as an ordinary number.

______________ electrostatic units

b) A starfish lays an average of 2.48×10^{6} eggs annually. Express that number as an ordinary number.

______________ eggs

a) .00000000048 electrostatic units

b) 2,480,000 eggs

68. a) The diameter of a large molecule is .00000017 centimeter. Express the diameter in scientific notation.

_______ x _______ centimeter

b) One year contains 31,600,000 seconds. Express that number in scientific notation.

_______ x _______ seconds.

a) 1.7×10^{-7} centimeter b) 3.16×10^{7} seconds

11-12 CALCULATORS AND SCIENTIFIC NOTATION

Many pocket calculators report very large and very small answers in scientific notation. We will discuss that use of scientific notation in this section.

69. When using a pocket calculator, you can get an answer that shows a number between 1 and 10 on the left side and another number on the right. For example:

Left Side	Right Side
6.27	8
1.96	-7

The answers above are really reported in scientific notation with the exponent of the power of ten on the right side. That is:

The top answer is 6.27×10^{8} or 627,000,000

The bottom answer is 1.96×10^{-7} or ______________

.000000196

70. When an answer is given in scientific notation, sometimes a "0" is printed in front of a one-digit exponent. For example:

Left Side	Right Side
2.54	08
9.33	-09

The "0" should be ignored when writing the answer in scientific notation. That is:

The top answer is 2.54×10^{8}. The bottom answer is _______ x _______.

9.33×10^{-9}

71. Two calculator answers are shown at the right. Express each in scientific notation and as an ordinary number.

Left Side	Right Side
3.8	10
7.5	-09

a) Top answer = ________ x ________ = ____________________

b) Bottom answer = ________ x ________ = ____________________

a) 3.8×10^{10} = 38,000,000,000   b) 7.5×10^{-9} = .0000000075

SELF-TEST 33 (pages 289-297)

Convert each to an ordinary number.

1. 5.6×10^4 = ____________
2. 8×10^7 = ____________
3. 3.77×10^{-1} = ____________
4. 1.5×10^{-5} = ____________

Write each number in scientific notation.

5. 17.4 = ____________
6. .586 = ____________
7. 29,300 = ____________
8. .0044 = ____________
9. .0000006 = ____________
10. 750,000,000 = ____________

11. A light-year (the distance light travels in one year) is 5.87×10^{12} miles. Express that distance as an ordinary number.

____________________ miles

12. One cycle of a television broadcast signal takes .00000000481 second. Express that time in scientific notation.

____________________ second

Write each calculator answer in scientific notation and then convert it to an ordinary number.

13.

Left Side	Right Side
2.67	-6

________ x ________ = ____________

14.

Left Side	Right Side
8.43	05

________ x ________ = ____________

ANSWERS:

1. 56,000
2. 80,000,000
3. .377
4. .000015
5. 1.74×10^1
6. 5.86×10^{-1}
7. 2.93×10^4
8. 4.4×10^{-3}
9. 6×10^{-7}
10. 7.5×10^8
11. 5,870,000,000,000 miles
12. 4.81×10^{-9} second
13. 2.67×10^{-6} = .00000267
14. 8.43×10^5 = 843,000

11-13 CONVERTING OTHER MULTIPLICATIONS TO SCIENTIFIC NOTATION

In this section, we will discuss the method for converting multiplications like 415×10^7 or $.0028 \times 10^{-5}$ to scientific notation.

---

72. The multiplication 387×10^4 is not in scientific notation because 387 is not a number between 1 and 10. We have converted 387×10^4 to scientific notation below.

$$387 \times 10^4 = 3.87 \times 10^2 \times 10^4 = 3.87 \times 10^6$$

Note: 1) First we converted 387 to scientific notation.

2) Then we combined 10^2 and 10^4 by multiplying.

Following the steps above, convert the multiplication below to scientific notation.

91.6×10^6 = ________ x ________ $\times 10^6$ = ________ x ________

$9.16 \times 10^1 \times 10^6 = 9.16 \times 10^7$

73. Following the example, convert the other multiplication to scientific notation.

$256 \times 10^{-5} = 2.56 \times 10^2 \times 10^{-5} = 2.56 \times 10^{-3}$

81.9×10^{-8} = ________ x ________ $\times 10^{-8}$ = ________ x ________

$8.19 \times 10^1 \times 10^{-8} = 8.19 \times 10^{-7}$

74. Following the example, convert the other multiplication to scientific notation.

$.329 \times 10^{-7} = 3.29 \times 10^{-1} \times 10^{-7} = 3.29 \times 10^{-8}$

$.08 \times 10^{12}$ = ________ x ________ $\times 10^{12}$ = ________ x ________

$8 \times 10^{-2} \times 10^{12} = 8 \times 10^{10}$

75. Convert each multiplication below to scientific notation.

a) $2,700 \times 10^8$ = ____________________

b) $.0059 \times 10^9$ = ____________________

c) $.07 \times 10^{-7}$ = ____________________

d) $.0003 \times 10^7$ = ____________________

a) 2.7×10^{11}

b) 5.9×10^6

c) 7×10^{-9}

d) 3×10^3

11-14 MULTIPLICATIONS INVOLVING NUMBERS IN SCIENTIFIC NOTATION

In this section, we will discuss the method for multiplying when one or both factors are numbers in scientific notation.

---

76. To multiply a number in scientific notation by an ordinary number, we multiply the two ordinary-number factors. For example:

$$4 \times (2 \times 10^5) = (4 \times 2) \times 10^5 = 8 \times 10^5$$

$$3 \times (3 \times 10^{-8}) = (3 \times 3) \times 10^{-8} = \_\_\_\_\_ \times \_\_\_\_\_$$

Answer: 9×10^{-8}

77. We usually convert the product to scientific notation if the first factor is not a number between 1 and 10. For example:

$$20 \times (7 \times 10^6) = 140 \times 10^6 = 1.4 \times 10^2 \times 10^6 = 1.4 \times 10^8$$

$$.13 \times (5 \times 10^{-5}) = .65 \times 10^{-5} = 6.5 \times 10^{-1} \times 10^{-5} = 6.5 \times 10^{-6}$$

Complete. Convert each product to scientific notation.

a) $31 \times (4 \times 10^7) =$ __________

b) $.02 \times (3 \times 10^{-8}) =$ __________

Answers: a) 1.24×10^9 b) 6×10^{-10}

78. Convert each product to scientific notation.

a) $8 \times (3.6 \times 10^4) =$ __________

b) $.04 \times (8.2 \times 10^{-3}) =$ __________

Answers: a) 2.88×10^5 b) 3.28×10^{-4}

79. To multiply two numbers in scientific notation, we multiply the two ordinary-number factors and the two powers of ten. For example:

$$(3 \times 10^4) \times (2 \times 10^5) = (3 \times 2) \times (10^4 \times 10^5) = 6 \times 10^9$$

$$(2 \times 10^{-3}) \times (4 \times 10^{-7}) = (2 \times 4) \times (10^{-3} \times 10^{-7}) = \_\_\_\_\_ \times \_\_\_\_\_$$

Answer: 8×10^{-10}

80. After multiplying below, we converted the product to scientific notation. Convert the other product to scientific notation.

$$(5 \times 10^{-2}) \times (3.1 \times 10^5) = 15.5 \times 10^3 = 1.55 \times 10^1 \times 10^3 = 1.55 \times 10^4$$

$(9.2 \times 10^{-7}) \times (8 \times 10^4) =$ __________

Answer: 7.36×10^{-2}

81. To get the product below, we multiplied 2.54 and 1.78 and got 4.5212. The product is in scientific notation.

$$(2.54 \times 10^6) \times (1.78 \times 10^8) = 4.5212 \times 10^{14}$$

Ordinarily, we round the first factor in scientific notation to hundredths (or 2 decimal places). That is:

$$4.52\!\downarrow\!12 \times 10^{14} \text{ is rounded to } 4.52 \times 10^{14}.$$

Continued on following page.

81. Continued

Round each first factor below to <u>hundredths</u> (or 2 decimal places).

a) 7.8549×10^{12} rounds to ________ $\times 10^{12}$.

b) 3.1488×10^{-10} rounds to ________ $\times 10^{-10}$.

---

82. Convert the product to scientific notation with the first factor rounded to hundredths (or 2 decimal places).

a) 7.85×10^{12}
b) 3.15×10^{-10}

$(8.11 \times 10^{9}) \times (6.55 \times 10^{-2}) =$

5.31×10^{8}

11-15 DIVISIONS INVOLVING NUMBERS IN SCIENTIFIC NOTATION

In this section, we will discuss the method for dividing when one or both terms are numbers in scientific notation.

---

83. To divide a number in scientific notation by an ordinary number, we divide the two ordinary numbers. For example:

$$\frac{6 \times 10^{7}}{2} = \frac{6}{2} \times 10^{7} = 3 \times 10^{7}$$

$$\frac{8 \times 10^{-5}}{4} = \frac{8}{4} \times 10^{-5} = \text{\_\_\_\_\_\_} \times \text{\_\_\_\_\_\_}$$

2×10^{-5}

---

84. Following the example, complete the other division.

$$\frac{5.4 \times 10^{10}}{3} = \frac{5.4}{3} \times 10^{10} = 1.8 \times 10^{10}$$

$$\frac{9.2 \times 10^{-8}}{4} = \frac{9.2}{4} \times 10^{-8} = \text{\_\_\_\_\_\_} \times \text{\_\_\_\_\_\_}$$

2.3×10^{-8}

---

85. Notice how we converted the quotient below to scientific notation. Convert the other quotient to scientific notation.

$$\frac{3.01 \times 10^{7}}{7} = .43 \times 10^{7} = 4.3 \times 10^{-1} \times 10^{7} = 4.3 \times 10^{6}$$

$$\frac{4.88 \times 10^{-6}}{8} = \text{\_\_\_\_\_\_\_\_\_\_\_\_\_\_\_\_}$$

6.1×10^{-7}

86. Convert each quotient to scientific notation.

a) $\frac{6.48 \times 10^4}{20} =$ ______________________

b) $\frac{1.75 \times 10^{-7}}{.05} =$ ______________________

---

a) 3.24×10^3

b) 3.5×10^{-6}

87. To divide two numbers in scientific notation, we divide the two ordinary numbers and the two powers of ten. For example:

$$\frac{8 \times 10^9}{2 \times 10^4} = \frac{8}{2} \times \frac{10^9}{10^4} = 4 \times 10^5$$

$$\frac{9 \times 10^{-2}}{3 \times 10^{-8}} = \frac{9}{3} \times \frac{10^{-2}}{10^{-8}} = \_\_\_\_\_\_ \times \_\_\_\_\_\_$$

3×10^6

88. Following the example, complete the other division.

$$\frac{2.48 \times 10^{10}}{1.24 \times 10^{20}} = \frac{2.48}{1.24} \times \frac{10^{10}}{10^{20}} = 2 \times 10^{-10}$$

$$\frac{8.88 \times 10^{-16}}{2.22 \times 10^{-12}} = \_\_\_\_\_\_\_\_\_\_\_\_$$

4×10^{-4}

11-16 APPLIED PROBLEMS INVOLVING NUMBERS IN SCIENTIFIC NOTATION

In this section, we will discuss some applied problems involving numbers in scientific notation.

---

89. If 1 mole of carbon dioxide contains 6.02×10^{23} molecules, how many molecules are there in 5 moles of carbon dioxide?

<u>Hint</u>: To solve the problem, you can solve the proportion below. Express the answer in scientific notation.

$$\frac{1 \text{ mole}}{6.02 \times 10^{23} \text{ molecules}} = \frac{5 \text{ moles}}{x \text{ molecules}} \quad \text{or} \quad \frac{1}{6.02 \times 10^{23}} = \frac{5}{x}$$

3.01×10^{24} molecules, from:

$x = 5 \times (6.02 \times 10^{23})$

90. The nearest star is 4.3 light-years from earth. If one light-year is 5.87×10^{12} miles, how many miles from earth is that star?

<u>Hint:</u> To solve the problem, you can solve the proportion below. Express the answer in scientific notation with the first factor rounded to hundredths.

$$\frac{1 \text{ light-year}}{5.87 \times 10^{12} \text{ miles}} = \frac{4.3 \text{ light-years}}{x \text{ miles}} \quad \text{or} \quad \frac{1}{5.87 \times 10^{12}} = \frac{4.3}{x}$$

91. If 22.4 liters of a gas contain 6.02×10^{23} molecules, how many molecules are there in 2 liters of the same gas?

Hint: To solve the problem, you can solve the proportion below. Express the answer in scientific notation with the first factor rounded to hundredths.

$$\frac{22.4 \text{ liters}}{6.02 \times 10^{23} \text{ molecules}} = \frac{2 \text{ liters}}{8 \text{ molecules}} \quad \text{or} \quad \frac{22.4}{6.02 \times 10^{23}} = \frac{2}{x}$$

> 2.52×10^{13} miles, from:
>
> $x = 4.3 \times (5.87 \times 10^{12})$

---

92. Using the formula below, we found V_2 when $C_1 = 4 \times 10^{-3}$, $V_1 = 100$, and $C_2 = 2 \times 10^{-4}$.

$$V_2 = \frac{C_1V_1}{C_2} = \frac{(4 \times 10^{-3})(100)}{2 \times 10^{-4}} = \frac{400 \times 10^{-3}}{2 \times 10^{-4}} = 200 \times 10^1 = 2,000$$

Note: 1) First we multiplied (4×10^{-3}) by 100.

2) Then we divided (400×10^{-3}) by (2×10^{-4}).

3) We wrote the answer as an ordinary number.

Using the same formula and the same steps, find V_2 when $C_1 = 2 \times 10^{-3}$, $V_1 = 50$, and $C_2 = 4 \times 10^{-4}$.

$$V_2 = \frac{C_1V_1}{C_2} = \underline{\hspace{8cm}}$$

> 5.38×10^{22} molecules, from:
>
> $x = \frac{12.04 \times 10^{23}}{22.4}$

> $V_2 = 250$, from:
>
> $\frac{100 \times 10^{-3}}{4 \times 10^{-4}}$

SELF-TEST 34 (pages 298-303)

Convert each multiplication to scientific notation.

1. $.259 \times 10^8 =$ ____________

2. $914 \times 10^{-7} =$ ____________

Write each product and quotient in scientific notation.

3. $5 \times (7 \times 10^6) =$

4. $(8 \times 10^{-7}) \times (9 \times 10^{10}) =$

5. $\dfrac{2 \times 10^{-8}}{5} =$

6. $\dfrac{8 \times 10^{12}}{2 \times 10^4} =$

We performed the multiplication and division below. Write each answer in scientific notation with the first factor rounded to hundredths.

7. $5.8 \times (2.96 \times 10^{10}) = 17.168 \times 10^{10} =$ ____________

8. $\dfrac{1.29 \times 10^{20}}{2.4 \times 10^{12}} = .5375 \times 10^8 =$ ____________

9. A star is 10 light years from earth. If one light-year is 5.87×10^{12} miles, how far is that star from earth? (Note: Express the answer in scientific notation.)

ANSWERS:

1. 2.59×10^7
2. 9.14×10^{-5}
3. 3.5×10^7
4. 7.2×10^4
5. 4×10^{-9}
6. 4×10^8
7. 1.72×10^{11}
8. 5.38×10^7
9. 5.87×10^{13} miles

Unit 12 PERCENTS AND EQUATIONS

In this unit, we will review the relationship between percents, fractions, and decimal numbers. The equation-solving method is used to solve the three types of percent problems. The formula for simple interest is discussed and used.

12-1 CONVERTING PERCENTS TO FRACTIONS AND DECIMAL NUMBERS

In this section, we will define "percents" and show the methods for converting percents to fractions and decimal numbers. A table of common percent-fraction-decimal equivalents is given.

1. The word "percent" means "hundredth". Therefore, any percent is equal to a "hundredths" decimal fraction with the percent-value as its numerator. For example:

$$83\% = \frac{83}{100} \qquad 12\frac{1}{2}\% = \frac{12\frac{1}{2}}{100} \qquad 5.75\% = \frac{5.75}{100}$$

$$100\% = \frac{100}{100} \qquad 240\% = \frac{240}{100} \qquad .16\% = \frac{.16}{100}$$

Convert each percent to a "hundredths" decimal fraction.

a) $1\% =$ _____ b) $66\frac{2}{3}\% =$ _____ c) $150\% =$ _____ d) $.5\% =$ _____

2. Some whole-number percents can be converted to simple fractions by reducing the "hundredths" decimal fraction to lowest terms. For example:

$$50\% = \frac{50}{100} = \frac{1}{2} \qquad 80\% = \frac{80}{100} = \frac{4}{5}$$

Convert each percent to a simple fraction. a) $25\% =$ _____ b) $10\% =$ _____

Answers to 1: a) $\frac{1}{100}$ b) $\frac{66\frac{2}{3}}{100}$ c) $\frac{150}{100}$ d) $\frac{.5}{100}$

3. Some mixed-number percents can be converted to simple fractions by converting the numerator to an improper fraction and then dividing that fraction by 100. For example:

$$33\frac{1}{3}\% = \frac{33\frac{1}{3}}{100} = \frac{\frac{100}{3}}{100} = \frac{\overset{1}{\cancel{100}}}{3} \times \frac{1}{\underset{1}{\cancel{100}}} = \frac{1}{3}$$

$$66\frac{2}{3}\% = \frac{66\frac{2}{3}}{100} = \frac{\frac{200}{3}}{100} = \frac{200}{3} \times \frac{1}{100} = \underline{\qquad}$$

Answers to 2: a) $\frac{1}{4}$ b) $\frac{1}{10}$

Answer to 3: $\frac{2}{3}$

4. Any percent that is a multiple of 100% equals a whole number. That is:

$100\% = \frac{100}{100} = 1$ $\qquad$ $200\% = \frac{200}{100} = 2$ $\qquad$ $500\% = \frac{500}{100} =$ ______

5

5. To convert each percent below to a decimal number, we converted it to a decimal fraction and then divided the numerator by 100. (Note: To divide by 100, we can simply shift the decimal point two places to the left.)

$83\% = \frac{83}{100} = 83. = .83$

$172\% = \frac{172}{100} = 172. = 1.72$

$6.5\% = \frac{6.5}{100} = 06.5 = .065$

$.19\% = \frac{.19}{100} = 00.19 = .0019$

The same conversions can be made directly <u>by simply shifting the decimal point two places to the left and dropping the percent sign</u>. That is:

$83\% = 83. = .83$ $\qquad$ $6.5\% = 06.5 = .065$

$172\% = 172. = 1.72$ $\qquad$ $.19\% = 00.19 = .0019$

Convert each percent below to a decimal number.

a) 7% = ______ b) 215% = ______ c) 17.5% = ______ d) .05% = ______

a) .07
b) 2.15
c) .175
d) .0005

6. After converting some percents to decimal numbers, we can drop one or two final 0's after the decimal point. For example:

$30\% = 30. = .30$ or $.3$

$180\% = 180. = 1.80$ or 1.8

$400\% = 400. = 4.00$ or 4

Convert each percent to a decimal number.

a) 90% = ______ b) 200% = ______ c) 2.7% = ______ d) .01% = ______

a) .9
b) 2
c) .027
d) .0001

7. To convert a mixed-number percent to a decimal number, we convert it to a decimal-number percent first. For example:

$12\frac{1}{2}\% = 12.5\% = .125$ $\qquad$ $9\frac{3}{4}\% = 9.75\% = .0975$

Convert each percent to a decimal number.

a) $11\frac{3}{4}\% =$ ______ b) $5\frac{1}{2}\% =$ ______

a) .1175
b) .055

8. Percents smaller than 1% are sometimes written as fractions. To convert them to decimal numbers, we convert them to decimal-number percents first. For example:

$\frac{1}{2}\% = .5\% = .005$ $\frac{3}{4}\% = .75\% = .0075$

Convert each percent to a decimal number.

a) $\frac{1}{4}\% =$ ______ b) $\frac{4}{5}\% =$ ______

---

a) .0025
b) .008

9. To avoid confusing percents smaller than 1% with percents larger than 1%, we name them in a special way. That is:

.1% is called "one tenth of 1%"

$\frac{3}{4}\%$ is called "three-fourths of 1%"

Write the percent corresponding to each word name.

a) fifteen hundredths of 1% = ______ b) one half of 1% = ______

---

a) .15%
b) $\frac{1}{2}\%$

10. Some common percent-fraction-decimal equivalents are given in the table below. Because of their usefulness, they should be memorized.

$25\% = \frac{1}{4} = .25$	$33\frac{1}{3}\% = \frac{1}{3} = .33$	$20\% = \frac{1}{5} = .2$	$10\% = \frac{1}{10} = .1$
$50\% = \frac{1}{2} = .5$	$66\frac{2}{3}\% = \frac{2}{3} = .67$	$40\% = \frac{2}{5} = .4$	$30\% = \frac{3}{10} = .3$
$75\% = \frac{3}{4} = .75$		$60\% = \frac{3}{5} = .6$	$70\% = \frac{7}{10} = .7$
		$80\% = \frac{4}{5} = .8$	$90\% = \frac{9}{10} = .9$

Note: The decimal equivalents for $33\frac{1}{3}\%$ and $66\frac{2}{3}\%$ are rounded to the nearest hundredth.

From memory, convert each percent to a fraction.

a) 25% = ______ b) 80% = ______ c) $66\frac{2}{3}\%$ = ______ d) 30% = ______

---

a) $\frac{1}{4}$ b) $\frac{4}{5}$ c) $\frac{2}{3}$ d) $\frac{3}{10}$

12-2 CONVERTING DECIMAL NUMBERS TO PERCENTS

In this section, we will discuss a method for converting decimal numbers to percents.

11. Because "percent" means "hundredth", any "hundredths" decimal fraction can be converted directly to a percent. For example:

$$\frac{9}{100} = 9\% \qquad \frac{60}{100} = 60\% \qquad \frac{200}{100} = 200\%$$

Convert to a percent. a) $\frac{59}{100}$ = ________ b) $\frac{1}{100}$ = ________ c) $\frac{100}{100}$ = ________

a) 59%
b) 1%
c) 100%

12. Any "hundredths" decimal number can be converted to a percent by converting it to a "hundredths" decimal fraction first. For example:

$$.05 = \frac{5}{100} = 5\% \qquad 1.37 = \frac{137}{100} = 137\%$$

The same conversions can be made directly <u>by simply shifting the decimal point two places to the right and adding the percent sign</u>. That is:

$$.05 = .05_{\curvearrowright} = 5\% \qquad 1.37 = 1.37_{\curvearrowright} = 137\%$$

Convert to a percent. a) .83 = ________ b) .01 = ________ c) 2.96 = ________

a) 83%
b) 1%
c) 296%

13. Any "tenths" decimal number can be converted to a percent by converting it to a "hundredths" decimal fraction first. For example:

$$.4 = \frac{4}{10} = \frac{40}{100} = 40\% \qquad 1.8 = \frac{18}{10} = \frac{180}{100} = 180\%$$

The decimal-point-shift method can be used for the same conversions. That is:

$$.4 = .40_{\curvearrowright} = 40\% \qquad 1.8 = 1.80_{\curvearrowright} = 180\%$$

Convert to a percent. a) .1 = ________ b) .9 = ________ c) 2.3 = ________

a) 10%
b) 90%
c) 230%

14. Any whole-number can be converted to a percent by converting it to a "hundredths" decimal fraction first. For example:

$$1 = \frac{100}{100} = 100\% \qquad 3 = \frac{300}{100} = 300\%$$

The decimal-point-shift method can be used for the same conversions. That is:

$$1 = 1.00_{\curvearrowright} = 100\% \qquad 3 = 3.00_{\curvearrowright} = 300\%$$

Convert to a percent. a) 2 = ________ b) 4 = ________ c) 7 = ________

a) 200%
b) 400%
c) 700%

15. The decimal-point-shift method can be used to convert "thousandths" and "ten-thousandths" decimal numbers to percents. For example:

.497 = .49 7 = 49.7% .0975 = .09 75 = 9.75%

.003 = .00 3 = .3% .0018 = .00 18 = .18%

Convert each decimal number to a percent.

a) .185 = ______ b) .001 = ______ c) .0688 = ______ d) .0005 = ______

a) 18.5% b) .1% c) 6.88% d) .05%

12-3 CONVERTING FRACTIONS TO PERCENTS

In this section, we will discuss the methods for converting fractions to percents. Though the decimal-fraction method is shown, the division method is emphasized because it is the general method. A table of common fraction-percent equivalents is given.

---

16. Some fractions can be converted to percents by converting them to equivalent "hundredths" decimal fractions. For example:

$$\frac{1}{2} = \frac{50}{100} = 50\% \qquad \frac{3}{4} = \frac{75}{100} = 75\% \qquad \frac{2}{5} = \frac{40}{100} = 40\%$$

Using the decimal-fraction method, complete these conversions.

a) $\frac{1}{10} = \frac{\square}{100} =$ ______ b) $\frac{11}{20} = \frac{\square}{100} =$ ______ c) $\frac{22}{25} = \frac{\square}{100} =$ ______

---

a) $\frac{10}{100} = 10\%$

b) $\frac{55}{100} = 55\%$

c) $\frac{88}{100} = 88\%$

17. Each fraction below has been converted to a percent by the decimal-fraction method.

$$\frac{1}{4} = \frac{25}{100} = 25\% \qquad \frac{3}{5} = \frac{60}{100} = 60\%$$

The same conversions can be made by converting the fractions to decimal numbers first. To do so, we divide. An example is shown. Complete the other conversion.

$$\frac{1}{4} = 4\overline{)1.00} = 25\%$$

```
   .25
4)1.00
  -8
   20
  -20
```

$$\frac{3}{5} = 5\overline{)3.} = \text{______}$$

.6 or .60

$5\overline{)3.0}$ = 60%

18. The "division" method is the general method for converting a fraction to a percent. We used it for one conversion below. Use it for the other conversion.

$$\frac{12}{25} = 25\overline{)12.00} = 48\%$$

```
     .48
25)12.00
  -100
    200
   -200
```

$$\frac{3}{20} = 20\overline{)3.} = \text{______}$$

```
    .15
20)3.00 = 15%
  -20
   100
  -100
```

19. We used the division method to convert $\frac{1}{3}$ to a mixed-number percent below. Notice that we terminated the division by writing the "hundredths" remainder as a fraction in lowest terms. Convert $\frac{2}{3}$ to a mixed-number percent.

$$\frac{1}{3} = 3\overline{)1.00}\;\; (.33\tfrac{1}{3}) = 33\tfrac{1}{3}\%$$

$$\begin{array}{r} -9 \\ \hline 10 \\ -9 \\ \hline 1 \end{array}$$

$\frac{2}{3} = 3\overline{)2.}$ = ________

Answer: $3\overline{)2.00}$ ($.66\frac{2}{3}$) $= 66\frac{2}{3}\%$

20. Some common fraction-percent equivalents are shown in the table below. Because of their usefulness, they should be memorized.

$\frac{1}{4} = 25\%$	$\frac{1}{3} = 33\frac{1}{3}\%$	$\frac{1}{5} = 20\%$	$\frac{1}{10} = 10\%$
$\frac{1}{2} = 50\%$	$\frac{2}{3} = 66\frac{2}{3}\%$	$\frac{2}{5} = 40\%$	$\frac{3}{10} = 30\%$
$\frac{3}{4} = 75\%$		$\frac{3}{5} = 60\%$	$\frac{7}{10} = 70\%$
		$\frac{4}{5} = 80\%$	$\frac{9}{10} = 90\%$

From memory, convert each fraction to a percent.

a) $\frac{3}{4}$ = ________ b) $\frac{2}{5}$ = ________ c) $\frac{1}{3}$ = ________ d) $\frac{9}{10}$ = ________

Answers: a) 75% b) 40% c) $33\frac{1}{3}\%$ d) 90%

21. Many fractions that are not in lowest terms can be converted to percents by reducing them to fractions whose equivalents are known. For example:

Since $\frac{30}{60}$ can be reduced to $\frac{1}{2}$, $\frac{30}{60}$ = 50%

Since $\frac{50}{75}$ can be reduced to $\frac{2}{3}$, $\frac{50}{75}$ = ________

Answer: $66\frac{2}{3}\%$

22. Instead of converting to a mixed-number percent, we can continue the division process beyond the "hundredths" place. An example of a terminating division is shown below. Use the same method for the other conversion.

$$\frac{3}{16} = 16\overline{)3.0000}\;\;(.1875) = 18.75\%$$

$$\begin{array}{r} -16 \\ \hline 140 \\ -128 \\ \hline 120 \\ -112 \\ \hline 80 \\ -80 \\ \hline \end{array}$$

$\frac{5}{8} = 8\overline{)5.}$ = ________

Answer: $8\overline{)5.000}$ (.625) = 62.5%

23. When the division is non-terminating, we can round to the nearest whole-number percent or to the nearest "tenth" or "hundredth" of a percent.

To round to the nearest whole-number percent, we divide until we get three decimal places and then round back to two decimal places. For example:

$$\frac{5}{6} = 6\overline{)5.000} \quad .833 \text{ or } .83 \quad = 83\%$$

To round to the nearest "tenth" of a percent, we divide until we get four decimal places and then round back to three decimal places. For example:

$$\frac{1}{7} = 7\overline{)1.0000} \quad .1428 \text{ or } .143 \quad = 14.3\%$$

To round to the nearest "hundredth" of a percent, we divide until we get five decimal places and then round back to four decimal places. For example:

$$\frac{2}{9} = 9\overline{)2.00000} \quad .22222 \text{ or } .2222 \quad = 22.22\%$$

Convert the fraction at the right to the nearest whole-number percent.

$$\frac{17}{60} = 60\overline{)17}$$

Answer: $60\overline{)17.000}$ .282 or .28 = 28%

24. When the quotient is larger than "1", the percent is larger than 100%. An example is shown. Complete the other conversion.

$$\frac{31}{25} = 25\overline{)31.00} = 124\% \quad (\text{quotient } 1.24;\ -25,\ 60,\ -50,\ 100,\ -100)$$

$$\frac{47}{20} = 20\overline{)47}$$

Answer: $20\overline{)47.00}$ 2.35 = 235%

25. When the first non-zero digit of the quotient is beyond the "hundredths" place, the percent is smaller than 1%. An example is shown. Complete the other conversion.

$$\frac{3}{400} = 400\overline{)3.0000} = .75\% \quad (\text{quotient } .0075;\ -2800,\ 2000,\ -2000)$$

$$\frac{1}{500} = 500\overline{)1}$$

Answer: $500\overline{)1.000}$ .002 = .2%

26. Attempting to reduce a fraction to lowest terms before converting to a percent is useful in two situations:

1) When the fraction obtained is one whose percent equivalent is known. For example:

$$\frac{32}{40} = \frac{4}{5} = 80\% \qquad \frac{33}{99} = \frac{1}{3} = 33\frac{1}{3}\%$$

2) When the division required is simplified. For example:

$$\frac{15}{90} = \frac{1}{6} = 6\overline{)1.000} \;\; (.166 \text{ or } .17) = 17\%$$

However, since any fraction can be converted directly to a percent by the division method, <u>only obvious and easy reductions should be attempted</u>.

We converted the fraction below directly to a percent without attempting to reduce to lowest terms. Convert the other fraction directly to a percent.

$$\frac{45}{180} = 180\overline{)45.00} \;(.25) = 25\% \qquad \frac{140}{200} = 200\overline{)140}$$

```
     .2 5
180)4 5.0 0
   -3 6 0
      9 0 0
     -9 0 0
```

Answer: $200\overline{)140.0}$; .7 or .70 ; -1400 ; = 70%

---

27. When both terms of the fraction contain one or more 0's on the right side, an easy type of reduction is possible. That is, we can drop as many 0's as possible from the right side of both terms <u>as long as we drop the same number of 0's from each term</u>. For example:

$$\frac{2,700}{6,200} = \frac{27}{62}$$ (We dropped <u>two</u> 0's from each term.)

$$\frac{490,000}{800,000} = \frac{49}{80}$$ (We dropped <u>four</u> 0's from each term.)

Convert the fraction below <u>to the nearest whole-number</u> percent by dropping as many 0's as possible from each term and then using the division method.

$$\frac{13,000}{60,000} =$$

Answer: $\frac{13}{60} = 60\overline{)13.000}$ (.216 or .22) = 22%

12-4 OPERATIONS INVOLVING PERCENTS

In this section, we will discuss adding and subtracting percents and multiplying or dividing an ordinary number by a percent.

---

28. Percents can be added or subtracted just like ordinary numbers. For example:

25% + 50% = 75% 100% - 40% = ______

Answer: 60%

29. To multiply an ordinary number by a percent, we must convert the percent to a fraction or decimal number. For example:

$$20\% \times 35 = \frac{1}{5} \times 35 = 7$$

$$20\% \times 35 = .2 \times 35 = 7.0 \text{ or } 7$$

Use either method to complete these:

a) $70\% \times 40 =$ ________ b) $50\% \times 84 =$ ________

---

a) 28

b) 42

30. Since most percents are not equal to simple fractions, we ordinarily convert to a decimal number to multiply. An example is shown. Complete the other multiplication.

27.5% x 3,500 .6% x 91

```
   3500
   .275
  17500
 24500
7000
962.500  or  962.5
```

```
   91
x .006
  .546
```

31. To divide an ordinary number by a percent, we must also convert the percent to a fraction or decimal number. For example:

$$\frac{45}{30\%} = \frac{45}{\frac{3}{10}} = \overset{15}{\cancel{45}} \times \frac{10}{\cancel{3}} = 150$$

$$\frac{45}{30\%} = \frac{45}{.3} = .3_{\wedge}\overline{)4\,5.0_{\wedge}}\ (150) = 150$$

Use either method to complete these:

a) $\frac{27}{10\%} =$ ________________ b) $\frac{75}{50\%} =$ ________________

---

a) 270

b) 150

32. Since most percents are not equal to simple fractions, we ordinarily convert to a decimal number to divide. An example is shown. Complete the other division.

$$\frac{700}{35\%} = \frac{700}{.35} = .35_{\wedge}\overline{)7\,0\,0.0\,0_{\wedge}}\ (2{,}000)$$

```
 -70
   0
```

$$\frac{84}{6\%} =$$

$$\frac{84}{.06} = .06_{\wedge}\overline{)8\,4.0\,0_{\wedge}}\ (1{,}400)$$

SELF-TEST 35 (pages 304-313)

From memory, convert each percent to a fraction and each fraction to a percent.

1. 60% =	2. $\frac{1}{3}$ =	3. 10% =	4. $\frac{3}{4}$ =

Convert each percent to a whole number or decimal number.

5. $7\frac{1}{2}$% =	6. 300% =	7. .18% =	8. 146% =

Convert each whole number or decimal number to a percent.

9. .056 =	10. 1 =	11. .0002 =	12. 2.83 =

Convert each fraction to the nearest whole-number percent.

13. $\frac{15}{25}$ =	14. $\frac{6,500}{4,000}$ =

Convert each fraction to the nearest tenth of a percent.

15. $\frac{1}{12}$ =	16. $\frac{2}{500}$ =

17. 25% + 40% =

18. 100% - 90% =

19. 1.2% x 4,000 =

20. $\frac{64}{40\%}$ =

ANSWERS:

1. $\frac{3}{5}$	5. .075	9. 5.6%	13. 60%	17. 65%
2. $33\frac{1}{3}$%	6. 3	10. 100%	14. 163%	18. 10%
3. $\frac{1}{10}$	7. .0018	11. .02%	15. 8.3%	19. 48
4. 75%	8. 1.46	12. 283%	16. .4%	20. 160

12-5 THE THREE TYPES OF PERCENT PROBLEMS

In this section, we will discuss the three basic types of percent problems. We will show how they can be solved by converting to equations and then solving the equations.

33. We translated each problem below to an equation. Notice that "is" means "=" and "of" means "x" or "times".

What is 50% of 46 ?	80% of 20 is what?
↓ ↓ ↓ ↓ ↓	↓ ↓ ↓ ↓ ↓
n = 50% x 46	80% x 20 = n

Translate each problem below to an equation.

a) What is 25% of 60 ? b) $33\frac{1}{3}$% of 96 is what?

Answer: a) n = 25% x 60 b) $33\frac{1}{3}$% x 96 = n

34. We translated one problem below to an equation. Translate the other problem to an equation.

Find 40% of 25 .	Find 75% of 88 .
↓ ↓ ↓	
40% x 25 = n	

Answer: 75% x 88 = n

35. To solve the problem below, we translated to an equation and then solved for "n" by converting $33\frac{1}{3}$% <u>to a fraction</u> to multiply. Use the same method to solve the other problem.

What is $33\frac{1}{3}$% of 66 ? 25% of 48 is what?

$n = 33\frac{1}{3}\% \times 66$

$n = \frac{1}{3}(66) = 22$

Answer: 25% x 48 = n; $\frac{1}{4}(48) = n$; n = 12

36. Ordinarily we have to convert <u>to a decimal number</u> when multiplying by a percent. An example is shown. Use the same method for the other problem.

Find 2.4% of 120. What is 120% of 330 ?

2.4% x 120 = n

.024(120) = n

```
  120
 .024
  480
 240
 2.880  or 2.88
```

Therefore, n = 2.88

Answer: n = 120% x 330; n = 1.2(330); n = 396

37. Two examples of the second type of percent problems are shown below. We translated each to an equation.

17 is what percent of 80 ?	What percent of 200 is 159 ?
↓ ↓ ↓ ↓ ↓	↓ ↓ ↓ ↓ ↓
17 = n% x 80	n% x 200 = 159

Translate each problem below to an equation.

a) 20 is what percent of 50 ?

b) What percent of 90 is 60 ?

---

a) $20 = n\% \times 50$

b) $n\% \times 90 = 60$

38. To solve equations like those obtained in the last frame, we divide the other number by the coefficient of "n%" and then convert the fraction to a percent. An example is shown. Solve the other equation.

$$n\% \times 90 = 60$$

$$n\% = \frac{60}{90} \text{ or } \frac{2}{3}$$

$$n\% = 66\frac{2}{3}\%$$

$$20 = n\% \times 50$$

$n\% = 40\%$, from: $\frac{20}{50} = \frac{2}{5}$

39. Ordinarily we have to use the division method to solve the second type of percent problem. An example is shown. Round the other answer to the nearest whole-number percent.

What percent of 70 is 97 ?

$$n\% \times 70 = 97$$

$$n\% = \frac{97}{70} = 139\%$$

```
       1.3 8 5 or 1.39
  70)9 7.0 0 0
    -7 0
     2 7 0
    -2 1 0
       6 0 0
      -5 6 0
         4 0 0
        -3 5 0
```

49 is what percent of 300 ?

$n\% = 16\%$, from $\frac{49}{300}$

40. Two examples of the third type of percent problems are shown below. We translated each to an equation.

25 is 12% of what?	80% of what is 62 ?
↓ ↓ ↓ ↓ ↓	↓ ↓ ↓ ↓ ↓
25 = 12% x n	80% x n = 62

Translate each problem below to an equation.

a) 40 is $66\frac{2}{3}\%$ of what?

b) 50% of what is 20 ?

---

a) $40 = 66\frac{2}{3}\% \times n$

b) $50\% \times n = 20$

41. To solve equations like those obtained in the last frame, we can sometimes convert the percent to a fraction. An example is shown. Solve the other equation.

$$40 = 66\frac{2}{3}\% \times n$$

$$40 = \frac{2}{3}n$$

$$\frac{3}{\not{2}}\overset{20}{(\not{40})} = \frac{3}{2}\left(\frac{2}{3}n\right)$$

$$60 = 1n$$

$$n = 60$$

$$50\% \times n = 20$$

---

n = 40, from:

$$\frac{1}{2}n = 20$$

$$2\left(\frac{1}{2}n\right) = 2(20)$$

$$n = 40$$

42. Usually we have to convert the percent to a decimal number. An example is shown. Solve the other problem. Round to the nearest whole number.

12 is 6.5% of what?

$$12 = 6.5\% \times n$$

$$12 = .065n$$

$$n = \frac{12}{.065} = 185$$

```
          184.6  or 185
.065‸)12.000‸
      -65
       550
      -520
        300
       -260
         400
        -390
```

12% of what is 200 ?

---

n = 1,667 , from:

$$.12n = 200$$

$$n = \frac{200}{.12}$$

12-6 APPLIED PROBLEMS INVOLVING PERCENTS

In this section, we will discuss some applied problems involving percents. We will show how they can be solved by rewording, translating to an equation, and then solving the equation.

---

43. To solve the problem below, we reworded it, translated the rewording to an equation, and then solved the equation. Use the same method for the problem on the next page.

How many pounds of zinc are there in 50 pounds of an alloy if the alloy is 30% zinc?

Rewording: What is 30% of 50 ?

Equation: $n = 30\% \times 50$

$n = .3(50) = 15.0$ or 15

Therefore, there are 15 pounds of zinc in the alloy.

Continued on following page.

43. Continued

On a 40-item test, a student had 90% of the items correct. How many did she have correct?

Rewording: ____________________

Equation: ____________________

Answer: ____________________

What is 90% of 40 ?
n = 90% x 40
n = 36 items correct

44. We used the rewording-equation method to solve the following problem. Use the same method for the problem below.

If 60 of the 80 graduates from a small high school went on to college, what percent went on to college?

Rewording: 60 is what percent of 80 ?

Equation: $60 = n\% \times 80$

$$n\% = \frac{60}{80} = \frac{3}{4} = 75\%$$

Therefore, 75% of the graduates went on to college.

Out of a shipment of 50 cars to a dealer, 10 had a defective part. What percent had a defective part?

Rewording: ____________________

Equation: ____________________

Answer: ____________________

---

10 is what percent of 50 ?
10 = n% x 50
n% = 20% defective

45. We used the rewording-equation method to solve the following problem. Use the same method for the problem on the next page.

On an airline flight, 60% of the seats were filled. If 48 seats were filled, what was the seating capacity?

Rewording: 48 is 60% of what?

Equation: $48 = 60\% \times n$

$48 = .6n$

$$n = \frac{48}{.6} = .6_{\wedge}\overline{)48.0_{\wedge}} = 80$$

Therefore, the seating capacity was 80.

Continued on following page.

45. Continued

In a class election, Joe received 72 votes. If he received 40% of the votes cast, how many votes were cast?

Rewording: ______________

Equation: ______________

Answer: ______________

---

72 is 40% of what?

72 = 40% x n

n = 180 votes cast

46. Use the "rewording" hints to solve the problems below.

a) A real estate salesman's commission is 7%. What is his commission for selling a home for $45, 900 ?

(Hint: What is 7% of $45, 900 ?)

b) The sales tax on a $400 sofa is $18. What is the sales tax rate?

(Hint: $18 is what percent of $400 ?)

---

a) $3, 213

b) 4.5% or $4\frac{1}{2}\%$

47. Use the "rewording" hints to solve the problems below.

a) If an ore contains 15% iron, how many tons of ore are needed to get 30 tons of iron?

(Hint: 30 is 15% of what?)

b) Of 500 people tested, only 3 had diabetes. What percent had diabetes?

(Hint: 3 is what percent of 500 ?)

---

a) 200 tons

b) .6%

48. Make up your own "rewording" hints to solve these.

a) On a 32-item test, a student got 26 items correct. What percent did he get correct? (Round to the nearest whole-number percent.)

b) If the sales tax rate is 5%, how much tax is paid for a TV set that costs $425.95 ? (Round to the nearest cent.)

49. Make up your own "rewording" hints to solve these.

a) A family spends \$320 per month for food. If that amount is 20% of their monthly income, what is their monthly income?

b) A water supply contained .06% pollutants. Find the amount of pollutants in 5,000 kilograms of the water.

a) 81%, from: 26 = n% x 32

b) \$21.30, from: n = 5% x \$425.95

a) \$1,600, from: \$320 = 20% x n

b) 3 kilograms, from: n = .06% x 5,000

SELF-TEST 36 (pages 314-319)

1. What is 70% of 90 ?	2. 10 is what percent of 40 ?	3. 20 is $33\frac{1}{3}$% of what?
4. If 433 of the 600 teachers in a school system are female, what percent are female? (Round to the nearest whole-number percent.)	5. If 8% of 1,400 people polled gave no opinion, how many gave no opinion?	6. Mary sold her car for 75% of what she paid for it. If she sold it for \$3,600 , how much had she paid for it?

ANSWERS: 1. 63 2. 25% 3. 60 4. 72% 5. 112 people 6. \$4,800

12-7 PERCENT INCREASE OR DECREASE

In this section, we will discuss problems that involve a percent increase or a percent decrease.

50. The problem below involves a percent increase. The two steps needed to solve it are shown.

The number of major crimes in a city increased 200% from 1970 to 1975. If there were 150 major crimes in 1970, how many were there in 1975?

1) Finding the amount of the increase.

Rewording: What is 200% of 150 ?

Equation: $n = 200\% \times 150$

$n = 2(150) = 300$ increase from 1970 to 1975

2) Finding the number of major crimes in 1975.

$150 + 300 = 450$ major crimes

The problem below involves a percent decrease. The same two steps are needed to solve it.

Find the sale price of a $150 suit if it is sold at "$33\frac{1}{3}\%$ off".

a) Find the amount of the reduction.

b) Find the sale price.

---

a) $50, from: $n = 33\frac{1}{3}\% \times \$150$

b) $100, from: $150 - $50

---

51. The problem below involves a percent increase. The two steps needed to solve it are shown.

A woman received a 6% raise. If the raise amounted to $540 per year, what was her original annual salary and what is her new annual salary?

1) Finding her original annual salary.

Rewording: $540 is 6% of what?

Equation: $\$540 = 6\% \times n$

$\$540 = .06n$

$$n = \frac{\$540}{.06} = .06_\wedge\overline{)\$540.00_\wedge} \quad \$9{,}000.$$

2) Finding her new annual salary.

$9, 000 + $540 = $9, 540

Continued on following page.

51. Continued

The problem below involves a percent decrease. The same two steps are needed to solve it.

The value of a home decreased 10% in 5 years. If the amount of the decrease was \$3,200, find the original value and the new value of the home.

a) Find the original value.

b) Find the new value.

Answers
a) \$32,000, from: $\$3,200 = 10\% \times n$
b) \$28,800, from: \$32,000 - \$3,200

52. When computing a percent increase or percent decrease, we always compare the amount of the increase or decrease with the original amount. For example:

If a 100-pound girl gained 13 pounds and now weighs 113 pounds, the percent increase in weight can be found with the following formula.

$$\text{Percent Increase} = \frac{\text{Amount of Increase}}{\text{Original Amount}} = \frac{13}{100} = 13\%$$

If the price of a \$50 watch is reduced \$10 and sells for \$40, the percent decrease in cost can be found with the following formula.

$$\text{Percent Decrease} = \frac{\text{Amount of Decrease}}{\text{Original Amount}} = \frac{\quad}{\quad} = \underline{\qquad}$$

Answer
$\frac{10}{50} = 20\%$

53. The problem below involves a percent decrease. The two steps needed to solve it are shown.

In a 10-year period, the population of a city decreased from 600,000 to 540,000. Find the percent decrease.

1) Finding the amount of the decrease.

$600,000 - 540,000 = 60,000$

2) Finding the percent decrease.

$$\text{Percent Decrease} = \frac{\text{Amount of Decrease}}{\text{Original Amount}} = \frac{60,000}{600,000} = \frac{6}{60} = \frac{1}{10} = 10\%$$

Continued on following page.

53. Continued

The problem below involves a percent increase. The same two steps are needed to solve it.

In a 15-year period, the price of coffee increased from \$ .39 per pound to \$1.79 per pound. Find the percent increase to the nearest whole-number percent.

a) Find the amount of the increase.

________ - ________ = ________

b) Find the percent increase.

$$\text{Percent Increase} = \frac{\text{Amount of Increase}}{\text{Original Amount}}$$

a) \$1.40, from: \$1.79 - \$ .39

b) 359%, from: $\frac{\$1.40}{\$.39}$

12-8 SIMPLE INTEREST

In this section, we will discuss the formula for simple interest and use it to solve applied problems.

54. To find the simple interest (I) for a loan for a 1-year period, we multiply the amount or principle (P) of the loan by the interest rate (R). That is:

$$I = PR$$

We used the formula to solve one problem below. Use it to solve the other problem.

Find the interest for a \$200 loan at 9% for 1 year.

I = PR = (\$200)(9%)

```
 $200
 x.09
$18.00 or $18
```

Find the interest for a \$1,000 loan at 7% for 1 year.

\$70, from: (7%)(\$1,000)

55. Since interest rates are stated for a 1-year period, the amount of interest is frequently computed for a 1-year period. However, we can compute the amount of interest for other periods of time. For example:

If the interest for 1 year is \$50, the interest for 2 years is 2(\$50) = \$100.

If the interest for 1 year is \$80, the interest for $\frac{1}{2}$ year is $\frac{1}{2}$(\$80) = ________

\$40

56. Because the amount of interest is computed for periods of time other than 1 year, the general formula for interest contains a time factor "T". That is:

$$\boxed{I = PRT}$$

We used the general formula to solve one problem below. Use it to solve the other problem.

Find the interest for a \$300 loan at 8% for 2 years.

$$I = PRT = (\$300)(8\%)(2) = (\$300)(.08)(2)$$
$$= (\$24)\quad(2) = \$48$$

Find the interest for a \$600 loan at 10% for $\frac{1}{2}$ year.

$$I = PRT =$$

\$30, from: ($600)(10%)$\left(\frac{1}{2}\right)$

57. When using I = PRT, time (T) must be stated in terms of a year. If time is stated in months, we must convert the months to a fractional part of a year. For example:

6 months = $\frac{1}{2}$ year $\left(\text{from } \frac{6}{12}\right)$

3 months = $\frac{1}{4}$ year $\left(\text{from } \frac{3}{12}\right)$

Similarly: a) 4 months = ______ year b) 9 months = ______ year

---

a) $\frac{1}{3}$ year $\left(\text{from } \frac{4}{12}\right)$
b) $\frac{3}{4}$ year $\left(\text{from } \frac{9}{12}\right)$

58. Following the example, solve the other problem.

Find the interest for a \$500 loan at 6% for 2 months.

$$I = PRT = (\$500)(6\%)\left(\frac{1}{6}\right) = (\$500)(.06)\left(\frac{1}{6}\right)$$
$$= (\$30)\quad\left(\frac{1}{6}\right) = \$5$$

Find the interest for a \$2, 000 loan at 10% for 6 months.

$$I = PRT =$$

\$100, from: ($2, 000)(10%)$\left(\frac{1}{2}\right)$

59. The general formula can be used for 1-year time periods. However, since $T = 1$, the formula equivalently reduces to $I = PR$. An example is shown. Complete the other problem.

Find the interest for a $3,000 loan at 5% for 1 year.

$$I = PRT = (\$3,000)(5\%)(1) = (\$3,000)(5\%) = \$150.00 \text{ or } \$150$$

Find the interest for a $100 loan at 12% for 1 year.

$$I = PRT =$$

$12, from: ($100)(12%)(1)

60. Since the amount of a savings deposit or other investment is like the principal (P) of a loan, the same formula can be used to compute the interest for savings deposits and other investments. An example is shown. Solve the other problem.

Find the interest earned for a $500 savings deposit at 5% for 2 years.

$$I = PRT = (\$500)(5\%)(2) = \underline{(\$500)(.05)}(2)$$
$$= (\$25)\quad(2) = \$50$$

Find the interest earned for $10,000 invested at 9% for 4 months.

$$I = PRT =$$

$300, from: $(\$10,000)(9\%)\left(\frac{1}{3}\right)$

61. Following the example, solve the other problem.

Find the interest for a $1,000 loan at $8\frac{3}{4}\%$ for 6 months.

Find the interest for a $4,000 investment at $6\frac{1}{2}\%$ for 2 years.

$$I = PRT = (\$1,000)\left(8\frac{3}{4}\%\right)\left(\frac{1}{2}\right)$$
$$= (\$500)(.0875)$$

```
  $500
  .0875
  2500
 3500
4000
$43.7500  or  $43.75
```

$520, from: $(\$4,000)\left(6\frac{1}{2}\%\right)(2)$

62. Following the example, solve the other problem.

Find the interest for a \$300 savings deposit at 5.13% for 1 year.

$I = PRT = (\$300)(5.13\%)(1)$

$$\begin{array}{r} \$300 \\ \underline{.0513} \\ 900 \\ 300 \\ \underline{1500} \\ \$15.3900 \end{array} \text{ or } \$15.39$$

Find the interest for a \$5,000 savings deposit at 6.29% for 1 year.

\$314.50, from:

(\$5,000)(6.29%)

---

63. The formula was used to solve the problem below. Two methods are shown.

If \$16 interest is paid for an 8% loan for 3 months, what is the principal of the loan?

Equation-Solving Method

$$I = PRT$$

$$16 = P(8\%)\left(\frac{1}{4}\right)$$

$$16 = P(.08)\left(\frac{1}{4}\right)$$

$$16 = P(.02)$$

$$P = \frac{16}{.02} = \$800, \text{ from:}$$

$$.02_\wedge\overline{)16.00_\wedge} = 800.$$

Rearrangement Method

$$I = PRT$$

$$P = \frac{I}{RT}$$

$$= \frac{16}{(8\%)\left(\frac{1}{4}\right)}$$

$$= \frac{16}{(.08)\left(\frac{1}{4}\right)} = \frac{16}{.02} = \$800, \text{ from:}$$

$$.02_\wedge\overline{)16.00_\wedge} = 800.$$

Using the formula and either method above, solve this problem.

If \$210 is earned by a 7% investment in a 3-year period, how much was invested?

\$1,000 , from:

$$P = \frac{\$210}{.21}$$

64. The formula was used to solve the problem below. Two methods are shown.

If the interest for a \$400 loan for 2 years is \$56, what is the interest ratc?

Equation-Solving Method

$$I = PRT$$

$$56 = (400)(R)(2)$$

$$56 = 800R$$

$$R = \frac{56}{800} = 7\%, \text{ from:}$$

$$\begin{array}{r} .07 \\ 800\overline{)56.00} \\ -5600 \\ \hline \end{array}$$

Rearrangement Method

$$I = PRT$$

$$R = \frac{I}{PT}$$

$$R = \frac{56}{(400)(2)}$$

$$= \frac{56}{800} = 7\%, \text{ from:}$$

$$\begin{array}{r} .07 \\ 800\overline{)56.00} \\ -5600 \\ \hline \end{array}$$

Using the formula and either method above, solve this problem.

If the interest for 1 year for a \$3,000 investment is \$285, what is the interest rate?

9.5% or $9\frac{1}{2}\%$, from:

$$R = \frac{285}{3,000}$$

SELF-TEST 37 (pages 319-327)

1. A man is making \$7.50 per hour. If he receives an 8% raise, what is his new hourly rate?	2. After an engine tune-up, the "miles per gallon" for a car increased from 16 to 20. Find the percent increase.
3. The price of a car is reduced \$1,000 for a sale. If that amounts to a 20% reduction, what is the sale price of the car?	4. Find the interest on a \$2,000 loan at $9\frac{1}{2}\%$ for 6 months.
5. If the interest for a \$4,000 loan for 2 years is \$600, what is the interest rate?	6. A man can invest some money at 10%. In order to make \$500 interest in 1 year, how much must he invest?

ANSWERS: 1. \$8.10 2. 25% 3. \$4,000 4. \$95 5. 7.5% or $7\frac{1}{2}\%$ 6. \$5,000

Unit 13 GRAPHING

In this unit, we will introduce the coordinate system that is used for graphing equations. Though emphasis is given to graphing linear equations containing "x" and "y", some linear formulas are graphed and some curvilinear graphs are shown.

13-1 SOLUTIONS OF EQUATIONS CONTAINING BOTH "x" AND "y"

In this section, we will discuss the meaning of "solutions" for equations that contain both "x" and "y".

---

1. When an equation contains both "x" and "y", it is "true" when some pairs of values are substituted for the letters and "false" when other pairs of values are substituted for the letters. For example:

$y = 2x$ is true for ($x = 4$ and $y = 8$) but false for ($x = 1$ and $y = 3$) since:

$$\begin{aligned} y &= 2x \\ 8 &= 2(4) \\ 8 &= 8 \end{aligned} \qquad \begin{aligned} y &= 2x \\ 3 &\neq 2(1) \\ 3 &\neq 2 \end{aligned}$$

Is $y = 2x$ true for: a) $x = 3$ and $y = 5$? ________

b) $x = 10$ and $y = 20$? ________

---

2. The equation $y = 3x + 1$ is true for ($x = 2$ and $y = 7$) but false for ($x = 4$ and $y = 10$) since:

$$\begin{aligned} y &= 3x + 1 \\ 7 &= 3(2) + 1 \\ 7 &= 7 \end{aligned} \qquad \begin{aligned} y &= 3x + 1 \\ 10 &\neq 3(4) + 1 \\ 10 &\neq 13 \end{aligned}$$

Is $y = 3x + 1$ true for: a) $x = 3$ and $y = 8$? ________

b) $x = 5$ and $y = 16$? ________

> a) No, since: $5 \neq 2(3)$
>
> b) Yes, since: $20 = 2(10)$

---

3. The equation $x + y = 10$ is true for ($x = 3$ and $y = 7$) but false for ($x = 5$ and $y = 4$) since:

$$\begin{aligned} x + y &= 10 \\ 3 + 7 &= 10 \\ 10 &= 10 \end{aligned} \qquad \begin{aligned} x + y &= 10 \\ 5 + 4 &\neq 10 \\ 9 &\neq 10 \end{aligned}$$

Is $x + y = 10$ true for: a) $x = 1$ and $y = 9$? ________

b) $x = 6$ and $y = 2$? ________

> a) No, since: $8 \neq 3(3) + 1$
>
> b) Yes, since: $16 = 3(5) + 1$

> a) Yes, since: $1 + 9 = 10$
>
> b) No, since: $6 + 2 \neq 10$

4. The equation $2x - y = 7$ is <u>true</u> for $(x = 5$ and $y = 3)$ but <u>false</u> for $(x = 7$ and $y = 5)$ since:

$$\begin{aligned} 2x - y &= 7 \\ 2(5) - 3 &= 7 \\ 10 - 3 &= 7 \\ 7 &= 7 \end{aligned} \qquad \begin{aligned} 2x - y &= 7 \\ 2(7) - 5 &\neq 7 \\ 14 - 5 &\neq 7 \\ 9 &\neq 7 \end{aligned}$$

Is $2x - y = 7$ true for: a) $x = 6$ and $y = 1$? ________

b) $x = 10$ and $y = 13$? ________

---

a) No, since: $2(6) - 1 \neq 7$

b) Yes, since: $2(10) - 13 = 7$

5. When a pair of values for "x" and "y" makes an equation true, we say that the pair "satisfies" the equation. For example:

$(x = 4,\ y = 12)$ satisfies $y = 3x$, since $12 = 3(4)$

$(x = 2,\ y = 9)$ satisfies $y = 5x - 1$, since $9 = 5(2) - 1$

a) Does $(x = 3,\ y = 15)$ satisfy $y = 5x$? ________

b) Does $(x = 2,\ y = 8)$ satisfy $y = 3x + 4$? ________

---

a) Yes, since: $15 = 5(3)$

b) No, since: $8 \neq 3(2) + 4$

6. When a pair of values for "x" and "y" satisfies an equation, the pair is called a "<u>solution</u>" of the equation. For example:

$(x = 5,\ y = 2)$ is a solution of $x + y = 7$, since $5 + 2 = 7$

$(x = 4,\ y = 3)$ is a solution of $2x - y = 5$, since $2(4) - 3 = 5$

a) Is $(x = 10,\ y = 4)$ a solution of $x - y = 3$? ________

b) Is $(x = 1,\ y = 5)$ a solution of $3x + y = 8$? ________

---

a) No, since: $10 - 4 \neq 3$

b) Yes, since: $3(1) + 5 = 8$

7. The solution of an equation containing "x" and "y" (an "xy" equation) can include one or two negative values. For example:

$(x = -5,\ y = 7)$ is a solution of $x + y = 2$, since $(-5) + 7 = 2$

$(x = -2,\ y = -8)$ is a solution of $y = 4x$, since $-8 = 4(-2)$

a) Is $(x = -4,\ y = 1)$ a solution of $y = x + 5$? ________

b) Is $(x = -3,\ y = -6)$ a solution of $2x - y = 5$? ________

---

a) Yes, since: $1 = (-4) + 5$

b) No, since: $2(-3) - (-6) \neq 5$

8. The solution of an "xy" equation can include one or two fractions. For example:

$(x = \frac{1}{2},\ y = 4)$ is a solution of $y = 8x$, since $4 = 8\left(\frac{1}{2}\right)$

$(x = \frac{5}{4},\ y = -\frac{1}{4})$ is a solution of $x + y = 1$, since $\frac{5}{4} + \left(-\frac{1}{4}\right) = 1$

a) Is $(x = \frac{2}{3},\ y = 8)$ a solution of $y = 6x$? ________

b) Is $(x = \frac{3}{2},\ y = -\frac{1}{2})$ a solution of $x - y = 2$? ________

9. The solution of an "xy" equation can include one or two decimal numbers. For example:

$(x = 2.5,\ y = 1.5)$ is a solution of $x - y = 1$, since $2.5 - 1.5 = 1$

$(x = .2,\ y = 8)$ is a solution of $y = 40x$, since $8 = 40(.2)$

a) Is $(x = 3.4,\ y = 2.6)$ a solution of $x + y = 5$? ________

b) Is $(x = -1.4,\ y = -7)$ a solution of $y = 5x$? ________

> a) No, since: $8 \neq 6\left(\frac{2}{3}\right)$
>
> b) Yes, since: $\frac{3}{2} - \left(-\frac{1}{2}\right) = 2$

---

10. Some solutions for the equation below are listed at the right.

$$y = 2x$$

$x = 5,\ y = 10$
$x = 1.5,\ y = 3$
$x = \frac{1}{2},\ y = 1$
$x = 0,\ y = 0$
$x = -1,\ y = -2$
$x = -\frac{5}{2},\ y = -5$
$x = -3.5,\ y = -7$
$x = -40,\ y = -80$

How many more solutions are there for the same equation? ________________

> a) No, since: $3.4 + 2.6 \neq 5$
>
> b) Yes, since: $-7 = 5(-1.4)$

> An infinite number

13-2 SOLUTION-TABLES FOR EQUATIONS CONTAINING BOTH "x" AND "y"

When graphing an "xy" equation, we begin by making up a table that contains some of the solutions for the equation. We will discuss tables of that type in this section.

---

11. To find solutions for an "xy" equation, we substitute values for x and then find the corresponding values for y. For example, we found five solutions for the equation below by substituting 2, 1, 0, -1, and -2 for x and then evaluating for the corresponding y's. The five solutions can be summarized in a table as we have done below.

$$y = 5x$$

If $x = 2,\ y = 5(2) = 10.$
If $x = 1,\ y = 5(1) = 5$
If $x = 0,\ y = 5(0) = 0$
If $x = -1,\ y = 5(-1) = -5$
If $x = -2,\ y = 5(-2) = -10$

$$y = 5x$$

x	y
2	10
1	5
0	0
-1	-5
-2	-10

Are all of the possible solutions for $y = 5x$ listed in the table? ________

---

12. Using the same method, we found the values of y corresponding to $x = 3$, $x = 1$, and $x = 0$ in the table below. Complete the table by finding the values of y corresponding to $x = -1$ and $x = -3$.

> No, since there are an infinite number of possible solutions.

$$y = x + 1$$

If $x = 3,\ y = 3 + 1 = 4$
If $x = 1,\ y = 1 + 1 = 2$
If $x = 0,\ y = 0 + 1 = 1$

$$y = x + 1$$

x	y
3	4
1	2
0	1
-1	
-3	

13. When making up a table of solutions for an "xy" equation in order to graph it, we begin by picking some values of x to substitute. The following guidelines can be used.

1) Pick some positive values, "0", and some negative values.

2) Pick smaller whole-number values to make the evaluations for y simpler.

Complete the table of solutions for each equation below.

a) $y = 2x + 3$

x	y
4	11
2	7
0	
-2	
-4	

b) $y = x - 1$

x	y
5	4
3	2
0	
-3	
-5	

x	y
3	4
1	2
0	1
-1	0
-3	-2

---

14. In a solution-table, the values of x should be listed in order as they are below. Complete the tables.

a) $y = 10x$

x	y
2	
1	
0	
-1	
-2	

b) $y = 10 - x$

x	y
4	
2	
0	
-2	
-4	

a)

x	y
4	11
2	7
0	3
-2	-1
-4	-5

b)

x	y
5	4
3	2
0	-1
-3	-4
-5	-6

---

15. In $x + y = 10$, "y" is not solved-for. Therefore, if we substitute values for x in the equation as it stands, we have to solve an equation to find the corresponding value of "y". An example is shown. Complete the other evaluation.

Find y when x is 3.

$$x + y = 10$$
$$3 + y = 10$$
$$(-3) + 3 + y = 10 + (-3)$$
$$y = 7$$

Find y when x is -1.

$$x + y = 10$$

a)

x	y
2	20
1	10
0	0
-1	-10
-2	-20

b)

x	y
4	6
2	8
0	10
-2	12
-4	14

$y = 11$, from: $(-1) + y = 10$

16. In each case below, find the corresponding value of y by substituting in the equation as it stands and then solving for y.

a) Find y when x is 4.

$2x + y = 7$

b) Find y when x is -2.

$x - y = 1$

---

17. When making up a solution-table for graphing, several pairs of values for x and y are needed. Therefore, we must substitute several times for x and evaluate to find the corresponding value of y. It is easier to evaluate for y if y is solved-for in an equation because no equation-solving is then needed. Therefore, when y is not solved-for in the original equation, we ordinarily solve for it before making up a solution-table. An example is shown. Solve for y in the other equation.

$$\begin{aligned} y - x &= 1 \\ y + (-x) &= 1 \\ y + (-x) + x &= 1 + x \\ y &= x + 1 \end{aligned}$$

$y - 2x = 5$

a) $y = -1$, from: $2(4) + y = 7$

b) $y = -3$, from: $(-2) - y = 1$

---

18. When solving for "y" below, we converted the addition of a negative term (-x) back to subtraction in the final step. Solve for "y" in the other equation.

$$\begin{aligned} x + y &= 10 \\ (-x) + x + y &= 10 + (-x) \\ y &= 10 + (-x) \\ y &= 10 - x \end{aligned}$$

$2x + y = 12$

$y = 2x + 5$
or
$y = 5 + 2x$

---

19. When solving for "y" below, we used the oppositing principle for equations in the final step to get rid of the "-" in front of "y". Solve for "y" in the other equation.

$$\begin{aligned} 2x - y &= 15 \\ 2x + (-y) &= 15 \\ (-2x) + 2x + (-y) &= 15 + (-2x) \\ -y &= 15 + (-2x) \\ -y &= 15 - 2x \\ y &= 2x - 15 \end{aligned}$$

$x - y = 3$

$y = 12 - 2x$

$y = x - 3$

20. We solved for "y" in each equation below. Using that form of the equation, complete the solution-tables.

a) $2x + y = 10$

or

$y = 10 - 2x$

x	y
3	
1	
0	
-1	
-3	

b) $x - y = 3$

or

$y = x - 3$

x	y
4	
2	
0	
-2	
-4	

a)

x	y
3	4
1	8
0	10
-1	12
-3	16

b)

x	y
4	1
2	-1
0	-3
-2	-5
-4	-7

SELF-TEST 38 (pages 328-334)

1. Which pairs of values are solutions of the equation $y = 3x + 1$? ________

a) $x = 4,\ y = 13$

b) $x = 1,\ y = 3$

c) $x = 0,\ y = 1$

d) $x = -1,\ y = -1$

e) $x = -2,\ y = -7$

f) $x = -3,\ y = -8$

Complete the table of solutions for each equation below.

2. $y = 3x$

x	y
3	
1	
0	
-1	
-3	

3. $y = x + 2$

x	y
5	
2	
0	
-2	
-5	

Continued on the following page.

SELF-TEST 38 (pages 328-334) - Continued

4. $2x + y = 6$

x	y
2	
1	
0	
-1	
-2	

5. $x - y = 1$

x	y
4	
2	
0	
-2	
-4	

ANSWERS: 1. a), c), f)

2.

x	y
3	9
1	3
0	0
-1	-3
-3	-9

3.

x	y
5	7
2	4
0	2
-2	0
-5	-3

4.

x	y
2	2
1	4
0	6
-1	8
-2	10

5.

x	y
4	3
2	1
0	-1
-2	-3
-4	-5

13-3 READING POINTS ON THE COORDINATE SYSTEM

In this section, we will introduce the coordinate system that is used to graph "xy" equations. Then we will discuss the method for reading the coordinates of points on the coordinate system.

21. The coordinate system is shown at the right. It consists of a horizontal and a vertical number line.

The horizontal number line is called the horizontal axis.

The vertical number line is called the vertical axis.

The two number lines together are called coordinate axes.

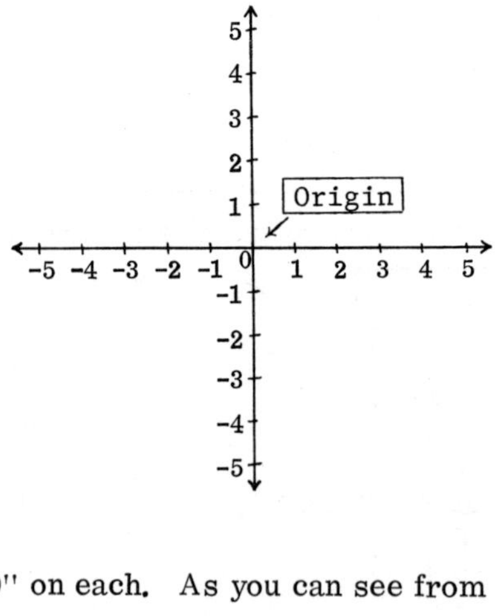

The two coordinate axes (number lines) intersect at "point 0" on each. As you can see from the figure, the special name for that point is the ____________.

origin

22. In the coordinate system at the right, we have included some additional horizontal and vertical lines that form a "gridwork". The gridwork makes it easier to use the coordinate system.

Each point in the coordinate system represents a pair of values, one related to the horizontal axis and one related to the vertical axis.

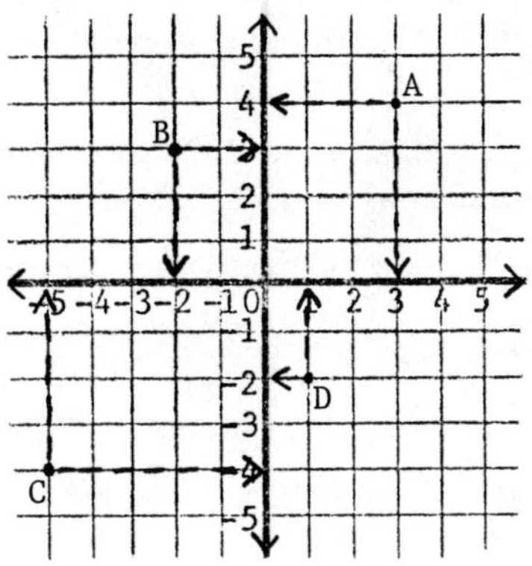

To find the pair of values for point A:

1) We drew an arrow down to the horizontal axis. The arrow points to "3".

2) We drew an arrow over to the vertical axis. The arrow points to "4".

3) Therefore, point A represents 3 on the horizontal axis and 4 on the vertical axis.

Similarly: Point B represents ______ on the horizontal axis and ______ on the vertical axis.

Point C represents ______ on the horizontal axis and ______ on the vertical axis.

Point D represents ______ on the horizontal axis and ______ on the vertical axis.

B) -2 ... 3

C) -5 ... -4

D) 1 ... -2

23. Mathematicians use the two axes to represent the values of "x" and "y", the letters ordinarily used in two-letter equations. They put:

x-values on the horizontal axis
y-values on the vertical axis

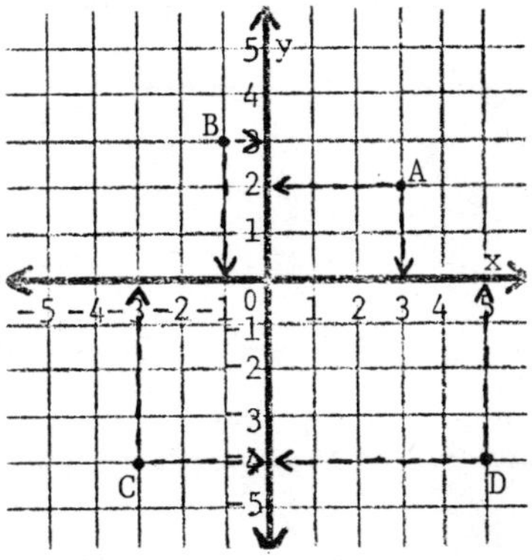

They label the axes "x" and "y" as we have done at the right.

On the coordinate system shown:

Point A represents: x = 3, y = 2.

Point B represents: x = ______, y = ______

Point C represents: x = ______, y = ______

Point D represents: x = ______, y = ______

B (x = -1, y = 3)

C (x = -3, y = -4)

D (x = 5, y = -4)

24. Since the axes are used to represent x-values and y-values, they are called the "x-axis" and the "y-axis".

a) The "x-axis" is another name for the ____________ (horizontal, vertical) axis.

b) The "y-axis" is another name for the ____________ (horizontal, vertical) axis.

---

a) horizontal

b) vertical

25. Point A at the right represents: x = 4, y = 2.

"4" and "2" are called the coordinates of point A.

"4" is called the x-coordinate.

"2" is called the y-coordinate.

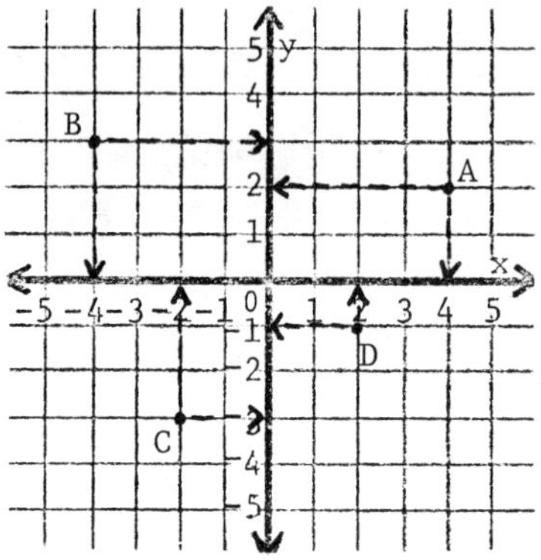

The coordinates of point A are usually written as we have done below. That is, they are enclosed in parentheses and separated by a comma with the x-coordinate written first.

(4, 2)

Write the coordinates of points B, C, and D below. Be sure to write the x-coordinate first.

B (,) C (,) D (,)

B (-4, 3)

C (-2, -3)

D (2, -1)

26. When writing the coordinates of a point, remember:

1) The coordinates are always enclosed in parentheses and separated by a comma.

2) The x-coordinate is always written first.

If a point represents: x = -2, y = 5, its coordinates are (-2, 5).

If a point represents: x = 4, y = -1, its coordinates are ____________.

(4, -1)
Don't forget the parentheses and comma.

27. If the coordinates of a point are (2, 8), the point represents: x = 2, y = 8.

If the coordinates of a point are (-5, -1), the point represents:

x = ______, y = ______.

---

x = -5, y = -1

28. When writing a pair of coordinates in parentheses, the order in which they are written makes a difference. That is:

(4, 7) means: x = 4, y = 7

(7, 4) means: x = 7, y = 4

Since the order in which they are written makes a difference, a pair of coordinates is called an "ordered pair".

a) In the ordered pair (-1, -2), the x-coordinate is ______.

b) In the ordered pair (-5, -3), the y-coordinate is ______.

a) -1

b) -3

29. Though we generally use smaller $\underline{x}$-values when graphing equations, we can get larger $\underline{y}$-values. Therefore, we sometimes use a different scale on the y-axis. For example, we counted "by 2's" to scale the y-axis at the right.

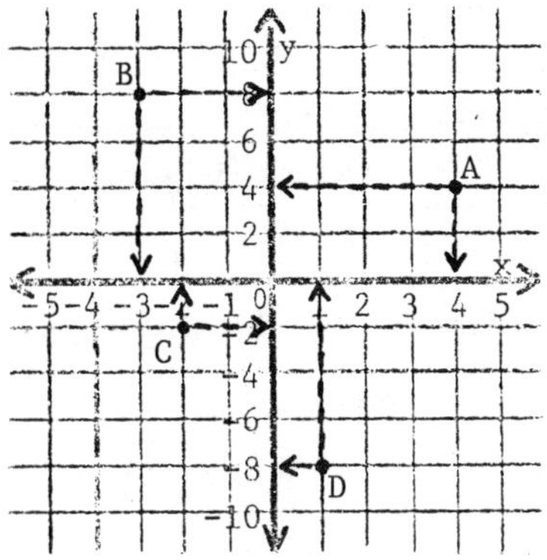

The coordinates of point A are (4, 4).

The coordinates of point B are (,).

The coordinates of point C are (,).

The coordinates of point D are (,).

B (-3, 8)

C (-2, -2)

D (1, -8)

30. To scale the y-axis at the right, we counted "by 10's". Write the coordinates of points A, B, C, and D below.

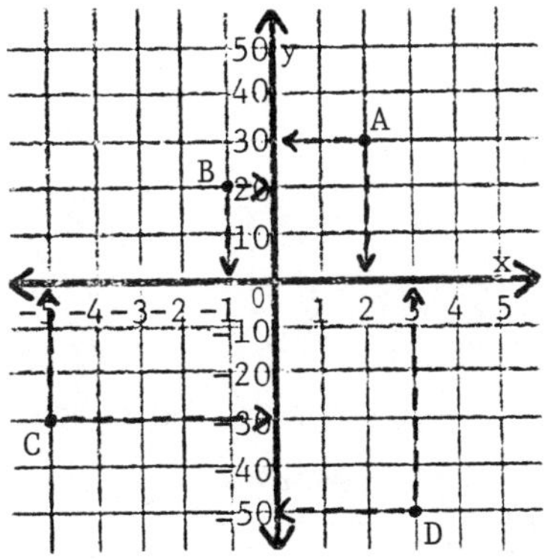

A ____________

B ____________

C ____________

D ____________

A (2, 30)

B (-1, 20)

C (-5, -30)

D (3, -50)

31. Only part of the scale is shown on each axis at the right. Therefore, you have to fill in some missing numbers to read the scales. For example:

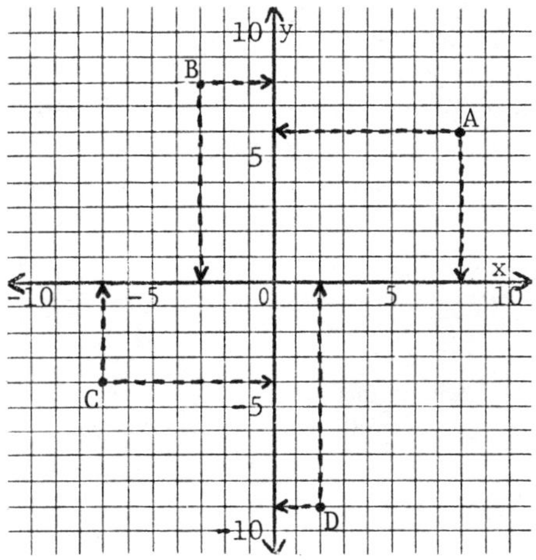

The coordinates of A are (8, 6). Write the coordinates of B, C, and D below.

B ____________

C ____________

D ____________

B (-3, 8)

C (-7, -4)

D (2, -9)

13-4 PLOTTING POINTS ON THE COORDINATE SYSTEM

If we know the coordinates of a point, we can locate or "plot" the point on the coordinate system. We will discuss the method in this section.

---

32. The coordinate axes divide the coordinate system into four parts. These four parts are called quadrants. We have labeled the four quadrants in Figure 1 below. Notice that they are numbered in a counter-clockwise direction, beginning with the upper right quadrant.

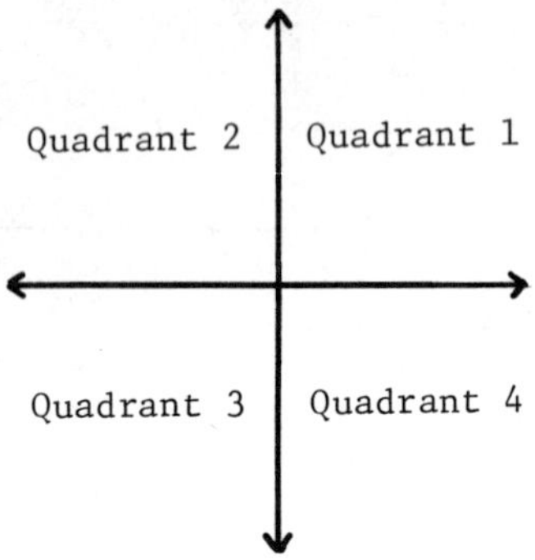

Figure 1

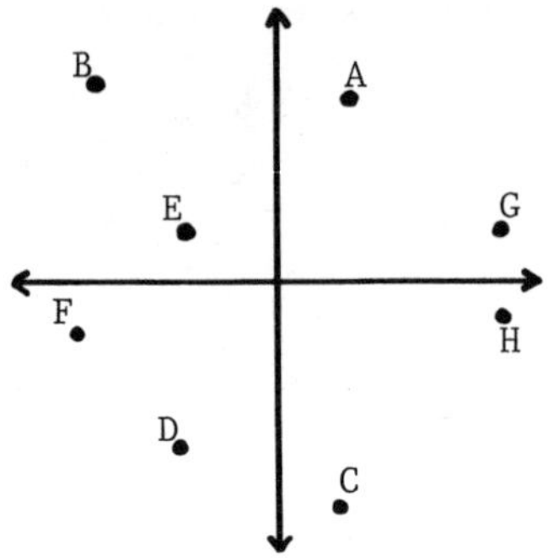

Figure 2

On the coordinate system in Figure 2 above, there are two points labeled in each quadrant.

a) Points B and E lie in Quadrant ______.   b) Points C and H lie in Quadrant ______.

---

33. To locate or "plot" a point on the coordinate system, we simply reverse the procedure for reading the coordinates of a point. For example, we have plotted points A and B at the right by drawing arrows from the axes.

A (2, 4)
B (-4, -3)

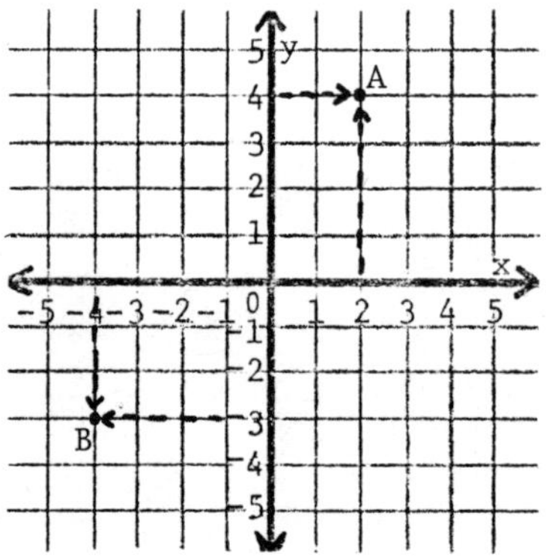

Plot and label the four points below on the same coordinate system.

C (3, 1)   E (-2, 5)
D (3, -2)   F (-2, -4)

> a) Quadrant 2
> b) Quadrant 4

---

34. The correct plotting of the points from the last frame is shown at the right.

Point C is in quadrant ______.

Point D is in quadrant ______.

Point E is in quadrant ______.

Point F is in quadrant ______.

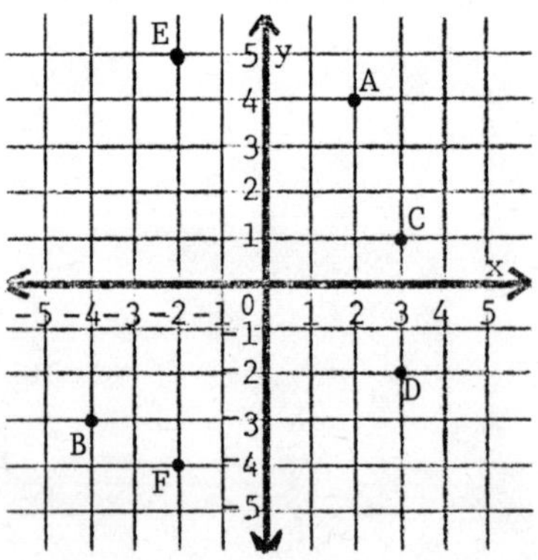

> See next frame for answer.

35. When the scales on the axes are different, be careful when plotting points. For example, we plotted points A and B on the coordinate system at the right.

A (-5, 20)
B (1, -30)

Plot and label the four points below on the same coordinate system.

C (-3, -10) E (1, 50)
D (-2, 40) F (5, -20)

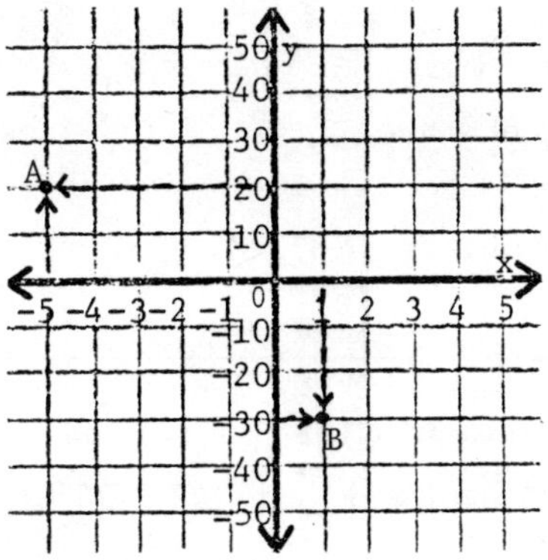

C (quadrant 1)
D (quadrant 4)
E (quadrant 2)
F (quadrant 3)

---

36. The correct plotting of the points from the last frame is shown at the right.

C is in quadrant ______.

D is in quadrant ______.

E is in quadrant ______.

F is in quadrant ______.

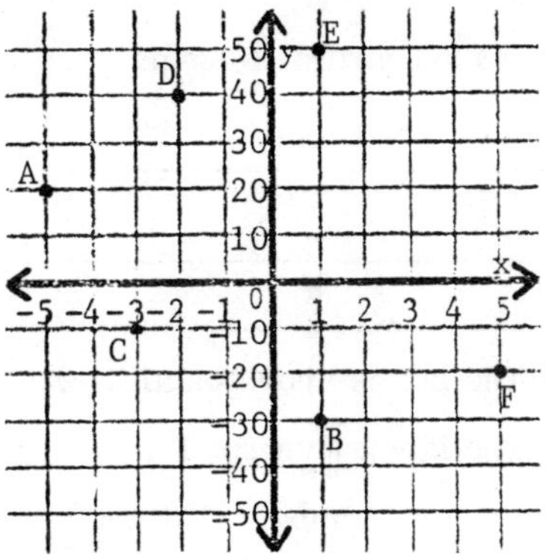

See next frame for answer.

---

37. Only part of the scale is shown on each axis at the right. We plotted the two points below.

A (3, 7)
B (-8, -4)

Plot and label the four points below on the same coordinate system.

C (9, 2) E (-2, -6)
D (-6, 5) F (4, -9)

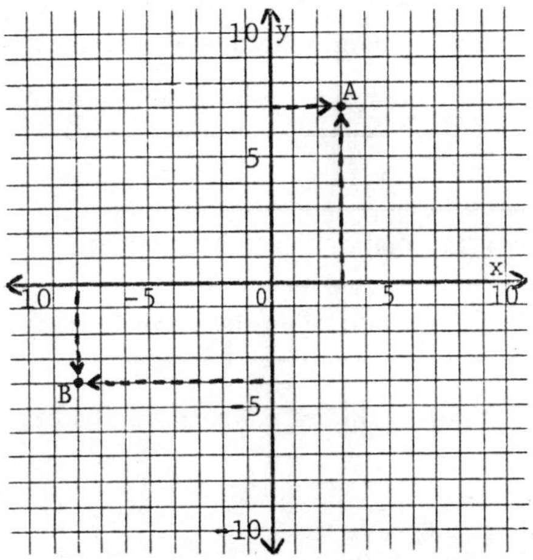

C (quadrant 3)
D (quadrant 2)
E (quadrant 1)
F (quadrant 4)

---

38. The correct plotting of the points from the last frame is shown at the right.

Two other names are used for the coordinates of a point. That is:

The x-coordinate is called the "abscissa".

The y-coordinate is called the "ordinate".

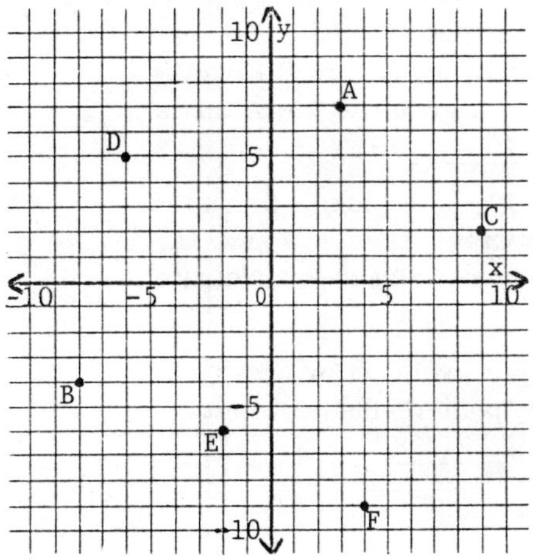

See next frame for answer.

Continued on the following page.

38. Continued.

Since the coordinates of A are (3, 7): its abscissa is 3.
its ordinate is 7.

Since the coordinates of B are (-8, -4): a) its abscissa is ______.

b) its ordinate is ______.

---

a) -8

b) -4

39. Write the coordinates of a point:

a) if its abscissa is 5 and its ordinate is 3. (,)

b) if its ordinate is -1 and its abscissa is -4. (,)

---

a) (5, 3)

b) (-4, -1)

40. a) Another name for the x-coordinate is the ____________________.

b) Another name for the y-coordinate is the ____________________.

---

a) abscissa b) ordinate

Note: The labels "horizontal", "x", and "abscissa" go together.

The labels "vertical", "y", and "ordinate" go together.

To help you remember what goes together, notice that alphabetically:

horizontal comes before vertical

x comes before y

abscissa comes before ordinate

13-5 READING AND PLOTTING POINTS ON THE AXES

In this section, we will read and plot points on the coordinate axes.

41. We have plotted various points on the coordinate system below. Points I, J, K, L lie on the horizontal axis.

a) Points A, B, C, and D have the same y-coordinate (or ordinate). It is ______.

b) Points E, F, G, and H have the same y-coordinate (or ordinate). It is ______.

c) Points I, J, K, and L lie on the horizontal axis. They have the same y-coordinate (or ordinate). It is ______.

d) The y-coordinate (or ordinate) of any point on the horizontal axis is ______.

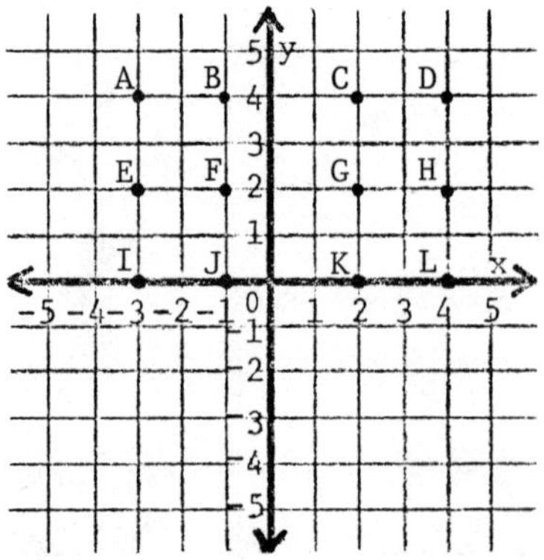

42. Eight points have been plotted on the coordinate system at the right. Write the coordinates of each point:

A ________ E ________

B ________ F ________

C ________ G ________

D ________ H ________

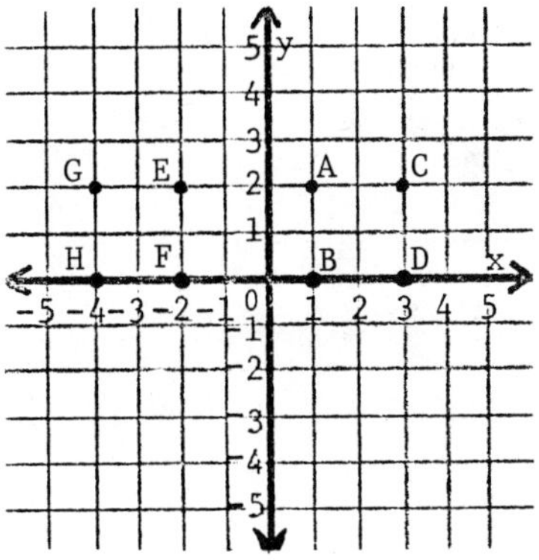

a) 4
b) 2
c) 0
d) 0

---

43. We have plotted various points on the coordinate system below. Some of the points lie on the vertical axis.

a) Points A, B, C, and D have the same x-coordinate (or abscissa). It is ______.

b) Points E, F, G, and H have the same x-coordinate (or abscissa). It is ______.

c) Points I, J, K, and L lie on the vertical axis. Each point has the same x-coordinate (or abscissa). It is ______.

d) The x-coordinate (or abscissa) of any point on the vertical axis is ______.

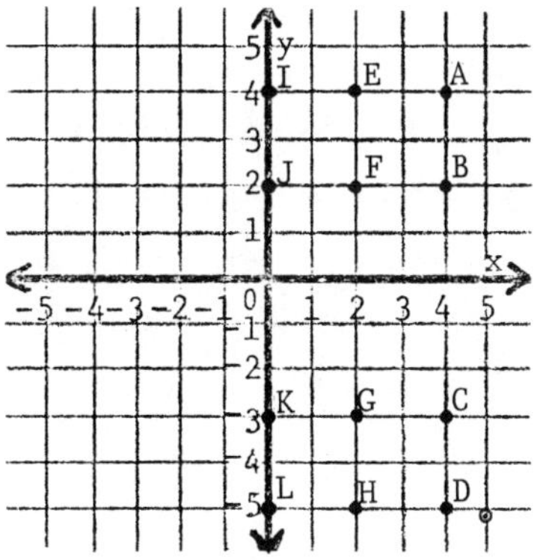

A (1, 2)
B (1, 0)

C (3, 2)
D (3, 0)

E (-2, 2)
F (-2, 0)

G (-4, 2)
H (-4, 0)

---

44. Various points have been plotted on the coordinate system at the right. Write the coordinates of each point:

A ________ E ________

B ________ F ________

C ________ G ________

D ________ H ________

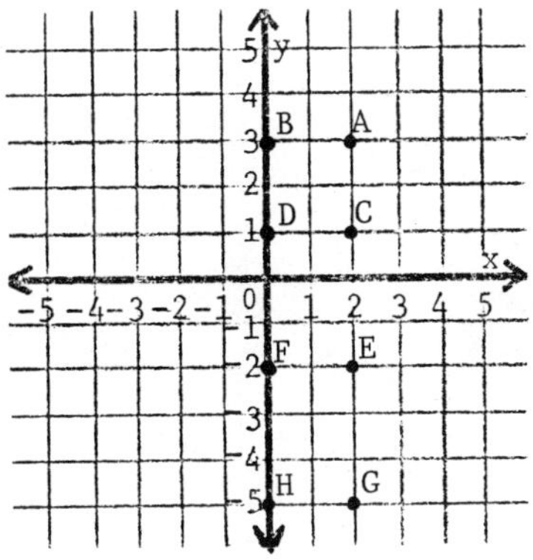

a) 4
b) 2
c) 0
d) 0

A (2, 3)	C (2, 1)	E (2, -2)	G (2, -5)
B (0, 3)	D (0, 1)	F (0, -2)	H (0, -5)

45. Four points have been plotted on the coordinate system at the right. Write the coordinates of each point:

A ________

B ________

C ________

D ________

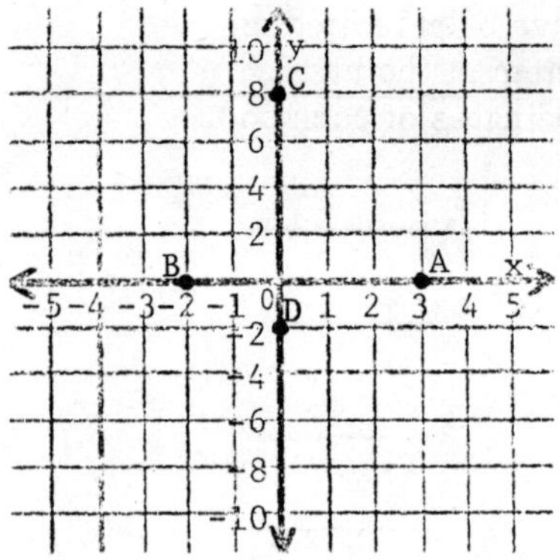

---

46. Plot the following points on the coordinate system at the right:

A (4, 0)

B (0, 4)

C (-3, 0)

D (0, -3)

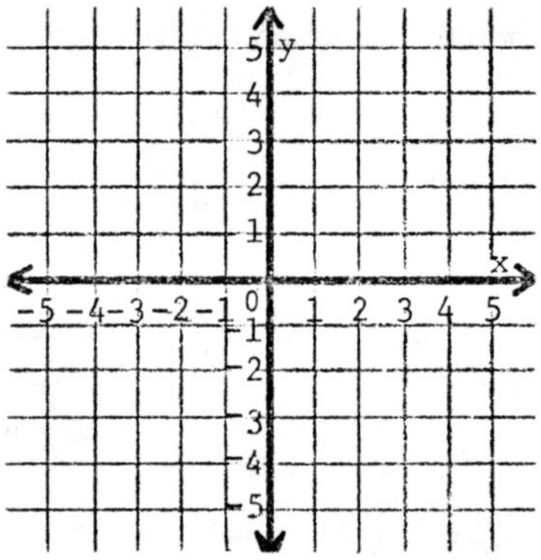

A (3, 0)
B (-2, 0)
C (0, 8)
D (0, -2)

---

47. The correct plotting of the points from the last frame is shown at the right.

As you can see from the graph, the point where the two axes intersect is called the "origin". Since the origin lies on both axes:

a) its x-coordinate is ______.

b) its y-coordinate is ______.

c) its coordinates are (,)

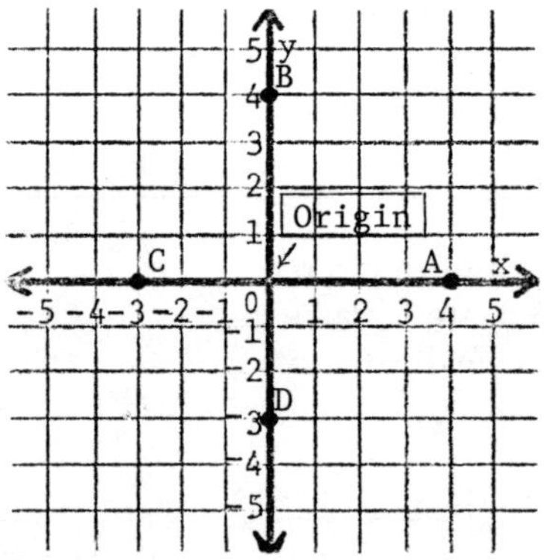

See next frame for answer.

---

a) 0

b) 0

c) (0, 0)

SELF-TEST 39 (pages 334-343)

1. Write the coordinates of the following points on the coordinate system at the right.

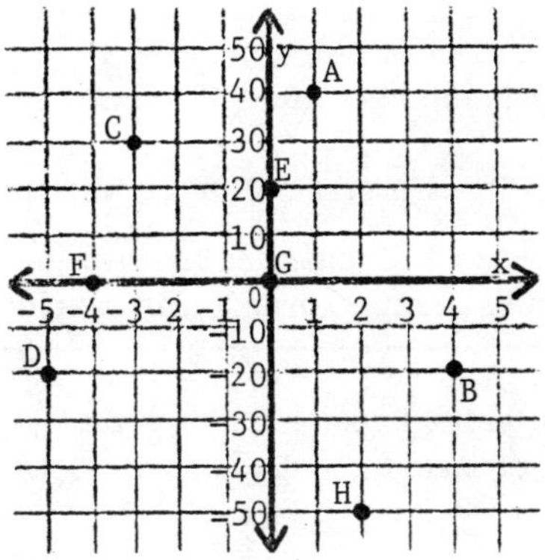

A ________ E ________

B ________ F ________

C ________ G ________

D ________ H ________

2. List the quadrant (1, 2, 3, or 4) in which each point below lies.

A ______ B ______ C ______ D ______

3. Plot and label the following points on the coordinate system at the right.

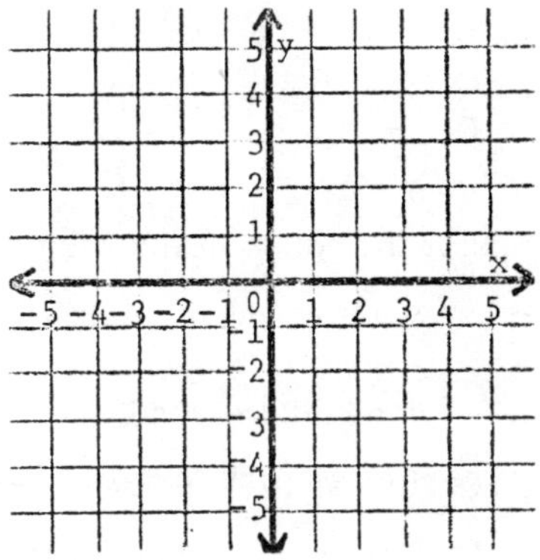

A (4, 3) E (3, 0)

B (-3, -5) F (0, 0)

C (2, -4) G (0, -4)

D (-2, 2) H (-2, -1)

ANSWERS:

1. A (1, 40) E (0, 20)
   B (4, -20) F (-4, 0)
   C (-3, 30) G (0, 0)
   D (-5, -20) H (2, -50)

2. A) 1
   B) 4
   C) 2
   D) 3

3.

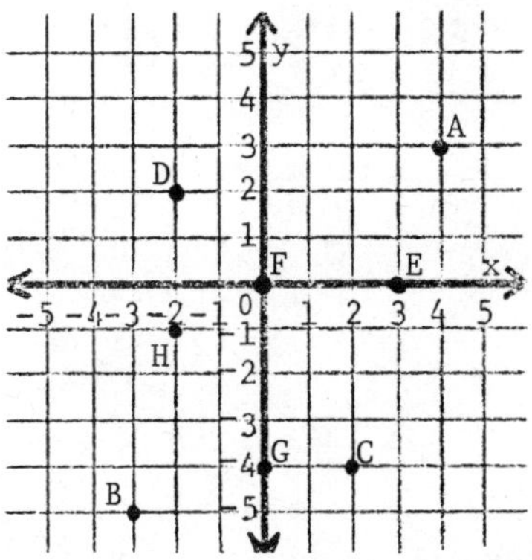

13-6 GRAPHING EQUATIONS CONTAINING "x" AND "y"

Any equation containing both "x" and "y" can be graphed on the coordinate system. To do so, the following three steps are used:

1. Make up a table of solutions.
2. Plot the points representing these solutions.
3. Draw a straight or curved line through the plotted points.

In this section, we will discuss the method for graphing "xy" equations. Though emphasis is given to straight-line graphs, some curved-line graphs are shown.

---

48. The three steps needed to graph the following equation are discussed below.

$$y = x + 1$$

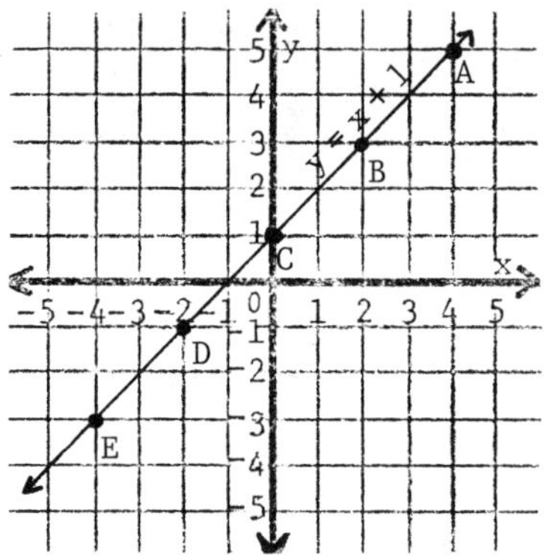

Step 1: Make up a solution-table.

	x	y
A	4	5
B	2	3
C	0	1
D	-2	-1
E	-4	-3

Step 2: Plot the points in the solution-table.

Step 3: Draw a straight line through the plotted points.

The straight line shown is the graph of $y = x + 1$.

a) Does the straight line (or graph) pass through the origin? ________

b) The straight line (or graph) passes through what three quadrants? ________

---

a) No

b) 1, 2, 3

49. When making up a solution-table for graphing, we always use "0" and a few positive and negative values for x. Even though different values are used for x, we get the same graph. For example, we used two different tables to graph $y = 2x$ below. Notice that the graphed lines are identical.

$$y = 2x$$

	x	y
A	2	4
B	1	2
C	0	0
D	-1	-2
E	-2	-4

$$y = 2x$$

	x	y
A	5	10
B	3	6
C	0	0
D	-3	-6
E	-5	-10

Continued on the following page.

49. Continued.

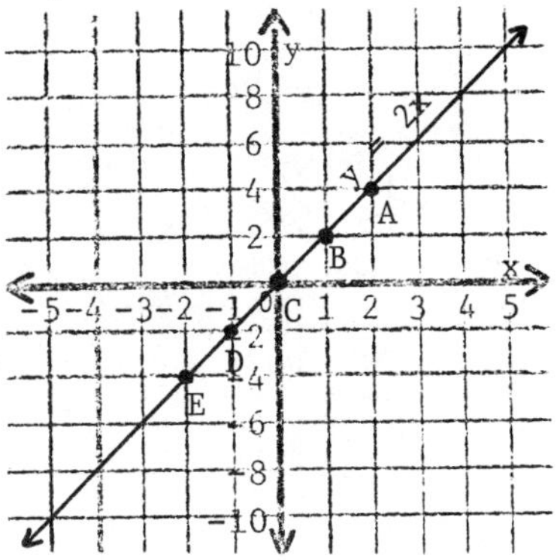

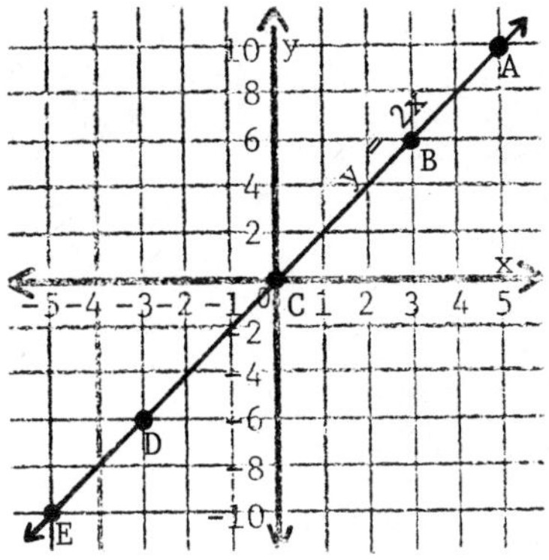

a) Does the graphed line pass through the origin? ________

b) The graphed line passes through what two quadrants? ________

---

a) Yes
b) 1 and 3

50. At the right we have graphed the equation:

$$x + y = 3$$

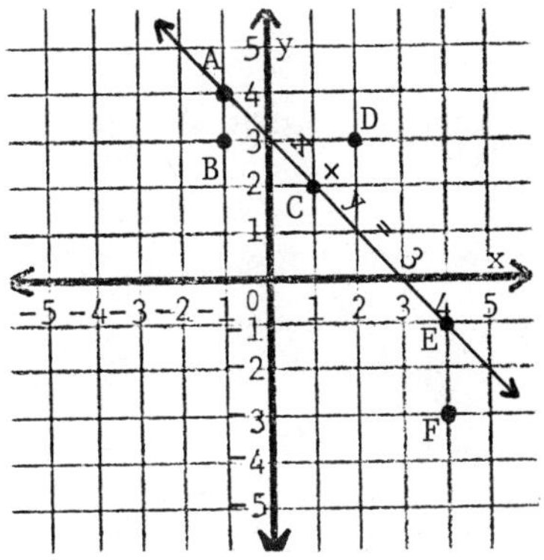

Any point on the graphed line represents one solution of the equation. Therefore, its coordinates satisfy the equation.

Any point off the graphed line does not represent a solution of the equation. Therefore, its coordinates do not satisfy the equation.

Let's examine the points on the line first.

a) Write the coordinates of the three points on the graphed line:

A (,) C (,) E (,)

b) Do the coordinates of these points satisfy the equation $x + y = 3$? ______

Now let's examine the points off the line.

c) Write the coordinates of the three points off the graphed line:

B (,) D (,) F (,)

d) Do the coordinates of these points satisfy the equation $x + y = 3$? ________

---

a) A (-1, 4)
C (1, 2)
E (4, -1)

b) Yes, since in each case the sum of the coordinates is "3".

c) B (-1, 3)
D (2, 3)
F (4, -3)

d) No, since in each case the sum of the coordinates is not "3".

51. On the two coordinate systems below, we have graphed the equation $x + y = 1$ from the table of solutions shown:

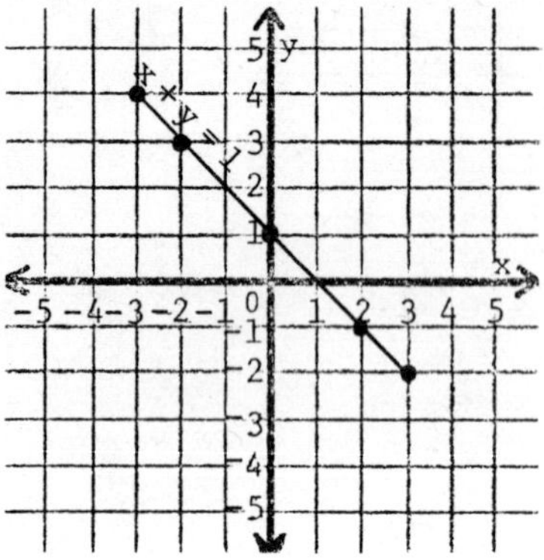

Figure 1

$x + y = 1$	
x	y
3	-2
2	-1
0	1
-2	3
-3	4

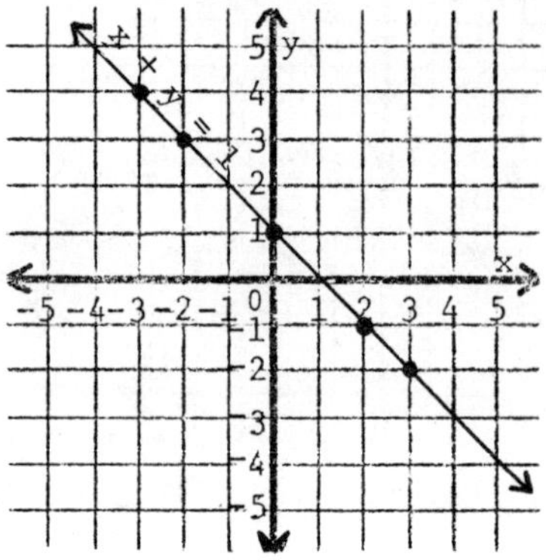

Figure 2

Note: In Figure 1 we stopped the graphed line at the last plotted point on each end.

In Figure 2 we extended the graphed line to the edge of the coordinate system shown and put arrowheads at each end of it.

Figure 2 is the correct graphing of $x + y = 1$ for these reasons:

1) The line should be extended to the edge of the coordinate system to show that there are other solutions, like (5, -4) and (-4, 5), beyond -3 and 3 on that part of the line.

2) Arrowheads should be put at each end of the graphed line to show that there are solutions, like (10, -9) and (-20, 21), beyond the edge of that part of the coordinate system.

---

52. Using the tables provided, graph each equation. Be sure to extend each line to the edge of the coordinate system and put arrowheads on each end of the line.

$y = 2x + 1$	
x	y
2	5
1	3
0	1
-1	-1
-2	-3

$x + y = 2$	
x	y
4	-2
2	0
0	2
-2	4
-4	6

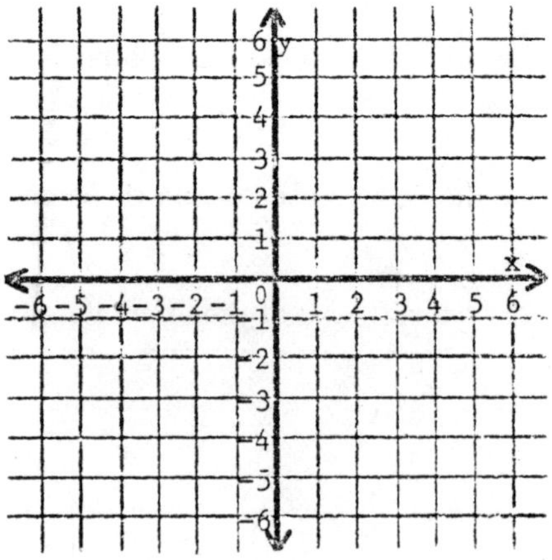

See next frame for answer.

53. The correct graphs of the equation from the last frame are shown at the right.

a) Does either graphed line pass through the origin? ______

b) The graph of $x + y = 2$ lies in what three quadrants? ______

c) The graph of $y = 2x + 1$ lies in what three quadrants? ______

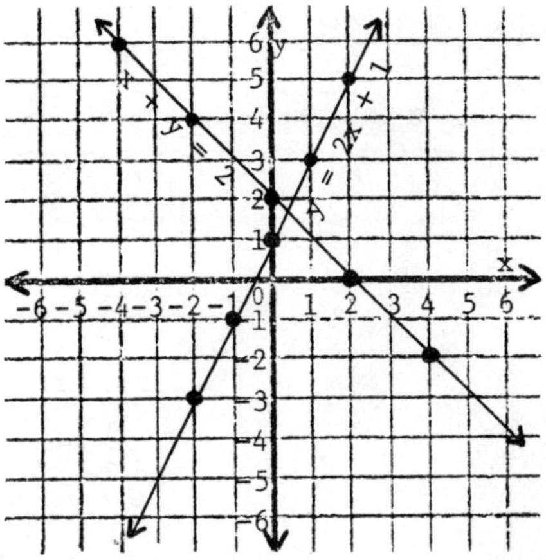

a) No

b) 1, 2, 4

c) 1, 2, 3

---

54. Complete each solution-table and then graph both equations on the coordinate system provided.

$y = 3x$

x	y
3	
1	
0	
-1	
-3	

$2x + y = 1$

x	y
4	
2	
0	
-2	
-4	

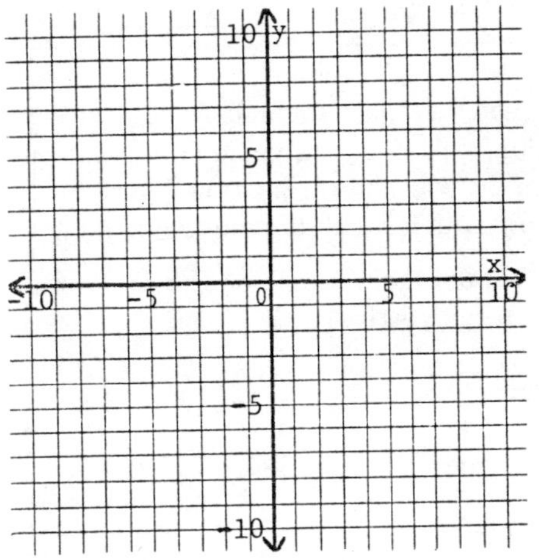

x	y
3	9
1	3
0	0
-1	-3
-3	-9

x	y
4	-7
2	-3
0	1
-2	5
-4	9

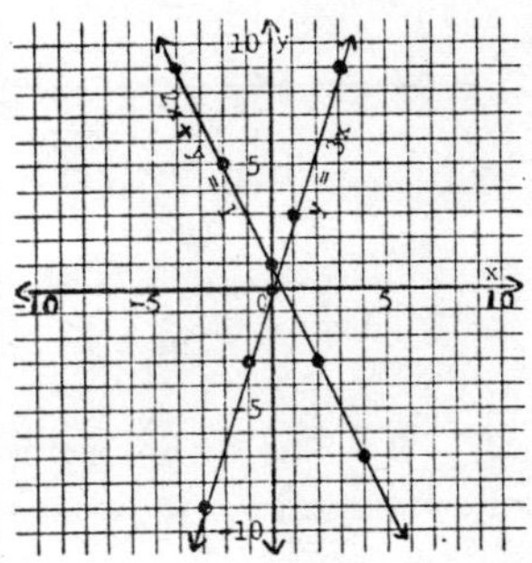

---

55. The graph of an equation can be a curved line rather than a straight line. For example, some sketches of the graphs of three equations are shown below. Each graph is a curved line.

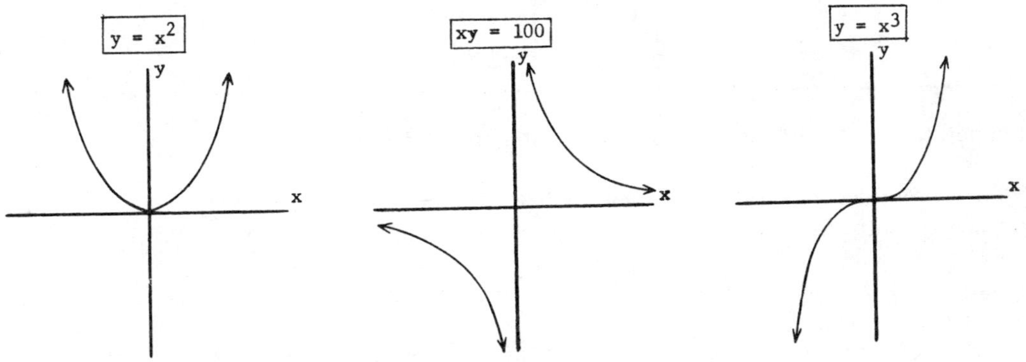

Note: We will not discuss the method for constructing curved graphs in this book.

13-7 GRAPHING FORMULAS

Any two-letter formula can be graphed on the coordinate system. We will discuss the method in this section.

---

56. The formula below can be used to convert temperature from the Celsius (C) scale to the Fahrenheit (F) scale.

$$F = \frac{9}{5}C + 32°$$

For example: If C = 10°, $F = \frac{9}{5}(10°) + 32°$

$= 18° + 32° = 50°$

If C = -20°, $F = \frac{9}{5}(-20°) + 32°$

$=$ ______ $+ 32° =$ ______

(-36°) + 32° = -4°

57. The method used to graph a formula is the same as the method used to graph an "xy" equation. That is, we make up a solution-table, plot the points in the table, and then draw a line through the plotted points. As an example, we graphed the Celsius-Fahrenheit formula below.

Note: We put the C-scale on the horizontal axis and the F-scale on the vertical axis.

$$F = \frac{9}{5}C + 32°$$

C	F
40°	104°
20°	68°
0°	32°
-20°	-4°
-40°	-40°

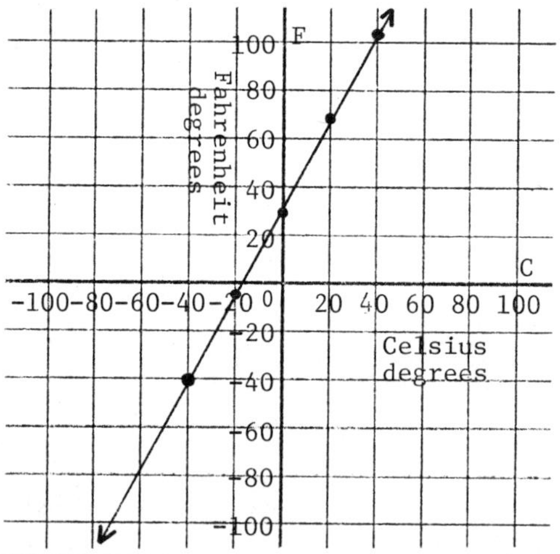

---

58. When measuring temperatures, negative values are used for both C (degrees - Celsius) and F (degrees - Fahrenheit). However, when measuring other quantities (like distance weight, pressure, volume, and so on) negative values are not ordinarily used. Therefore, only quadrant 1 is used when graphing almost all other formulas. An example is discussed below.

The following formula represents the distance traveled in a given period of time at a constant velocity (or speed) of 50 miles per hour.

$$d = 50t$$

where d is distance traveled (in miles)
and t is time (in hours)

Continued on the following page.

58. Continued.

Since negative values are not ordinarily used for either distance (d) or time (t), only quadrant 1 is used when graphing the formula. Complete the table below and then graph d = 50t.

$d = 50t$

t (in hours)	d (in miles)
0	
1	
2	
3	
4	

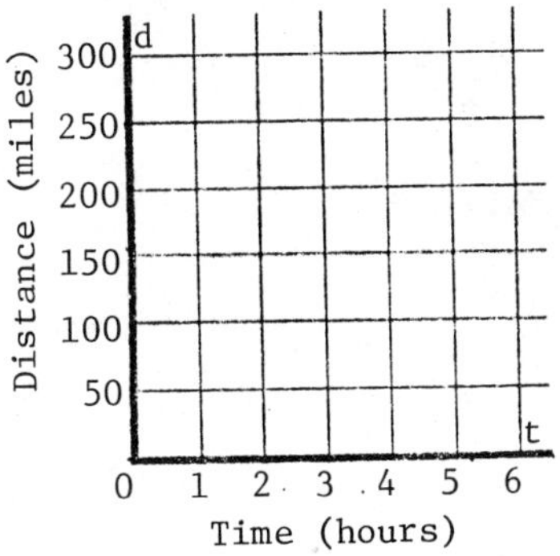

t	d
0	0
1	50
2	100
3	150
4	200

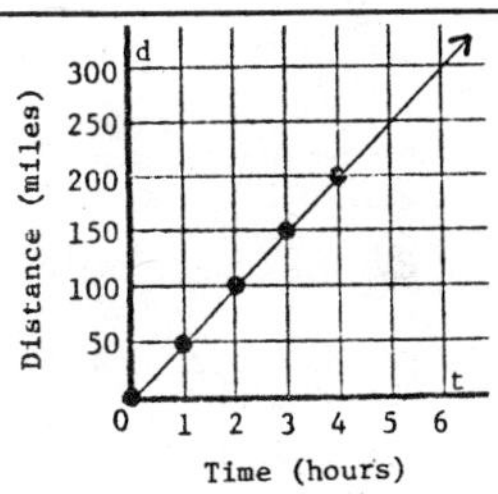

59. The graphs of formulas can also be curved lines. Two examples are sketched below. Notice again that the graphs appear only in quadrant 1.

The formula below represents the distance a dropped object will fall in a given period of time.

$s = 16t^2$

The formula below represents the relationship between the pressure and volume of a specific gas at a constant temperature.

$PV = 100$

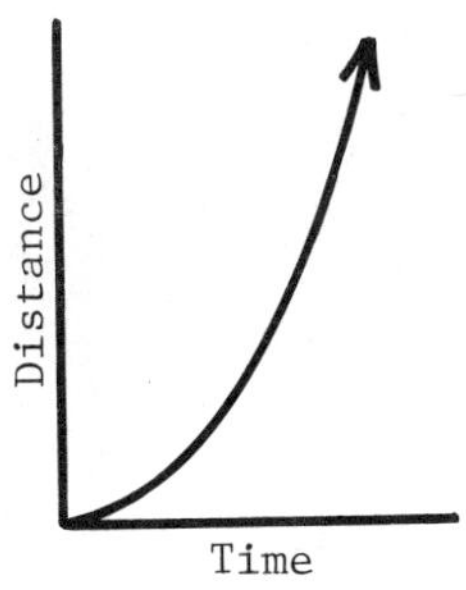

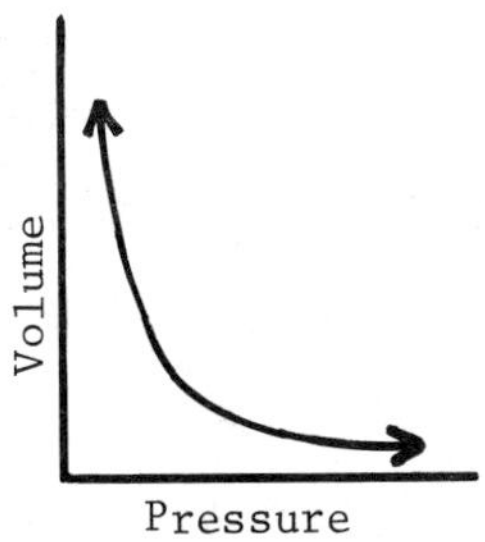

13-8 USING GRAPHS TO SOLVE APPLIED PROBLEMS

Graphs can be used to solve applied problems. We will discuss the method in this section.

60. The graph below shows the relationship between "distance traveled" and "time" at an average velocity (or speed) of 100 miles per hour.

We can use the graph to solve the following problem.

How far will a plane travel in 3 hours at an average speed of 100 miles per hour?

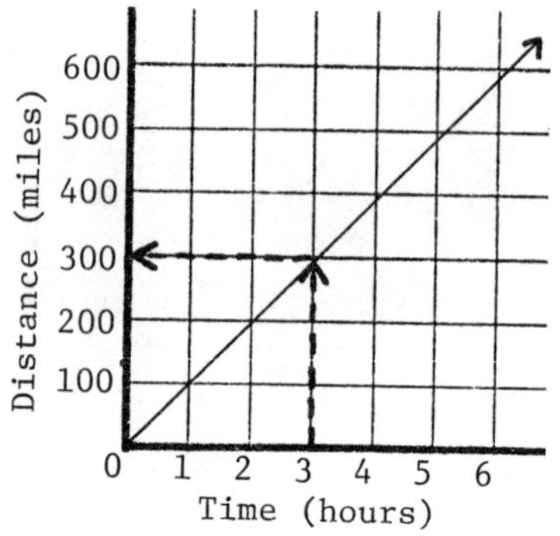

To solve the problem, we use these steps:

1) Draw an arrow up from "3" on the "time" axis to the graphed line.

2) Then draw an arrow over from that point to the "distance" axis.

3) Since the horizontal arrow points to "300" on the "distance" axis, we know that the plane will travel 300 miles in 3 hours at that speed.

Use the graph to answer these:

a) How far would the plane travel in 2 hours? ______________

b) How far would the plane travel in 5 hours? ______________

a) 200 miles

b) 500 miles

61. The graph below shows the relationship between "distance traveled" and "time" at an average velocity (or speed) of 50 miles per hour.

We can use the graph to solve this problem.

How long would it take a car to drive 150 miles at an average speed of 50 miles per hour?

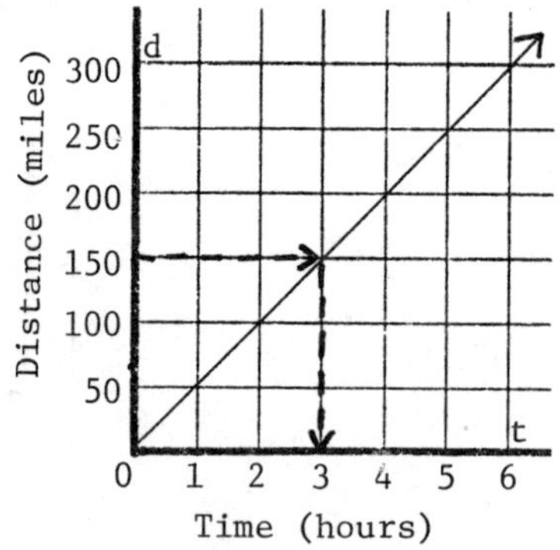

To solve the problem, we use these steps:

1) Draw an arrow from 150 on the "distance" axis to the graphed line.

2) Then draw an arrow from that point of intersection to the "time" axis.

3) Since the vertical arrow points to "3" on the "time" axis, we know that it would take the car 3 hours to drive 150 miles at that speed.

Use the graph to answer these:

a) How long would it take the car to drive 100 miles? ______________

b) How long would it take the car to drive 250 miles? ______________

62. If a force is applied to a spring, the spring stretches. The amount of stretch depends on the amount of force applied and also on the type of spring. The graph below shows the amount of stretch (in inches) for different amounts of force applied (in pounds) to a specific spring.

a) 2 hours
b) 5 hours

We can use the graph to solve this problem:

If 4 pounds of force are applied, how far will the spring stretch?

By looking at the arrows, you can see this answer:

The spring stretches 2 inches when a 4-pound force is applied.

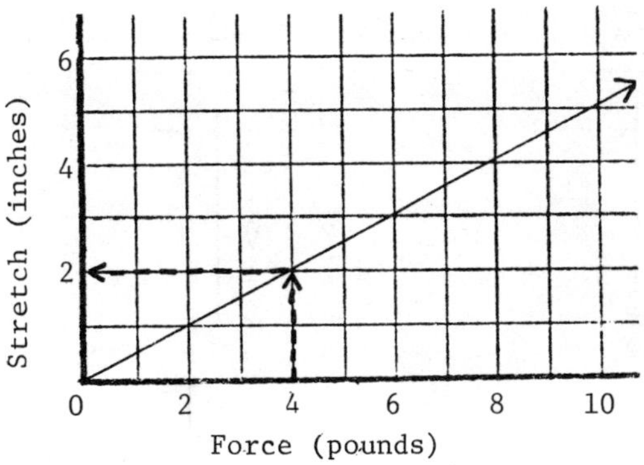

Use the graph to answer these:

a) How far would the spring stretch if a 6-pound force was applied? _______________

b) How far would the spring stretch if a 10-pound force was applied? _______________

---

a) 3 inches
b) 5 inches

63. The relationship between stretch (in inches) and force applied (in pounds) for a different spring is shown in the graph below.

We can use the graph to solve this problem:

How much force is needed to get a 10-inch stretch?

By looking at the arrows, you can see this answer:

A 400-pound force is needed to get a 10-inch stretch.

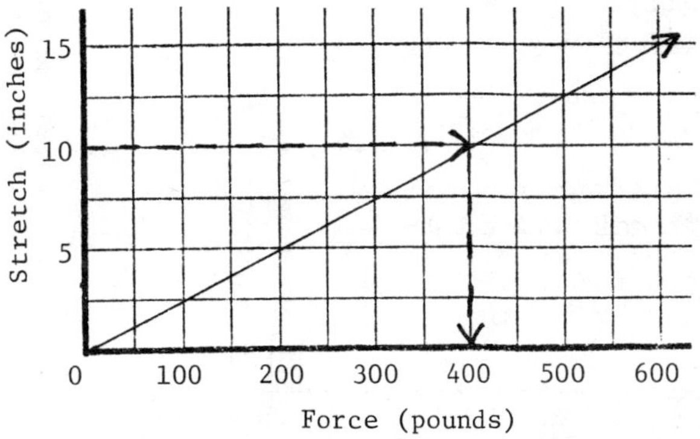

Use the graph to answer these:

a) How much force is needed to get a 15-inch stretch? _______________

b) How much force is needed to get a 5-inch stretch? _______________

---

a) 600 pounds
b) 200 pounds

SELF-TEST 40 (pages 344-352)

Complete each table and then graph each equation on the coordinate system provided.

1. $x - y = 0$

x	y
5	
3	
0	
-3	
-5	

2. $2x + y = 1$

x	y
2	
1	
0	
-1	
-2	

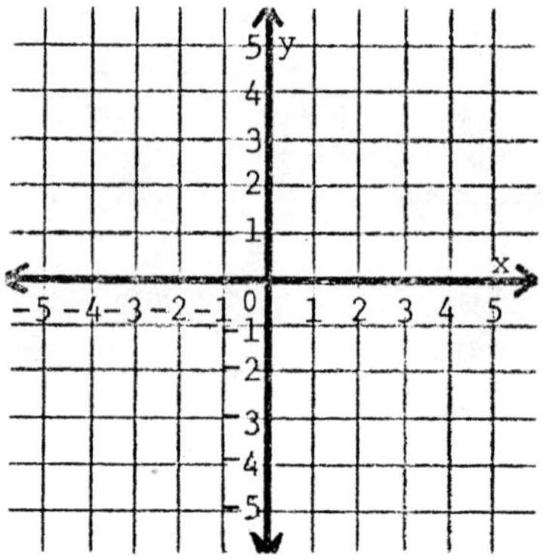

A student with a part-time job is paid on an hourly basis. The relationship between his wages (in dollars) and the amount of hours worked is shown in the graph below. Use the graph to answer these questions.

3. How much would he earn if he worked 12 hours? ________

4. How many hours would he have to work to earn $20? ________

5. How much would he earn if he worked 16 hours? ________

6. How many hours would he have to work to earn $50? ________

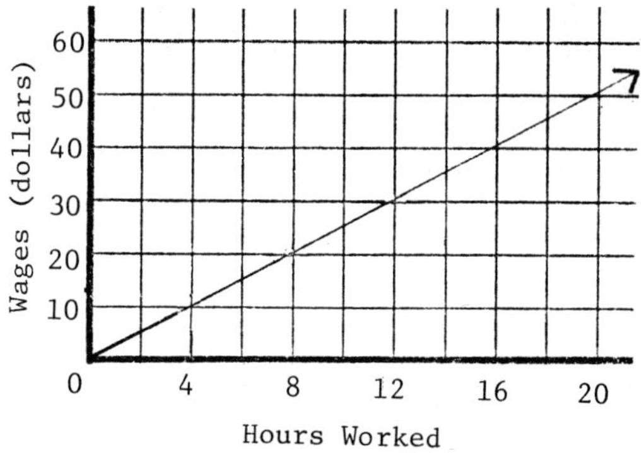

ANSWERS:

1.

x	y
5	5
3	3
0	0
-3	-3
-5	-5

2.

x	y
2	-3
1	-1
0	1
-1	3
-2	5

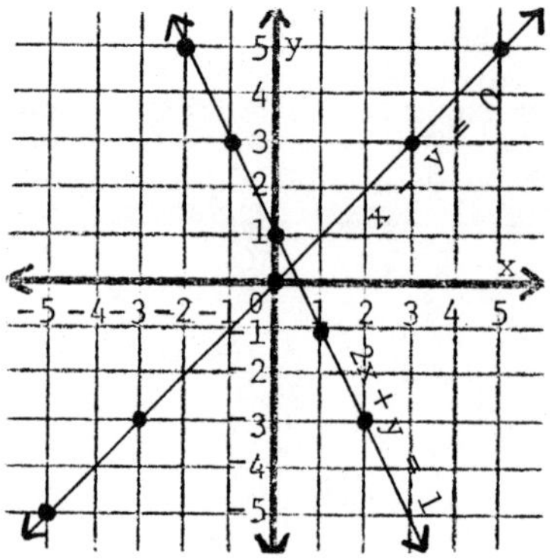

3. $30
4. 8 hours
5. $40
6. 20 hours

Unit 14 SYSTEMS OF EQUATIONS

In this unit, we will define what is meant by a "system" of equations and then solve systems of equations by both the graphing method and the addition method. Some applied problems are included.

14-1 THE MEANING OF A SYSTEM OF EQUATIONS

In this section, we will define what is meant by a system of equations and the solution of a system of equations.

1. It is sometimes easier to translate an applied problem to mathematical language if we use two letters and two equations. Two examples are shown.

Example 1: There are 27 students in a class. If there are twice as many females as males, how many females and males are there?

There are two statements in the problem. Each leads to a different equation involving number of females (f) and number of males (m). That is:

There are 27 students in a class.

$$f + m = 27$$

There are twice as many females as males.

$$f = 2m$$

Therefore, the problem has been translated to the following pair of equations.

$$f + m = 27$$
$$f = 2m$$

Example 2: The sum of two numbers is 21. The difference between the larger number and smaller number is 5. Find the two numbers.

Using "x" for the larger number and "y" for the smaller number, we can set up two different equations. That is:

The sum of two numbers is 21.

$$x + y = 21$$

The difference between the larger and smaller number is 5.

$$x - y = 5$$

Therefore, the problem has been translated to the following pair of equations.

$$x + y = 21$$
$$x - y = 5$$

2. In the last frame, we translated the problems to the following pairs of equations.

$f + m = 27$ $f = 2m$	$x + y = 21$ $x - y = 5$

Pairs of equations like those above are called "systems" of equations. The solution of a system of equations is a pair of values that satisfies both equations. For example, for the system at the left above:

($f = 20$, $m = 7$) is not a solution because it only satisfies the top equation. That is:

$f + m = 27$	$f = 2m$
$20 + 7 = 27$	$20 \neq 2(7)$
$27 = 27$	$20 \neq 14$

($f = 16$, $m = 8$) is not a solution because it only satisfies the bottom equation. That is:

$f + m = 27$	$f = 2m$
$16 + 8 \neq 27$	$16 = 2(8)$
$24 \neq 27$	$16 = 16$

($f = 18$, $m = 9$) is a solution because it satisfies both equations. That is:

$f + m = 27$	$f = 2m$
$18 + 9 = 27$	$18 = 2(9)$
$27 = 27$	$18 = 18$

Which of the following pairs of values is a solution of the other system above because it satisfies both $x + y = 21$ and $x - y = 5$? ______________

a) $x = 15$, $y = 6$ b) $x = 10$, $y = 5$ c) $x = 13$, $y = 8$

---

c) $x = 13$, $y = 8$

3. Remember that a pair of values is a solution of a system of equations only if it satisfies both equations.

Which pair of values below satisfies the system at the right? ______________

$p = q + 1$
$p + q = 5$

a) $p = 1$, $q = 0$ b) $p = 3$, $q = 2$ c) $p = 4$, $q = 1$

---

b) $p = 3$, $q = 2$

4. Which pair of values below is the solution of the system of equations at the right? ______________

$x + y = 5$
$y = 4x$

a) $x = 1$, $y = 4$ b) $x = 0$, $y = 5$ c) $y = 0$, $x = 5$

---

a) $x = 1$, $y = 4$

14-2 THE GRAPHING METHOD FOR SOLVING SYSTEMS

Various methods can be used to solve systems of equations. In this section, we will discuss the graphing method.

---

5. We have graphed both equations in the system below at the right.

$x - y = 1$
$x + y = 5$

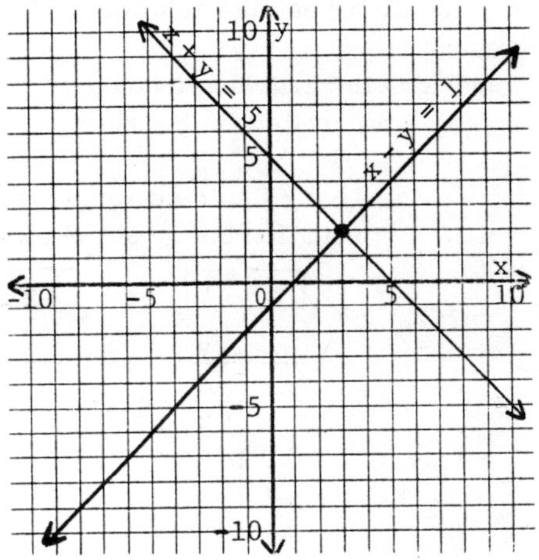

The point where the two graphed lines cross is called the "point of intersection". Since that point lies on both lines, its coordinates satisfy both equations. Therefore, its coordinates are the solution of the system. Check that fact by answering the questions below.

a) The coordinates of the point of intersection are:

x = ______ , y = ______

b) Do those coordinates satisfy both equations? ________

c) Therefore, the solution of the system is:

x = ______ , y = ______

---

a) $x = 3$, $y = 2$

b) Yes.

c) $x = 3$, $y = 2$

6. When the graphs of both equations are straight lines and the lines intersect, there is only one point of intersection. Therefore, there is only one solution of the system of equations.

Two systems of equations have been graphed below. Using the coordinates of the point of intersection, write the one solution of each system.

a)

$x + y = 3$
$y = 2x$

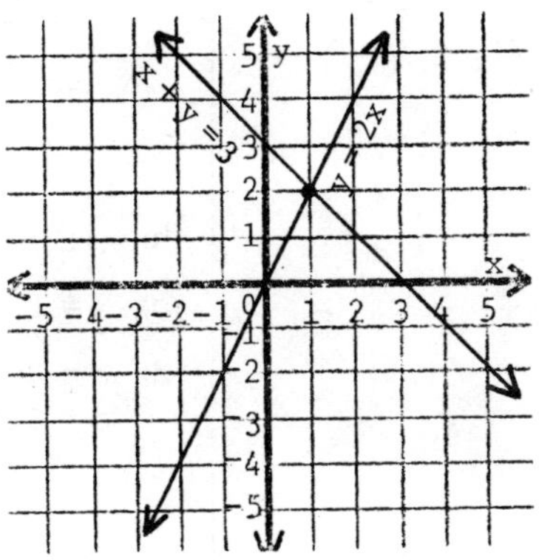

The solution is: x = _____ , y = _____

b)

$2x - y = 3$
$x + y = 0$

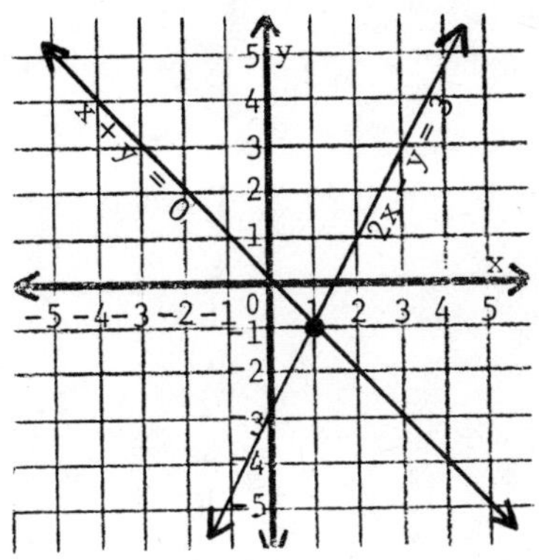

The solution is: x = _____ , y = _____

7. To solve a system of equations by the graphing method, we use these steps:

1. Graph both equations on the same set of axes.
2. Use the coordinates of the point of intersection as the solution of the system.
3. Check the solution in both equations of the system.

a) $x = 1,\ y = 2$

b) $x = 1,\ y = -1$

Let's use the graphing method to solve the system at the right.

$$\begin{aligned} y &= 3x \\ x + y &= 4 \end{aligned}$$

a) Using the tables provided, graph each equation on the set of axes.

$y = 3x$

x	y
2	6
1	3
0	0
-1	-3
-2	-6

$x + y = 4$

x	y
5	-1
3	1
0	4
-3	7
-5	9

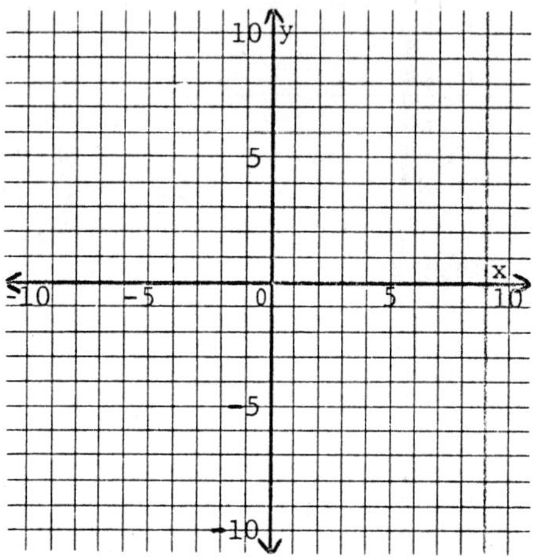

b) The coordinates of the point of intersection are (,).

c) Therefore, the solution of the system is:

x = ______ , y = ______

d) Check the solution in both original equations.

$y = 3x$ $x + y = 4$

a)

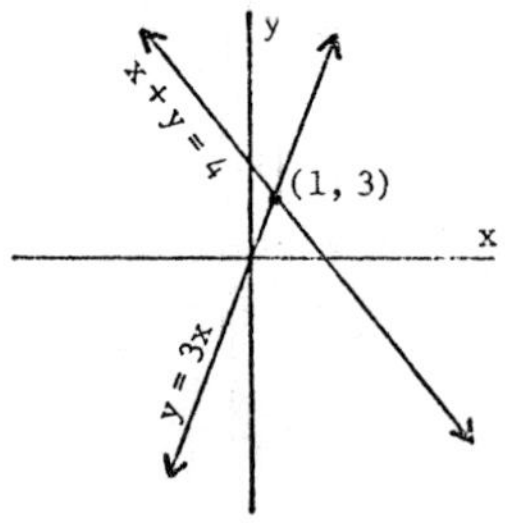

b) (1, 3)

c) $x = 1,\ y = 3$

d) $3 = 3(1)$ $1 + 3 = 4$
$3 = 3$ $4 = 4$

8. Let's use the graphing method to solve the system below.

$$y = 2x$$
$$y - x = 3$$

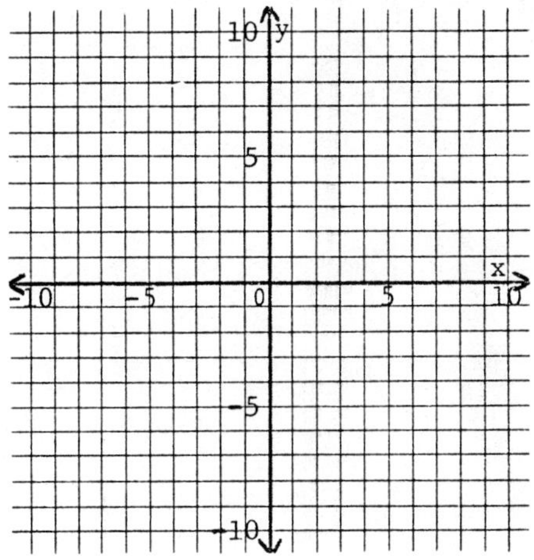

a) Complete each table and graph each equation.

$y = 2x$

x	y
5	
3	
0	
-3	
-5	

$y - x = 3$

x	y
4	
2	
0	
-2	
-4	

b) The coordinates of the point of intersection are (,).

c) The solution of the system is: x = ______ , y = ______

d) Check the solution in both original equations.

$y = 2x$ $\qquad$ $y - x = 3$

---

a)

x	y
5	10
3	6
0	0
-3	-6
-5	-10

x	y
4	7
2	5
0	3
-2	1
-4	-1

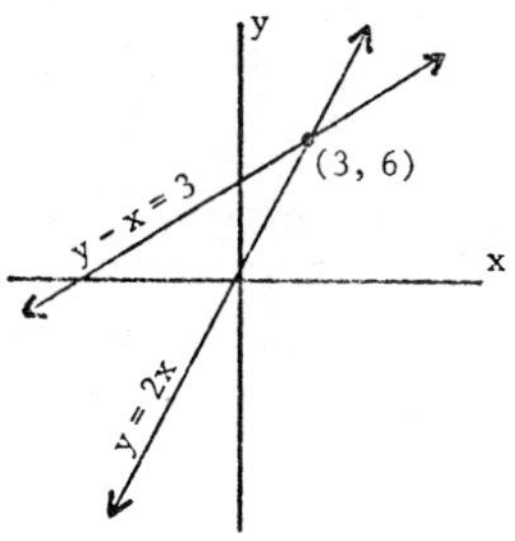

b) (3, 6)

c) x = 3, y = 6

d) $6 = 2(3)$ $\qquad$ $6 - 3 = 3$
$\quad$ $6 = 6$ $\qquad$ $3 = 3$

9. Using the tables provided, graph the following system on the axes below.

$$\begin{aligned} 2x - y &= 6 \\ x + 2y &= 13 \end{aligned}$$

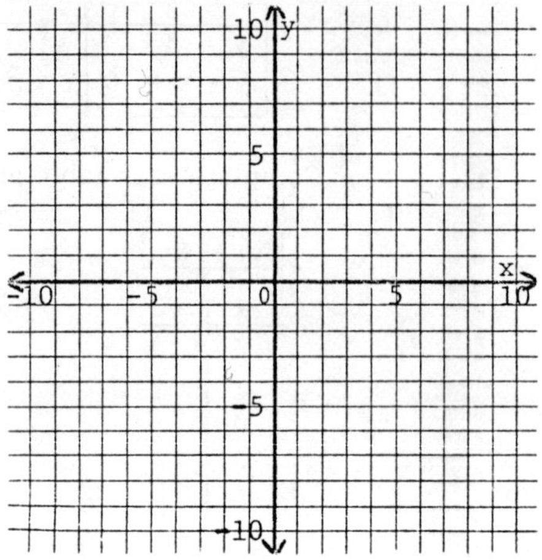

$2x - y = 6$

x	y

$x + 2y = 13$

x	y

a) The coordinates of the point of intersection are (,).

b) Therefore, the solution of the system is: x = ______ , y = ______

a) (5, 4)

b) x = 5, y = 4

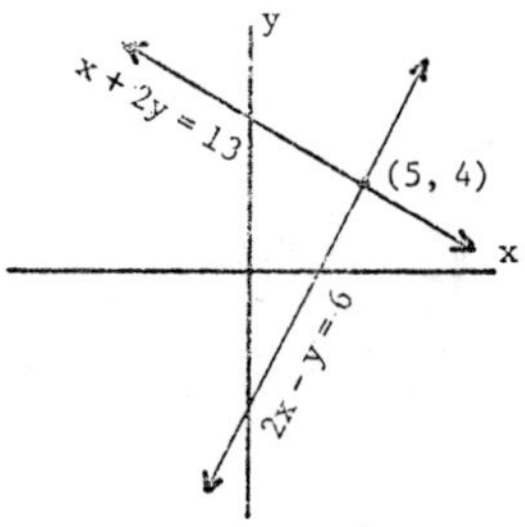

14-3 THE ADDITION METHOD FOR SOLVING SYSTEMS

Because the graphing method for solving systems of equations is time-consuming and inaccurate when the solutions are not whole numbers, an algebraic method is ordinarily used. In this section, we will discuss an algebraic method called the "addition" method for solving systems.

10. In the addition method for solving systems, the key step is adding the two equations. To do so, we add the like terms in columns as we have done below.

$$\begin{aligned} 2x + 5y &= 7 \\ 3x - 2y &= 4 \\ \hline 5x + 3y &= 11 \end{aligned}$$

<u>Note</u>: To get $5x$, we added $2x$ and $3x$.

To get $3y$, we added $5y$ and $-2y$.

To get 11, we added 7 and 4.

Use the same method to add each pair of equations below.

a) $$\begin{aligned} 4x + 7y &= 5 \\ 3x - 4y &= 10 \\ \hline \end{aligned}$$

b) $$\begin{aligned} 9s + 4t &= 3 \\ 5s - 2t &= 7 \\ \hline \end{aligned}$$

a) $7x + 3y = 15$

b) $14s + 2t = 10$

11. When adding a pair of equations, remember that the coefficient of a letter is "1" or "-1" if the coefficient is not explicitly shown. That is:

$$\begin{array}{r} x + 3y = 5 \\ \underline{2x - y = 7} \\ 3x + 2y = 12 \end{array}$$

Note: To get 3x, we added 1x and 2x.

To get 2y, we added 3y and -1y.

Use the same method to add each pair of equations below.

a) $\begin{array}{r} 3x + 4y = 3 \\ \underline{x - y = 1} \end{array}$ b) $\begin{array}{r} m + 7f = 9 \\ \underline{m - f = 0} \end{array}$

---

a) 4x + 3y = 4
b) 2m + 6f = 9

12. When the "y" terms were added below, we got "0y". Since "0y" can be dropped, we eliminated "y" by performing the addition. Complete the other additions.

$\begin{array}{r} x + 2y = 3 \\ \underline{x - 2y = 5} \\ 2x + 0y = 8 \end{array}$ a) $\begin{array}{r} x + 4y = 0 \\ \underline{2x - 4y = 7} \end{array}$ b) $\begin{array}{r} a + b = 10 \\ \underline{a - b = 2} \end{array}$

or

$2x = 8$

---

a) 3x + 0y = 7
or 3x = 7
b) 2a + 0b = 12
or 2a = 12

13. In the addition method for solving a system, we add the equations <u>in order to eliminate a letter</u>. An example is shown. Complete the other additions.

$\begin{array}{r} 2x + 3y = 0 \\ \underline{4x - 3y = 6} \\ 6x \quad = 6 \end{array}$ a) $\begin{array}{r} m + 5t = 7 \\ \underline{3m - 5t = 5} \end{array}$ b) $\begin{array}{r} p - q = 4 \\ \underline{p + q = 0} \end{array}$

---

a) 4m = 12
b) 2p = 4

14. We used the addition method below to solve the system at the right. The two steps are described.

x + y = 5
x - y = 1

1. <u>Finding the value of "x"</u>.

By adding the two equations, we can eliminate "y".

$$\begin{array}{r} x + y = 5 \\ \underline{x - y = 1} \\ 2x \quad = 6 \end{array}$$

Since there is no "y" term in the new equation, we can solve for "x".

$$2x = 6$$

$$x = 3$$

Continued on following page.

14. Continued

2. Finding the value of "y".

We can now find the corresponding value of "y" by substituting "3" for "x" in either of the original equations.

$$\begin{aligned} x + y &= 5 \\ 3 + y &= 5 \\ y &= 2 \end{aligned} \qquad\qquad \begin{aligned} x - y &= 1 \\ 3 - y &= 1 \\ y &= 2 \end{aligned}$$

The obtained solution is $(x = 3\ ,\ y = 2)$. Check that solution in each original equation below.

a) $x + y = 5$ b) $x - y = 1$

a) $3 + 2 = 5$
$5 = 5$

b) $3 - 2 = 1$
$1 = 1$

---

15. We used the addition method below to solve the system at the right.

$$\boxed{\begin{aligned} s + 2t &= 5 \\ 3s - 2t &= 7 \end{aligned}}$$

1. Finding the value of "s".

By adding the equations, we can eliminate "t" and solve for "s".

$$\begin{aligned} s + 2t &= 5 \\ 3s - 2t &= 7 \\ \hline 4s \quad &= 12 \\ s &= 3 \end{aligned}$$

2. Finding the value of "t".

To find the corresponding value of "t", we substituted "3" for "s" in the top equation.

$$\begin{aligned} s + 2t &= 5 \\ 3 + 2t &= 5 \\ 2t &= 2 \\ t &= 1 \end{aligned}$$

Note: We could also have substituted "3" for "s" in the bottom equation.

The obtained solution is $(s = 3\ ,\ t = 1)$. Show that the solution satisfies each of the original equations.

a) $s + 2t = 5$ b) $3s - 2t = 7$

a) $3 + 2(1) = 5$
$3 + 2 = 5$
$5 = 5$

b) $3(3) - 2(1) = 7$
$9 - 2 = 7$
$7 = 7$

16. Use the addition method to solve each system below.

a) $2x + y = 10$
$2x - y = 6$

b) $p - 4q = 16$
$p + 4q = 0$

a) $x = 4, \ y = 2$

b) $p = 8, \ q = -2$

SELF-TEST 41 (pages 353-362)

1. Which one of the following pairs of values is the solution of the system at the right? ________

$r + 2s = 7$
$5r - s = 2$

a) $r = 3$, $s = 2$ b) $r = -1$, $s = 4$ c) $r = 1$, $s = 3$ d) $r = 5$, $s = 1$

Given this system:

$x - y = 8$
$2x + y = 1$

2. Graph both equations on the axes at the right.

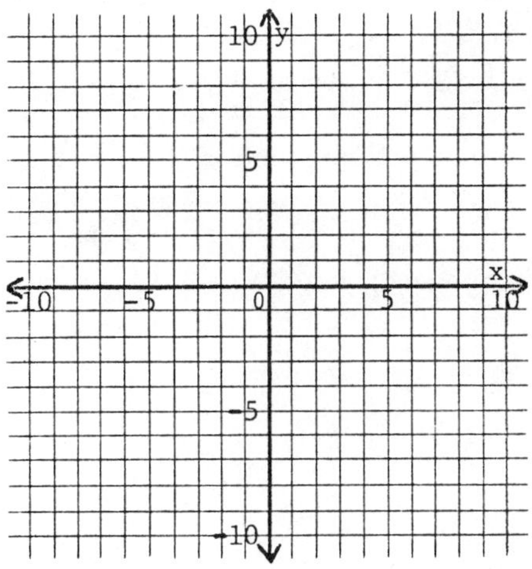

3. From the graph, the solution of the system is:

x = ______

y = ______

Continued on following page.

SELF-TEST 41 - Continued

Use the addition method to solve each system.

4. $3x + y = 6$
$x - y = 2$

5. $a - 3b = 8$
$2a + 3b = 7$

ANSWERS:

1. (c) $r = 1$
$s = 3$

2. Your graph should look like the graph at the right.

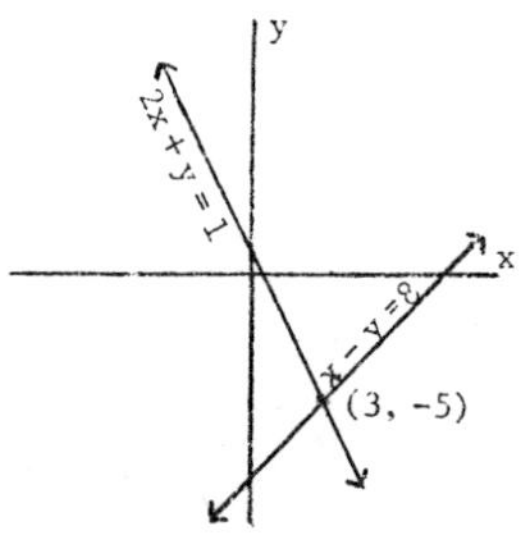

3. $x = 3$
$y = -5$

4. $x = 2$, $y = 0$

5. $a = 5$, $b = -1$

14-4 MULTIPLYING AN EQUATION BY A NUMBER

When using the addition method to solve a system, we sometimes have to multiply an equation by a number. We will discuss that multiplication process in this section.

17. We multiplied $x + 4y = 5$ by 2 below. To do so, we used the distributive principle on the left side.

Original: $x + 4y = 5$

$2(x + 4y) = 2(5)$

$2(x) + 2(4y) = 2(5)$

New: $2x + 8y = 10$

Continued on following page.

17. Continued

Notice that each term in the new equation is 2 times the corresponding term in the original equation. That is:

$$2x = 2(x)$$

$$8y = 2(4y)$$

$$10 = 2(5)$$

Complete these by simply multiplying each term by the number.

a) Multiply this equation by 3.

$$3x + y = 6$$

b) Multiply this equation by 4.

$$p + 2q = 0$$

---

a) $9x + 3y = 18$

b) $4p + 8q = 0$

18. We multiplied $2p - q = 4$ by 3 below. To do so, we again used the distributive principle on the left side.

Original: $2p - q = 4$

$$3(2p - q) = 3(4)$$

$$3(2p) - 3(q) = 3(4)$$

New: $6p - 3q = 12$

Notice that each term in the new equation is 3 times the corresponding term in the original equation. That is:

$$6p = 3(2p)$$

$$-3q = 3(-q)$$

$$12 = 3(4)$$

Complete these by simply multiplying each term by the number.

a) Multiply this equation by 2.

$$x - 3y = 7$$

b) Multiply this equation by 5.

$$a - b = 3$$

a) $2x - 6y = 14$

b) $5a - 5b = 15$

19. We multiplied $2x + 4y = 1$ by -3 below. Notice that we converted the addition to subtraction in the last step.

Original:

$$\begin{aligned} 2x + 4y &= 1 \\ -3(2x + 4y) &= -3(1) \\ -3(2x) + (-3)(4y) &= -3(1) \\ -6x + (-12y) &= -3 \end{aligned}$$

New:

$$-6x - 12y = -3$$

Notice that each term in the new equation is -3 times the corresponding term in the original equation. That is:

$$\begin{aligned} -6x &= -3(2x) \\ -12y &= -3(4y) \\ -3 &= -3(1) \end{aligned}$$

Complete these by simply multiplying each term by the number.

a) Multiply this equation by -2.

$$3x + y = 0$$

b) Multiply this equation by -4.

$$2s + 5t = 3$$

---

a) $-6x - 2y = 0$
b) $-8s - 20t = -12$

20. We multiplied $3a - b = 5$ by -2 below. Notice that we converted the subtraction to addition in the last step.

Original:

$$\begin{aligned} 3a - b &= 5 \\ -2(3a - b) &= -2(5) \\ -2(3a) - (-2)(b) &= -2(5) \\ -6a - (-2b) &= -10 \end{aligned}$$

New:

$$-6a + 2b = -10$$

Notice that each term in the new equation is -2 times the corresponding term in the original equation. That is:

$$\begin{aligned} -6a &= -2(3a) \\ 2b &= -2(-b) \\ -10 &= -2(5) \end{aligned}$$

Complete these by simply multiplying each term by the number.

a) Multiply this equation by -5.

$$m - 4t = 2$$

b) Multiply this equation by -3.

$$2x - 3y = 7$$

21. Multiply each equation below by the circled number at its left.

a) (5) $x + y = 3$

b) (2) $4p - 3q = 7$

c) (-4) $a + b = 1$

d) (-3) $2m - 5f = 4$

a) $-5m + 20t = -10$
b) $-6x + 9y = -21$

---

22. We multiplied both $3x + y = 2$ and $p - 2q = 3$ by -1 below.

Original: $3x + y = 2$

$-1(3x + y) = -1(2)$

$-1(3x) + (-1)(y) = -1(2)$

$-3x + (-y) = -2$

New: $-3x - y = -2$

Original: $p - 2q = 3$

$-1(p - 2q) = -1(3)$

$-1(p) - (-1)(2q) = -1(3)$

$-p - (-2q) = -3$

New: $-p + 2q = -3$

Notice that each new equation can be obtained by simply changing the sign of each term in the original equation. Use that method to multiply each equation below by -1.

a) $5x + 4y = 9$ b) $2x - 3y = 7$ c) $p + q = 1$ d) $a - b = 0$

a) $5x + 5y = 15$
b) $8p - 6q = 14$
c) $-4a - 4b = -4$
d) $-6m + 15f = -12$

---

23. Multiply each equation below by the circled number at its left.

a) (-1) $4t - 3s = 5$

b) (-2) $3a - 2b = 1$

c) (-1) $p - q = 5$

d) (-3) $x - y = 0$

a) $-5x - 4y = -9$
b) $-2x + 3y = -7$
c) $-p - q = -1$
d) $-a + b = 0$

a) $-4t + 3s = -5$
b) $-6a + 4b = -2$
c) $-p + q = -5$
d) $-3x + 3y = 0$

24. When an equation is multiplied by a number, the new equation is equivalent to the original equation. That is, both equations have the same solutions. An example is discussed below.

We multiplied the equation below by 3.

Original: $x + y = 2$

New: $3x + 3y = 6$

A table of solutions for each equation is given below.

$x + y = 2$

x	y
2	0
1	1
0	2
-1	3
-2	4

$3x + 3y = 6$

x	y
2	0
1	1
0	2
-1	3
-2	4

Since the solution-tables are identical, both equations have the same solutions. Therefore, we say that they are ________________ equations.

equivalent

14-5 MULTIPLICATION IN THE ADDITION METHOD

When using the addition method for solving systems, we sometimes have to multiply an equation by a number before adding the equations. We will discuss systems of that type in this section. The discussion is limited to systems that can be solved by multiplying one equation by a number.

---

25. If we added the equations in the system at the right, neither letter would be eliminated.

$$x + 2y = 7$$
$$2x - y = 4$$

However, if the "-y" in the bottom equation were a "-2y", we could eliminate "y". We can get a "-2y" in that equation by multiplying it by 2. We did so at the right.

$$2(2x - y) = 4$$
$$4x - 2y = 8$$

Substituting the new equation for $2x - y = 4$, we get the new system at the right. We added the equations and solved for "x".

$$\begin{aligned} x + 2y &= 7 \\ 4x - 2y &= 8 \\ \hline 5x &= 15 \\ x &= 3 \end{aligned}$$

Substituting 3 for "x" in one of the original equations, we can find the corresponding value for "y".

$$\begin{aligned} x + 2y &= 7 \\ 3 + 2y &= 7 \\ 2y &= 4 \\ y &= 2 \end{aligned}$$

Check $(x = 3,\ y = 2)$ in each original equation below.

a) $x + 2y = 7$

b) $2x - y = 4$

a) $3 + 2(2) = 7$
$3 + 4 = 7$
$7 = 7$

b) $2(3) - 2 = 4$
$6 - 2 = 4$
$4 = 4$

26. We could eliminate "b" in the system at the right if the "2b" in the top equation were a "-6b".

$$a + 2b = 0$$
$$5a + 6b = 4$$

To get a "-6b" in the top equation, we multiplied it by -3 at the right and then solved for "a".

$$\begin{aligned} -3a - 6b &= 0 \\ \underline{5a + 6b} &= \underline{4} \\ 2a &= 4 \\ a &= 2 \end{aligned}$$

Substituting "2" for "a" in the bottom equation, we found the corresponding value of "b" at the right.

$$\begin{aligned} 5a + 6b &= 4 \\ 5(2) + 6b &= 4 \\ 10 + 6b &= 4 \\ 6b &= -6 \\ b &= -1 \end{aligned}$$

Check (a = 2, b = -1) in each original equation below.

a) $a + 2b = 0$ b) $5a + 6b = 4$

Answers:

a) $2 + 2(-1) = 0$
$2 + (-2) = 0$
$0 = 0$

b) $5(2) + 6(-1) = 4$
$10 + (-6) = 4$
$4 = 4$

27. We could eliminate "y" in the system at the right if the "3y" were "-3y" in either equation.

$$2x + 3y = 8$$
$$x + 3y = 7$$

To get a "-3y" in the bottom equation, we multiplied it by "-1" at the right and then solved for "x".

$$\begin{aligned} 2x + 3y &= 8 \\ \underline{-x - 3y} &= \underline{-7} \\ x &= 1 \end{aligned}$$

To find the corresponding value for "y", we substituted in the original bottom equation at the right.

$$\begin{aligned} x + 3y &= 7 \\ 1 + 3y &= 7 \\ 3y &= 6 \\ y &= 2 \end{aligned}$$

Check (x = 1, y = 2) in each original equation below.

a) $2x + 3y = 8$ b) $x + 3y = 7$

Answers:

a) $2(1) + 3(2) = 8$
$2 + 6 = 8$
$8 = 8$

b) $1 + 3(2) = 7$
$1 + 6 = 7$
$7 = 7$

28. Let's solve the system at the right by eliminating "t". To do so, we must multiply the top equation by 4.

$$\begin{aligned} 2s - t &= 7 \\ s + 4t &= 8 \end{aligned}$$

a) Write the new system obtained if the top equation is multiplied by 4.

b) Solve the system.

---

a) $$\begin{aligned} 8s - 4t &= 28 \\ s + 4t &= 8 \end{aligned}$$

b) $s = 4, \ t = 1$

29. Let's solve the system at the right by eliminating "x". To do so, we must multiply the bottom equation by -2.

$$\begin{aligned} 4x + y &= 1 \\ 2x + 3y &= 13 \end{aligned}$$

a) Write the new system obtained if the bottom equation is multiplied by -2.

b) Solve the system.

a) $$\begin{aligned} 4x + y &= 1 \\ -4x - 6y &= -26 \end{aligned}$$

b) $x = -1, \ y = 5$

30. Let's solve the system at the right by eliminating "p". To do so, we can multiply either equation by -1. Let's multiply the bottom equation by -1.

$$\begin{aligned} p + q &= 7 \\ p + 3q &= 15 \end{aligned}$$

a) Write the new system obtained if the bottom equation is multiplied by -1.

b) Solve the system.

---

a) $$\begin{aligned} p + q &= 7 \\ -p - 3q &= -15 \end{aligned}$$

b) $p = 3$, $q = 4$

31. When you have to multiply an equation in a system by a number, it is important that you know what number to use. For example:

a) To eliminate "t" in the system at the right, we multiply the bottom equation by ______.

$$\begin{aligned} s + 5t &= 9 \\ 2s - t &= 0 \end{aligned}$$

b) To eliminate "y" in the system at the right, we multiply the top equation by ______.

$$\begin{aligned} x + 2y &= 0 \\ 3x - 4y &= 5 \end{aligned}$$

c) To eliminate "b" in the system at the right, we multiply the bottom equation by ______.

$$\begin{aligned} 2a + 3b &= 5 \\ 3a + b &= 9 \end{aligned}$$

d) To eliminate "q" in the system at the right, we can multiply either the top equation or the bottom equation by ______.

$$\begin{aligned} 3p + 4q &= 7 \\ p + 4q &= 1 \end{aligned}$$

a) 5

b) 2

c) -3

d) -1

32. When solving a system, we can eliminate either letter and still get the same solution. As an example, we have solved the system below in two different ways.

$$2x + y = 2$$
$$x + 4y = 1$$

To eliminate "x", we multiplied the bottom equation by -2 and then completed the solution.

$$\begin{aligned} 2x + y &= 2 \\ \underline{-2x - 8y} &= \underline{-2} \\ -7y &= 0 \\ y &= 0 \end{aligned}$$

$$\begin{aligned} x + 4y &= 1 \\ x + 4(0) &= 1 \\ x + 0 &= 1 \\ x &= 1 \end{aligned}$$

To eliminate "y", we multiplied the top equation by -4 and then completed the solution.

$$\begin{aligned} -8x - 4y &= -8 \\ \underline{x + 4y} &= \underline{1} \\ -7x \quad &= -7 \\ x &= 1 \end{aligned}$$

$$\begin{aligned} x + 4y &= 1 \\ 1 + 4y &= 1 \\ 4y &= 0 \\ y &= 0 \end{aligned}$$

We obtained the same solution ($x = 1$, $y = 0$) both ways. Check that solution in each original equation below.

a) $2x + y = 2$

b) $x + 4y = 1$

---

a) $2(1) + 0 = 2$
$2 + 0 = 2$
$2 = 2$

b) $1 + 4(0) = 1$
$1 + 0 = 1$
$1 = 1$

---

33. To solve the system at the right:

$$x + 4y = 0$$
$$3x + 2y = 5$$

a) we can eliminate "x" by multiplying the top equation by ______.

b) we can eliminate "y" by multiplying the bottom equation by ______.

To solve the system at the right:

$$2a - b = 1$$
$$a + 5b = 9$$

c) we can eliminate "a" by multiplying the bottom equation by ______.

d) we can eliminate "b" by multiplying the top equation by ______.

To solve the system at the right:

$$2s + 3t = 7$$
$$2s - t = 3$$

e) we can eliminate "t" by multiplying the bottom equation by ______.

f) we can eliminate "s" by multiplying either equation by ______.

a) -3
b) -2

c) -2
d) 5

e) 3
f) -1

34. By eliminating either letter, solve each system below.

a) $x + 4y = 10$
$3x - 2y = 16$

b) $2k + m = 0$
$2k + 3m = 8$

a) $x = 6, \ y = 1$
b) $k = -2, \ m = 4$

SELF-TEST 42 (pages 362-371)

Multiply each equation by the circled number at its left.

1. (4) $2x - 3y = 1$

2. (-2) $x + y = 5$

3. (-5) $2a - b = 0$

4. (-1) $5s - 4y = 7$

Solve each system.

5. $4x + y = 14$
$2x - 3y = 0$

6. $k + m = 2$
$2k + m = 8$

ANSWERS:

1. $8x - 12y = 4$
2. $-2x - 2y = -10$
3. $-10a + 5b = 0$
4. $-5s + 4y = -7$
5. $x = 3, \ y = 2$
6. $k = 6, \ m = -4$

14-6 SOLVING SYSTEMS BY CONVERTING TO STANDARD FORM

To solve a system by the addition method, both equations must be in standard form. When an equation is not in standard form, we must convert it to standard form first. We will discuss solutions of that type in this section.

35. In the two systems below, the equations are in standard form because both letter-terms are on the left side and the number-term is on the right side.

$$x + y = 5 \qquad\qquad 3a + b = 7$$
$$x - y = 0 \qquad\qquad a + 5b = 9$$

In the two systems below, one equation in each is not in standard form because it has a letter-term on the right side.

$$5x + y = 4 \qquad\qquad s - 2t = 0$$
$$y = 2x \qquad\qquad t = s + 6$$

At the left, $y = 2x$ is not in standard form.

At the right, $t = s + 6$ is not in standard form.

In each system below, identify the equation that is not in standard form because it has a letter-term on the right side.

a) $3k - m = 7$, $m = k$

b) $x + y = 10$, $y = 3x - 1$

c) $2t = 5s + 1$, $s - 4t = 0$

Answers: a) $m = k$ b) $y = 3x - 1$ c) $2t = 5s + 1$

36. When an equation in a system is not in standard form, we must convert it to standard form before using the addition method to solve the system. Two examples are shown. Notice these two points:

1. We used the addition axiom to get the letter-term to the left side.
2. We lined up the terms with the same letters on the left side.

Original System	Original System
$x + y = 10$	$a = b + 7$
$y = 2x + 3$	$2a + b = 5$
Conversion	**Conversion**
$y = 2x + 3$	$a = b + 7$
$(-2x) + y = 2x + (-2x) + 3$	$a + (-b) = b + (-b) + 7$
$-2x + y = 3$	$a - b = 7$
Standard System	**Standard System**
$x + y = 10$	$a - b = 7$
$-2x + y = 3$	$2a + b = 5$

Continued on following page.

36. Continued

Convert each system below to one with two standard-form equations.

a) $2p + q = 9$
$q = 4p + 5$

b) $t = 2d - 1$
$t + d = 10$

---

a) $2p + q = 9$
$-4p + q = 5$

b) $t - 2d = -1$
$t + d = 10$

---

37. Two more examples of conversions to standard form are shown below. Notice in each that we lined up the terms with the same letters on the left side.

Original System

$p - q = 7$

$q = 2p$

Conversion

$q = 2p$

$(-2p) + q = 2p + (-2p)$

$-2p + q = 0$

Standard System

$p - q = 7$

$-2p + q = 0$

Original System

$x = y$

$2x + y = 3$

Conversion

$x = y$

$x + (-y) = y + (-y)$

$x - y = 0$

Standard System

$x - y = 0$

$2x + y = 3$

Convert each system below to one with two standard-form equations.

a) $b = a$
$2a - 3b = 8$

b) $t + 6m = 1$
$t = 3m$

---

a) $-a + b = 0$
$2a - 3b = 8$

b) $t + 6m = 1$
$t - 3m = 0$

38. In the system at the right, $y = 3x - 8$ is not in standard form

$$x - y = 2$$
$$y = 3x - 8$$

To solve the system, we begin by converting $y = 3x - 8$ to standard form.

$$(-3x) + y = 3x + (-3x) - 8$$
$$-3x + y = -8$$

Then we can use the addition method to solve the standard system at the right.

$$x - y = 2$$
$$-3x + y = -8$$

a) Solve the standard system.

b) Check your solution in each original equation.

$x - y = 2$ $\qquad\qquad$ $y = 3x - 8$

Answers:

a) $x = 3, \ y = 1$

b) $3 - 1 = 2$
$2 = 2$

$1 = 3(3) - 8$
$1 = 9 - 8$
$1 = 1$

---

39. Following the steps in the last frame, solve each system.

a)
$$y = x + 2$$
$$x + y = 6$$

b)
$$2p + q = 30$$
$$p = 2q$$

Answers:

a) $x = 2, \ y = 4$

b) $p = 12, \ q = 6$

14-7 APPLIED PROBLEMS

In this section, we will discuss some applied problems that can be solved by setting up and solving a system of equations.

---

40. The following problem can be solved by setting up and solving a system of equations.

> There are 31 students in a class. If there are 5 more males than females, how many males and females are there?

There are two statements in the problem. Each leads to an equation involving number of males (m) and number of females (f). That is:

There are 31 students in a class.

$$m + f = 31$$

There are 5 more males than females.

$$m = f + 5$$

or

$$m - f = 5$$

Therefore, the problem can be solved by solving the system at the right.

$$\begin{aligned} m + f &= 31 \\ m - f &= 5 \end{aligned}$$

a) Solve the system.

b) Therefore, there are ______ males and ______ females in the class.

---

a) $m = 18,\ f = 13$
b) 18 males and 13 females

41. Here is another problem that can be solved by setting up and solving a system of equations.

> A portable radio costs twice as much as a digital clock. If the total cost of the radio and clock is \$72, how much does each cost?

The two statements in the problem lead to two equations involving the cost of the radio (r) and the cost of the clock (c). That is:

A radio costs twice as much as a clock.

$$r = 2c$$

The total cost of the radio and clock is \$72.

$$r + c = \$72$$

Continued on following page.

41. Continued

Therefore, the problem can be solved by solving the system at the right.

$$r = 2c$$
$$r + c = \$72$$

a) Solve the system.

b) Therefore, the radio costs ________ and the clock costs ________.

---

a) $r = \$48$, $c = \$24$

b) radio costs \$48 and clock costs \$24

---

42. Here is a third problem that can be solved by setting up and solving a system of equations.

The sum of two numbers is 50. The difference between the larger and smaller numbers is 14. Find the two numbers.

Using "x" for the larger number and "y" for the smaller number, we can set up two equations.

The sum of two numbers is 50.

$$x + y = 50$$

The difference between the larger and smaller numbers is 14.

$$x - y = 14$$

Therefore, the problem can be solved by solving the system at the right.

$$x + y = 50$$
$$x - y = 14$$

a) Solve the system.

b) Therefore, the larger number is ________ and the smaller number is ________.

---

a) $x = 32$, $y = 18$

b) larger number is 32 and smaller number is 18

43. Systems of equations can be used to solve various problems involving geometric figures. An example is given.

A 10-foot board is cut into two parts. The larger part is 2 feet longer than the smaller. How long is each part?

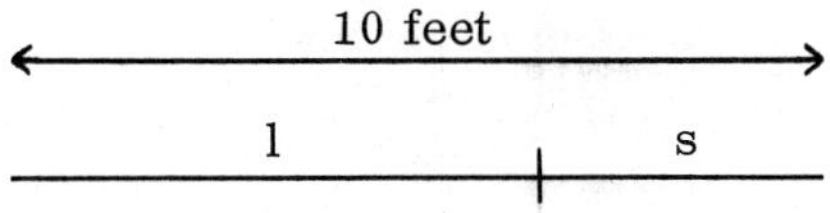

Using "l" for the larger part and "s" for the smaller part, we can set up two equations.

The sum of the two parts is 10 feet.

$$l + s = 10$$

The larger part is 2 feet longer than the smaller part.

$$l = s + 2$$

or

$$l - s = 2$$

Therefore, we can solve the problem by solving the system at the right.

$$l + s = 10$$
$$l - s = 2$$

a) Solve the system.

b) Therefore, the larger part is ______ feet and the smaller part is ______ feet.

---

a) $l = 6$, $s = 4$

b) larger is 6 feet, smaller is 4 feet

---

44. The geometric problem below can be solved by means of a system.

The perimeter of a rectangle is 180 centimeters. If its length is 20 centimeters longer than its width, what are its length and width?

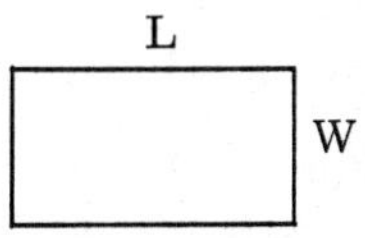

Using "L" for length and "W" for width, we can set up two equations.

The perimeter of a rectangle is 180 centimeters.

$$2L + 2W = 180$$

The length is 20 centimeters longer than the width.

$$L = W + 20$$

or

$$L - W = 20$$

Continued on following page.

44. Continued

Therefore, we can solve the problem by solving the system at the right.

$$2L + 2W = 180$$
$$L - W = 20$$

a) Solve the system.

b) Therefore, its length is ________ centimeters and its width is ________ centimeters.

---

a) $L = 55$, $W = 35$

b) length is 55 centimeters, width is 35 centimeters

---

45. When using a system of equations to solve an applied problem, the key step is setting up the two equations in the system. Therefore, we will give practice in setting up systems in the following frames.

Without solving, set up the system for each problem.

a) There are 36 adults at a party. If there are 3 times as many non-smokers (n) as smokers (s), how many non-smokers and smokers are there?

b) House A has 3 more rooms than house B. If the two houses together contain 17 rooms, how many rooms are in each house?

---

a) $n + s = 36$
$n = 3s$

b) $A = B + 3$
$A + B = 17$
or
$A - B = 3$
$A + B = 17$

---

46. Without solving, set up the system for each problem.

a) The difference between two numbers is 6. The larger number (x) is four times the smaller number (y). Find the two numbers.

b) The sum of two numbers is 40. If the larger (x) is 20 more than the smaller (y), find the two numbers.

---

a) $x - y = 6$
$x = 4y$

b) $x + y = 40$
$x = y + 20$
or
$x + y = 40$
$x - y = 20$

47. Without solving, set up the system for each problem.

a) A piece of wire 28 centimeters long is cut into two parts. If the larger part (x) is three times the smaller part (y), how long is each part?

b) A metal rod 12 feet long is cut into two parts. If the larger part (x) is 3 feet longer than the smaller part (y), how long is each part?

---

a) $x + y = 28$
$x = 3y$

b) $x + y = 12$
$x = y + 3$
or
$x + y = 12$
$x - y = 3$

48. Without solving, set up the system for each problem.

a) The perimeter of a rectangle is 48 inches. If the length (L) is five times the width (W), find the length and width.

b) The perimeter of a rectangle is 100 centimeters. If the length (L) is 10 centimeters longer than the width (W), find the length and width.

---

a) $2L + 2W = 48$
$L = 5W$

b) $2L + 2W = 100$
$L = W + 10$
or
$2L + 2W = 100$
$L - W = 10$

49. Without solving, set up the system for each problem.

a) A student paid a total of $10.50 for two tickets, one for a rock concert (c) and one for a baseball game (b). If the ticket for the rock concert cost twice as much as the ticket for the baseball game, how much was paid for each?

b) A husband (h) weighs 60 pounds more than his wife (w). Together they weigh 300 pounds. Find the weight of each.

---

a) $b + c = \$10.50$
$c = 2b$

b) $h = w + 60$
$h + w = 300$
or
$h - w = 60$
$h + w = 300$

SELF-TEST 43 (pages 372-380)

Solve each system.

1. $2x - y = 7$
   $y = x - 3$

2. $a = 2b$
   $a + 4b = 24$

Solve each problem.

3. The sum of two numbers is 64. The larger (x) is 18 more than the smaller (y). Find the two numbers.

4. The perimeter of a rectangular lot is 360 meters. If the length (L) is twice the width (W), find the length and width.

ANSWERS: 1. $x = 4$, $y = 1$ 2. $a = 8$, $b = 4$ 3. $x = 41$, $y = 23$ 4. $L = 120m$, $W = 60m$

Unit 15 SPECIAL PRODUCTS AND FACTORING

In this unit, we will discuss some special products and factorings that occur frequently. The "special products" include multiplication of two monomials, of a monomial and a binomial, and of two binomials. The "special factorings" include factoring of monomials, binomials, and trinomials.

15-1 MONOMIALS, BINOMIALS, TRINOMIALS

In this section, we will define what is meant by a monomial, a binomial, and a trinomial.

---

1. An algebraic expression can contain one or more terms. For example:

   Each expression below contains <u>one</u> term.

   $5 \qquad 3x \qquad y^2$

   Each expression below contains <u>two</u> terms.

   $2x + 7 \qquad y - 1 \qquad 4x^2 + 5$

   Each expression below contains <u>three</u> terms.

   $x^2 + 5x + 6 \qquad 2y^2 - 4y + 7$

   How many terms are there in each expression below?

   a) $7y$ ______ c) 12 ______

   b) $9m + 10$ ______ d) $p^2 - 2p + 8$ ______

---

2. When an expression contains only <u>one</u> term, it is called a "<u>monomial</u>". Which of the following are monomials? ______

   a) 7 b) $x^2 - x$ c) $3y^2$ d) $m^2 - 3m + 4$ e) $R + 10$

Answers to 1: a) one b) two c) one d) three

---

3. When an expression contains <u>two</u> terms, it is called a "<u>binomial</u>". Which of the following are binomials? ______

   a) $2m$ b) $3y^2 + y$ c) $x + 5$ d) $3b^2 - b + 5$ e) $p^2 - 5p$

Answer to 2: (a) and (c)

---

4. When an expression contains <u>three</u> terms, it is called a "<u>trinomial</u>". Which of the following are trinomials? ______

   a) $b^2 - 3b$ b) $r^2 + 6r + 9$ c) $2x + 7$ d) $5a^2 - 7a - 12$

Answer to 3: (b), (c), and (e)

Answer to 4: (b) and (d)

15-2 MULTIPLYING TWO MONOMIALS

In this section, we will discuss multiplications in which both factors are monomials.

---

5. To multiply a number and a letter, we simply write the number as the coefficient of the letter. That is:

$2(x) = 2x$ $y(5) = 5y$ a) $7(m) =$ ______ b) $t(10) =$ ______

---

a) $7m$
b) $10t$

6. To multiply a number and a letter with a coefficient, we multiply the two numerical factors. That is:

$3(4x) = (3)(4)(x) = 12x$ $(5y)(2) = (5)(2)(y) = 10y$

Following the examples, complete each multiplication below.

a) $5(6t) =$ ______ b) $(3d)(6) =$ ______ c) $(9m)(7) =$ ______

---

a) $30t$
b) $18d$
c) $63m$

7. When one factor is "1", the product is identical to the other factor. That is:

$1(x) = x$ $1(3t) = 3t$ a) $(y)(1) =$ ______ b) $(7p)(1) =$ ______

---

a) y
b) $7p$

8. Multiplying a letter by itself is the same as squaring the letter. Therefore:

$(x)(x) = x^2$ $(y)(y) = y^2$ $(t)(t) =$ ______

---

t^2

9. To multiply a letter and the same letter with a coefficient, we square the letter. For example:

$x(5x) = (5)(x)(x) = 5x^2$ $(2y)(y) = (2)(y)(y) = 2y^2$

Following the examples, complete each multiplication below.

a) $t(7t) =$ ______ b) $p(10p) =$ ______ c) $(6d)(d) =$ ______

---

a) $7t^2$
b) $10p^2$
c) $6d^2$

10. To perform each multiplication below, we multiply the two numerical factors and square the letter.

$(2x)(5x) = (2)(5)(x)(x) = 10x^2$ $(7y)(6y) = (7)(6)(y)(y) = 42y^2$

Following the examples, complete each multiplication below.

a) $(3t)(4t) =$ ______ b) $(8p)(2p) =$ ______ c) $(9d)(4d) =$ ______

---

a) $12t^2$
b) $16p^2$
c) $36d^2$

15-3 MULTIPLYING A MONOMIAL AND A BINOMIAL

In this section, we will show that the distributive principle is used to multiply a monomial and a binomial.

11. We used the distributive principle for each multiplication below.

$$3(x + 5) = 3(x) + 3(5) = 3x + 15$$

$$7(4y - 1) = 7(4y) - 7(1) = 28y - 7$$

Following the examples, complete these multiplications.

a) $10(3b + 2) =$ ________ b) $6(p - 5) =$ ________

---

a) $30b + 20$
b) $6p - 30$

12. We used the distributive principle for each multiplication below.

$$x(x + 2) = x(x) + x(2) = x^2 + 2x$$

$$y(y - 7) = y(y) - y(7) = y^2 - 7y$$

Following the examples, complete these multiplications.

a) $m(m + 5) =$ ________ b) $t(t - 6) =$ ________

---

a) $m^2 + 5m$
b) $t^2 - 6t$

13. Notice that we did not write the coefficient "1" in each product below.

$$x(x + 1) = x(x) + x(1) = x^2 + x$$

$$y(y - 1) = y(y) - y(1) = y^2 - y$$

Following the examples, complete these multiplications.

a) $p(p + 1) =$ ________ b) $t(t - 1) =$ ________

---

a) $p^2 + p$
b) $t^2 - t$

14. Following the examples, complete the other multiplications.

$$b(3b + 4) = b(3b) + b(4) = 3b^2 + 4b$$

$$t(5t - 1) = t(5t) - t(1) = 5t^2 - t$$

a) $x(7x + 1) =$ ________ b) $y(2y - 8) =$ ________

---

a) $7x^2 + x$
b) $2y^2 - 8y$

15. Following the examples, complete the other multiplications.

$$2x(x + 5) = 2x(x) + 2x(5) = 2x^2 + 10x$$

$$6y(y - 1) = 6y(y) - 6y(1) = 6y^2 - 6y$$

a) $3m(m + 1) =$ ________ b) $7d(d - 4) =$ ________

---

a) $3m^2 + 3m$
b) $7d^2 - 28d$

16. Following the examples, complete the other multiplications.

$$4x(2x + 1) = 4x(2x) + 4x(1) = 8x^2 + 4x$$

$$3y(5y - 6) = 3y(5y) - 3y(6) = 15y^2 - 18y$$

a) $7x(3x + 8) =$ ________ b) $9y(4y - 1) =$ ________

17. Complete each multiplication.

a) $t(t + 1) =$ ______

b) $y(2y - 5) =$ ______

c) $4m(m + 3) =$ ______

d) $9a(2a - 7) =$ ______

a) $21x^2 + 56x$
b) $36y^2 - 9y$

a) $t^2 + t$ b) $2y^2 - 5y$ c) $4m^2 + 12m$ d) $18a^2 - 63a$

15-4 FACTORING MONOMIALS

In this section, we will discuss the factoring process for monomials.

18. Factoring a number is the same as writing the number as a multiplication. For example:

15 can be factored into (3)(5)

42 can be factored into (7)(6)

The two numbers in the multiplication are called "factors" of the original number. That is:

"3" and "5" are called "factors" of 15.

"7" and "6" are called "______" of 42.

factors

19. A first number is a factor of a second number if the second is divisible by the first. By "divisible", we mean that the division comes out even with no remainder. For example:

8 is a factor of 32, since $32 \div 8 = 4$

5 is not a factor of 17, since $17 \div 5 = 3$ remainder 2

a) Is 3 a factor of 36? ______

b) Is 7 a factor of 52? ______

c) Is 6 a factor of 18? ______

d) Is 9 a factor of 40? ______

a) Yes
b) No
c) Yes
d) No

20. Since any number is divisible by "1" and itself, "1" and the number itself are factors of any number. That is:

Two factors of 8 are "1" and "8".

To determine the other factors of 8, we test whole numbers up to 4 (one-half of 8). That is:

Is 8 divisible by 2? Yes
Is 8 divisible by 3? No
Is 8 divisible by 4? Yes

Therefore, the four factors of 8 are ______, ______, ______, and ______.

1, 2, 4, 8

21. List all of the factors for each number below.

a) 7 ______

b) 25 ______

c) 32 ______

d) 36 ______

a) 1, 7 b) 1, 5, 25 c) 1, 2, 4, 8, 16, 32 d) 1, 2, 3, 4, 6, 9, 12, 18, 36

22. Some numbers can be factored into only one pair of factors. For example:

3 = (1)(3) or (3)(1)   7 = (1)(7) or (7)(1)

Most numbers, however, can be factored into more than one pair of factors. For example:

18 can be factored into three pairs of factors. They are:

18 = (1)(18) or (18)(1)

18 = (2)(9) or (9)(2)

18 = (3)(6) or (6)(3)

24 can be factored into four pairs of factors. Do so below.

24 = (   )(   )

24 = (   )(   )

24 = (   )(   )

24 = (   )(   )

---

(1)(24) or (24)(1)
(2)(12) or (12)(2)
(3)(8) or (8)(3)
(4)(6) or (6)(4)

23. To factor a letter with a coefficient, we can write the factors explicitly in either order. That is:

3x = (3)(x) or (x)(3)   7y = (   )(   ) or (   )(   )

---

(7)(y) or (y)(7)

24. When the coefficient is not explicitly shown, the coefficient is "1". Therefore:

x = (1)(x) or (x)(1)   t = (   )(   ) or (   )(   )

---

(1)(t) or (t)(1)

25. We have factored 12x and 18y in various ways below.

12x = (2)(6x)   18y = (2)(9y)
12x = (3)(4x)   18y = (3)(6y)
12x = (4)(3x)   18y = (6)(3y)
12x = (6)(2x)   18y = (9)(2y)

Write the missing factor in each blank.

a) 6t = (2)(   )   b) 10t = (   )(2t)   c) 16x = (4)(   )

---

a) (2)(3t)
b) (5)(2t)
c) (4)(4x)

26. We have factored 16x and 20y in various ways below.

16x = (2x)(8)   20y = (2y)(10)
16x = (4x)(4)   20y = (4y)(5)
16x = (8x)(2)   20y = (5y)(4)
                  20y = (10y)(2)

Write the missing factor in each blank.

a) 30d = (5d)(   )   b) 14m = (   )(2)   c) 32p = (4p)(   )

---

a) (5d)(6)
b) (7m)(2)
c) (4p)(8)

27. Any letter with a coefficient can be factored into itself and "1". That is:

3x = (3x)(1) or (1)(3x)   8t = (   )(   ) or (   )(   )

(8t)(1) or (1)(8t)

28. Any squared letter can be factored by multiplying the letter by itself. That is:

$x^2 = (x)(x)$ $y^2 = (y)(y)$ $p^2 = (\quad)(\quad)$

(p)(p)

29. We have factored "$3x^2$" and "$7y^2$" in two different ways below.

$3x^2 = (3x)(x)$ $7y^2 = (7y)(y)$

$3x^2 = (x)(3x)$ $7y^2 = (y)(7y)$

Write the missing factor in each blank.

a) $2t^2 = (2t)(\quad)$ b) $5d^2 = (d)(\quad)$ c) $8m^2 = (\quad)(m)$

a) (2t)(t)
b) (d)(5d)
c) (8m)(m)

30. We have factored $12x^2$ and $18y^2$ in various ways below.

$12x^2 = (2x)(6x)$ $18y^2 = (2y)(9y)$

$12x^2 = (3x)(4x)$ $18y^2 = (3y)(6y)$

$12x^2 = (4x)(3x)$ $18y^2 = (6y)(3y)$

$12x^2 = (6x)(2x)$ $18y^2 = (9y)(2y)$

Write the missing factor in each blank.

a) $20t^2 = (2t)(\quad)$ b) $36d^2 = (9d)(\quad)$ c) $48p^2 = (\quad)(6p)$

a) (2t)(10t) b) (9d)(4d) c) (8p)(6p)

15-5 FACTORING BINOMIALS OF THE FORM: ax + b

In this section, we will show how the distributive principle can be used to factor some binomials of the form: ax + b.

31. We used the distributive principle for each multiplication below.

$2(x + 3) = 2(x) + 2(3) = 2x + 6$

$5(y - 4) = 5(y) - 5(4) = 5y - 20$

By simply reversing the process, we can use the distributive principle to factor each binomial product. That is:

$2x + 6 = 2(x) + 2(3) = 2(x + 3)$

$5y - 20 = 5(y) - 5(4) = 5(y - 4)$

Complete each of these factorings.

a) $3t + 18 = 3(t) + 3(6) =$ ________ b) $4d - 36 = 4(d) - 4(9) =$ ________

a) 3(t + 6)
b) 4(d - 9)

32. Two more factorings of binomials are shown below.

$$8x + 10 = 2(4x) + 2(5) = 2(4x + 5)$$

$$15y - 10 = 5(3y) - 5(2) = 5(3y - 2)$$

Complete each of these factorings.

a) $9t + 12 = 3(3t) + 3(4) =$ ____________ b) $35d - 30 = 5(7d) - 5(6) =$ ____________

---

a) $3(3t + 4)$

b) $5(7d - 6)$

33. When factoring a binomial of the form: $ax + b$, we "factor out" a number that is a common factor of the coefficient and the number. For example:

In $3x + 6$, the common factor is "3". Therefore:

$$3x + 6 = 3(x + 2)$$

In $6y + 4$, the common factor is "2". Therefore:

$$6y + 4 = 2(3y + 2)$$

Factor out the common factor from each binomial below.

a) $7x + 14 =$ ____________ c) $6a + 21 =$ ____________

b) $5y - 45 =$ ____________ d) $10t - 8 =$ ____________

---

a) $7(x + 2)$

b) $5(y - 9)$

c) $3(2a + 7)$

d) $2(5t - 4)$

34. When factoring a binomial, we can sometimes "factor out" more than one number. For example, we can factor out 2, 3, or 6 from the binomial below.

$$6x + 12 = 2(3x) + 2(6) = 2(3x + 6)$$

$$6x + 12 = 3(2x) + 3(4) = 3(2x + 4)$$

$$6x + 12 = 6(x) + 6(2) = 6(x + 2)$$

Ordinarily, we factor out <u>the largest possible number</u>. Therefore, the ordinary factoring for $6x + 12$ is ____________.

$6(x + 2)$

35. Factoring out the largest possible number from a binomial is called "<u>factoring completely</u>". Factor each binomial below completely.

a) $4x + 12 =$ ____________ c) $12t + 18 =$ ____________

b) $9b - 45 =$ ____________ d) $16y - 12 =$ ____________

---

a) $4(x + 3)$

b) $9(b - 5)$

c) $6(2t + 3)$

d) $4(4y - 3)$

36. When the largest possible factor is identical to the number-term, we get "1" as the second term within the parentheses when factoring. For example:

$$6x + 2 = 2(3x) + 2(1) = 2(3x + 1)$$

$$5y - 5 = 5(y) - 5(1) = 5(y - 1)$$

Factor each binomial below completely.

a) $3m + 3 =$ ____________ b) $20p - 10 =$ ____________

---

a) $3(m + 1)$

b) $10(2p - 1)$

37. To check a factoring of a binomial, we simply multiply by the distributive principle. That is:

$$8x + 12 = 4(2x + 3), \quad \text{since} \quad 4(2x + 3) = 8x + 12$$

$$14y - 7 = 7(2y - 1), \quad \text{since} \quad 7(2y - 1) = 14y - 7$$

Factor each binomial completely and check your results.

a) $16x + 24 =$ ________________ c) $9m + 9 =$ ________________

b) $36y - 12 =$ ________________ d) $15d - 3 =$ ________________

---

a) $8(2x + 3)$
b) $12(3y - 1)$
c) $9(m + 1)$
d) $3(5d - 1)$

38. A binomial containing a letter-term and a number-term can be factored only if the coefficient and the number-term have a common factor. Therefore, the binomials below cannot be factored.

$$3x + 7 \qquad 9y - 5 \qquad 12t + 17$$

Factor each binomial below completely if possible.

a) $7t + 9 =$ ________________ c) $4y + 2 =$ ________________

b) $6d - 6 =$ ________________ d) $5m - 2 =$ ________________

a) Not possible b) $6(d - 1)$ c) $2(2y + 1)$ d) Not possible

15-6 FACTORING BINOMIALS OF THE FORM: $ax^2 + bx$

In this section, we will show how the distributive principle can be used to factor all binomials of the form: $ax^2 + bx$.

---

39. We used the distributive principle for each multiplication below.

$$x(x + 5) = x(x) + x(5) = x^2 + 5x$$

$$y(y - 3) = y(y) - y(3) = y^2 - 3y$$

By simply reversing the process, we can use the distributive principle to factor each binomial product. That is:

$$x^2 + 5x = x(x) + x(5) = x(x + 5)$$

$$y^2 - 3y = y(y) - y(3) = y(y - 3)$$

Complete each of these factorings.

a) $m^2 + 2m = m(m) + m(2) =$ ____________ b) $t^2 - 10t = t(t) - t(10) =$ ____________

---

a) $m(m + 2)$
b) $t(t - 10)$

40. Notice how we got a "1" as the second term within the parentheses in each factoring below.

$$x^2 + x = x(x) + x(1) = x(x + 1)$$

$$y^2 - y = y(y) - y(1) = y(y - 1)$$

Continued on following page.

40. Continued

Following the examples, complete these factorings.

a) $t^2 + t = t(t) + t(1) =$ ________ b) $m^2 - m = m(m) - m(1) =$ ________

---

a) $t(t + 1)$
b) $m(m - 1)$

41. Two more examples of "factoring out" a letter from a binomial are shown below.

$$3b^2 + 4b = b(3b) + b(4) = b(3b + 4)$$

$$5t^2 - t = t(5t) - t(1) = t(5t - 1)$$

Following the examples, complete these factorings.

a) $7x^2 + x = x(7x) + x(1) =$ ________ b) $2y^2 - 5y = y(2y) - y(5) =$ ________

---

a) $x(7x + 1)$
b) $y(2y - 5)$

42. To check a factoring, we simply multiply by the distributive principle. That is:

$t^2 + t = t(t + 1)$ is correct, since $t(t + 1) = t^2 + t$

$2p^2 - 5p = p(2p - 5)$ is correct, since $p(2p - 5) = 2p^2 - 5p$

Factor out a letter from each binomial below and check your results.

a) $b^2 + 9b =$ ________ c) $8p^2 + p =$ ________

b) $d^2 - d =$ ________ d) $7x^2 + 8x =$ ________

---

a) $b(b + 9)$
b) $d(d - 1)$
c) $p(8p + 1)$
d) $x(7x + 8)$

43. When the coefficients of the terms have a common factor, we can factor out a number and a letter. For example:

$$3x^2 + 6x = 3x(x) + 3x(2) = 3x(x + 2)$$

$$4y^2 - 10y = 2y(2y) - 2y(5) = 2y(2y - 5)$$

Following the examples, complete these factorings.

a) $7t^2 + 21t = 7t(t) + 7t(3) =$ ________ b) $8m^2 - 6m = 2m(4m) - 2m(3) =$ ________

---

a) $7t(t + 3)$
b) $2m(4m - 3)$

44. Sometimes we get a "1" as the second term in the parentheses when factoring out a number and a letter. That is:

$4x^2 + 2x = 2x(2x) + 2x(1) = 2x(2x + 1)$ $15y^2 - 5y = 5y(3y) - 5y(1) =$ ________

$5y(3y - 1)$

45. In order to factor completely, we must factor out the largest possible number. Factor these completely and check your results.

a) $5x^2 + 10x =$ ________ c) $12t^2 - 18t =$ ________

b) $15y^2 - 3y =$ ________ d) $16m^2 - 8m =$ ________

a) $5x(x + 2)$
b) $3y(5y - 1)$
c) $6t(2t - 3)$
d) $8m(2m - 1)$

46. Factor each binomial completely and check your results.

a) $2p^2 - p$ = ________ c) $14m^2 + 7m$ = ________

b) $4d^2 + 3d$ = ________ d) $18h^2 - 24h$ = ________

a) $p(2p - 1)$ b) $d(4d + 3)$ c) $7m(2m + 1)$ d) $6h(3h - 4)$

SELF-TEST 44 (pages 381-390)

Perform each multiplication.

1. $(4p)(5)$ = ________ 2. $(m)(m)$ = ________ 3. $x(3x)$ = ________ 4. $(7y)(2y)$ = ________

Perform each multiplication.

5. $b(b + 1)$ = ________

6. $p(3p - 2)$ = ________

7. $4x(x - 1)$ = ________

8. $6y(2y + 3)$ = ________

Write the missing factor.

9. $10x = (2)(\quad)$ 10. $m^2 = (\quad)(m)$ 11. $4t^2 = (t)(\quad)$ 12. $16y^2 = (\quad)(8y)$

Factor completely.

13. $3y + 9$ = ________

14. $12x - 6$ = ________

15. $t^2 + 8t$ = ________

16. $p^2 - p$ = ________

17. $4x^2 + 12x$ = ________

18. $10y^2 - 5y$ = ________

ANSWERS:

1. $20p$
2. m^2
3. $3x^2$
4. $14y^2$
5. $b^2 + b$
6. $3p^2 - 2p$
7. $4x^2 - 4x$
8. $12y^2 + 18y$
9. $(2)(5x)$
10. $(m)(m)$
11. $(t)(4t)$
12. $(2y)(8y)$
13. $3(y + 3)$
14. $6(2x - 1)$
15. $t(t + 8)$
16. $p(p - 1)$
17. $4x(x + 3)$
18. $5y(2y - 1)$

15-7 MULTIPLYING TWO BINOMIALS: THE DISTRIBUTIVE-PRINCIPLE METHOD

In this section, we will show how the distributive principle can be used to multiply two binomials.

47. Before showing how the distributive principle can be used to multiply two binomials, we will show how it can be used to multiply a monomial and binomial when the binomial is on the left side. An example is discussed below.

$$(x + 4)(2) = x(2) + 4(2) = 2x + 8$$

<u>Note</u>: The arrows show that "2" is multiplied by each term ("x" and "4") in the binomial.

Following the example above, complete each multiplication below.

a) $(y + 3)(5)$ = ______________________

b) $(2m + 1)(7)$ = ______________________

---

a) $5y + 15$
b) $14m + 7$

48. Another example of multiplying a monomial by a binomial is shown below. The arrows again show that "x" is multiplied by each term ("x" and "8") in the binomial.

$$(x + 8)(x) = x(x) + 8(x) = x^2 + 8x$$

Following the example above, complete each multiplication below.

a) $(y + 1)(5y)$ = ______________________

b) $(2t + 3)(4t)$ = ______________________

---

a) $5y^2 + 5y$
b) $8t^2 + 12t$

49. The same form of the distributive principle can be used to multiply a binomial by a binomial. An example is shown. The arrows show that $(x + 3)$ is multiplied by each term ("x" and "2") in the first binomial.

$$(x + 2)(x + 3) = x(x + 3) + 2(x + 3)$$

Notice these two points about using the distributive principle above:

1. The first binomial $(x + 2)$ is broken up.
2. The right side contains two terms, $x(x + 3)$ and $2(x + 3)$, each of which is a multiplication of a binomial by a monomial.

Using the same pattern, complete these.

a) $(y + 4)(y + 5) = (\quad)(y + 5) + (\quad)(y + 5)$

b) $(2p + 3)(p + 7) = (\quad)(p + 7) + (\quad)(p + 7)$

c) $(3t + 1)(4t + 9) = (\quad)(4t + 9) + (\quad)(4t + 9)$

---

a) $(y)(y + 5) + (4)(y + 5)$ b) $(2p)(p + 7) + (3)(p + 7)$ c) $(3t)(4t + 9) + (1)(4t + 9)$

50. Complete these by applying the distributive principle.

a) $(m + 4)(m + 6) = (\quad)(\qquad) + (\quad)(\qquad)$

b) $(x + 7)(3x + 5) = (\quad)(\qquad) + (\quad)(\qquad)$

c) $(4y + 3)(2y + 7) = (\quad)(\qquad) + (\quad)(\qquad)$

---

a) $(m)(m + 6) + (4)(m + 6)$

b) $(x)(3x + 5) + (7)(3x + 5)$

c) $(4y)(2y + 7) + (3)(2y + 7)$

51. The three steps needed to multiply two binomials are shown and discussed below.

Step 1: $(x + 2)(x + 3) = x(x + 3) + 2(x + 3)$

Step 2: $= x^2 + 3x + 2x + 6$

Step 3: $= x^2 + 5x + 6$

In Step 1, we applied the distributive principle to break up the first binomial.

In Step 2, we performed the two multiplications of a binomial by a monomial.

In Step 3, we combined 3x and 2x to get 5x.

Following the steps above, complete this multiplication.

$(y + 5)(y + 4) = y(y + 4) + 5(y + 4)$

$= \_\_\_\_\_\_\_\_$

$= \_\_\_\_\_\_\_\_$

$y^2 + 4y + 5y + 20$

$y^2 + 9y + 20$

52. Use the distributive-principle method to complete this multiplication.

$(a + 6)(a + 4) = \_\_\_\_\_\_\_\_$

$= \_\_\_\_\_\_\_\_$

$= \_\_\_\_\_\_\_\_$

$a(a + 4) + 6(a + 4)$

$a^2 + 4a + 6a + 24$

$a^2 + 10a + 24$

53. Use the distributive-principle method to complete this multiplication.

$(2x + 1)(x + 5) = \_\_\_\_\_\_\_\_$

$= \_\_\_\_\_\_\_\_$

$= \_\_\_\_\_\_\_\_$

$2x(x + 5) + 1(x + 5)$

$2x^2 + 10x + x + 5$

$2x^2 + 11x + 5$

15-8 MULTIPLYING TWO BINOMIALS: THE FOIL METHOD

In this section, we will show a shorter method for multiplying two binomials. The shorter method is called the "FOIL" method.

54. The first two steps used to multiply $(x + 2)$ and $(x + 6)$ by the distributive-principle method are shown below.

$$(x + 2)(x + 6) = x(x + 6) + 2(x + 6)$$
$$= x^2 + 6x + 2x + 12$$

In the second step above, there are four terms in the product. We can skip the first step and write those four terms directly as we have done below.

$$\underset{\text{I L}}{\overset{\text{F O}}{(x + 2)(x + 6)}} = \underset{F}{x^2} + \underset{O}{6x} + \underset{I}{2x} + \underset{L}{12}$$

Note: 1. To get F (or "x^2"), we multiplied the first terms of the binomials.

2. To get O (or "6x"), we multiplied the outside terms of the binomials.

3. To get I (or "2x"), we multiplied the inside terms of the binomials.

4. To get L (or "12"), we multiplied the last terms of the binomials.

Using the FOIL method, write the four terms of each product below.

a) $\underset{\text{I L}}{\overset{\text{F O}}{(y + 3)(y + 5)}}$ = ___ (F) + ___ (O) + ___ (I) + ___ (L)

b) $\underset{\text{I L}}{\overset{\text{F O}}{(m + 7)(m + 4)}}$ = ___ (F) + ___ (O) + ___ (I) + ___ (L)

---

a) $y^2 + 5y + 3y + 15$

b) $m^2 + 4m + 7m + 28$

55. Using the FOIL method, write the four terms of each product below.

a) $\underset{\text{I L}}{\overset{\text{F O}}{(5x + 1)(x + 3)}}$ = ___ (F) + ___ (O) + ___ (I) + ___ (L)

b) $\underset{\text{I L}}{\overset{\text{F O}}{(2t + 3)(4t + 1)}}$ = ___ (F) + ___ (O) + ___ (I) + ___ (L)

---

a) $5x^2 + 15x + x + 3$

b) $8t^2 + 2t + 12t + 3$

56. In the FOIL method, we multiply both terms in the second binomial:

1. by the first term in the first binomial
2. by the second term in the first binomial

Using the FOIL method, write each four-term product.

a) $(x + 7)(x + 5)$ = ______ + ______ + ______ + ______

b) $(y + 6)(y + 1)$ = ______ + ______ + ______ + ______

---

a) $x^2 + 5x + 7x + 35$
b) $y^2 + y + 6y + 6$

57. Using the FOIL method, write each four-term product.

a) $(3t + 2)(t + 4)$ = ______ + ______ + ______ + ______

b) $(5x + 1)(2x + 8)$ = ______ + ______ + ______ + ______

---

a) $3t^2 + 12t + 2t + 8$
b) $10x^2 + 40x + 2x + 8$

58. After writing the four-term product below, we combined the "like" terms (4x and 8x) to get a trinomial product.

$$(x + 8)(x + 4) = x^2 + 4x + 8x + 32$$
$$= x^2 + 12x + 32$$

Following the example, write each four-term product and then combine the "like" terms to get a trinomial product.

a) $(y + 5)(y + 9)$ = ______ + ______ + ______ + ______

= ____________________

b) $(2t + 1)(4t + 5)$ = ______ + ______ + ______ + ______

= ____________________

---

a) $y^2 + 9y + 5y + 45$
$y^2 + 14y + 45$

b) $8t^2 + 10t + 4t + 5$
$8t^2 + 14t + 5$

59. Find each trinomial product.

a) $(x + 4)(x + 4)$ = ____________________

b) $(3y + 1)(3y + 1)$ = ____________________

a) $x^2 + 8x + 16$
b) $9y^2 + 6y + 1$

15-9 MULTIPLICATIONS INVOLVING "DIFFERENCE" BINOMIALS

In this section, we will discuss multiplications in which one or both binomials is a difference.

60. The FOIL method can also be used when one or both binomials is a difference. An example is shown below.

F O   F O I L

$$(x + 5)(x - 2) = x^2 - 2x + 5x - 10$$

I L

Note: 1. To get O (or $-2x$), we multiplied "x" and "-2".

2. To get L (or -10), we multiplied "5" and "-2".

Using the FOIL method, write each four-term product below.

a) $(y + 3)(y - 4)$ = ______________________

b) $(2t + 1)(t - 6)$ = ______________________

---

a) $y^2 - 4y + 3y - 12$

b) $2t^2 - 12t + t - 6$

61. The FOIL method was used to get the four-term product below.

F O   F O I L

$$(x - 4)(x + 9) = x^2 + 9x - 4x - 36$$

I L

Note: 1. To get I (or $-4x$), we multiplied "-4" and "x".

2. To get L (or -36), we multiplied "-4" and "9".

Using the FOIL method, write each four-term product below.

a) $(y - 7)(y + 2)$ = ______________________

b) $(t - 1)(4t + 5)$ = ______________________

---

a) $y^2 + 2y - 7y - 14$

b) $4t^2 + 5t - 4t - 5$

62. The FOIL method was used to get the four-term product below.

F O   F O I L

$$(m - 2)(m - 3) = m^2 - 3m - 2m + 6$$

I L

Note: 1. To get O (or $-3m$), we multiplied "m" and "-3".

2. To get I (or $-2m$), we multiplied "-2" and "m".

3. To get L (or $+6$), we multiplied "-2" and "-3".

Using the FOIL method, write each four-term product below.

a) $(p - 8)(p - 1)$ = ______________________

b) $(2t - 3)(2t - 4)$ = ______________________

---

a) $p^2 - p - 8p + 8$

b) $4t^2 - 8t - 6t + 12$

63. After writing the four-term product below, we combined the like terms $(-5x$ and $3x)$ to get a trinomial product.

$$(x + 3)(x - 5) = x^2 - 5x + 3x - 15$$
$$= x^2 - 2x - 15$$

Following the example, write each four-term product and then combine the like terms to get a trinomial product.

a) $(y + 2)(y - 8) =$ ____________________
$=$ ____________________

b) $(5m + 1)(m - 4) =$ ____________________
$=$ ____________________

Answers:
a) $y^2 - 8y + 2y - 16$
$y^2 - 6y - 16$

b) $5m^2 - 20m + m - 4$
$5m^2 - 19m - 4$

64. After writing the four-term product, we combined like terms $(6y$ and $-4y)$ to get a trinomial product.

$$(y - 4)(y + 6) = y^2 + 6y - 4y - 24$$
$$= y^2 + 2y - 24$$

Following the example, write each four-term product and then combine like terms to get a trinomial product.

a) $(x - 1)(x + 2) =$ ____________________
$=$ ____________________

b) $(d - 5)(3d + 4) =$ ____________________
$=$ ____________________

Answers:
a) $x^2 + 2x - x - 2$
$x^2 + x - 2$

b) $3d^2 + 4d - 15d - 20$
$3d^2 - 11d - 20$

65. After writing the four-term product below, we combined like terms $(-4t$ and $-3t)$ to get a trinomial product.

$$(t - 3)(t - 4) = t^2 - 4t - 3t + 12$$
$$= t^2 - 7t + 12$$

Following the example, write each four-term product and then combine like terms to get a trinomial product.

a) $(x - 9)(x - 1) =$ ____________________
$=$ ____________________

b) $(3y - 1)(2y - 4) =$ ____________________
$=$ ____________________

Answers:
a) $x^2 - x - 9x + 9$
$x^2 - 10x + 9$

b) $6y^2 - 12y - 2y + 4$
$6y^2 - 14y + 4$

66. Find each trinomial product.

a) $(h + 7)(h - 5) =$ ______

b) $(t - 6)(2t + 1) =$ ______

---

a) $h^2 + 2h - 35$

b) $2t^2 - 11t - 6$

67. Find each trinomial product.

a) $(x - 2)(x - 2) =$ ______

b) $(2y - 5)(2y - 5) =$ ______

---

a) $x^2 - 4x + 4$

b) $4y^2 - 20y + 25$

15-10 MULTIPLYING THE SUM AND DIFFERENCE OF TWO TERMS

In this section, we will discuss multiplications of the sum and difference of two terms.

68. In the multiplication below, both binomials contain an "x" and a "3". One binomial is a sum; the other binomial is a difference. Since $-3x + 3x = 0$, the product reduces to a binomial.

$$(x + 3)(x - 3) = x^2 - 3x + 3x - 9$$
$$= x^2 - 9$$

Following the example, complete this multiplication.

$(y + 6)(y - 6) =$ ______

$=$ ______

$y^2 - 6y + 6y - 36$

$y^2 - 36$

69. In the multiplication below, both binomials contain a "2t" and a "5". One binomial is a sum; the other binomial is a difference. Since $-10t + 10t = 0$, the product reduces to a binomial.

$$(2t + 5)(2t - 5) = 4t^2 - 10t + 10t - 25$$
$$= 4t^2 - 25$$

Following the example, complete this multiplication.

$(3m + 2)(3m - 2) =$ ______

$=$ ______

$9m^2 - 6m + 6m - 4$

$9m^2 - 4$

70. Here is a multiplication performed earlier.

$$(y + 6)(y - 6) = y^2 - 36$$

Notice these points about the product $y^2 - 36$.

1. y^2 is the square of "y".
2. 36 is the square of "6".
3. The two squares are <u>subtracted</u>.

Following the pattern above, write each binomial product directly.

a) $(x + 2)(x - 2) =$ ______ - ______ b) $(p + 9)(p - 9) =$ ______ - ______

---

a) $x^2 - 4$
b) $p^2 - 81$

71. Here is another multiplication performed earlier.

$$(3m + 2)(3m - 2) = 9m^2 - 4$$

Notice these points about the product $9m^2 - 4$.

1. $9m^2$ is the square of 3m.
2. 4 is the square of 2.
3. The two squares are <u>subtracted</u>.

Following the pattern above, write each binomial product directly.

a) $(4x + 1)(4x - 1) =$ ________ b) $(5t + 7)(5t - 7) =$ ________

---

a) $16x^2 - 1$
b) $25t^2 - 49$

72. Write each binomial product directly.

a) $(x + 1)(x - 1) =$ ________ c) $(2t + 9)(2t - 9) =$ ________

b) $(y + 10)(y - 10) =$ ________ d) $(6p + 1)(6p - 1) =$ ________

---

a) $x^2 - 1$ b) $y^2 - 100$ c) $4t^2 - 81$ d) $36p^2 - 1$

SELF-TEST 45 (pages 391-399)

Find each trinomial product.

1. $(x + 3)(x + 3) =$	2. $(2y + 5)(3y + 1) =$
3. $(t + 8)(t - 1) =$	4. $(5p - 4)(3p + 2) =$

Continued on following page.

SELF-TEST 45 - Continued

5. $(a - 2)(a - 9) =$	6. $(4b - 1)(4b - 1) =$

Find each binomial product.

7. $(x + 4)(x - 4) =$	8. $(5y - 1)(5y + 1) =$

ANSWERS:

1. $x^2 + 6x + 9$
2. $6y^2 + 17y + 5$
3. $t^2 + 7t - 8$
4. $15p^2 - 2p - 8$
5. $a^2 - 11a + 18$
6. $16b^2 - 8b + 1$
7. $x^2 - 16$
8. $25y^2 - 1$

15-11 FACTORING THE DIFFERENCE OF TWO PERFECT SQUARES

In this section, we will discuss the procedure for factoring the difference of two perfect squares.

73. A whole number whose principal square root is a whole number is a perfect square. For example:

25 is a perfect square, since $\sqrt{25} = 5$

a) 81 is a perfect square, since $\sqrt{81} =$ ______

b) 1 is a perfect square, since $\sqrt{1} =$ ______

a) 9
b) 1

74. Any squared letter is a "perfect square". For example:

x^2 is a perfect square, since $\sqrt{x^2} = x$

y^2 is a perfect square, since $\sqrt{y^2} =$ ______

y

75. Any term containing a squared letter with a perfect-square coefficient is a "perfect square". For example:

$4x^2$ is a perfect square, since $\sqrt{4x^2} = 2x$

$9m^2$ is a perfect square, since $\sqrt{9m^2} = 3m$

$64y^2$ is a perfect square, since $\sqrt{64y^2} =$ ______

8y

76. In the multiplication below, the product is the difference of two perfect squares.

$$(x + 4)(x - 4) = x^2 - 16$$

By simply reversing the two sides above, we can factor $x^2 - 16$. That is:

$$x^2 - 16 = (x + 4)(x - 4)$$

Notice these points about the factoring:

1. "x" is the principal square root of "x^2".
2. "4" is the principal square root of "16".
3. One factor is a sum; the other factor is a difference.

Following the pattern above, complete these factorings.

a) $y^2 - 36 = (y + 6)(\qquad)$ b) $m^2 - 1 = (\qquad)(m - 1)$

Answers: a) $(y + 6)(y - 6)$ b) $(m + 1)(m - 1)$

77. In the multiplication below, the product is the difference of two perfect squares.

$$(3y + 2)(3y - 2) = 9y^2 - 4$$

By simply reversing the two sides, we can factor $9y^2 - 4$. That is:

$$9y^2 - 4 = (3y + 2)(3y - 2)$$

Notice these points about the factoring:

1. "3y" is the principal square root of "$9y^2$".
2. "2" is the principal square root of "4".
3. One factor is a sum; the other factor is a difference.

Following the pattern above, complete these factorings.

a) $25t^2 - 64 = (5t + 8)(\qquad)$ b) $49y^2 - 1 = (\qquad)(7y - 1)$

Answers: a) $(5t + 8)(5t - 8)$ b) $(7y + 1)(7y - 1)$

78. Following the pattern in the last two frames, factor these:

a) $p^2 - 16 =$ ________ c) $36x^2 - 81 =$ ________

b) $t^2 - 100 =$ ________ d) $9y^2 - 1 =$ ________

Answers: a) $(p + 4)(p - 4)$ b) $(t + 10)(t - 10)$ c) $(6x + 9)(6x - 9)$ d) $(3y + 1)(3y - 1)$

79. Using the same pattern, factor these:

a) $x^2 - 49 =$ ________ c) $49q^2 - 4 =$ ________

b) $y^2 - 1 =$ ________ d) $64h^2 - 9 =$ ________

Answers: a) $(x + 7)(x - 7)$ b) $(y + 1)(y - 1)$ c) $(7q + 2)(7q - 2)$ d) $(8h + 3)(8h - 3)$

15-12 FACTORING TRINOMIALS OF THE FORM: $x^2 + bx + c$

In this section, we will discuss the procedure for factoring trinomials in which the coefficient of the squared letter is "1".

---

80. The trinomial product for the multiplication below is given.

$$(x + 2)(x + 3) = x^2 + 5x + 6$$

Notice these points about the numbers in the product:

1. The number-term is "6". It is the product of 2 and 3.
2. The coefficient of "x" is "5". It is the sum of 2 and 3.

Using the facts above, we can directly write the trinomial product for this multiplication.

$$(y + 5)(y + 4)$$

a) Since $5 + 4 = 9$, the coefficient of the "y" term must be ______.

b) Since $(5)(4) = 20$, the number-term must be ______.

c) Therefore: $(y + 5)(y + 4) =$ ______ + ______ + ______

---

a) 9
b) 20
c) $y^2 + 9y + 20$

81. Using the method in the last frame, write the trinomial product for these.

a) $(m + 1)(m + 7) =$ ______________

b) $(p + 6)(p + 8) =$ ______________

---

a) $m^2 + 8m + 7$
b) $p^2 + 14p + 48$

82. We used the same method to write the trinomial product below.

$$(x + 4)(x + 2) = x^2 + 6x + 8$$

To factor $x^2 + 6x + 8$, we can simply reverse the two sides of the equation. We get:

$$x^2 + 6x + 8 = (x + 4)(x + 2)$$

Notice the same two points about the numbers "4" and "2" in the binomial factors.

1. Their product is "8", the number-term of the trinomial.
2. Their sum is "6", the coefficient of the "x" term of the trinomial.

Let's check the same two points about the numbers in the binomials below to see whether the factoring is correct.

$$y^2 + 8y + 15 = (y + 3)(y + 5)$$

a) Does $(3)(5)$ equal the number-term of the trinomial? ______

b) Does $3 + 5$ equal the coefficient of the "y" term? ______

c) Can $y^2 + 8y + 15$ be factored into $(y + 3)(y + 5)$? ______

---

a) Yes b) Yes c) Yes

83. The steps needed to factor the following trinomial are discussed below.

$$x^2 + 5x + 6$$

1. The product of the numbers in the binomials must be 6. The possible pairs of factors for 6 are (1 and 6) and (2 and 3).
2. The sum of the numbers in the binomials must be 5. Therefore, the correct pair of factors is (2 and 3) since $2 + 3 = 5$.

Therefore: $x^2 + 5x + 6 = (\quad)(\quad)$

Answer: $(x + 2)(x + 3)$

84. Let's factor $y^2 + 5y + 4$. The possible pairs of factors of "4" are (1 and 4) and (2 and 2).

a) The sum of which pair of factors is 5? ________

b) Therefore: $y^2 + 5y + 4$ = ________

Answer: a) 1 and 4 b) $(y + 1)(y + 4)$

85. Let's factor $t^2 + 8t + 12$. The possible pairs of factors of "12" are (1 and 12), (2 and 6), and (3 and 4).

a) The sum of which pair of factors is 8? ________

b) Therefore: $t^2 + 8t + 12$ = ________

Answer: a) 2 and 6 b) $(t + 2)(t + 6)$

86. Using the same method, factor each trinomial below.

a) $x^2 + 6x + 5$ = ________ b) $p^2 + 10p + 16$ = ________

Answer: a) $(x + 1)(x + 5)$ b) $(p + 2)(p + 8)$

87. The trinomial product for the multiplication below is given.

$$(x - 3)(x - 5) = x^2 - 8x + 15$$

Notice these points about the numbers in the product.

1. The number-term is "15". It is the product of -3 and -5.
2. The coefficient of "x" is -8. It is the sum of -3 and -5.

Using the facts above, we can directly write the trinomial product for this multiplication.

$$(y - 8)(y - 4)$$

a) Since $(-8) + (-4) = -12$, the coefficient of the "y" term must be ________.

b) Since $(-8)(-4) = 32$, the number-term must be ________.

c) Therefore: $(y - 8)(y - 4)$ = ________

Answer: a) -12 b) 32 c) $y^2 - 12y + 32$

88. Using the method in the last frame, write each trinomial product.

a) $(t - 1)(t - 2)$ = ____________ b) $(m - 7)(m - 5)$ = ____________

---

a) $t^2 - 3t + 2$

b) $m^2 - 12m + 35$

89. In the trinomial below, the coefficient of the "x" term is negative. Let's factor the trinomial.

$$x^2 - 7x + 10$$

Since the product of the numbers in the binomials must be +10 and their sum must be -7, both numbers must be negative. The possible pairs of negative factors for +10 are (-1 and -10) and (-2 and -5).

a) The sum of which pair of factors is -7? ____________

b) Therefore: $x^2 - 7x + 10 = (\quad)(\quad)$

---

a) -2 and -5

b) $(x - 2)(x - 5)$

90. When factoring the trinomial below, the numbers in both binomials must also be negative.

$$y^2 - 7y + 6$$

The possible pairs of negative factors for +6 are (-1 and -6) and (-2 and -3).

a) The sum of which pair of factors is -7? ____________

b) Therefore: $y^2 - 7y + 6 = (\quad)(\quad)$

---

a) -1 and -6 b) $(y - 1)(y - 6)$

91. Using the same method, factor each trinomial below.

a) $t^2 - 3t + 2$ = ____________ b) $m^2 - 8m + 16$ = ____________

---

a) $(t - 1)(t - 2)$

b) $(m - 4)(m - 4)$

92. When the number-term of a trinomial is positive, the numbers in the binomial factors are either both positive or both negative. To decide whether they are both positive or both negative, we look at the sign of the letter-term (or middle term).

If the sign of the letter-term is positive, both numbers must be positive. That is:

$$x^2 + 5x + 6 = (x + 2)(x + 3)$$

If the sign of the letter-term is negative, both numbers must be negative. That is:

$$x^2 - 5x + 6 = (x - 2)(x - 3)$$

Using the facts above, factor each trinomial below.

a) $y^2 + 6y + 5$ = ____________ c) $m^2 + 9m + 14$ = ____________

b) $y^2 - 6y + 5$ = ____________ d) $m^2 - 9m + 14$ = ____________

93. The trinomial product for the multiplication below is given.

$$(x + 2)(x - 5) = x^2 - 3x - 10$$

Notice these points about the numbers in the product.

1. The number-term is -10. It is the <u>product</u> of 2 and -5.
2. The coefficient of "x" is -3. It is the <u>sum</u> of 2 and -5.

Using the facts above, we can directly write the trinomial product for the multiplication below.

$$(y - 3)(y + 4)$$

a) Since $(-3) + 4 = 1$, the coefficient of the "y" term must be ________.

b) Since $(-3)(4) = -12$, the number-term must be ________.

c) Therefore: $(y - 3)(y + 4)$ = ________________

Answers to 92:
a) $(y + 1)(y + 5)$
b) $(y - 1)(y - 5)$
c) $(m + 2)(m + 7)$
d) $(m - 2)(m - 7)$

---

94. Using the same method, write each trinomial product.

a) $(x + 7)(x - 1)$ = ________________

b) $(y + 3)(y - 5)$ = ________________

c) $(m - 3)(m + 6)$ = ________________

d) $(p - 2)(p + 1)$ = ________________

Answers to 93: a) 1 b) -12 c) $y^2 + y - 12$

---

95. In the trinomial below, the number-term is negative. Let's factor the trinomial.

$$x^2 + 5x - 6$$

Since the product of the numbers in the binomials must be <u>negative</u>, <u>one number must be positive and the other negative</u>. The possible pairs of factors for -6 are (-1 and 6), (-6 and 1), (-2 and 3), and (-3 and 2).

a) The sum of which pair of factors is +5? ________

b) Therefore: $x^2 + 5x - 6$ = (   )(   )

Answers to 94:
a) $x^2 + 6x - 7$
b) $y^2 - 2y - 15$
c) $m^2 + 3m - 18$
d) $p^2 - p - 2$

---

96. In the trinomial below, the number-term is negative. Let's factor the trinomial.

$$y^2 - 2y - 8$$

Since the product of the numbers in the binomials must be -8, <u>one must be positive and the other negative</u>. The possible pairs of factors for -8 are (-1 and 8), (-8 and 1), (-2 and 4), and (2 and -4).

a) The sum of which pair of factors is -2? ________

b) Therefore: $y^2 - 2y - 8$ = (   )(   )

Answers to 95:
a) -1 and 6
b) $(x - 1)(x + 6)$

Answers to 96:
a) 2 and -4
b) $(y + 2)(y - 4)$

97. If the number-term of a trinomial is negative, the numbers in the two binomials must have different signs. Using that fact, factor each trinomial below.

a) $m^2 + 4m - 5$ = ________________

b) $t^2 - 2t - 3$ = ________________

c) $x^2 + 3x - 10$ = ________________

d) $y^2 - 5y - 14$ = ________________

---

a) $(m - 1)(m + 5)$
b) $(t + 1)(t - 3)$
c) $(x - 2)(x + 5)$
d) $(y + 2)(y - 7)$

98. To check the factoring of a trinomial, multiply the two binomial factors to see whether you obtain the original trinomial. For example:

$x^2 + 8x + 15 = (x + 3)(x + 5)$ is correct, since:

$(x + 3)(x + 5) = x^2 + 8x + 15$

$y^2 + 7y - 18 = (y - 2)(y + 9)$ is correct, since:

$(y - 2)(y + 9) = y^2 + 7y - 18$

Check each factoring. State whether it is "correct" or "incorrect".

a) $x^2 - 7x + 10 = (x - 2)(x - 5)$ ____________

b) $p^2 + 2p - 15 = (p - 3)(p + 5)$ ____________

c) $t^2 - 5t - 14 = (t - 2)(t + 7)$ ____________

---

a) correct

b) correct

c) incorrect, since:
$(t - 2)(t + 7) = t^2 + 5t - 14$

99. Remember that the key to factoring a trinomial is the number-term. That is:

1. If the number-term is positive, the numbers in the binomials have the same sign.

2. If the number-term is negative, the numbers in the binomials have different signs.

Factor each trinomial below.

a) $x^2 + 9x + 20$ = ________________

b) $y^2 - 3y - 18$ = ________________

c) $t^2 - 9t + 8$ = ________________

d) $m^2 + 4m - 12$ = ________________

---

a) $(x + 4)(x + 5)$

b) $(y + 3)(y - 6)$

c) $(t - 1)(t - 8)$

d) $(m - 2)(m + 6)$

15-13 FACTORING TRINOMIALS OF THE FORM: $ax^2 + bx + c$

In this section, we will discuss the procedure for factoring trinomials in which the coefficient of the squared letter is a number other than "1".

100. The steps needed to multiply $(4x + 3)$ and $(x + 2)$ are shown below.

$$(4x + 3)(x + 2) = 4x^2 + 8x + 3x + 6 = 4x^2 + 11x + 6$$

Notice these points about the trinomial product:

"$4x^2$" is the product of "4x" and "x", the letter-terms.

" 6 " is the product of " 3 " and "2", the number-terms.

Using the facts above, if we multiplied $(2y + 4)$ and $(3y + 1)$:

a) The first term of the trinomial product would be ________.

b) The last term of the trinomial product would be ________.

---

a) $6y^2$, from $(2y)(3y)$

b) 4 , from $(4)(1)$

101. Here are the steps needed to factor the trinomial below.

$$3x^2 + 14x + 8$$

1. Since the product of the letter-terms of the binomials must be "$3x^2$", the only possible pair of letter-terms is (3x and x).
2. Since the product of the number-terms of the binomials must be 8, the possible pairs of number-terms are (1 and 8) and (2 and 4).
3. Using the factors in all combinations, we get four possible pairs of binomials:

A: $(3x + 1)(x + 8)$

B: $(3x + 8)(x + 1)$

C: $(3x + 2)(x + 4)$

D: $(3x + 4)(x + 2)$

4. Only one pair of binomials is correct. It is the pair that produces "14x" as the middle term of the trinomial product.

a) Which pair of binomial factors has "14x" as the middle term of its product? Pair ______

b) Therefore: $3x^2 + 14x + 8 = (\quad\quad)(\quad\quad)$

---

a) Pair C

b) $(3x + 2)(x + 4)$

102. Let's factor the trinomial: $2y^2 - 13y + 15$

1. The only possible pair of letter-terms is (2y and y).
2. The possible pairs of number-terms are (-1 and -15) and (-3 and -5).
3. The possible pairs of binomial factors are:

A: $(2y - 1)(y - 15)$

B: $(2y - 15)(y - 1)$

C: $(2y - 3)(y - 5)$

D: $(2y - 5)(y - 3)$

The correct pair of binomial factors is the one with "-13y" as the middle term of its product.

a) Which pair has "-13y" as the middle-term of its product? Pair ______

b) Therefore: $2y^2 - 13y + 15 = (\qquad)(\qquad)$

---

a) Pair C

b) $(2y - 3)(y - 5)$

103. Let's factor the trinomial: $5t^2 + 2t - 3$

1. The only possible pair of letter-terms is (5t and t).
2. The possible pairs of number-terms are (1 and -3) and (-1 and 3).
3. The possible pairs of binomial factors are:

A: $(5t + 1)(t - 3)$

B: $(5t - 3)(t + 1)$

C: $(5t - 1)(t + 3)$

D: $(5t + 3)(t - 1)$

a) Which pair of factors has a product whose middle term is "2t"? Pair ______

b) Therefore: $5t^2 + 2t - 3 = (\qquad)(\qquad)$

---

a) Pair B

b) $(5t - 3)(t + 1)$

104. Let's factor the trinomial: $7t^2 - 2t - 5$

1. The only possible pair of letter-terms is (7t and t).
2. The possible pairs of number-terms are (1 and -5) and (-1 and 5).
3. The possible pairs of binomial factors are:

A: $(7t + 1)(t - 5)$

B: $(7t - 5)(t + 1)$

C: $(7t - 1)(t + 5)$

D: $(7t + 5)(t - 1)$

a) Which pair of factors has a product whose middle term is "-2t"? Pair ______

b) Therefore: $7t^2 - 2t - 5 = (\qquad)(\qquad)$

a) Pair D

b) $(7t + 5)(t - 1)$

105. When the coefficient of the squared letter is a number other than "1", factoring is a process of trial and error. Each possible pair of factors has to be checked by multiplying the binomial factors. That is:

$$3x^2 - 11x + 6 = (3x - 2)(x - 3) \text{ is correct, since:}$$
$$(3x - 2)(x - 3) = 3x^2 - 11x + 6$$

$$2y^2 - y - 1 = (2y + 1)(y - 1) \text{ is correct, since:}$$
$$(2y + 1)(y - 1) = 2y^2 - y - 1$$

Factor each trinomial below and check your results.

a) $5m^2 + 13m + 6 =$ ____________________

b) $7m^2 - 11m + 4 =$ ____________________

---

106. Factor each trinomial below and check your results.

a) $11m^2 + 4m - 7 =$ ____________________

b) $2t^2 - 5t - 3 =$ ____________________

Answers to 105:
a) $(5m + 3)(m + 2)$
b) $(7m - 4)(m - 1)$

Answers to 106:
a) $(11m - 7)(m + 1)$
b) $(2t + 1)(t - 3)$

SELF-TEST 46 (pages 399-408)

Factor these binomials.

1. $x^2 - 25 =$	2. $4y^2 - 81 =$

Factor these trinomials.

3. $t^2 + 7t + 12 =$	4. $p^2 + 10p + 25 =$
5. $m^2 - 9m + 8 =$	6. $b^2 - 2b + 1 =$
7. $h^2 - 6h - 16 =$	8. $m^2 + m - 2 =$
9. $5x^2 - 8x + 3 =$	10. $3y^2 - 2y - 5 =$

ANSWERS:

1. $(x + 5)(x - 5)$	3. $(t + 3)(t + 4)$	5. $(m - 1)(m - 8)$	7. $(h + 2)(h - 8)$	9. $(5x - 3)(x - 1)$
2. $(2y + 9)(2y - 9)$	4. $(p + 5)(p + 5)$	6. $(b - 1)(b - 1)$	8. $(m - 1)(m + 2)$	10. $(3y - 5)(y + 1)$

Unit 16 QUADRATIC EQUATIONS

Any one-letter equation that has two roots is called a "quadratic" equation. There are three types of quadratic equations - "pure", "incomplete", and "complete". In this unit, we will discuss some of the methods available for solving each type. Some applied problems are included.

16-1 THE THREE TYPES OF QUADRATIC EQUATIONS

There are three types of quadratic equations - "pure", "incomplete", and "complete". We will discuss the three types in this section.

1. The equation below is a "quadratic" equation because it has two roots, +6 and -6.

$$x^2 - 36 = 0$$

We have checked +6 as its root below. Check -6 as its root.

$$\begin{aligned} x^2 - 36 &= 0 \\ (6)^2 - 36 &= 0 \\ 36 - 36 &= 0 \\ 0 &= 0 \end{aligned} \qquad\qquad x^2 - 36 = 0$$

$$\begin{aligned} (-6)^2 - 36 &= 0 \\ 36 - 36 &= 0 \\ 0 &= 0 \end{aligned}$$

2. The equation below is a "quadratic" equation because it has two roots, 0 and 5.

$$y^2 - 5y = 0$$

We have checked "0" as its root below. Check 5 as its root.

$$\begin{aligned} y^2 - 5y &= 0 \\ (0)^2 - 5(0) &= 0 \\ 0 - 0 &= 0 \\ 0 &= 0 \end{aligned} \qquad\qquad y^2 - 5y = 0$$

$$\begin{aligned} (5^2) - 5(5) &= 0 \\ 25 - 25 &= 0 \\ 0 &= 0 \end{aligned}$$

3. The equation below is a "quadratic" equation because it has two roots, 2 and 3.

$$m^2 - 5m + 6 = 0$$

We have checked 2 as its root below. Check 3 as its root.

$$m^2 - 5m + 6 = 0 \qquad\qquad m^2 - 5m + 6 = 0$$

$$(2)^2 - 5(2) + 6 = 0$$

$$4 - 10 + 6 = 0$$

$$10 - 10 = 0$$

$$0 = 0$$

$$(3^2) - 5(3) + 6 = 0$$

$$9 - 15 + 6 = 0$$

$$15 - 15 = 0$$

$$0 = 0$$

4. We have seen that each equation below is a "quadratic" equation because each has two roots.

$$x^2 - 36 = 0 \qquad y^2 - 5y = 0 \qquad m^2 - 5m + 6 = 0$$

Notice that each equation contains a squared letter in one term. The squared letter can be used as a clue to identify quadratic equations. Which of the following are quadratic equations? ______________

a) $b^2 - 81 = 0$ b) $3x + 8 = 0$ c) $m^2 + 7m = 0$ d) $y^2 - 7y + 12 = 0$

(a), (c), and (d)

5. Each quadratic equation below contains a squared-letter term, a letter-term, and a number-term. Equations of that type are called "complete" quadratic equations.

$$x^2 - 6x + 8 = 0 \qquad 3y^2 + 5y + 2 = 0$$

Which of the following are "complete" quadratic equations? ______________

a) $y^2 - 9 = 0$ b) $t^2 + 7t + 10 = 0$ c) $x^2 + 9x = 0$

Only (b)

6. Each quadratic equation below contains a squared-letter term and a letter-term, but no number-term. Equations of that type are called "incomplete" quadratic equations.

$$m^2 - 3m = 0 \qquad 4d^2 + 20d = 0$$

Which of the following are "incomplete" quadratic equations? ______________

a) $x^2 - 49 = 0$ b) $b^2 + 9b + 14 = 0$ c) $y^2 + 6y = 0$

Only (c)

7. Each quadratic equation below contains a squared-letter term and a number-term, but no letter-term. Equations of that type are called "pure" quadratic equations.

$$t^2 - 64 = 0 \qquad 16p^2 - 81 = 0$$

Which of the following are "pure" quadratic equations? ______________

a) $x^2 - 1 = 0$ b) $y^2 - 8y + 12 = 0$ c) $d^2 + 10d = 0$

Only (a)

8. Here is a summary of the three types of quadratic equations.

Complete:	$x^2 + 3x - 10 = 0$
Incomplete:	$x^2 + 3x = 0$
Pure:	$x^2 - 10 = 0$

The "complete" quadratic has a trinomial on the left side. The trinomial contains a squared-letter term, a letter-term, and a number-term.

a) The "incomplete" quadratic has a binomial on the left side. It does not contain a ______________-term.

b) The "pure" quadratic has a binomial on the left side. It does not contain a ______________-term.

---

a) number-term

b) letter-term

9. For each quadratic equation, state whether it is "complete", "incomplete", or "pure".

______________ a) $m^2 - 25 = 0$

______________ b) $t^2 - 6t + 5 = 0$

______________ c) $p^2 - 8p = 0$

---

a) pure

b) complete

c) incomplete

10. For each quadratic equation, state whether it is "complete", "incomplete", or "pure".

______________ a) $x^2 + 12x = 0$

______________ b) $V^2 - 100 = 0$

______________ c) $3y^2 + 14y + 8 = 0$

---

a) incomplete

b) pure

c) complete

16-2 SOLVING PURE QUADRATICS

In an earlier unit, we discussed a method for solving "pure" quadratic equations. We will review that method in this section.

---

11. In the standard form of a pure quadratic equation, both the squared-letter term and the number-term are on the same side. For example:

$$x^2 - 49 = 0 \qquad 16y^2 - 9 = 0 \qquad \frac{1}{2}t^2 - 8 = 0$$

However, pure quadratic equations frequently appear in an equivalent form with the squared-letter term on one side and the number term on the other side. For example:

$$x^2 = 49 \qquad 16y^2 = 9 \qquad \frac{1}{2}t^2 = 8$$

Continued on following page.

11. Continued

Similarly: an equivalent form for $y^2 - 1 = 0$ is $y^2 = 1$.

a) an equivalent form for $4p^2 - 25 = 0$ is ________ = ________

b) an equivalent form for $\frac{1}{3}d^2 - 27 = 0$ is ________ = ________

---

a) $4p^2 = 25$

b) $\frac{1}{3}d^2 = 27$

12. When a pure quadratic contains a squared letter without a coefficient on one side and a number on the other side, <u>its two roots are the two square roots of the number</u>. That is:

For $x^2 = 49$, $x = \pm\sqrt{49} = +7$ and -7

For $y^2 = 1$, $y = \pm\sqrt{1} =$ ______ and ______

+1 and -1

13. When the number is a perfect-square fraction, the two roots of the equation are the two square roots of the fraction. That is:

For $x^2 = \frac{36}{25}$, $x = \pm\sqrt{\frac{36}{25}} = \frac{6}{5}$ and $-\frac{6}{5}$

For $y^2 = \frac{1}{4}$, $y = \pm\sqrt{\frac{1}{4}} =$ ______ and ______

$\frac{1}{2}$ and $-\frac{1}{2}$

14. To solve the standard-form equation below, we used the addition axiom to get "t^2" on one side and 9 on the other side. Solve the other equation.

$$t^2 - 9 = 0$$
$$t^2 = 9$$
$$t = \pm\sqrt{9} = 3 \text{ and } -3$$

$$m^2 - 100 = 0$$

$m = 10$ and -10

15. To solve the equation below, we used the multiplication axiom to isolate "t^2" and then found both square roots of $\frac{9}{16}$. Solve the other equation.

$$16t^2 = 9$$
$$t^2 = \frac{9}{16}$$
$$t = \frac{3}{4} \text{ and } -\frac{3}{4}$$

$$49y^2 = 100$$

$y = \frac{10}{7}$ and $-\frac{10}{7}$

16. To solve the standard-form equation below, we used both axioms to isolate "x^2" and then found both square roots of $\frac{81}{64}$. Solve the other equation.

$$64x^2 - 81 = 0$$
$$64x^2 = 81$$
$$x^2 = \frac{81}{64}$$
$$x = \pm\sqrt{\frac{81}{64}} = \frac{9}{8} \text{ and } -\frac{9}{8}$$

$$9b^2 - 1 = 0$$

$b = \pm\sqrt{\frac{1}{9}} = \frac{1}{3}$ and $-\frac{1}{3}$

17. To solve the equation below, we used the multiplication axiom to isolate "x^2". Solve the other equation.

$$\frac{1}{2}x^2 = 8$$
$$2\left(\frac{1}{2}x^2\right) = 2(8)$$
$$x^2 = 16$$
$$x = 4 \text{ and } -4$$

$$12 = \frac{1}{3}y^2$$

$y = 6$ and -6

18. We used both axioms to solve the equation below. Solve the other equation.

$$\frac{1}{2}t^2 - 32 = 0$$
$$\frac{1}{2}t^2 = 32$$
$$t^2 = 64$$
$$t = \pm\sqrt{64} = 8 \text{ and } -8$$

$$\frac{1}{4}m^2 - 25 = 0$$

$m = 10$ and -10

19. When the number-term is not a perfect square, we use the square-root table on page 190. That is:

For $x^2 = 69$, $x = \pm\sqrt{69}$ = 8.307 and -8.307

For $y^2 = 175$, $y = \pm\sqrt{175}$ = ________ and ________

13.229 and -13.229

20. Use the square-root table to find the roots of each equation below.

a) $t^2 - 94 = 0$

b) $\frac{1}{4}m^2 = 30$

---

a) $t = \pm\sqrt{94} = \pm 9.695$

b) $m = \pm\sqrt{120} = \pm 10.954$

21. In solving the equation below, we obtained a fraction that is not a perfect square. Therefore, we performed a division and then used the square-root table. Solve the other equation.

$$2p^2 = 30$$
$$p^2 = \frac{30}{2}$$
$$p^2 = 15$$
$$p = \pm 3.873$$

$$5d^2 - 975 = 0$$

Answer: $d = \pm\sqrt{195} = \pm 13.964$

22. In solving the equation below, we obtained a negative number on the right side.

$$x^2 + 25 = 0$$
$$x^2 = -25$$
$$x = \pm\sqrt{-25}$$

However, we have already seen that no ordinary signed number is the square root of a negative number like -25. That is:

Since $(+5)(+5) = +25$, +5 is not a square root of -25.

Since $(-5)(-5) = +25$, -5 is not a square root of -25.

Since no ordinary signed number is the square root of -25, does $x^2 + 25 = 0$ have two ordinary numbers as its roots? ________

Answer: No

23. In solving the equation at the right, we get a negative fraction on the right side.

$$9y^2 + 4 = 0$$
$$9y^2 = -4$$
$$y^2 = -\frac{4}{9}$$
$$y = \pm\sqrt{-\frac{4}{9}}$$

a) Is $+\frac{2}{3}$ a square root of $-\frac{4}{9}$? ________

b) Is $-\frac{2}{3}$ a square root of $-\frac{4}{9}$? ________

c) Does $9y^2 + 4 = 0$ have two ordinary numbers as its roots? ________

Answers:

a) No, since: $\left(+\frac{2}{3}\right)\left(+\frac{2}{3}\right) = +\frac{4}{9}$

b) No, since: $\left(-\frac{2}{3}\right)\left(-\frac{2}{3}\right) = +\frac{4}{9}$

c) No

24. Which of the following pure quadratics have two ordinary numbers as their roots? ________

a) $x^2 = 16$

b) $y^2 = -64$

c) $t^2 + 1 = 0$

d) $b^2 - 9 = 0$

e) $4d^2 - 1 = 0$

f) $64h^2 + 9 = 0$

Answer: (a), (d), and (e)

16-3 THE PRINCIPLE OF ZERO PRODUCTS

All incomplete and some complete quadratics can be solved by the factoring method. The basic principle used in the factoring method is the "principle of zero products". We will discuss that principle in this section.

---

25. If one factor in a multiplication is "0", the product is "0". For example:

$$(5)(0) = 0 \qquad (0)(9) = 0$$

Therefore, <u>if the product in a multiplication is "0", one of the factors must be "0"</u>. That is:

If $ab = 0$, either "a" or "b" must be "0".

If $xy = 0$, either "x" or "y" must be ______.

Answer: 0

26. The equation below involves a multiplication. The factors are $(x - 2)$ and $(x - 3)$.

$$(x - 2)(x - 3) = 0$$

Since the product is "0", the equation is true when either $(x - 2)$ or $(x - 3)$ is "0". Therefore, we can find the two roots of the equation by solving the two equations below as we have done.

$$x - 2 = 0 \qquad x - 3 = 0$$
$$x = 2 \qquad x = 3$$

We checked 2 as one root of the original equation. Check 3 as the second root.

$$(x - 2)(x - 3) = 0$$
$$(2 - 2)(2 - 3) = 0$$
$$(0)(-1) = 0$$
$$0 = 0$$

$$(x - 2)(x - 3) = 0$$

Answer:

$$(3 - 2)(3 - 3) = 0$$
$$(1)(0) = 0$$
$$0 = 0$$

27. The equation below is true if either $(y + 4)$ or $(y - 5)$ is "0".

$$(y + 4)(y - 5) = 0$$

Therefore, to find the two roots, we solved the equations below.

$$y + 4 = 0 \qquad y - 5 = 0$$
$$y = -4 \qquad y = 5$$

The two roots of the original equation are ______ and ______.

Answer: -4 and 5

28. We solved the equation below by setting both factors equal to "0". Solve the other equation.

$$(t + 2)(t + 8) = 0$$

$t + 2 = 0$	$t + 8 = 0$
$t = -2$	$t = -8$

$$(m - 1)(m + 10) = 0$$

Answer: $m = 1$ and -10

29. The equation below is true if either $(3x - 2)$ or $(2x + 5)$ is "0".

$$(3x - 2)(2x + 5) = 0$$

Therefore to find the two roots, we solved the equations below.

$$3x - 2 = 0 \qquad 2x + 5 = 0$$

$$3x = 2 \qquad 2x = -5$$

$$x = \frac{2}{3} \qquad x = -\frac{5}{2}$$

The two roots of the original equation are ________ and ________.

$\frac{2}{3}$ and $-\frac{5}{2}$

30. We solved the equation below by setting both factors equal to "0". Solve the other equation.

$$(3p + 1)(4p - 7) = 0 \qquad (5t - 9)(6t - 1) = 0$$

$$3p + 1 = 0 \qquad 4p - 7 = 0$$

$$3p = -1 \qquad 4p = 7$$

$$p = -\frac{1}{3} \qquad p = \frac{7}{4}$$

$t = \frac{9}{5}$ or $\frac{1}{6}$

31. The equation below is true if either (x) or $(x - 3)$ is "0".

$$x(x - 3) = 0$$

Therefore to find the two roots, we solved the equations below.

$$x = 0 \qquad x - 3 = 0$$

$$x = 3$$

The two roots of the original equation are ______ and ______.

---

0 and 3

32. In the last frame, we saw that "0" is one of the roots of $x(x - 3) = 0$.
The "0" root is frequently overlooked.

We solved the equation below by setting both factors equal to "0". Solve the other equation.

$$m(m + 6) = 0 \qquad y(y - 12) = 0$$

$$m = 0 \qquad m + 6 = 0$$

$$m = -6$$

$y = 0$ and 12

33. We solved the equation below by setting both factors equal to "0". Solve the other equation.

$$b(2b - 1) = 0 \qquad\qquad x(5x + 6) = 0$$

$b = 0$	$2b - 1 = 0$
	$2b = 1$
	$b = \frac{1}{2}$

$x = 0$ and $-\frac{6}{5}$

34. We solved one equation below. Solve the other equation.

$$4y(y + 3) = 0 \qquad\qquad 7t(4t - 3) = 0$$

$4y = 0$	$y + 3 = 0$
$y = \frac{0}{4}$	$y = -3$
$y = 0$	

$t = 0$ and $\frac{3}{4}$

35. The principle of "setting the factors equal to zero" to solve an equation applies <u>only when the product is "0"</u>. Therefore, it could not be used with either equation below.

$$(x - 3)(x - 9) = \overset{\downarrow}{5} \qquad\qquad y(y + 8) = \overset{\downarrow}{9}$$

In which equations below does the principle apply? ______________

a) $(3x - 1)(2x + 3) = 7$ b) $d(d - 1) = 0$ c) $p(p + 4) = 12$

Only (b)

SELF-TEST 47 (pages 409-417)

Solve each equation. Use the square-root table when necessary.

1. $x^2 - 16 = 0$

2. $9 = \frac{1}{4}y^2$

3. $25t^2 - 1 = 0$

4. $4d^2 = 480$

Continued on following page.

SELF-TEST 47 - Continued

Solve each equation.

5. $(x + 1)(x - 4) = 0$	6. $(2y - 1)(3y + 6) = 0$
7. $m(m + 5) = 0$	8. $3t(4t - 1) = 0$

ANSWERS:

1. $x = \pm 4$	3. $t = \pm\frac{1}{5}$	5. $x = -1$ and 4	7. $m = 0$ and -5
2. $y = \pm 6$	4. $d = \pm 10.954$	6. $y = \frac{1}{2}$ and -2	8. $t = 0$ and $\frac{1}{4}$

16-4 SOLVING INCOMPLETE QUADRATICS BY THE FACTORING METHOD

In this section, we will discuss the factoring method for solving incomplete quadratic equations.

36. To solve the incomplete quadratic at the right, we factored the left side and then set both factors equal to "0"

$$x^2 - 5x = 0$$
$$x(x - 5) = 0$$
$$x = 0 \quad \Big| \quad x - 5 = 0$$
$$x = 5$$

The two roots are "0" and "5". We checked "0" as a root of the original equation below. Check "5" as its root.

$$x^2 - 5x = 0$$
$$(0)^2 - 5(0) = 0$$
$$0 - 0 = 0$$
$$0 = 0$$

$$x^2 - 5x = 0$$

Answer:

$$(5)^2 - 5(5) = 0$$
$$25 - 25 = 0$$
$$0 = 0$$

37. Use the steps below to solve the incomplete quadratic at the right.

$$y^2 + 8y = 0$$

a) Factor the left side.

b) Solve the new equation by setting both factors equal to "0".

c) The two roots of the original equation are ________ and ________.

38. We used the same steps to solve one equation below. Solve the other equation.

$$m^2 + m = 0 \qquad\qquad t^2 - t = 0$$

$$m(m + 1) = 0$$

$$m = 0 \quad \Big| \quad m + 1 = 0$$

$$m = -1$$

a) $y(y + 8) = 0$

b) $y = 0$ | $y + 8 = 0$, $y = -8$

c) 0 and -8

---

39. We used the factoring method to solve one equation below. Solve the other equation.

$$2x^2 - 3x = 0 \qquad\qquad 7y^2 + 5y = 0$$

$$x(2x - 3) = 0$$

$$x = 0 \quad \Big| \quad 2x - 3 = 0$$

$$2x = 3$$

$$x = \frac{3}{2}$$

t = 0 and 1, from:

$$t(t - 1) = 0$$

$t = 0$ | $t - 1 = 0$, $t = 1$

---

40. When using the factoring method, we can sometimes factor out a number as well as the letter. An example is shown. Solve the other equation.

$$2t^2 - 6t = 0 \qquad\qquad 5p^2 + 10p = 0$$

$$2t(t - 3) = 0$$

$$2t = 0 \quad \Big| \quad t - 3 = 0$$

$$t = 0 \quad \Big| \quad t = 3$$

$y = 0$ and $-\frac{5}{7}$, from:

$$y(7y + 5) = 0$$

$y = 0$ | $7y + 5 = 0$, $7y = -5$, $y = -\frac{5}{7}$

---

41. When using the factoring method, we usually factor completely. Solve the other equation.

$$8x^2 + 12x = 0 \qquad\qquad 12y^2 - 6y = 0$$

$$4x(2x + 3) = 0$$

$$4x = 0 \quad \Big| \quad 2x + 3 = 0$$

$$x = 0 \quad \Big| \quad 2x = -3$$

$$x = -\frac{3}{2}$$

p = 0 and -2, from:

$$5p(p + 2) = 0$$

$5p = 0$, $p = 0$ | $p + 2 = 0$, $p = -2$

$y = 0$ and $\frac{1}{2}$, from:

$$6y(2y - 1) = 0$$

$6y = 0$, $y = 0$ | $2y - 1 = 0$, $2y = 1$, $y = \frac{1}{2}$

42. Following the example, solve the other equation.

$$6b^2 - 5b = 0$$

$$b(6b - 5) = 0$$

$b = 0$ | $6b - 5 = 0$

$$6b = 5$$

$$b = \frac{5}{6}$$

$$2d^2 + 7d = 0$$

---

43. When factoring an incomplete quadratic, we always get a letter (with or without a coefficient) as one of the factors. For example:

$m^2 + 9m = 0$ $\qquad$ $4p^2 - 12p = 0$

$m(m + 9) = 0$ $\qquad$ $4p(p - 3) = 0$

Since we always set a letter (with or without a coefficient) equal to "0", what number is always one of the roots of an incomplete quadratic? ______

Answer: $d = 0$ and $-\frac{7}{2}$, from:

$$d(2d + 7) = 0$$

$d = 0$ | $2d + 7 = 0$

$$2d = -7$$

$$d = -\frac{7}{2}$$

---

44. Though the equations below are not in standard form, each is an incomplete quadratic because it contains a squared-letter term and a letter-term.

$x^2 = 10x$ $\qquad$ $2R^2 = 8R$

To solve the equations, we use the addition axiom to put them in standard form first. We get:

$x^2 - 10x = 0$ $\qquad$ $2R^2 - 8R = 0$

Then we can solve them in the usual way. Complete each solution below.

a) $x(x - 10) = 0$ $\qquad$ b) $2R(R - 4) = 0$

Answer: 0

---

45. Using the same steps, solve each equation at the right.

a) $m^2 = 8m$ $\qquad$ b) $5t^2 = -3t$

Answer:

a) $x = 0$ and 10

b) $R = 0$ and 4

Answer:

a) $m = 0$ and 8, from:

$$m^2 - 8m = 0$$

b) $t = 0$ and $-\frac{3}{5}$, from:

$$5t^2 + 3t = 0$$

16-5 THE FACTORING METHOD AND COMPLETE QUADRATICS OF THE FORM: $x^2 + bx + c = 0$

In this section, we will discuss the factoring method for solving complete quadratic equations in which the coefficient of the squared letter is "1".

---

46. To solve the complete quadratic at the right, we factored the left side and then set both factors equal to "0".

$$x^2 + 3x + 2 = 0$$
$$(x + 1)(x + 2) = 0$$

$x + 1 = 0$	$x + 2 = 0$
$x = -1$	$x = -2$

The two roots are "-1" and "-2". We checked "-1" as a root of the original equation below. Check "-2" as its root.

$$x^2 + 3x + 2 = 0$$
$$(-1)^2 + 3(-1) + 2 = 0$$
$$1 + (-3) + 2 = 0$$
$$3 + (-3) = 0$$
$$0 = 0$$

$$x^2 + 3x + 2 = 0$$

Answer:

$$(-2)^2 + 3(-2) + 2 = 0$$
$$4 + (-6) + 2 = 0$$
$$6 + (-6) = 0$$
$$0 = 0$$

47. Use the steps below to solve the complete quadratic at the right.

$$y^2 - 5y + 6 = 0$$

a) Factor the left side.

b) Solve the new equation by setting both factors equal to "0".

c) The two roots of the original equation are ______ and ______.

Answer:

a) $(y - 2)(y - 3) = 0$

b) $y - 2 = 0$, $y = 2$ $y - 3 = 0$, $y = 3$

c) 2 and 3

48. We used the same steps to solve one equation below. Solve the other equation.

$$t^2 + 3t - 10 = 0$$
$$(t - 2)(t + 5) = 0$$

$t - 2 = 0$	$t + 5 = 0$
$t = 2$	$t = -5$

$$m^2 - 8m - 9 = 0$$

Answer:

$m = -1$ and 9, from:

$$(m + 1)(m - 9) = 0$$

$m + 1 = 0$, $m = -1$ $m - 9 = 0$, $m = 9$

49. Use the same steps to solve each equation below.

a) $x^2 + 11x + 10 = 0$

b) $y^2 - 2y - 8 = 0$

Answer:

a) $x = -1$ and -10

b) $y = -2$ and 4

50. We solved the equation at the left below. Notice that both roots are -3. A solution of that type is called a <u>double root</u>. Solve the other equation.

$$x^2 + 6x + 9 = 0 \qquad\qquad y^2 - 4y + 4 = 0$$

$$(x + 3)(x + 3) = 0$$

$$x + 3 = 0 \quad \Big| \quad x + 3 = 0$$

$$x = -3 \quad \Big| \quad x = -3$$

Both roots are +2.

---

51. Though the equations below are not in standard form, each is a complete quadratic because it contains a squared-letter term, a letter-term, and a number-term.

$$x^2 + 3x = 4 \qquad\qquad m^2 = 7m - 8$$

We can use the addition axiom to put them in standard form. We get:

$$x^2 + 3x - 4 = 0 \qquad\qquad m^2 - 7m + 8 = 0$$

Use the addition axiom to put each complete quadratic below in standard form.

a) $t^2 + 6 = 5t$ b) $d^2 = d + 2$

a) $t^2 - 5t + 6 = 0$

b) $d^2 - d - 2 = 0$

---

52. When a complete quadratic is not in standard form, the first step of the solution is putting it in standard form. An example is shown. Solve the other quadratic.

$$y^2 = y + 6 \qquad\qquad p^2 + 15 = 8p$$

$$y^2 + (-y) + (-6) = y + 6 + (-y) + (-6)$$

$$y^2 - y - 6 = 0$$

$$(y + 2)(y - 3) = 0$$

$$y + 2 = 0 \quad \Big| \quad y - 3 = 0$$

$$y = -2 \quad \Big| \quad y = 3$$

$p = 3$ and 5, from:

$$p^2 - 8p + 15 = 0$$

$$(p - 3)(p - 5) = 0$$

16-6 THE FACTORING METHOD AND COMPLETE QUADRATICS OF THE FORM: $ax^2 + bx + c = 0$

In this section, we will discuss the factoring method for solving complete quadratic equations in which the coefficient of the squared letter is a number other than "1".

53. The same method is used to solve a complete quadratic when the coefficient of the squared letter is a number other than "1". That is, we factor the left side and then set both factors equal to "0". An example is shown. Solve the other equation.

$$3x^2 + 5x + 2 = 0$$
$$(3x + 2)(x + 1) = 0$$

$3x + 2 = 0$	$x + 1 = 0$
$3x = -2$	$x = -1$
$x = -\frac{2}{3}$	

$$2y^2 - 7y + 6 = 0$$

---

54. Following the example, solve the other equation.

$$2p^2 - 5p - 12 = 0$$
$$(2p + 3)(p - 4) = 0$$

$2p + 3 = 0$	$p - 4 = 0$
$2p = -3$	$p = 4$
$p = -\frac{3}{2}$	

$$5d^2 + 3d - 2 = 0$$

Answer to 53: $y = \frac{3}{2}$ and 2, from:

$$(2y - 3)(y - 2) = 0$$

$2y - 3 = 0$	$y - 2 = 0$
$2y = 3$	$y = 2$
$y = \frac{3}{2}$	

---

55. To solve the equation below, we began by putting it in standard form first. Solve the other equation.

$$7x^2 = 3x + 4$$
$$7x^2 - 3x - 4 = 0$$
$$(7x + 4)(x - 1) = 0$$

$7x + 4 = 0$	$x - 1 = 0$
$7x = -4$	$x = 1$
$x = -\frac{4}{7}$	

$$3m^2 + 2 = 7m$$

Answer to 54: $d = \frac{2}{5}$ and -1, from:

$$(5d - 2)(d + 1) = 0$$

$5d - 2 = 0$	$d + 1 = 0$
$5d = 2$	$d = -1$
$d = \frac{2}{5}$	

Answer to 55: $m = \frac{1}{3}$ and 2, from:

$$3m^2 - 7m + 2 = 0$$
$$(3m - 1)(m - 2) = 0$$

SELF-TEST 48 (pages 418-424)

Solve each equation.

1. $x^2 - 7x = 0$	2. $3y^2 + y = 0$	3. $6m^2 = 8m$

Solve each equation.

4. $m^2 + m - 2 = 0$	5. $t^2 - 9t + 18 = 0$
6. $x^2 - 12 = x$	7. $5w^2 + 8w + 3 = 0$

ANSWERS:

1. $x = 0$ and 7
2. $y = 0$ and $-\frac{1}{3}$
3. $m = 0$ and $\frac{4}{3}$
4. $m = 1$ and -2
5. $t = 3$ and 6
6. $x = -3$ and 4
7. $w = -\frac{3}{5}$ and -1

16-7 LIMITATIONS OF THE FACTORING METHOD

Some complete quadratic equations are not ordinarily solved by the factoring method either because they cannot be factored into binomials containing whole numbers or because they are too difficult to factor. We will give examples of both types in this section.

56. There are some trinomials that cannot be factored into binomials in which the numbers are whole numbers. An example is given below.

$$x^2 + 4x + 6$$

The pairs of factors for 6 are (1 and 6) and (2 and 3). Neither pair has 4 as its sum.

Continued on following page.

56. Continued

Which trinomials below cannot be factored into binomials containing whole numbers? ____________

a) $y^2 + 9y + 14$ b) $t^2 - 11t + 16$ c) $m^2 + m - 7$

Both (b) and (c).

(a) can be factored:

$(x + 2)(x + 7)$

57. The trinomial below also cannot be factored into binomials containing whole numbers.

$$2m^2 + 6m + 3$$

The possible pairs of factors for the trinomial are:

$(2m + 3)(m + 1)$
$(2m + 1)(m + 3)$

Neither pair has "6m" as the middle term of its product.

Which trinomials below cannot be factored into binomials containing whole numbers? ____________

a) $3x^2 - 8x + 5$ b) $5t^2 - 6t - 2$ c) $2d^2 + 5d - 7$

Only (b)

(a) can be factored: $(3x - 5)(x - 1)$

(c) can be factored: $(2d + 7)(d - 1)$

58. When a complete quadratic contains a trinomial that cannot be factored into binomials containing whole numbers, we do not ordinarily solve it by the factoring method.

Which of the following quadratics would not ordinarily be solved by the factoring method? ______________

a) $t^2 + 7t + 3 = 0$ b) $d^2 + d - 12 = 0$ c) $3x^2 - x - 5 = 0$

Both (a) and (c).

(b) can be factored:

$(d + 4)(d - 3) = 0$

59. When the numbers in a quadratic are large, we do not ordinarily use the factoring method to solve it because the factoring is too difficult. An example is shown. Complete the other solution.

$50y^2 - 495y + 144 = 0$	$y^2 - 44y + 288 = 0$
$(5y - 48)(10y - 3) = 0$	$(y - 8)(y - 36) = 0$

$5y - 48 = 0$	$10y - 3 = 0$
$5y = 48$	$10y = 3$
$y = \frac{48}{5}$	$y = \frac{3}{10}$

y = 8 and 36

---

60. Even though the numbers in the quadratic below are quite small, the number of possible pairs of factors is quite large as you can see. Therefore, the equation is difficult to solve by the factoring method.

$$6x^2 + 5x - 14 = 0$$

$(6x + 1)(x - 14) = 0$	$(3x + 1)(2x - 14) = 0$
$(6x - 1)(x + 14) = 0$	$(3x - 1)(2x + 14) = 0$
$(6x + 14)(x - 1) = 0$	$(3x + 14)(2x - 1) = 0$
$(6x - 14)(x + 1) = 0$	$(3x - 14)(2x + 1) = 0$
$(6x + 2)(x - 7) = 0$	$(3x + 2)(2x - 7) = 0$
$(6x - 2)(x + 7) = 0$	$(3x - 2)(2x + 7) = 0$
$(6x + 7)(x - 2) = 0$	$(3x + 7)(2x - 2) = 0$
$(6x - 7)(x + 2) = 0$	$(3x - 7)(2x + 2) = 0$

a) By checking the factorings above, find the correct factoring for the equation. ______________

b) Therefore, the roots of the equation are ________ and ________.

a) $(6x - 7)(x + 2) = 0$

b) $x = \frac{7}{6}$ and -2

16-8 IDENTIFYING "a", "b", AND "c" IN THE STANDARD FORM OF QUADRATIC EQUATIONS

When a quadratic equation cannot be easily solved by the factoring method, we use the quadratic formula to solve it. In order to use the quadratic formula, we must be able to identify "a", "b", and "c" in the standard form of quadratic equations. We will discuss the method in this section.

---

61. Any quadratic equation can be written in the following standard form:

$$ax^2 + bx + c = 0$$

In the standard form, "a" is the coefficient of the squared-letter, "b" is the coefficient of the letter, and "c" is the number term. Two examples are shown.

$2x^2 + 5x + 3 = 0$	$5y^2 - 4y - 1 = 0$
$a = 2$	$a = 5$
$b = 5$	$b = -4$
$c = 3$	$c = -1$

Continued on following page.

61. Continued

Identify "a", "b", and "c" in each equation below.

a) $3m^2 + 4m - 9 = 0$ a = ______ b = ______ c = ______

b) $7t^2 - 2t + 8 = 0$ a = ______ b = ______ c = ______

---

a) a = 3
b = 4
c = -9

b) a = 7
b = -2
c = 8

62. When "a" or "b" is either "1" or "-1", the "1" or "-1" are not ordinarily shown. For example:

$x^2 + x - 6 = 0$ $y^2 - y + 12 = 0$

It is helpful to write each "1" or "-1" before identifying "a", "b", and "c". We get:

$1x^2 + 1x - 6 = 0$ $1y^2 - 1y + 12 = 0$

a = 1
b = 1
c = -6

a = 1
b = -1
c = 12

Identify "a", "b", and "c" in each equation below.

a) $t^2 + t - 30 = 0$ a = ______ b = ______ c = ______

b) $d^2 - d - 2 = 0$ a = ______ b = ______ c = ______

---

a) a = 1
b = 1
c = -30

b) a = 1
b = -1
c = -2

63. Neither equation below is in standard form because there are some non-zero terms on the right side.

$x^2 + 6 = 5x$ $2d^2 = d + 3$

In order to identify "a", "b", and "c", we must put them in standard form first. We get:

$x^2 - 5x + 6 = 0$ $2d^2 - d - 3 = 0$

a) For the left equation: a = ______ b = ______ c = ______

b) For the right equation: a = ______ b = ______ c = ______

---

a) a = 1
b = -5
c = 6

b) a = 2
b = -1
c = -3

64. Put each equation in standard form and then identify "a", "b", and "c".

a) $2p^2 - 1 = p$ ____________________

a = ______ b = ______ c = ______

b) $d^2 = 3d - 10$ ____________________

a = ______ b = ______ c = ______

---

a) $2p^2 - p - 1 = 0$
a = 2, b = -1, c = -1

b) $d^2 - 3d + 10 = 0$
a = 1, b = -3, c = 10

65. In the standard form of pure quadratics, the letter term is not usually shown because its coefficient is "0". For example:

$x^2 - 16 = 0$ $\qquad$ $2y^2 - 80 = 0$

However, when identifying "a", "b", and "c", it is helpful to insert the letter term with its "0" coefficient. We get:

$x^2 + 0x - 16 = 0$ $\qquad$ $2y^2 + 0y - 80 = 0$

a) For the left equation: a = ______ b = ______ c = ______

b) For the right equation: a = ______ b = ______ c = ______

Answers to 65: a) a = 1, b = 0, c = −16; b) a = 2, b = 0, c = −80

66. In the standard form of incomplete quadratics, the number term is not usually shown because it is "0". For example:

$x^2 - 6x = 0$ $\qquad$ $3y^2 + 12y = 0$

However, when identifying "a", "b", and "c", it is helpful to insert "0" as the number term. We get:

$x^2 - 6x + 0 = 0$ $\qquad$ $3y^2 + 12y + 0 = 0$

a) For the left equation: a = ______ b = ______ c = ______

b) For the right equation: a = ______ b = ______ c = ______

Answers to 66: a) a = 1, b = −6, c = 0; b) a = 3, b = 12, c = 0

67. Identify "a", "b", and "c" for each equation below.

a) $m^2 - 1 = 0$ a = ______ b = ______ c = ______

b) $5p^2 - p = 0$ a = ______ b = ______ c = ______

Answers to 67:

a) a = 1, b = 0, c = −1 from: $m^2 + 0m - 1 = 0$

b) a = 5, b = −1, c = 0, from: $5p^2 - p + 0 = 0$

68. Neither equation below is in standard form because there are non-zero terms on the right side.

$3x^2 = 75$ $\qquad$ $y^2 = 10y$

In order to identify "a", "b", and "c", we must put them in standard form first. We get:

$3x^2 - 75 = 0$ $\qquad$ $y^2 - 10y = 0$

$3x^2 + 0x - 75 = 0$ $\qquad$ $y^2 - 10y + 0 = 0$

a) For the left equation: a = ______ b = ______ c = ______

b) For the right equation: a = ______ b = ______ c = ______

Answers to 68: a) a = 3, b = 0, c = −75; b) a = 1, b = −10, c = 0

69. Write each equation in standard form and then identify "a", "b", and "c".

a) $t^2 = 147$ ________________

a = ________ b = ________ c = ________

b) $8q^2 = q$ ________________

a = ________ b = ________ c = ________

a) $t^2 + 0t - 147 = 0$ $a = 1,\ b = 0,\ c = -147$	b) $8q^2 - q + 0 = 0$ $a = 8,\ b = -1,\ c = 0$

16-9 THE QUADRATIC FORMULA

Though the quadratic formula is a general method that can be used to solve any type of quadratic equation, it is ordinarily used only to solve complete quadratics that are not easily solved by the factoring method. We will discuss the quadratic formula in this section.

70. The quadratic formula is given below. It contains "a", "b", and "c" from the standard form of quadratic equations. Since any quadratic equation has two roots, there are two forms of the formula. The only difference between them is the "+" and "-" in front of the radical in the numerator (see the arrows).

$$\text{First root} = \frac{(-b) \overset{\downarrow}{+} \sqrt{b^2 - 4ac}}{2a}$$

$$\text{Second root} = \frac{(-b) \overset{\downarrow}{-} \sqrt{b^2 - 4ac}}{2a}$$

If we use the factoring method to solve $x^2 + 3x - 10 = 0$ we get 2 and -5 as its roots. Let's use the quadratic formula to solve the same equation.

$$x^2 + 3x - 10 = 0$$

$$a = 1, \quad b = 3, \quad c = -10$$

First Root

$$x_1 = \frac{(-b) + \sqrt{b^2 - 4ac}}{2a} = \frac{(-3) + \sqrt{(-3)^2 - 4(1)(-10)}}{2(1)} = \frac{(-3) + \sqrt{9 - (-40)}}{2} = \frac{(-3) + \sqrt{49}}{2} = \frac{(-3) + 7}{2} = \frac{4}{2} = 2$$

Second Root

$$x_2 = \frac{(-b) - \sqrt{b^2 - 4ac}}{2a} = \frac{(-3) - \sqrt{(-3)^2 - 4(1)(-10)}}{2(1)} = \frac{(-3) - \sqrt{9^2 - (-40)}}{2} = \frac{(-3) - \sqrt{49}}{2} = \frac{(-3) - 7}{2} = \frac{-10}{2} = -5$$

Did we get the same roots that would be obtained by the factoring method? ____________

Yes

71. Since the formulas for the two roots are so similar, we usually write one formula with a "±" symbol. The "±" symbol means "add the radical for the first root" and "subtract the radical for the second root".

$$\text{Two roots} = \frac{(-b) \pm \sqrt{b^2 - 4ac}}{2a}$$

Let's use the single formula to solve the equation below.

$$y^2 + 6y - 16 = 0$$

$$a = 1, \quad b = 6, \quad c = -16$$

$$\text{Two roots} = \frac{(-6) \pm \sqrt{(6)^2 - 4(1)(-16)}}{2(1)}$$

$$= \frac{(-6) \pm \sqrt{36 - (-64)}}{2}$$

$$= \frac{(-6) \pm \sqrt{100}}{2}$$

a) The first root $= \frac{(-6) + 10}{2} =$ ________________

b) The second root $= \frac{(-6) - 10}{2} =$ ________________

---

a) 2, from $\frac{4}{2}$

b) -8, from $\frac{-16}{2}$

72. There are a few points you should notice about plugging values into the quadratic formula.

$$\text{Two roots} = \frac{(-b) \pm \sqrt{b^2 - 4ac}}{2a}$$

(1) (-b) means "the opposite of b". If $b = 4$, $-b = -4$
If $b = -5$, $-b = +5$

(2) b^2 (under the radical) is always a positive number. If $b = 4$, $b^2 = (4)^2 = +16$
If $b = -5$, $b^2 = (-5)^2 = +25$

(3) Watch your signs with $b^2 - 4ac$ under the radical.

"4ac" is a set of three factors. Perform the multiplication before handling the subtraction. That is:

If $a = 1$, $b = -10$, $c = -6$

$$b^2 - 4ac = (-10)^2 - [4(1)(-6)]$$
$$= 100 - [-24]$$
$$= 100 + [+24] = 124$$

If $a = -3$, $b = 12$, $c = -9$

$$b^2 - 4ac = (12)^2 - [4(-3)(-9)]$$
$$= 144 - [+108]$$
$$= 144 + [-108]$$
$$= 36$$

---

73. For the equation: $2x^2 - 5x - 3 = 0$

a) $-b =$ ________ b) $b^2 =$ ________ c) $b^2 - 4ac =$ ________

74. For the equation: $7m^2 - 6m - 1 = 0$

a) $-b =$ ________ b) $b^2 =$ ________ c) $b^2 - 4ac =$ ________

a) +5
b) 25
c) 49

---

75. <u>The quadratic formula should be memorized</u>. Therefore, study the formula as it is given in frame 72 and then write it from memory below.

Two roots =

a) +6
b) 36
c) 64

---

76. Let's use the formula to solve the equation below.

$$3t^2 - 2t - 5 = 0$$

$$a = 3, \quad b = -2, \quad c = -5$$

$$\text{Two roots} = \frac{(2) \pm \sqrt{(-2)^2 - 4(3)(-5)}}{2(3)}$$

$$= \frac{2 \pm \sqrt{4 - (-60)}}{6}$$

$$= \frac{2 \pm \sqrt{64}}{6}$$

a) The first root = ________ b) The second root = ________

$$\frac{-b \pm \sqrt{b^2 - 4ac}}{2a}$$

---

77. Though we ordinarily solve incomplete quadratics by the factoring method, we can use the quadratic formula to solve them. An example is shown.

$$3x^2 + 4x = 0$$

$$a = 3, \quad b = 4, \quad c = 0$$

$$\text{Two roots} = \frac{(-4) \pm \sqrt{(4)^2 - 4(3)(0)}}{2(3)}$$

$$= \frac{(-4) \pm \sqrt{16 - 0}}{6}$$

$$= \frac{(-4) \pm \sqrt{16}}{6}$$

a) The first root is ________ b) The second root is ________

a) $\frac{5}{3}$, from $\frac{10}{6}$

b) -1, from $\frac{-6}{6}$

---

a) 0, from $\frac{0}{6}$

b) $-\frac{4}{3}$, from $\frac{-8}{6}$

78. Though we ordinarily solve pure quadratics by isolating the squared letter, we can use the quadratic formula to solve them. An example is shown.

$$y^2 - 9 = 0$$

$$a = 1, \quad b = 0, \quad c = -9$$

$$\text{Two roots} = \frac{0 \pm \sqrt{(0)^2 - 4(1)(-9)}}{2(1)}$$

$$= \frac{0 \pm \sqrt{0 - (-36)}}{2}$$

$$= \frac{0 \pm \sqrt{36}}{2}$$

a) The first root is ________ b) The second root is ________

Answers to 78: a) 3, from $\frac{6}{2}$ b) -3, from $\frac{-6}{2}$

79. Because we cannot factor into binomials containing whole numbers, we would use the formula to solve the equation below.

$$y^2 + 9y + 7 = 0$$

$$a = 1, \quad b = 9, \quad c = 7$$

$$\text{Two roots} = \frac{(-9) \pm \sqrt{(9)^2 - 4(1)(7)}}{2(1)}$$

$$= \frac{(-9) \pm \sqrt{81 - 28}}{2}$$

$$= \frac{(-9) \pm \sqrt{53}}{2}$$

When the radicand is not a perfect square (like 53), the roots are sometimes left in the form above. However, we can also use the square-root table to find $\sqrt{53}$ and then report the roots as decimal numbers. We get:

$$\text{The first root} = \frac{(-9) + 7.28}{2} = \frac{-1.72}{2} = -.86$$

$$\text{The second root} = \frac{(-9) - 7.28}{2} = \underline{\qquad\qquad} = \underline{\qquad\qquad}$$

Answer to 79: $\frac{-16.28}{2} = -8.14$

80. Because we cannot factor into binomials containing whole numbers, we would use the formula to solve the equation at the right. Complete the solution. Use the square-root table and round each root to the nearest hundredth.

$$3m^2 - 10m + 6 = 0$$

$$\text{Two roots} = \frac{10 \pm \sqrt{(-10)^2 - 4(3)(6)}}{2(3)}$$

81. When the radical contains a negative number, the equation does not have ordinary numbers as its roots. For example:

$$x^2 - 2x + 7 = 0$$

$$\text{Two roots} = \frac{2 \pm \sqrt{(-2)^2 - 4(1)(7)}}{2(1)}$$

$$= \frac{2 \pm \sqrt{4 - 28}}{2}$$

$$= \frac{2 \pm \sqrt{-24}}{2}$$

We will avoid equations like those above in this book.

m = 2.55 and .78, from: $\frac{10 \pm \sqrt{28}}{6}$

---

82. The following strategy can be used to solve quadratic equations.

Pure quadratics - Isolate the squared letter and then find both square roots of the number on the other side.

Incomplete quadratics - Use the factoring method.

Complete quadratics - Try the factoring method first. If the factoring is too difficult or if you can't find factors in which the numbers are whole numbers, switch immediately to using the formula. Don't waste a lot of time trying to factor.

16-10 APPLIED PROBLEMS

In this section, we will discuss a few applied problems involving quadratic equations.

---

83. Many applied problems involving formulas require solving a pure quadratic equation. An example is given.

Find the side of a square if its area is 144 square centimeters.

$$A = S^2$$

$$144 = S^2$$

$$S = \pm\sqrt{144} = \pm 12 \text{ centimeters}$$

Of the two answers (+12 centimeters and -12 centimeters), only one makes sense. Which one? ____________

+12 centimeters

84. Here is another formula problem that leads to a pure quadratic equation.

The distance (s) that an object falls from rest in "t" seconds is given by the formula:

$$s = 16t^2$$

How many seconds does it take an object to fall 64 feet?

$$64 = 16t^2$$

$$t^2 = \frac{64}{16} = 4$$

$$t = \pm 2 \text{ seconds}$$

Of the two answers (+2 seconds and -2 seconds), only one makes sense. Which one? ________________

+2 seconds

85. To solve the problem below, we have to solve a pure quadratic equation.

The length of a rectangle is 3 times its width. If the area of the rectangle is 48 square feet, what are its dimensions?

Let: x = width of the rectangle
$3x$ = length of the rectangle

Then:

$$A = LW$$

$$48 = (3x)(x)$$

$$48 = 3x^2$$

$$x^2 = 16$$

$$x = \pm 4 \text{ feet}$$

For the problem, -4 feet does not make sense as a value for width. Therefore:

a) the width of the rectangle is ________ feet.

b) the length of the rectangle is ________ feet.

---

a) 4 feet
b) 12 feet

86. To solve the problem below, we have to solve a complete quadratic equation.

The length of a rectangle is 3 meters more than its width. If the area of the rectangle is 54 square meters, what are its dimensions?

Let: x = width of the rectangle
$x + 3$ = length of the rectangle

Then:

$$A = LW$$

$$54 = (x + 3)(x)$$

$$54 = x^2 + 3x$$

$$x^2 + 3x - 54 = 0$$

$$(x - 6)(x + 9) = 0$$

$$x = +6 \text{ or } -9 \text{ meters}$$

For the problem, -9 meters does not make sense as a value for width. Therefore:

a) the width of the rectangle is ________ meters.

b) the length of the rectangle is ________ meters.

---

a) 6 meters
b) 9 meters

SELF-TEST 49 (pages 424-435)

Identify "a", "b", and "c" in each equation below.

1. $6x^2 - 8x - 1 = 0$

a = ______

b = ______

c = ______

2. $y^2 - 5y + 4 = 0$

a = ______

b = ______

c = ______

3. $t^2 = 1$

a = ______

b = ______

c = ______

Use the quadratic formula to solve each equation. Round to the nearest tenth for decimal roots.

4. $4t^2 - 5t - 6 = 0$

5. $m^2 - 7m + 2 = 0$

ANSWERS:

1. $a = 6$
   $b = -8$
   $c = -1$

2. $a = 1$
   $b = -5$
   $c = 4$

3. $a = 1$
   $b = 0$
   $c = -1$

4. $t = -\frac{3}{4}$ and 2

5. $m = 6.7$ and .3

Unit 17 EXPONENTS AND POLYNOMIALS

In this unit, we will discuss powers, some laws of exponents, and the four basic operations with polynomials. The polynomials are limited to those with only one letter.

17-1 THE MEANING OF POWERS

In this section, we will discuss the meaning of powers with positive whole-number exponents.

1. In an earlier unit, we discussed base-exponent expressions in which the base is "10". However, any number can be the base. For example:

In 2^3: the base is 2 ; the exponent is 3

In 7^5: the base is ___; the exponent is ___

> 7 5

2. A base-exponent expression is a short way of writing a multiplication of identical factors. For example:

$5^3 = (5)(5)(5)$ (The exponent tells us to multiply three 5's.)

$2^4 = (2)(2)(2)(2)$ (The exponent tells us to multiply four 2's.)

Write each expression as a multiplication of identical factors.

a) 7^2 = ________ b) 8^3 = ________ c) 4^5 = ________

> a) (7)(7)
> b) (8)(8)(8)
> c) (4)(4)(4)(4)(4)

3. In a base-exponent expression, the base can be a letter. For example:

In x^2, the base is "x". In y^4, the base is "y".

Expressions of that type also stand for a multiplication of identical factors. That is:

$x^2 = (x)(x)$ (The exponent tells us to multiply two x's.)

$y^4 = (y)(y)(y)(y)$ (The exponent tells us to multiply four y's.)

Write each expression as a multiplication of identical factors.

a) t^3 = ________ b) m^2 = ________ c) d^5 = ________

> a) (t)(t)(t)
> b) (m)(m)
> c) (d)(d)(d)(d)(d)

4. Any multiplication of identical factors can be written in base-exponent form. For example:

$(5)(5)(5) = 5^3$ (Since there are three 5's, the exponent is 3.)

$(x)(x)(x)(x) = x^4$ (Since there are four x's, the exponent is 4.)

Write each multiplication in base-exponent form.

a) $(7)(7) =$ ______ b) $(3)(3)(3)(3) =$ ______ c) $(y)(y)(y)(y)(y)(y) =$ ______

Answers: a) 7^2 b) 3^4 c) y^6

5. Any base-exponent expression is called a "power" of the base. For example:

6^2, 6^3, and 6^5 are called "powers of 6".

x^3, x^4, and x^7 are called "powers of x".

a) 9^6 is a power of ______ b) y^3 is a power of ______

Answers: a) 9 b) y

6. The following type of language is used to distinguish the powers of the same number or letter.

4^2 is called "4 to the second power".

4^3 is called "4 to the third power".

4^5 is called "4 to the fifth power".

Similarly: a) 2^4 is called "2 to the ______ power".

b) x^7 is called "x to the ______ power".

Answers: a) fourth b) seventh

7. The words "squared" and "cubed" are frequently used for the second and third power of any base. That is:

5^2 is called "5 to the second power" or "5 squared".

x^3 is called "x to the third power" or "x cubed".

Write the base-exponent expression corresponding to these:

a) 9 squared = ______ b) y squared = ______ c) t cubed = ______

Answers: a) 9^2 b) y^2 c) t^3

8. When the base is a number, we can convert the power to an ordinary number by performing the multiplication. That is:

$4^3 = (4)(4)(4) = 64$

a) $3^4 = (3)(3)(3)(3) =$ ______ b) $2^5 = (2)(2)(2)(2)(2) =$ ______

Answers: a) 81 b) 32

9. When the base is a letter, we can only convert the power to an ordinary number by substituting a number for the letter. That is:

If $y = 5$, $y^3 = 5^3 = (5)(5)(5) = 125$

If $t = 2$, $t^4 = 2^4 = (2)(2)(2)(2) =$ ______

16

10. Any power whose exponent is "1" is equal to its base. That is:

$5^1 = 5 \qquad x^1 = x$

The definition above fits the pattern for powers. That is:

$5^3 = (5)(5)(5) \qquad x^3 = (x)(x)(x)$

$5^2 = (5)(5) \qquad x^2 = (x)(x)$

$5^1 = (5) \text{ or } 5 \qquad x^1 = (x) \text{ or } x$

Using the definition above, complete these:

a) $7^1 =$ ______ b) $2^1 =$ ______ c) $y^1 =$ ______

---

11. By reversing the definition in the last frame, we can write any number or letter as a power. That is:

a) 7
b) 2
c) y

$3 = 3^1 \qquad m = m^1$ a) $8 =$ ______ b) $t =$ ______

a) 8^1	b) t^1

17-2 MULTIPLYING AND DIVIDING POWERS

In this section, we will discuss the laws of exponents for multiplying and dividing powers with the same base.

---

12. Each expression below means "multiply 7^2 and 7^4".

$7^2 \times 7^4 \qquad 7^2 \cdot 7^4 \qquad (7^2)(7^4)$

To perform the multiplication, we can write each factor as a multiplication of 7's and then count the total number of 7's to get the exponent of the product. That is:

$7^2 \cdot 7^4 = (7)(7) \cdot (7)(7)(7)(7) = 7^6$

Count the total number of x's to write the product below as a power.

$x^5 \cdot x^3 = (x)(x)(x)(x)(x) \cdot (x)(x)(x) =$ ______

---

13. The two multiplications from the last frame are shown below.

x^8

$7^2 \cdot 7^4 = 7^6 \qquad x^5 \cdot x^3 = x^8$

In each, <u>the exponent of the product is the sum of the exponents of the factors</u>. That is:

$7^2 \cdot 7^4 = 7^{2+4} = 7^6 \qquad x^5 \cdot x^3 = x^{5+3} = x^8$

Therefore we can use the following law of exponents to multiply powers with the same base.

$$\boxed{a^m \cdot a^n = a^{m+n}}$$

Using the law of exponents for multiplication, complete these:

a) $5^4 \cdot 5^3 =$ ______ b) $9^1 \cdot 9^2 =$ ______ c) $y^5 \cdot y^5 =$ ______

a) 5^7
b) 9^3
c) y^{10}

14. The law of exponents for multiplication applies <u>only if the powers have the same base</u>. For example, it does not apply to the multiplication below.

$$2^3 \cdot 5^2 = (2)(2)(2) \cdot (5)(5)$$

<u>Note:</u> The product cannot be written as a single power because all of the factors are not identical.

If it applies, use the law of exponents to write the products.

a) $3^5 \cdot 3^3 =$ ______ b) $5^3 \cdot 6^4 =$ ______ c) $x^2 \cdot y^6 =$ ______

---

a) 3^8
b) Does not apply
c) Does not apply

15. To use the law of exponents for the multiplications below, we substituted 4^1 and x^1 for 4 and x.

$$4 \cdot 4^3 = 4^1 \cdot 4^3 = 4^4 \qquad x^5 \cdot x = x^5 \cdot x^1 = x^6$$

Use the same method to complete these:

a) $2 \cdot 2^5 =$ ______ b) $y \cdot y^4 =$ ______ c) $m^2 \cdot m =$ ______

---

a) 2^6
b) y^5
c) m^3

16. The law of exponents can also be used for multiplications containing more than two factors. For example:

$$x^3 \cdot x^2 \cdot x^5 \cdot x = x^{3+2+5+1} = x^{11}$$

Following the example, complete these:

a) $y^4 \cdot y^6 \cdot y^2 =$ ______ b) $m^3 \cdot m \cdot m^3 \cdot m^2 =$ ______

---

a) y^{12}
b) m^9

17. To perform the division below, we began by writing each term as a multiplication of 4's. Then we factored out $\left(\frac{4 \cdot 4 \cdot 4}{4 \cdot 4 \cdot 4}\right)$ which equals "1".

$$\frac{4^5}{4^3} = \frac{4 \cdot 4 \cdot 4 \cdot 4 \cdot 4}{4 \cdot 4 \cdot 4} = \left(\frac{4 \cdot 4 \cdot 4}{4 \cdot 4 \cdot 4}\right)(4 \cdot 4) = (1)(4 \cdot 4) = 4 \cdot 4 = 4^2$$

Write the quotient below as a power.

$$\frac{x^4}{x^1} = \frac{x \cdot x \cdot x \cdot x}{x} = \left(\frac{x}{x}\right)(x \cdot x \cdot x) = (1)(x \cdot x \cdot x) = x \cdot x \cdot x = \_\_\_\_\_\_$$

---

x^3

18. The two divisions from the last frame are shown below.

$$\frac{4^5}{4^3} = 4^2 \qquad \frac{x^4}{x^1} = x^3$$

In each, <u>the exponent of the quotient can be obtained by subtracting the exponent in the denominator from the exponent in the numerator</u>.

$$\frac{4^5}{4^3} = 4^{5-3} = 4^2 \qquad \frac{x^4}{x^1} = x^{4-1} = x^3$$

Continued on following page.

18. Continued

Therefore, we can use the following law of exponents to divide powers with the same base.

$$\frac{a^m}{a^n} = a^{m-n}$$

Using the law of exponents for division, complete these:

a) $\frac{5^6}{5^2} =$ ______ b) $\frac{9^5}{9^4} =$ ______ c) $\frac{t^8}{t^1} =$ ______

---

a) 5^4
b) 9^1 or 9
c) t^7

19. The law of exponents for division applies <u>only if the powers have the same base</u>. For example, it does not apply to the division below since we cannot factor out a fraction that equals "1".

$$\frac{6^4}{5^2} = \frac{6 \cdot 6 \cdot 6 \cdot 6}{5 \cdot 5}$$

If it applies, use the law of exponents to write the quotient.

a) $\frac{2^5}{3^4} =$ ______ b) $\frac{x^2}{x^1} =$ ______ c) $\frac{m^8}{t^5} =$ ______

---

a) Does not apply
b) x^1 or x
c) Does not apply

20. To use the law of exponents for the divisions below, we substituted 3^1 and y^1 for 3 and y.

$$\frac{3^3}{3} = \frac{3^3}{3^1} = 3^2 \qquad \frac{y^6}{y} = \frac{y^6}{y^1} = y^5$$

Use the same method to complete these:

a) $\frac{7^5}{7} =$ ______ b) $\frac{4^2}{4} =$ ______ c) $\frac{a^{10}}{a} =$ ______

---

a) 7^4
b) 4^1 or 4
c) a^9

21. To multiply powers, we <u>add</u> exponents. To divide powers, we <u>subtract</u> exponents. Complete these:

a) $3^7 \cdot 3^5 =$ ______ c) $x^6 \cdot x^2 =$ ______

b) $\frac{3^7}{3^5} =$ ______ d) $\frac{x^6}{x^2} =$ ______

---

a) 3^{12}
b) 3^2
c) x^8
d) x^4

22. Complete:

a) $9^2 \cdot 9 =$ ______ c) $b^7 \cdot b =$ ______

b) $\frac{9^2}{9} =$ ______ d) $\frac{b^7}{b} =$ ______

---

a) 9^3
b) 9^1 or 9
c) b^8
d) b^6

17-3 ZERO POWERS

In this section, we will define the meaning of any power with a "0" exponent.

---

23. When the exponent of a power is "0", the power equals "+1". That is:

$5^0 = 1$ $\quad x^0 = 1$ $\quad$ a) $9^0 =$ ______ $\quad$ b) $t^0 =$ ______

Answers: a) 1 b) 1

24. The definition of a power with a "0" exponent is consistent with the law of exponents for multiplication. To show that fact, we performed the same multiplications in two different ways below.

1. Substituting the definition of each power for the power.

$5^0 \cdot 5^2 = (1) \cdot (5)(5) = 5^2$ $\qquad x^3 \cdot x^0 = (x)(x)(x) \cdot (1) = x^3$

2. Using the law of exponents for multiplication.

$5^0 \cdot 5^2 = 5^{0+2} = 5^2$ $\qquad x^3 \cdot x^0 = x^{3+0} = x^3$

Use the law of exponents to complete these:

a) $3^0 \cdot 3^4 =$ ______ $\quad$ b) $8^6 \cdot 8^0 =$ ______ $\quad$ c) $m^0 \cdot m^5 =$ ______

Answers: a) 3^4 b) 8^6 c) m^5

25. The definition of a power with a "0" exponent is also consistent with the law of exponents for division. To show that fact, we performed the same divisions in two different ways.

1. Using this fact: If any quantity is divided by "1", the quotient is the original quantity.

$\frac{3^5}{3^0} = \frac{3^5}{1} = 3^5$ $\qquad \frac{t^4}{t^0} = \frac{t^4}{1} = t^4$

2. Using the law of exponents for division.

$\frac{3^5}{3^0} = 3^{5-0} = 3^5$ $\qquad \frac{t^4}{t^0} = t^{4-0} = t^4$

Use the law of exponents to complete these:

a) $\frac{7^2}{7^0} =$ ______ $\quad$ b) $\frac{2^1}{2^0} =$ ______ $\quad$ c) $\frac{m^8}{m^0} =$ ______

Answers: a) 7^2 b) 2^1 or 2 c) m^8

26. To show that the definition of a zero power is consistent with the law of exponents for division, we performed two more divisions in two different ways below.

1. Using this fact: If any quantity is divided by itself, the quotient is "1".

$\frac{4^3}{4^3} = 1$ $\qquad \frac{x^5}{x^5} = 1$

2. Using the law of exponents for division.

$\frac{4^3}{4^3} = 4^{3-3} = 4^0$ $\qquad \frac{x^5}{x^5} = x^{5-5} = x^0$

Use the law of exponents to complete these:

a) $\frac{6^4}{6^3} =$ ______ $\quad$ b) $\frac{6^4}{6^4} =$ ______ $\quad$ c) $\frac{y^7}{y^6} =$ ______ $\quad$ d) $\frac{y^6}{y^6} =$ ______

Answers: a) 6^1 or 6 b) 6^0 or 1 c) y^1 or y d) y^0 or 1

27. Complete: a) $2 \cdot 2^0 =$ ______ c) $x \cdot x^0 =$ ______

b) $\frac{2}{2^0} =$ ______ d) $\frac{x}{x^0} =$ ______

a) 2^1 or 2 b) 2^1 or 2 c) x^1 or x d) x^1 or x

17-4 RAISING A POWER TO A POWER

In this section, we will discuss the procedure for raising a power to a power.

28. In a base-exponent expression, the base can be a power. For example:

In $(5^3)^2$: the base is 5^3 ; the exponent is "2".

In $(y^2)^4$: the base is y^2 ; the exponent is "4".

Expressions like those above are also a short way of writing a multiplication of identical factors. That is:

$(5^3)^2 = 5^3 \cdot 5^3$ (There are two 5^3's.)

$(y^2)^4 = y^2 \cdot y^2 \cdot y^2 \cdot y^2$ (There are four y^2's.)

Write each power as a multiplication of identical factors.

a) $(4^5)^3 =$ ______ b) $(m^4)^5 =$ ______

a) $4^5 \cdot 4^5 \cdot 4^5$

b) $m^4 \cdot m^4 \cdot m^4 \cdot m^4 \cdot m^4$

29. To raise a power to a power, we can convert to a multiplication and then use the law of exponents for multiplication. That is:

$(3^5)^2 = 3^5 \cdot 3^5 = 3^{5+5} = 3^{10}$

$(y^3)^4 = y^3 \cdot y^3 \cdot y^3 \cdot y^3 = y^{3+3+3+3} =$ ______

y^{12}

30. The two problems from the last frame are shown below.

$(3^5)^2 = 3^{10}$ $(y^3)^4 = y^{12}$

As you can see, we can raise a power to a power by simply multiplying the exponents. That is:

$(3^5)^2 = 3^{(5)(2)} = 3^{10}$ $(y^3)^4 = y^{(3)(4)} = y^{12}$

Therefore, we can use the following law of exponents to raise a power to a power.

$$(a^m)^n = a^{mn}$$

Using the law of exponents for raising a power to a power, complete these:

a) $(8^3)^2 =$ ______ b) $(p^2)^4 =$ ______ c) $(m^5)^6 =$ ______

a) 8^6 b) p^8 c) m^{30}

31. Any power raised to the first power equals the original power. That is:

$(2^4)^1 = 2^{(4)(1)} = 2^4$ $\quad$ $(y^2)^1 = y^{(2)(1)} = $ ______

---

y^2

32. Any power raised to the zero power equals "1". That is:

$(8^3)^0 = 8^{(3)(0)} = 8^0 = 1$ $\quad$ $(m^5)^0 = m^{(5)(0)} = m^0 = $ ______

1

33. The three basic laws of exponents are reviewed below.

$a^m \cdot a^n = a^{m+n}$	$\dfrac{a^m}{a^n} = a^{m-n}$	$(a^m)^n = a^{mn}$

Using the laws complete these:

a) $7^4 \cdot 7^2 = $ ______ $\quad$ d) $x^5 \cdot x^3 = $ ______

b) $\dfrac{7^4}{7^2} = $ ______ $\quad$ e) $\dfrac{x^5}{x^3} = $ ______

c) $(7^4)^2 = $ ______ $\quad$ f) $(x^5)^3 = $ ______

a) 7^6 $\quad$ b) 7^2 $\quad$ c) 7^8 $\quad$ d) x^8 $\quad$ e) x^2 $\quad$ f) x^{15}

SELF-TEST 50 (pages 436-443)

Convert each power to an ordinary number.

1. $7^0 = $ ______	2. $5^1 = $ ______	3. $3^3 = $ ______	4. $2^6 = $ ______

Complete each multiplication.

5. $x^6 \cdot x^3 = $ ______	6. $8^2 \cdot 8^2 = $ ______	7. $y \cdot y^4 = $ ______	8. $4^2 \cdot 4^0 = $ ______

Complete each division.

9. $\dfrac{7^6}{7^2} = $ ______	10. $\dfrac{y^4}{y^0} = $ ______	11. $\dfrac{3}{3^0} = $ ______	12. $\dfrac{m^5}{m^5} = $ ______

Raise each power to the indicated power.

13. $(5^3)^2 = $ ______	14. $(x^4)^6 = $ ______	15. $(3^5)^1 = $ ______	16. $(a^4)^0 = $ ______

ANSWERS:

1. 1	5. x^9	9. 7^4	13. 5^6
2. 5	6. 8^4	10. y^4	14. x^{24}
3. 27	7. y^5	11. 3^1 or 3	15. 3^5
4. 64	8. 4^2	12. m^0 or 1	16. a^0 or 1

17-5 POLYNOMIALS

In this section, we will define "polynomials" and introduce some terminology used in discussing them.

---

34. The terms below are called "power-terms". They contain a coefficient and a power.

$3x^4$ $-2y^5$ $6t^3$ $-9d^2$

To show that x^5 and $-y^3$ are power-terms, we can write their coefficients explicitly. That is:

x^5 can be written $1x^5$ $-y^3$ can be written ________

$-1y^3$

35. To show that 5x and -7t are power-terms, we can write the exponent "1" of the power explicitly. That is:

5x can be written $5x^1$ -7t can be written ________

$-7t^1$

36. To show that "t" and "-p" are power-terms, we can write both the coefficient and the exponent of the letter explicitly. That is:

t can be written $1t^1$ -p can be written ________

$-1p^1$

37. To show that numbers like 3 and -5 are power-terms, we can use the zero power of some letter. That is:

3 can be written $3x^0$, since $x^0 = 1$

-5 can be written $-5y^0$, since $y^0 =$ ______

1

38. Any power-term is of the form ax^n where "n" is a positive whole-number or "0". Each term in the expressions below can be written explicitly as a power-term. That is:

$2x^3 - x^2 + x - 1$ can be written $2x^3 - 1x^2 + 1x^1 - 1x^0$

$y^5 - 8y^3 - y + 5$ can be written ________________

$1y^5 - 8y^3 - 1y^1 + 5y^0$

39. However, when writing expressions containing power terms, we do not usually write "1" and "-1" coefficients, "1" exponents, or zero powers. That is:

$1t^3 - 1t^2 + 8t^1 + 3t^0$ is usually written $t^3 - t^2 + 8t + 3$

$2b^5 + 1b^4 - 1b^1 - 9b^0$ is usually written ________________

$2b^5 + b^4 - b - 9$

40. Any expression containing only power-terms is called a "polynomial". Some examples are shown.

$3x$ $y + 7$ $2a^2 - 5a + 6$ $t^4 - 4t^3 + t^2 - 1$

Continued on following page.

40. Continued

A polynomial can contain one or more power-terms. For example:

$3x$ contains <u>one</u> power-term.

$y + 7$ contains <u>two</u> power-terms.

$t^4 - 4t^3 + t^2 - 1$ contains ________ power-terms.

Answer: four

41. There are special names for polynomials containing one, two, or three terms.

A <u>one-term</u> polynomial is called a "<u>monomial</u>". Some examples are:

-5 $\quad x$ $\quad -3y$ $\quad -t^2$ $\quad 5d^4$

A <u>two-term</u> polynomial is called a "<u>binomial</u>". Some examples are:

$2x + 9$ $\quad y^4 - 5y$ $\quad 2t^5 + 3t^3$

A <u>three-term</u> polynomial is called a "<u>trinomial</u>". Some examples are:

$x^2 + 8x - 7$ $\quad 4p^4 - p^3 + 8p^2$

State whether each expression is a monomial, binomial, or trinomial.

a) 8 ________________

b) $x^4 - 1$ ________________

c) $2y^3 + y + 7$ ________________

d) $5m^6 - 6m^5$ ________________

e) $-p$ ________________

f) $a^4 - 3a^2 + 1$ ________________

Answers: a) monomial b) binomial c) trinomial d) binomial e) monomial f) trinomial

42. There are no special names for polynomials with more than three terms. That is:

$2x^3 - x^2 + 4x - 6$ is simply called a polynomial with <u>four</u> terms.

$x^4 - 5x^3 + 2x^2 - x + 9$ is simply called a polynomial with ________ terms.

Answer: five

43. Polynomials are usually written <u>in descending order</u>. That is, they are written so that the power-terms are in order, with the largest exponent at the left. For example:

$$3x^4 + 5x^3 + 6x^2 + 2x + 9$$

To put each polynomial below in descending order, we simply rearrange the terms. That is:

$2 + 5y^4 + 3y^2$ is rearranged to $5y^4 + 3y^2 + 2$

$8b + b^2 + 1 + 4b^3$ is rearranged to ____________________

Answer: $4b^3 + b^2 + 8b + 1$

17-6 MULTIPLYING AND DIVIDING MONOMIALS

In this section, we will discuss the procedure for multiplying and dividing monomials containing the same letter.

44. To multiply a monomial containing a power by a number, we simply multiply the number and the coefficient. For example:

$3(2x^4) = 6x^4$ a) $-4(5y^2) =$ ________ b) $8(t^5) =$ ________

45. To multiply two monomials containing powers of the same letter, we multiply their coefficients and use the law of exponents for multiplication to combine the powers into one. For example:

a) $-20y^2$
b) $8t^5$

$$(3x^2)(5x^4) = (3)(5)(x^2)(x^4) = 15x^6$$

$$(-2y^2)(4y^2) = (-2)(4)(y^2)(y^2) = -8y^4$$

Using the same method, perform each multiplication.

a) $(4t^3)(-3t^2) =$ ________ b) $(-2m^5)(-5m^5) =$ ________

46. When the exponent of a power is "1", it is helpful to insert the "1". For example:

a) $-12t^5$
b) $10m^{10}$

$$(2x)(4x) = (2)(4)(x^1)(x^1) = 8x^2$$

$$(-3y)(5y^4) = (-3)(5)(y^1)(y^4) = -15y^5$$

Using the same method, perform each multiplication.

a) $(2p^2)(-7p) =$ ________ b) $(-5t)(-4t^5) =$ ________

47. When the coefficient of the power is "1" or "-1", it is helpful to insert the "1" or "-1". For example:

a) $-14p^3$
b) $20t^6$

$$(x^2)(4x^5) = (1x^2)(4x^5) = 4x^7$$

$$(3y^3)(-y^4) = (3y^3)(-1y^4) = -3y^7$$

Using the same method, complete these:

a) $(2y)(y^2) =$ ________ b) $(-t^4)(7t) =$ ________

48. To divide a monomial containing a power by a number, we simply divide the coefficient by the number. That is:

a) $2y^3$
b) $-7t^5$

$\frac{12x^3}{6} = 2x^3$ a) $\frac{-6y}{2} =$ ________ b) $\frac{-20t^4}{-5} =$ ________

a) $-3y$
b) $4t^4$

49. To divide two monomials containing powers of the same letter, we divide their coefficients and use the law of exponents for division to combine the powers into one. For example:

$$\frac{10x^5}{5x^2} = \left(\frac{10}{5}\right)\left(\frac{x^5}{x^2}\right) = 2x^3 \qquad \frac{-8y^6}{2y^4} = \left(\frac{-8}{2}\right)\left(\frac{y^6}{y^4}\right) = -4y^2$$

Using the same method, perform each division.

a) $\frac{12t^7}{3t^3} =$ ________ b) $\frac{-12m^4}{-4m^2} =$ ________

---

a) $4t^4$
b) $3m^2$

50. When the coefficient of the quotient is "1" or "-1", the "1" or "-1" is not ordinarily written. For example:

$$\frac{4t^7}{4t^2} = 1t^5 \text{ or } t^5 \qquad \frac{-8m^5}{8m^3} = -1m^2 \text{ or } \_\_\_\_$$

$-m^2$

51. When a monomial contains a letter to the first power, it is helpful to write the "1" explicitly. For example:

$$\frac{6x^3}{2x} = \frac{6x^3}{2x^1} = 3x^2 \qquad \frac{-7y^5}{-7y} = \frac{-7y^5}{-7y^1} = \_\_\_\_$$

$1y^4$ or y^4

52. When the quotient contains a letter to the first power, the "1" exponent is not ordinarily written. For example:

$$\frac{10x^4}{5x^3} = 2x^1 \text{ or } 2x \qquad \frac{9y^2}{-3y} = -3y^1 \text{ or } \_\_\_\_$$

$-3y$

53. When the quotient contains a letter to the zero power, that power is not ordinarily written since it equals "1". For example:

$$\frac{12t^3}{4t^3} = 3t^0 \text{ or } 3(1) \text{ or } 3 \qquad \frac{-5d^4}{5d^4} = -1d^0 \text{ or } -1(1) \text{ or } \_\_\_\_$$

-1

54. When the coefficient of the denominator is "1" or "-1", it is helpful to write the "1" or "-1" explicitly. That is:

$$\frac{5x^3}{x} = \frac{5x^3}{1x} = 5x^2 \qquad \frac{7t^5}{-t^2} = \frac{7t^5}{-1t^2} = \_\_\_\_$$

$-7t^3$

55. Perform each multiplication and division.

a) $(-2y^3)(7y^4) =$ ________ c) $(t)(-5t^2) =$ ________ e) $(-d^2)(-d^2) =$ ________

b) $\frac{32m^6}{4m^5} =$ ________ d) $\frac{16x^3}{16x} =$ ________ f) $\frac{-2p^5}{p^5} =$ ________

---

a) $-14y^7$	c) $-5t^3$	e) d^4
b) $8m$	d) x^2	f) -2

17-7 MULTIPLYING OTHER POLYNOMIALS BY A MONOMIAL

In this section, we will show that the distributive principle can be used to multiply other polynomials by a monomial.

---

56. To multiply $4y^3 + 5$ by 2 below, we used the distributive principle. Notice that we multiplied each term in the binomial by 2. Complete the other multiplication.

$2(4y^3 + 5) = 2(4y^3) + 2(5)$

$= 8y^3 + 10$

$3(5x^4 - 1) = (\quad)(\quad) - (\quad)(\quad)$

$= \_\_\_\_\_ - \_\_\_\_\_$

$3(5x^4) - 3(1)$

$15x^4 - 3$

57. To multiply $2x^4 - x^2 + 7$ by 4 below, we multiplied each term in the trinomial by 4. Complete the other multiplication.

$4(2x^4 - x^2 + 7) = 4(2x^4) - 4(x^2) + 4(7)$

$= 8x^4 - 4x^2 + 28$

$5(y^3 + 2y^2 - y) = (\quad)(\quad) + (\quad)(\quad) - (\quad)(\quad)$

$= \_\_\_\_\_ + \_\_\_\_\_ - \_\_\_\_\_$

$(5)(y^3) + 5(2y^2) - 5(y)$

$5y^3 + 10y^2 - 5y$

58. To multiply $2x^2 + 5$ by $3x$ below, we multiplied each term of the binomial by $3x$. Complete the other multiplication.

$3x(2x^2 + 5) = 3x(2x^2) + 3x(5)$

$= 6x^3 + 15x$

$y^2(3y^3 - y) = (\quad)(\quad) - (\quad)(\quad)$

$= \_\_\_\_\_ - \_\_\_\_\_$

$(y^2)(3y^3) - (y^2)(y)$

$3y^5 - y^3$

59. Complete each multiplication.

a) $x^3(x^2 + x) = \_\_\_\_\_$

b) $y(y^6 - y^4) = \_\_\_\_\_$

---

a) $x^5 + x^4$

b) $y^7 - y^5$

60. To multiply $x^2 - 5x + 7$ by $4x$ below, we multiplied each term in the trinomial by $4x$.

$4x(x^2 - 5x + 7) = 4x(x^2) - 4x(5x) + 4x(7)$

$= 4x^3 - 20x^2 + 28x$

Following the example, complete this multiplication.

$y^3(2y^2 + y - 1) = (\quad)(\quad) + (\quad)(\quad) - (\quad)(\quad)$

$= \_\_\_\_\_ + \_\_\_\_\_ - \_\_\_\_\_$

$(y^3)(2y^2) + (y^3)(y) - (y^3)(1)$

$2y^5 + y^4 - y^3$

61. Complete:

a) $t(t^4 - 5t^2 - 8) = \_\_\_\_\_$

b) $3p^5(p^3 - p^2 + p) = \_\_\_\_\_$

---

a) $t^5 - 5t^3 - 8t$

b) $3p^8 - 3p^7 + 3p^6$

17-8 DIVIDING OTHER POLYNOMIALS BY A MONOMIAL

In this section, we will show the procedure for dividing other polynomials by a monomial.

62. To divide a binomial by a number, we simply divide each term in the binomial by the number. That is:

$$\frac{8x^2 + 6}{2} = \frac{8x^2}{2} + \frac{6}{2} = 4x^2 + 3 \qquad \frac{9y^4 - 12}{3} = \frac{9y^4}{3} - \frac{12}{3} = \_\_\_\_\_\_$$

Answer: $3y^4 - 4$

63. To divide a trinomial by a number, we divide each term in the trinomial by the number. That is:

$$\frac{5x^2 - 10x + 15}{5} = \frac{5x^2}{5} - \frac{10x}{5} + \frac{15}{5} = x^2 - 2x + 3$$

$$\frac{8p^4 + 6p^2 - 12}{2} = \frac{8p^4}{2} + \frac{6p^2}{2} - \frac{12}{2} = \_\_\_\_\_\_$$

Answer: $4p^4 + 3p^2 - 6$

64. To divide polynomials by other power-terms, we also divide each term in the polynomial by the power-term. That is:

$$\frac{6y^2 + 4y}{2y} = \frac{6y^2}{2y} + \frac{4y}{2y} = 3y + 2 \qquad \frac{x^8 - 5x^4}{x^2} = \frac{x^8}{x^2} - \frac{5x^4}{x^2} = \_\_\_\_\_\_$$

Answer: $x^6 - 5x^2$

65. Complete each division.

a) $\frac{3t^4 + t}{t} = \_\_\_\_\_\_$ b) $\frac{6m^5 - 9m^3}{3m^3} = \_\_\_\_\_\_$

Answer: a) $3t^3 + 1$ b) $2m^2 - 3$

66. Complete:

a) $\frac{2y^3 + 6y^2 - 4y}{2y} = \frac{2y^3}{2y} + \frac{6y^2}{2y} - \frac{4y}{2y} = \_\_\_\_\_\_$

b) $\frac{3x^6 - 7x^5 - x^4}{x^2} = \frac{3x^6}{x^2} - \frac{7x^5}{x^2} - \frac{x^4}{x^2} = \_\_\_\_\_\_$

Answer: a) $y^2 + 3y - 2$ b) $3x^4 - 7x^3 - x^2$

67. Complete:

a) $\frac{2x^5 - x^3 + 4x}{x} = \_\_\_\_\_\_$

b) $\frac{2p^5 + 4p^4 - 6p^3}{2p^3} = \_\_\_\_\_\_$

Answer: a) $2x^4 - x^2 + 4$ b) $p^2 + 2p - 3$

SELF-TEST 51 (pages 444-450)

Perform each multiplication.

1. $5(2x^5)$ = ______	2. $(t)(-4t^3)$ = ______	3. $(-2x)(x^5)$ = ______	4. $(-3y^4)(-6y^2)$ = ______

Perform each division.

5. $\frac{-10a^4}{5}$ = ______	6. $\frac{3x^5}{3x^2}$ = ______	7. $\frac{9d^3}{-d}$ = ______	8. $\frac{-16y^8}{-4y^8}$ = ______

Perform each multiplication.

9. $2x(x^5 - 3x^2)$ =	10. $y^3(y^2 - 7y + 1)$ =

Perform each division.

11. $\frac{10t^2 - 2}{2}$ =	12. $\frac{6m^5 - 5m^4 + m^3}{m^3}$ =

ANSWERS:

1. $10x^5$	5. $-2a^4$	9. $2x^6 - 6x^3$
2. $-4t^4$	6. x^3	10. $y^5 - 7y^4 + y^3$
3. $-2x^6$	7. $-9d^2$	11. $5t^2 - 1$
4. $18y^6$	8. 4	12. $6m^2 - 5m + 1$

17-9 ADDING AND SUBTRACTING MONOMIALS

In this section, we will define what is meant by monomials that are "like" terms and show how such monomials can be combined by addition or subtraction.

68. Two monomials are called "like" terms if they contain the same letter to the same power. For example:

$2x$ and $3x$ are "like" terms because they both contain "x" or "x^1".

$5y^3$ and $4y^3$ are "like" terms because they both contain "y^3".

$8t^5$ and $6t^5$ are "like" terms because they both contain ______.

t^5

69. Two monomials are called "unlike" terms if they contain the same letter to different powers. For example:

$2x^2$ and $5x^3$ are "unlike" terms.

$3y$ and $7y^4$ are "unlike" terms.

Which of the following are pairs of "like" terms? ______

a) $5m^4$ and $2m^4$ b) $3p^5$ and $4p^2$ c) $6b$ and $4b$ d) $5x$ and $2x^2$

(a) and (c)

70. Two monomials are also called "unlike" terms <u>if they contain different letters, even when the letters are to the same power</u>. For example:

$2x$ and $3y$ are "unlike" terms.

$5t^3$ and $4m^3$ are "unlike" terms.

Which of the following are pairs of "like" terms? ________

a) 6m and 3b b) 5d and 8d c) $3d^2$ and $4d^2$ d) $7x^3$ and $2y^3$

(b) and (c)

71. When the terms in an addition are "like" terms, we can factor by the distributive principle to combine them. Doing so is the same as adding their coefficients. That is:

$$2x + 3x = (2 + 3)x = 5x$$

$$5y^3 + 4y^3 = (5 + 4)y^3 = 9y^3$$

Perform each addition by simply adding the coefficients.

a) $6y + 2y$ = ________ b) $3t^2 + 7t^2$ = ________ c) $3m^5 + 3m^5$ = ________

---

a) 8y
b) $10t^2$
c) $6m^5$

72. When the coefficient of a term is "1", the "1" is not ordinarily written. However, it is helpful to insert the "1" when adding. For example:

$$5x^2 + x^2 = 5x^2 + 1x^2 = 6x^2$$

$$t^8 + t^8 = 1t^8 + 1t^8 = 2t^8$$

Perform each addition.

a) $y + 6y$ = ________ b) $2m^4 + m^4$ = ________ c) $b^3 + b^3$ = ________

---

a) 7y
b) $3m^4$
c) $2b^3$

73. When the terms in a subtraction are "like" terms, we can also factor by the distributive principle to combine them. Doing so is the same as subtracting their coefficients. That is:

$$7x - 4x = (7 - 4)x = 3x$$

$$3y^3 - 5y^3 = (3 - 5)y^3 = -2y^3$$

Perform each subtraction by simply subtracting the coefficients.

a) $2m - 6m$ = ________ b) $8t^4 - 3t^4$ = ________ c) $4x^2 - 5x^2$ = ________

---

a) -4m
b) $5t^4$
c) $-1x^2$ or $-x^2$

74. For each subtraction below, it is helpful to insert "1" as a coefficient. That is:

$$4x^2 - x^2 = 4x^2 - 1x^2 = 3x^2$$

$$p^5 - 7p^5 = 1p^5 - 7p^5 = -6p^5$$

Perform each subtraction.

a) $8y - y$ = ________ b) $2t^3 - t^3$ = ________ c) $d^6 - 2d^6$ = ________

---

a) 7y
b) $1t^3$ or t^3
c) $-1d^6$ or $-d^6$

75. When the coefficients in a subtraction are identical, we get "0" as the difference. That is:

$$3y^2 - 3y^2 = (3 - 3)y^2 = 0y^2 = 0$$

$$t^5 - t^5 = (1 - 1)t^5 = 0t^5 = 0$$

Therefore:

a) $5x^3 - 5x^3 =$ ______ b) $3t^4 - 7t^4 =$ ______ c) $d^2 - d^2 =$ ______

Answers: a) 0 b) $-4t^4$ c) 0

76. When the terms in an addition or subtraction are "unlike", they cannot be combined into one. For example, **none** of the expressions below can be simplified by combining terms.

$2x + 3y$ $5m^2 + 6t^2$ $4b^4 - 7b^2$

If possible, simplify each expression by combining like terms.

a) $3p + q =$ ______ c) $x^3 + 6x^3 =$ ______ e) $3y^3 + y^4 =$ ______

b) $t - 5t =$ ______ d) $m^5 - 2p^5 =$ ______ f) $a^5 - a^5 =$ ______

Answers: a) Not possible b) $-4t$ c) $7x^3$ d) Not possible e) Not possible f) 0

77. An addition or subtraction of different powers of the same letter cannot be simplified because "unlike" terms cannot be combined. For example:

$x^2 + x^3$ cannot be simplified

$y^4 - y^3$ cannot be simplified

However, a multiplication of different powers of the same letter can be simplified. For example:

$$(x^2)(x^3) = (1x^2)(1x^3) = 1x^5 \text{ or } x^5$$

$$(y^4)(-y^3) = (1y^4)(-1y^3) = -1y^7 \text{ or } -y^7$$

Simplify if possible.

a) $t + t^4 =$ ______ c) $m^3 - m =$ ______

b) $(t)(t^4) =$ ______ d) $(m^3)(-m) =$ ______

Answers: a) Not possible b) t^5 c) Not possible d) $-m^4$

17-10 SIMPLIFYING POLYNOMIALS BY COMBINING LIKE TERMS

In this section, we will show how polynomials containing like terms can be simplified by combining the like terms.

---

78. In the polynomial below, there are two pairs of like terms: ($2x^2$ and $3x^2$) and (5 and 1). To identify the like terms, we drew the same type of line under each pair. We then simplified the polynomial by combining like terms.

$$\underline{2x^2} + \underline{\underline{5}} + \underline{3x^2} + \underline{\underline{1}} = 5x^2 + 6$$

Simplify the following polynomial by combining like terms.

$$\underline{4y^3} + \underline{\underline{2y}} + \underline{5y^3} + \underline{\underline{6y}} =$$

> $9y^3 + 8y$

79. Following the example, simplify the other polynomial.

$$\underline{5x^4} + \underline{\underline{3x^2}} - \underline{2x^4} - \underline{\underline{9x^2}} = 3x^4 - 6x^2 \qquad 7d^3 - 8 - 5d^3 + 2 =$$

> $2d^3 - 6$

80. Following the example, simplify the other polynomial.

$$\underline{3x^2} - \underline{\underline{5x}} - \underline{7x^2} - \underline{\underline{2x}} = -4x^2 - 7x \qquad 4y^4 - 3 - 6y^4 - 1 =$$

> $-2y^4 - 4$

81. When simplifying polynomials, it is helpful to write "1" and "-1" coefficients explicitly. Simplify these polynomials.

a) $x^3 - 5x + 3x^3 - x =$ b) $2y^4 - y^2 + y^4 - y^2 =$

> a) $4x^3 - 6x$
> b) $3y^4 - 2y^2$

82. When combining like terms to simplify a polynomial, we do not ordinarily write "1" and "-1" coefficients. Simplify these.

a) $5p^3 - 7p^2 - 4p^3 + 6p^2 =$ b) $2t^2 + t - t^2 - 2t =$

> a) $p^3 - p^2$
> b) $t^2 - t$

83. When combining $3x^2$ and $-3x^2$, we got "$0x^2$" or "0". Notice that the "0" is not written in the final simplification.

$$\underline{3x^2} + \underline{\underline{x}} - \underline{3x^2} - \underline{\underline{5x}} = 0x^2 - 4x = 0 - 4x = -4x$$

Simplify each polynomial.

a) $5y + 7 - 5y - 2 =$ b) $x^4 + x^2 + x^4 - x^2 =$

> a) 5
> b) $2x^4$

84. There are three pairs of like terms in the polynomial below. Notice how we simplified it.

$$\underline{3x^2} - \underline{\underline{5x}} + \underline{\underline{\underline{1}}} + \underline{2x^2} + \underline{\underline{5x}} - \underline{\underline{\underline{7}}} = 5x^2 + 0x - 6 = 5x^2 - 6$$

Following the example, simplify each polynomial.

a) $y^2 + 2y - 9 + y^2 - y - 3 =$

b) $t^4 - t^2 + 3 - t^4 + 4t^2 - 3 =$

---

a) $2y^2 + y - 12$

b) $3t^2$

85. There are only two pairs of like terms in the polynomial below. Notice how we wrote the terms in the simplified polynomial in descending order.

$$5a^3 + \underline{2a^2} - \underline{\underline{a}} + \underline{a^2} - \underline{\underline{3a}} + 8 = 5a^3 + 3a^2 - 4a + 8$$

Simplify this polynomial. Write the terms in descending order.

$3t^5 + t^4 - 1 - t^5 - t^4 + 4t =$

$2t^5 + 4t - 1$

17-11 ADDING POLYNOMIALS

In this section, we will discuss the procedure for adding polynomials both horizontally and in columns.

---

86. When writing an addition of polynomials horizontally, the polynomials are written in parentheses. For example:

Add $2x^2 + 5$ and $3x^2 - 1$ is written $(2x^2 + 5) + (3x^2 - 1)$

To perform the addition, we simply drop the parentheses and combine like terms. For example:

$$(2x^2 + 5) + (3x^2 - 1) = \underline{2x^2} + \underline{\underline{5}} + \underline{3x^2} - \underline{\underline{1}} = 5x^2 + 4$$

Following the example, perform the following addition.

$(y^3 - 2y) + (2y^3 - 3y) =$

$3y^3 - 5y$

87. You should be able to perform an addition without rewriting the addition with the parentheses dropped. That is:

$$(\underline{y^4} - \underline{\underline{2y^2}} + \underline{\underline{\underline{1}}}) + (\underline{y^4} + \underline{\underline{3y^2}} - \underline{\underline{\underline{9}}}) = 2y^4 + y^2 - 8$$

$(\underline{3t^2} - \underline{\underline{t}} + \underline{\underline{\underline{7}}}) + (\underline{t^2} - \underline{\underline{t}} - \underline{\underline{\underline{6}}}) =$

$4t^2 - 2t + 1$

88. Following the example, perform the other addition.

$$(\underline{x^2} + \underline{\underline{5}}) + (\underline{3x^2} - \underline{\underline{\underline{7x}}} + \underline{\underline{2}}) + (\underline{\underline{\underline{2x}}} - \underline{\underline{10}}) = 4x^2 - 5x - 3$$

$(y^3 - 2y^2 - y) + (y^3 + y) + (2y^2 + y) =$

$2y^3 + y$

89. Perform each addition. Write each sum in descending order.

a) $(y^3 - 2) + (3y^4 - y^2 - 1) + (y^4 + 5) =$

b) $(4t^2 - 3t + 1) + (t^3 - 1) + (t^3 - 4t^2) =$

---

90. To add polynomials vertically, we line up like terms in columns and then find the sum of the columns. An example is shown. Complete the other addition.

Answer (89): a) $4y^4 + y^3 - y^2 + 2$ b) $2t^3 - 3t$

Add $2x^2 - 1$ and $5x^2 - 3$.

$$\begin{array}{r} 2x^2 - 1 \\ \underline{5x^2 - 3} \\ 7x^2 - 4 \end{array}$$

Add $y^3 + 4y$ and $-2y^3 - y$.

---

91. Following the example, complete the other addition.

Answer (90):
$$\begin{array}{r} y^3 + 4y \\ \underline{-2y^3 - y} \\ -y^3 + 3y \end{array}$$

Add $2x^2 + 5x - 1$, $3x + 5$, and $x^3 - 4x^2 - 7$.

$$\begin{array}{rrrr} & 2x^2 & + 5x & - 1 \\ & & 3x & + 5 \\ \underline{x^3} & \underline{- 4x^2} & & \underline{- 7} \\ x^3 & - 2x^2 & + 8x & - 3 \end{array}$$

Add $y + 6$, $y^2 - 6y + 1$, and $y^3 + 5y - 5$.

Answer (91):
$$\begin{array}{rrrr} & & y & + 6 \\ & y^2 & - 6y & + 1 \\ \underline{y^3} & & \underline{+ 5y} & \underline{- 5} \\ y^3 & + y^2 & + 0y & + 2 \end{array}$$
or
$y^3 + y^2 + 2$

17-12 OPPOSITES OF POLYNOMIALS

In this section, we will discuss the opposites of monomials and other polynomials.

---

92. Two monomials are a pair of opposites if their sum is "0". For example:

Since $3x^2 + (-3x^2) = 0$: the opposite of $3x^2$ is $-3x^2$.
the opposite of $-3x^2$ is $3x^2$.

Since $y^4 + (-y^4) = 0$: a) the opposite of y^4 is _______.

b) the opposite of $-y^4$ is _______.

---

93. To find the opposite of a monomial, we simply change the sign of its coefficient. That is:

Answer (92): a) $-y^4$ b) y^4

The opposite of $5t$ is $-5t$. The opposite of $-a^5$ is a^5.

Write the opposite of each monomial.

a) $2b^2$ _______ b) $-7t^4$ _______ c) a^3 _______ d) $-x$ _______

Answer (93): a) $-2b^2$ b) $7t^4$ c) $-a^3$ d) x

94. Two polynomials are a pair of opposites if each pair of like terms is a pair of opposites. For example:

Since $(2x + 5) + (-2x - 5) = 0x + 0 = 0$:

the opposite of $2x + 5$ is $-2x - 5$.
the opposite of $-2x - 5$ is $2x + 5$.

Since $(y^2 - 2y - 7) + (-y^2 + 2y + 7) = 0y^2 + 0y + 0 = 0$:

a) the opposite of $y^2 - 2y - 7$ is ________________.

b) the opposite of $-y^2 + 2y + 7$ is ________________.

---

a) $-y^2 + 2y + 7$
b) $y^2 - 2y - 7$

95. To find the opposite of a polynomial, we simply change the sign of each term. That is:

The opposite of $2t^3 - 5t^2 - 7$ is $-2t^3 + 5t^2 + 7$.

Write the opposite of each polynomial.

a) $3y^2 - 4$ ________________ b) $x^3 - 7x + 1$ ________________

---

a) $-3y^2 + 4$
b) $-x^3 + 7x - 1$

96. Write the opposite of each polynomial.

a) $b^5 - b^4$ ______________ b) $-3b^4 + 8b^2 - b - 1$ ________________

a) $-b^5 + b^4$ b) $3b^4 - 8b^2 + b + 1$

17-13 SUBTRACTING POLYNOMIALS

In this section, we will discuss the procedure for subtracting polynomials both horizontally and in columns.

---

97. To subtract polynomials, we simply add the opposite of the second polynomial. That is:

Opposites

$$(3x^2 - 5) - (x^2 + 4) = (3x^2 - 5) + (-x^2 - 4) = 2x^2 - 9$$

$$(5y + 7) - (6y - 3) = (5y + 7) + (-6y + 3) = -y + 10$$

Following the examples, complete these subtractions.

a) $(y - 1) - (3y - 5) = (y - 1) + (\qquad) =$

b) $(2t^4 + t^2) - (t^4 + t^2) = (2t^4 + t^2) + (\qquad) =$

a) $+ (-3y + 5) = -2y + 4$
b) $+ (-t^4 - t^2) = t^4$

98. Another subtraction of polynomials is shown below. Notice again that we added the opposite of the second polynomial.

Opposite

$$(5y^3 - 2y^2 + 1) - (y^3 + 2y^2 + 1) = (5y^3 - 2y^2 + 1) + \overbrace{(-y^3 - 2y^2 - 1)} = 4y^3 - 4y^2$$

Following the example, complete this subtraction.

$$(m^2 - m + 4) - (m^2 - 2m - 1) = (m^2 - m + 4) + (\qquad\qquad) =$$

Answer: $+ (-m^2 + 2m + 1) = m + 5$

99. Perform each subtraction by adding the opposite of the second polynomial. Write each answer in descending order.

a) $(2x^2 - 5x) - (x^4 + 3x) =$

b) $(6y - 1) - (-3y^3 + 6y - 1) =$

Answer: a) $-x^4 + 2x^2 - 8x$ b) $3y^3$

100. To subtract polynomials vertically, we line up like terms in columns and then add the opposite of the bottom polynomial. An example is shown. Complete the other subtraction.

Subtract $2y - 7$ from $5y + 1$.

$$\begin{array}{r} 5y + 1 \\ -\ \ 2y - 7 \\ \hline \end{array} \quad \text{becomes} \quad \begin{array}{r} 5y + 1 \\ + -2y + 7 \\ \hline 3y + 8 \end{array}$$

Subtract $x^3 + x^2$ from $3x^3 - 2x^2$.

$$\begin{array}{r} 3x^3 - 2x^2 \\ -\ \ x^3 + x^2 \\ \hline \end{array} \quad \text{becomes}$$

Answer:
$$\begin{array}{r} 3x^3 - 2x^2 \\ + -x^3 - x^2 \\ \hline 2x^3 - 3x^2 \end{array}$$

101. Following the example, complete the other subtraction.

Subtract $2x^2 - 3x + 5$ from $7x^2 - x - 3$.

$$\begin{array}{r} 7x^2 - x - 3 \\ -\ \ 2x^2 - 3x + 5 \\ \hline \end{array} \quad \text{becomes} \quad \begin{array}{r} 7x^2 - x - 3 \\ + -2x^2 + 3x - 5 \\ \hline 5x^2 + 2x - 8 \end{array}$$

Subtract $t^4 - t^2 - 1$ from $t^4 + t^2 + 1$.

$$\begin{array}{r} t^4 + t^2 + 1 \\ -\ \ t^4 - t^2 - 1 \\ \hline \end{array} \quad \text{becomes}$$

Answer:
$$\begin{array}{r} t^4 + t^2 + 1 \\ + -t^4 + t^2 + 1 \\ \hline 2t^2 + 2 \end{array}$$

SELF-TEST 52 (pages 450-458)

Perform each addition and subtraction.

1. $3t^3 + 7t^3 =$	2. $m^5 + m^5 =$
3. $x^2 - 2x^2 =$	4. $5p^4 - p^4 =$

Perform each addition and subtraction horizontally. Write each answer in descending order.

5. $(3y^2 + 4) + (4y^2 - 3) =$

6. $(a^3 - 6a^2 + a) + (2a^4 + 6a^2 - a) =$

7. $(x^2 - 5) - (2x^2 - 8) =$

8. $(m^3 - 2m + 1) - (m^3 - 3m^2 + 4) =$

Perform each addition and subtraction in columns.

9. Add $3x^2 - 5x + 2$, $x + 7$, and $x^3 + 3x - 9$.	10. Subtract $y^2 - 10$ from $3y^2 + 2$.

ANSWERS:

1. $10t^3$	3. $-x^2$	5. $7y^2 + 1$	7. $-x^2 + 3$	9. $x^3 + 3x^2 - x$
2. $2m^5$	4. $4p^4$	6. $2a^4 + a^3$	8. $3m^2 - 2m - 3$	10. $2y^2 + 12$

Unit 18 RADICALS

In this unit, we will discuss the four basic operations with square root radicals and show how radicals are simplified. The discussion is limited to principal square roots. It is assumed that any letter in a radicand stands for a positive number.

18-1 MULTIPLYING RADICALS

In this section, we will discuss the procedure for multiplying square-root radicals.

---

1. As we saw in an earlier unit, the following language is used in connection with square-root radicals like $\sqrt{3}$.

   "$\sqrt{\ }$ " is called the <u>radical sign</u>.

   " 3 " is called the <u>radicand</u>.

   "$\sqrt{3}$ " is called a <u>square-root radical</u> or simply a <u>radical</u>.

   Write the square-root radical whose radicand is 5. ________

$\sqrt{5}$

2. To indicate a multiplication of radicals, we use a dot or simply write the radicals side by side. That is:

   Both $\sqrt{3}\cdot\sqrt{5}$ and $\sqrt{3}\sqrt{5}$ mean "multiply $\sqrt{3}$ and $\sqrt{5}$ ".

   To multiply two radicals, we simply multiply their radicands. That is:

   $\sqrt{3}\cdot\sqrt{5} = \sqrt{(3)(5)}$ $\qquad$ $\sqrt{7}\cdot\sqrt{x} = \sqrt{(7)(x)}$ $\qquad$ $\sqrt{y}\cdot\sqrt{y} = \sqrt{(\quad)(\quad)}$

$\sqrt{(y)(y)}$

3. To show that the multiplication procedure makes sense, two examples involving perfect-square radicands are shown below.

$$\begin{aligned} \sqrt{9}\cdot\sqrt{4} &= \sqrt{(9)(4)} \\ 3\cdot 2 &= \sqrt{36} \\ 6 &= 6 \end{aligned} \qquad\qquad \begin{aligned} \sqrt{16}\cdot\sqrt{25} &= \sqrt{(16)(25)} \\ 4\cdot 5 &= \sqrt{400} \\ 20 &= 20 \end{aligned}$$

   By simply multiplying the radicands, complete these:

   a) $\sqrt{7x}\cdot\sqrt{2x} = \sqrt{(\qquad)(\qquad)}$ $\qquad$ b) $\sqrt{3y^2}\cdot\sqrt{y^6} = \sqrt{(\qquad)(\qquad)}$

4. When multiplying radicals, we simplify the radicand of the product as much as possible. For example:

$$\sqrt{2} \cdot \sqrt{3} = \sqrt{(2)(3)} = \sqrt{6}$$
$$\sqrt{5} \cdot \sqrt{y^4} = \sqrt{(5)(y^4)} = \sqrt{5y^4}$$
$$\sqrt{3} \cdot \sqrt{7x} = \sqrt{(3)(7x)} = \sqrt{21x}$$

Following the examples, complete these:

a) $\sqrt{7} \cdot \sqrt{11} = \sqrt{\quad}$ b) $\sqrt{10} \cdot \sqrt{t} = \sqrt{\quad}$ c) $\sqrt{2} \cdot \sqrt{11b^6} = \sqrt{\quad}$

a) $\sqrt{(7x)(2x)}$
b) $\sqrt{(3y^2)(y^6)}$

---

5. More examples of simplifying the radicand of the product are shown below.

$$\sqrt{x^4} \cdot \sqrt{x} = \sqrt{(x^4)(x)} = \sqrt{x^5}$$
$$\sqrt{3y} \cdot \sqrt{5y} = \sqrt{(3y)(5y)} = \sqrt{15y^2}$$
$$\sqrt{2t^2} \cdot \sqrt{5t^2} = \sqrt{(2t^2)(5t^2)} = \sqrt{10t^4}$$

Following the examples, complete these:

a) $\sqrt{b} \cdot \sqrt{b} = \sqrt{\quad}$ b) $\sqrt{5d} \cdot \sqrt{d^3} = \sqrt{\quad}$ c) $\sqrt{3p^5} \cdot \sqrt{7p^2} = \sqrt{\quad}$

a) $\sqrt{77}$
b) $\sqrt{10t}$
c) $\sqrt{22b^6}$

---

6. To multiply three or more radicals, we simply multiply their radicands. For example:

$$\sqrt{2} \cdot \sqrt{3} \cdot \sqrt{5} = \sqrt{(2)(3)(5)} = \sqrt{30}$$
$$\sqrt{7} \cdot \sqrt{x} \cdot \sqrt{30} \cdot \sqrt{x^2} = \sqrt{(7)(x)(30)(x^2)} = \sqrt{210x^3}$$

Write each product.

a) $\sqrt{y} \cdot \sqrt{2y} \cdot \sqrt{3} = \sqrt{\quad}$ b) $\sqrt{3p} \cdot \sqrt{7} \cdot \sqrt{p} \cdot \sqrt{p^5} = \sqrt{\quad}$

a) $\sqrt{b^2}$
b) $\sqrt{5d^4}$
c) $\sqrt{21p^7}$

---

7. To indicate a multiplication of a non-radical and a radical, we write the non-radical in front of the radical. For example:

$2\sqrt{5}$ means "multiply 2 and $\sqrt{5}$".

$x\sqrt{3x}$ means "multiply ____ and ________".

a) $\sqrt{6y^2}$
b) $\sqrt{21p^7}$

---

8. To simplify the expressions below, we multiplied the non-radical factors.

$2 \cdot 4\sqrt{7} = 8\sqrt{7}$ $\qquad$ $y \cdot 6\sqrt{y^3} = 6y\sqrt{y^3}$

Following the examples, simplify these:

a) $3x \cdot 5\sqrt{x}$ = ____________ b) $t^2 \cdot t^3\sqrt{7}$ = ____________

x and $\sqrt{3x}$

---

a) $15x\sqrt{x}$
b) $t^5\sqrt{7}$

9. To simplify the expressions below, we multiplied the radical factors.

$7\sqrt{3}\cdot\sqrt{2} = 7\sqrt{6}$ $\qquad$ $x\sqrt{5}\cdot\sqrt{2x} = x\sqrt{10x}$

Following the examples, simplify these:

a) $9\sqrt{y}\cdot\sqrt{y} =$ ________ $\qquad$ b) $t^3\sqrt{t}\cdot\sqrt{3t} =$ ________

---

a) $9\sqrt{y^2}$
b) $t^3\sqrt{3t^2}$

10. To simplify the expression below, we multiplied both the non-radical and the radical factors.

$$2\sqrt{3}\cdot 5\sqrt{7} = (2\cdot 5)(\sqrt{3}\cdot\sqrt{7})$$
$$= 10\cdot\sqrt{21} \quad \text{or} \quad 10\sqrt{21}$$

Following the example, simplify these:

a) $3\sqrt{5x}\cdot x\sqrt{2} =$ ________ $\qquad$ b) $y\sqrt{y}\cdot y^3\sqrt{y^5} =$ ________

a) $3x\sqrt{10x}$
b) $y^4\sqrt{y^6}$

18-2 SIMPLIFYING RADICALS

In this section, we will show how radicals can sometimes be simplified by factoring out a perfect square.

---

11. By reversing the procedure for multiplying radicals, we can factor a single radical into two radicals. That is:

$\sqrt{6} = \sqrt{(2)(3)} = \sqrt{2}\cdot\sqrt{3}$ $\qquad$ $\sqrt{35} = \sqrt{(5)(7)} = \sqrt{\quad}\cdot\sqrt{\quad}$

$\sqrt{5}\cdot\sqrt{7}$

12. Following the example, complete the other factorings.

$\sqrt{10} = \sqrt{2}\cdot\sqrt{5}$ $\qquad$ a) $\sqrt{14} = \sqrt{\quad}\cdot\sqrt{\quad}$ $\qquad$ b) $\sqrt{15} = \sqrt{\quad}\cdot\sqrt{\quad}$

---

a) $\sqrt{2}\cdot\sqrt{7}$
b) $\sqrt{3}\cdot\sqrt{5}$

13. A radical is not factored when the radicand can only be factored into itself and "1". For example:

$\sqrt{2}$ is not factored, since 2 is only equal to (2)(1).

$\sqrt{11}$ is not factored, since 11 is only equal to (11)(1).

Factor each radical if possible.

a) $\sqrt{7} =$ ________ $\qquad$ c) $\sqrt{13} =$ ________

b) $\sqrt{21} =$ ________ $\qquad$ d) $\sqrt{34} =$ ________

a) Not factorable
b) $\sqrt{3}\cdot\sqrt{7}$
c) Not factorable
d) $\sqrt{2}\cdot\sqrt{17}$

14. When a radical factor contains a perfect square, we can simplify. For example:

$$\sqrt{12} = \sqrt{4}\cdot\sqrt{3} = 2\sqrt{3} \qquad \sqrt{50} = \sqrt{25}\cdot\sqrt{2} = 5\sqrt{2}$$

The process above is called "<u>simplifying by factoring out a perfect square</u>". Using that process, complete these:

a) $\sqrt{700} = \sqrt{100}\cdot\sqrt{7} =$ ________ b) $\sqrt{27} = \sqrt{9}\cdot\sqrt{3} =$ ________

---

a) $10\sqrt{7}$
b) $3\sqrt{3}$

15. When the radicand can be factored in various ways, we always look for a perfect-square factor in order to simplify. That is:

Instead of $\sqrt{18} = \sqrt{6}\cdot\sqrt{3}$, we use $\sqrt{18} = \sqrt{9}\cdot\sqrt{2} = 3\sqrt{2}$

Instead of $\sqrt{28} = \sqrt{2}\cdot\sqrt{14}$, we use $\sqrt{28} = \sqrt{4}\cdot\sqrt{7} = 2\sqrt{7}$

Simplify each of these by factoring out a perfect square.

a) $\sqrt{8} =$ ________ c) $\sqrt{45} =$ ________

b) $\sqrt{20} =$ ________ d) $\sqrt{75} =$ ________

---

a) $2\sqrt{2}$, from $\sqrt{4}\cdot\sqrt{2}$
b) $2\sqrt{5}$, from $\sqrt{4}\cdot\sqrt{5}$
c) $3\sqrt{5}$, from $\sqrt{9}\cdot\sqrt{5}$
d) $5\sqrt{3}$, from $\sqrt{25}\cdot\sqrt{3}$

16. Some numbers have more than one perfect-square factor. For example, 32 has 4 and 16; 72 has 4, 9, and 36. When a number of that type is the radicand, we can simplify the radical completely <u>in one step</u> only by factoring out the largest perfect square. We get:

$$\sqrt{32} = \sqrt{16}\cdot\sqrt{2} = 4\sqrt{2} \qquad \sqrt{72} = \sqrt{36}\cdot\sqrt{2} = 6\sqrt{2}$$

If we do not factor out the largest perfect square, the remaining radical (like $\sqrt{8}$ and $\sqrt{18}$ below) will still contain a perfect-square factor.

$$\sqrt{32} = \sqrt{4}\cdot\sqrt{8} = 2\sqrt{8} \qquad \sqrt{72} = \sqrt{4}\cdot\sqrt{18} = 2\sqrt{18}$$

Therefore, to simplify completely, we have to factor again. We get:

$$\sqrt{32} = 2\sqrt{8} = 2\sqrt{4}\cdot\sqrt{2} = 2\cdot 2\sqrt{2} = 4\sqrt{2}$$

$$\sqrt{72} = 2\sqrt{18} = 2\sqrt{9}\cdot\sqrt{2} = \text{\_\_\_\_\_\_\_\_}$$

$2\cdot 3\sqrt{2} = 6\sqrt{2}$

17. Simplify completely. Try to do so in one step by factoring out the largest perfect square.

a) $\sqrt{48} =$ ________ c) $\sqrt{80} =$ ________

b) $\sqrt{200} =$ ________ d) $\sqrt{300} =$ ________

---

a) $4\sqrt{3}$, from $\sqrt{16}\cdot\sqrt{3}$
b) $10\sqrt{2}$, from $\sqrt{100}\cdot\sqrt{2}$
c) $4\sqrt{5}$, from $\sqrt{16}\cdot\sqrt{5}$
d) $10\sqrt{3}$, from $\sqrt{100}\cdot\sqrt{3}$

18. When the radicand is a letter with a coefficient, we can factor the radical. For example:

$$\sqrt{5x} = \sqrt{(5)(x)} = \sqrt{5}\cdot\sqrt{x}$$

$$\sqrt{7y} = \sqrt{(7)(y)} = \sqrt{\quad}\cdot\sqrt{\quad}$$

$\sqrt{7}\cdot\sqrt{y}$

19. If the coefficient is a perfect square, we can simplify after factoring. For example:

$\sqrt{9t} = \sqrt{9}\cdot\sqrt{t} = 3\sqrt{t}$ $\sqrt{49d} = \sqrt{49}\cdot\sqrt{d} =$ ________

$7\sqrt{d}$

20. If the coefficient is not a perfect square, we factor out a perfect square if possible. The remaining factors are left under a single radical. For example:

$$\sqrt{12t} = \sqrt{4}\cdot\sqrt{3t} = 2\sqrt{3t}$$

$$\sqrt{18x} = \sqrt{9}\cdot\sqrt{2x} = 3\sqrt{2x}$$

Simplify completely by factoring out the largest possible perfect square.

a) $\sqrt{8y} =$ ________ c) $\sqrt{27p} =$ ________

b) $\sqrt{32b} =$ ________ d) $\sqrt{24m} =$ ________

---

a) $2\sqrt{2y}$
b) $4\sqrt{2b}$
c) $3\sqrt{3p}$
d) $2\sqrt{6m}$

21. If a radical does not contain a perfect-square factor, it cannot be simplified. For example:

$\sqrt{14}$ cannot be simplified

$\sqrt{33x}$ cannot be simplified

Simplify each radical if possible.

a) $\sqrt{40} =$ ________ c) $\sqrt{21y} =$ ________

b) $\sqrt{55} =$ ________ d) $\sqrt{45q} =$ ________

a) $2\sqrt{10}$
b) Not possible
c) Not possible
d) $3\sqrt{5q}$

18-3 SQUARE ROOTS OF POWERS

In this section, we will discuss the square roots of powers.

---

22. To square a power, we simply multiply the power by itself. For example:

$(x^3)^2 = x^3 \cdot x^3 = x^6$ $(y^4)^2 = y^4 \cdot y^4 = y^8$

You can see that squaring a power is the same as doubling its exponent. That is:

$(x^3)^2 = x^{(3)(2)} = x^6$ $(y^4)^2 = y^{(4)(2)} = y^8$

By simply doubling the exponent, complete each squaring.

a) $(t^1)^2 =$ ________ b) $(p^5)^2 =$ ________ c) $(x^6)^2 =$ ________ d) $(y^{10})^2 =$ ________

a) t^2
b) p^{10}
c) x^{12}
d) y^{20}

23. Since we square a power by doubling its exponent, the square of any power has an <u>even</u> exponent. For example:

$(x^1)^2 = x^2$ $\quad (x^2)^2 = x^4$ $\quad (x^5)^2 = x^{10}$

Therefore, powers with <u>even</u> exponents are called "<u>perfect-square</u>" powers. Which of these are perfect-square powers? ____________

a) y^2 b) y^3 c) y^4 d) y^7 e) y^{10}

(a), (c), and (e)

24. We can find the square root of any perfect-square power. That is:

Since $x^2 \cdot x^2 = x^4$, $\sqrt{x^4} = x^2$

Since $y^3 \cdot y^3 = y^6$, $\sqrt{y^6} = y^3$

Since $t^5 \cdot t^5 = t^{10}$, $\sqrt{t^{10}} =$ ______

---

t^5

25. The three square roots from the last frame are shown below.

$\sqrt{x^4} = x^2$ $\quad \sqrt{y^6} = y^3$ $\quad \sqrt{t^{10}} = t^5$

You can see that finding the square root of a perfect-square power is the same as taking half of its exponent. That is:

$\sqrt{x^4} = x^{\frac{4}{2}} = x^2$ $\quad \sqrt{y^6} = y^{\frac{6}{2}} = y^3$ $\quad \sqrt{t^{10}} = t^{\frac{10}{2}} = t^5$

By taking half of the exponent, find each square root.

a) $\sqrt{m^2} =$ ________ b) $\sqrt{p^8} =$ ________ c) $\sqrt{b^{20}} =$ ________

---

a) m^1 or m
b) p^4
c) b^{10}

26. By reversing the law of exponents for multiplication, we can factor a non-perfect square power into a perfect-square power and a first power. That is:

$x^3 = x^2 \cdot x^1$ $\quad y^5 = y^4 \cdot y^1$ $\quad t^7 =$ ________

$t^6 \cdot t^1$

27. Using the factoring process from the last frame, we can simplify each radical below by factoring out the largest perfect square. That is:

$\sqrt{x^3} = \sqrt{x^2 \cdot x^1} = \sqrt{x^2} \cdot \sqrt{x^1} = x\sqrt{x}$

$\sqrt{y^5} = \sqrt{y^4 \cdot y^1} = \sqrt{y^4} \cdot \sqrt{y^1} = y^2\sqrt{y}$

$\sqrt{p^{11}} = \sqrt{p^{10} \cdot p^1} = \sqrt{p^{10}} \cdot \sqrt{p^1} =$ ________

$p^5\sqrt{p}$

28. Simplify these by either finding the square root or factoring out the largest perfect square.

a) $\sqrt{p^2} =$ ________ c) $\sqrt{h^8} =$ ________

b) $\sqrt{p^3} =$ ________ d) $\sqrt{h^9} =$ ________

a) p
b) $p\sqrt{p}$
c) h^4
d) $h^4\sqrt{h}$

18-4 SIMPLIFYING RADICALS CONTAINING POWERS

In this section, we will show how radicals containing a power can be simplified by factoring out perfect squares.

29. A radical containing a power can also be simplified by factoring out perfect squares. For example:

$$\sqrt{4x^2} = \sqrt{4}\cdot\sqrt{x^2} = 2x$$
$$\sqrt{9y^6} = \sqrt{9}\cdot\sqrt{y^6} = 3y^3$$

Following the examples, complete these simplifications.

a) $\sqrt{25d^4} = \sqrt{25}\cdot\sqrt{d^4} =$ ________ b) $\sqrt{100t^8} = \sqrt{100}\cdot\sqrt{t^8} =$ ________

a) $5d^2$
b) $10t^4$

30. Each radical below has been simplified by factoring out a perfect square.

$$\sqrt{5x^4} = \sqrt{5}\cdot\sqrt{x^4} = x^2\sqrt{5}$$
$$\sqrt{7t^{10}} = \sqrt{7}\cdot\sqrt{t^{10}} = t^5\sqrt{7}$$

Following the examples, complete these simplifications.

a) $\sqrt{3y^8} = \sqrt{3}\cdot\sqrt{y^8} =$ ________ b) $\sqrt{2d^2} = \sqrt{2}\cdot\sqrt{d^2} =$ ________

a) $y^4\sqrt{3}$
b) $d\sqrt{2}$

31. Each radical below has been simplified by factoring out perfect squares.

$$\sqrt{12y^6} = \sqrt{4}\cdot\sqrt{3}\cdot\sqrt{y^6} = 2y^3\sqrt{3}$$
$$\sqrt{50x^2} = \sqrt{25}\cdot\sqrt{2}\cdot\sqrt{x^2} = 5x\sqrt{2}$$

Following the examples, complete these simplifications.

a) $\sqrt{18t^4} = \sqrt{9}\cdot\sqrt{2}\cdot\sqrt{t^4} =$ ________ b) $\sqrt{24m^{10}} = \sqrt{4}\cdot\sqrt{6}\cdot\sqrt{m^{10}} =$ ________

a) $3t^2\sqrt{2}$
b) $2m^5\sqrt{6}$

32. Simplify by factoring out as many perfect squares as possible.

a) $\sqrt{49x^2} =$ ________ c) $\sqrt{28a^2} =$ ________

b) $\sqrt{11y^8} =$ ________ d) $\sqrt{32b^6} =$ ________

a) $7x$
b) $y^4\sqrt{11}$
c) $2a\sqrt{7}$
d) $4b^3\sqrt{2}$

33. We simplified each radical below by factoring out perfect squares.

$$\sqrt{16x^3} = \sqrt{16}\cdot\sqrt{x^2}\cdot\sqrt{x^1} = 4x\sqrt{x}$$
$$\sqrt{25y^7} = \sqrt{25}\cdot\sqrt{y^6}\cdot\sqrt{y^1} = 5y^3\sqrt{y}$$

Following the examples, complete these:

a) $\sqrt{9t^5} = \sqrt{9}\cdot\sqrt{t^4}\cdot\sqrt{t^1} =$ ________ b) $\sqrt{81m^9} = \sqrt{81}\cdot\sqrt{m^8}\cdot\sqrt{m^1} =$ ________

a) $3t^2\sqrt{t}$
b) $9m^4\sqrt{m}$

34. We simplified below by factoring out a perfect square.

$$\sqrt{3x^5} = \sqrt{x^4}\cdot\sqrt{3x^1} = x^2\sqrt{3x}$$
$$\sqrt{5y^7} = \sqrt{y^6}\cdot\sqrt{5y^1} = y^3\sqrt{5y}$$

Following the examples, complete these:

a) $\sqrt{7t^3} = \sqrt{t^2}\cdot\sqrt{7t^1} =$ ________ b) $\sqrt{13d^9} = \sqrt{d^8}\cdot\sqrt{13d^1} =$ ________

Answers: a) $t\sqrt{7t}$ b) $d^4\sqrt{13d}$

35. We simplified below by factoring out perfect squares.

$$\sqrt{8x^3} = \sqrt{4}\cdot\sqrt{x^2}\cdot\sqrt{2x^1} = 2x\sqrt{2x}$$
$$\sqrt{45y^7} = \sqrt{9}\cdot\sqrt{y^6}\cdot\sqrt{5y^1} = 3y^3\sqrt{5y}$$

Following the examples, complete these:

a) $\sqrt{27p^5} = \sqrt{9}\cdot\sqrt{p^4}\cdot\sqrt{3p^1} =$ ________

b) $\sqrt{50h^9} = \sqrt{25}\cdot\sqrt{h^8}\cdot\sqrt{2h^1} =$ ________

Answers: a) $3p^2\sqrt{3p}$ b) $5h^4\sqrt{2h}$

36. Simplify by factoring out as many perfect squares as possible.

a) $\sqrt{64t^3} =$ ________ c) $\sqrt{40x^5} =$ ________

b) $\sqrt{14d^9} =$ ________ d) $\sqrt{75y^7} =$ ________

Answers: a) $8t\sqrt{t}$ b) $d^4\sqrt{14d}$ c) $2x^2\sqrt{10x}$ d) $5y^3\sqrt{3y}$

18-5 SIMPLIFYING RADICAL PRODUCTS

After multiplying radicals, we can sometimes simplify the product by factoring out perfect squares. We will discuss that procedure in this section.

---

37. When multiplying radicals, we simplify the product (if possible) by factoring out any perfect square. For example:

$\sqrt{2}\cdot\sqrt{6} = \sqrt{12} = 2\sqrt{3}$ $\sqrt{5}\cdot\sqrt{10} = \sqrt{50} =$ ________

Answer: $5\sqrt{2}$

38. Following the example, simplify the other product.

$\sqrt{3}\cdot\sqrt{6x} = \sqrt{18x} = 3\sqrt{2x}$ $\sqrt{10x}\cdot\sqrt{2} = \sqrt{20x} =$ ________

Answer: $2\sqrt{5x}$

39. Following the example, simplify the other product.

$\sqrt{2x}\cdot\sqrt{3x} = \sqrt{6x^2} = x\sqrt{6}$ $\sqrt{5y^2}\cdot\sqrt{6y^4} = \sqrt{30y^6} =$ ________

Answer: $y^3\sqrt{30}$

40. Following the example, simplify the other product.

$\sqrt{2t}\cdot\sqrt{12t} = \sqrt{24t^2} = 2t\sqrt{6}$ $\sqrt{5x}\cdot\sqrt{20x^3} = \sqrt{100x^4} =$ ______

$10x^2$

41. Following the example, simplify the other product.

$\sqrt{3x}\cdot\sqrt{6x^2} = \sqrt{18x^3} = 3x\sqrt{2x}$ $\sqrt{2t^3}\cdot\sqrt{8t^4} = \sqrt{16t^7} =$ ______

$4t^3\sqrt{t}$

42. When multiplying radicals, it is not always possible to simplify the product. Some examples are shown.

$\sqrt{2}\cdot\sqrt{3} = \sqrt{6}$ $\sqrt{3}\cdot\sqrt{7x} = \sqrt{21x}$

Perform each multiplication and simplify each product if possible.

a) $\sqrt{2}\cdot\sqrt{32} =$ ______ c) $\sqrt{3y}\cdot\sqrt{5y} =$ ______

b) $\sqrt{5x}\cdot\sqrt{2} =$ ______ d) $\sqrt{4d^3}\cdot\sqrt{8d^2} =$ ______

a) 8 b) $\sqrt{10x}$ c) $y\sqrt{15}$ d) $4d^2\sqrt{2d}$

SELF-TEST 53 (pages 459-467)

Perform each multiplication.

1. $\sqrt{7}\cdot\sqrt{11} =$ ______
2. $\sqrt{x}\cdot\sqrt{3} =$ ______
3. $\sqrt{7}\cdot\sqrt{3t} =$ ______
4. $\sqrt{2}\cdot\sqrt{3}\cdot\sqrt{5y} =$ ______
5. $6\cdot 2\sqrt{11} =$ ______
6. $5\sqrt{2}\cdot 3\sqrt{5} =$ ______

Simplify by factoring out perfect squares.

7. $\sqrt{44} =$ ______
8. $\sqrt{36x} =$ ______
9. $\sqrt{20y} =$ ______
10. $\sqrt{3t^2} =$ ______
11. $\sqrt{9p^3} =$ ______
12. $\sqrt{18m^7} =$ ______

Multiply and simplify each product.

13. $\sqrt{8}\cdot\sqrt{3} =$ ______
14. $\sqrt{2x}\cdot\sqrt{5x} =$ ______
15. $\sqrt{3y^2}\cdot\sqrt{12y^4} =$ ______
16. $\sqrt{t}\cdot\sqrt{8t^2} =$ ______

ANSWERS:

1. $\sqrt{77}$
2. $\sqrt{3x}$
3. $\sqrt{21t}$
4. $\sqrt{30y}$
5. $12\sqrt{11}$
6. $15\sqrt{10}$
7. $2\sqrt{11}$
8. $6\sqrt{x}$
9. $2\sqrt{5y}$
10. $t\sqrt{3}$
11. $3p\sqrt{p}$
12. $3m^3\sqrt{2m}$
13. $2\sqrt{6}$
14. $x\sqrt{10}$
15. $6y^3$
16. $2t\sqrt{2t}$

18-6 DIVIDING RADICALS

In this section, we will discuss the procedure for dividing square-root radicals. The divisions are limited to those with non-fractional radical quotients.

---

43. To indicate a division of radicals, we use a fraction. For example:

$$\frac{\sqrt{50}}{\sqrt{2}} \text{ means "divide } \sqrt{50} \text{ by } \sqrt{2}\text{".}$$

To divide radicals, we simply divide their radicands and then simplify. That is:

$$\frac{\sqrt{50}}{\sqrt{2}} = \sqrt{\frac{50}{2}} = \sqrt{25} = 5 \qquad \frac{\sqrt{27}}{\sqrt{3}} = \sqrt{\frac{27}{3}} = \sqrt{9} = \_\_\_\_$$

Answer: 3

44. To show that the division procedure makes sense, two examples involving perfect-square radicands are shown below.

$$\frac{\sqrt{36}}{\sqrt{9}} = \sqrt{\frac{36}{9}} \qquad \frac{\sqrt{64}}{\sqrt{4}} = \sqrt{\frac{64}{4}}$$

$$\frac{6}{3} = \sqrt{4} \qquad \frac{8}{2} = \sqrt{16}$$

$$2 = 2 \qquad 4 = 4$$

By simply dividing the radicands and simplifying, complete these:

a) $\frac{\sqrt{12}}{\sqrt{3}} = \sqrt{\frac{12}{3}} = \sqrt{\quad} = \_\_\_\_$ b) $\frac{\sqrt{144}}{\sqrt{4}} = \sqrt{\frac{144}{4}} = \sqrt{\quad} = \_\_\_\_$

Answers: a) $\sqrt{4} = 2$ b) $\sqrt{36} = 6$

45. When the radicand of the quotient is not a perfect square, we simplify it as much as possible by factoring out perfect squares. For example:

$$\frac{\sqrt{25}}{\sqrt{5}} = \sqrt{\frac{25}{5}} = \sqrt{5} \qquad \frac{\sqrt{36}}{\sqrt{3}} = \sqrt{\frac{36}{3}} = \sqrt{12} = 2\sqrt{3}$$

Simplify each radical quotient as much as possible.

a) $\frac{\sqrt{20}}{\sqrt{2}} = \sqrt{\frac{20}{2}} = \_\_\_\_$ b) $\frac{\sqrt{150}}{\sqrt{3}} = \sqrt{\frac{150}{3}} = \_\_\_\_$

Answers: a) $\sqrt{10}$ b) $5\sqrt{2}$, from $\sqrt{50}$

46. The same procedure is used for divisions of radicals containing letters. For example:

$$\frac{\sqrt{8x}}{\sqrt{2x}} = \sqrt{\frac{8x}{2x}} = \sqrt{4} = 2 \qquad \frac{\sqrt{54y^3}}{\sqrt{6y^3}} = \sqrt{\frac{54y^3}{6y^3}} = \sqrt{\quad} = \_\_\_\_$$

Answer: $\sqrt{9} = 3$

47. Following the example, complete the other division.

$$\frac{\sqrt{x^8}}{\sqrt{x^2}} = \sqrt{\frac{x^8}{x^2}} = \sqrt{x^6} = x^3 \qquad \frac{\sqrt{y^5}}{\sqrt{y}} = \sqrt{\frac{y^5}{y}} = \sqrt{\quad} = \_\_\_\_$$

Answer: $\sqrt{y^4} = y^2$

48. Following the example, complete the other division.

$$\frac{\sqrt{18x^4}}{\sqrt{2x^2}} = \sqrt{\frac{18x^4}{2x^2}} = \sqrt{9x^2} = 3x \qquad \frac{\sqrt{75y^7}}{\sqrt{3y^3}} = \sqrt{\frac{75y^7}{3y^3}} = \sqrt{\qquad} = \underline{\qquad}$$

$\sqrt{25y^4} = 5y^2$

---

49. When the radicand of the quotient is not a perfect square, we simplify it as much as possible. For example:

$$\frac{\sqrt{24x^2}}{\sqrt{6x}} = \sqrt{\frac{24x^2}{6x}} = \sqrt{4x} = 2\sqrt{x}$$

$$\frac{\sqrt{36y}}{\sqrt{2}} = \sqrt{\frac{36y}{2}} = \sqrt{18y} = 3\sqrt{2y}$$

Simplify each quotient as much as possible.

a) $\frac{\sqrt{16t^5}}{\sqrt{t^4}} = \sqrt{\frac{16t^5}{t^4}} = \underline{\qquad}$ b) $\frac{\sqrt{64p}}{\sqrt{2}} = \sqrt{\frac{64p}{2}} = \underline{\qquad}$

a) $4\sqrt{t}$, from $\sqrt{16t}$

b) $4\sqrt{2p}$, from $\sqrt{32p}$

---

50. Two more examples of simplifying radical quotients are shown below.

$$\frac{\sqrt{6x^6}}{\sqrt{2x^2}} = \sqrt{\frac{6x^6}{2x^2}} = \sqrt{3x^4} = x^2\sqrt{3}$$

$$\frac{\sqrt{40y^4}}{\sqrt{2y}} = \sqrt{\frac{40y^4}{2y}} = \sqrt{20y^3} = 2y\sqrt{5y}$$

Simplify each quotient as much as possible.

a) $\frac{\sqrt{48t^3}}{\sqrt{2t}} = \underline{\qquad}$ b) $\frac{\sqrt{15p^7}}{\sqrt{3p^2}} = \underline{\qquad}$

a) $2t\sqrt{6}$, from $\sqrt{24t^2}$ b) $p^2\sqrt{5p}$, from $\sqrt{5p^5}$

18-7 SIMPLIFYING RADICALS CONTAINING FRACTIONS

In this section, we will discuss the procedure for simplifying radicals containing fractions. The fractions are limited to those with perfect-square denominators.

---

51. To simplify a radical containing a fraction, we can reverse the procedure for dividing radicals. That is, we can convert the radical to a division of two radicals and then simplify both terms. For example:

$$\sqrt{\frac{9}{25}} = \frac{\sqrt{9}}{\sqrt{25}} = \frac{3}{5}$$

a) $\sqrt{\frac{81}{49}} = \frac{\sqrt{81}}{\sqrt{49}} = \underline{\qquad}$ b) $\sqrt{\frac{1}{4}} = \frac{\sqrt{1}}{\sqrt{4}} = \underline{\qquad}$

a) $\frac{9}{7}$

b) $\frac{1}{2}$

52. When the numerator of the radicand is not a perfect square, we simplify as much as possible. For example:

$$\sqrt{\frac{3}{4}} = \frac{\sqrt{3}}{\sqrt{4}} = \frac{\sqrt{3}}{2} \qquad \sqrt{\frac{8}{9}} = \frac{\sqrt{8}}{\sqrt{9}} = \frac{2\sqrt{2}}{3}$$

Following the examples, simplify these:

a) $\sqrt{\frac{17}{25}} = \frac{\sqrt{17}}{\sqrt{25}} =$ ______ b) $\sqrt{\frac{12}{49}} = \frac{\sqrt{12}}{\sqrt{49}} =$ ______

Answers: a) $\frac{\sqrt{17}}{5}$ b) $\frac{2\sqrt{3}}{7}$

53. When a fractional radicand contains a letter, we use the same procedure to simplify. For example:

$$\sqrt{\frac{2x}{25}} = \frac{\sqrt{2x}}{\sqrt{25}} = \frac{\sqrt{2x}}{5} \qquad \sqrt{\frac{9t}{16}} = \frac{\sqrt{9t}}{\sqrt{16}} = \frac{3\sqrt{t}}{4}$$

Following the examples, simplify these:

a) $\sqrt{\frac{y}{36}} = \frac{\sqrt{y}}{\sqrt{36}} =$ ______ b) $\sqrt{\frac{20m}{49}} = \frac{\sqrt{20m}}{\sqrt{49}} =$ ______

Answers: a) $\frac{\sqrt{y}}{6}$ b) $\frac{2\sqrt{5m}}{7}$

54. When a fractional radicand contains a power, we use the same procedure to simplify. For example:

$$\sqrt{\frac{x^8}{16}} = \frac{\sqrt{x^8}}{\sqrt{16}} = \frac{x^4}{4} \qquad \sqrt{\frac{9}{25y^2}} = \frac{\sqrt{9}}{\sqrt{25y^2}} = \frac{3}{5y}$$

Following the examples, simplify these:

a) $\sqrt{\frac{4m^4}{9}} =$ ______ b) $\sqrt{\frac{81}{t^6}} =$ ______

Answers: a) $\frac{2m^2}{3}$ b) $\frac{9}{t^3}$

55. Following the example, complete the other simplification.

$$\sqrt{\frac{18}{x^2}} = \frac{\sqrt{18}}{\sqrt{x^2}} = \frac{3\sqrt{2}}{x} \qquad \sqrt{\frac{32}{9y^4}} = \text{______}$$

Answer: $\frac{4\sqrt{2}}{3y^2}$

56. Two more examples of simplifying radicals are shown below.

$$\sqrt{\frac{32x^2}{25}} = \frac{\sqrt{32x^2}}{\sqrt{25}} = \frac{4x\sqrt{2}}{5} \qquad \sqrt{\frac{8y^5}{49}} = \frac{\sqrt{8y^5}}{\sqrt{49}} = \frac{2y^2\sqrt{2y}}{7}$$

Following the examples, simplify these:

a) $\sqrt{\frac{28m^4}{81}} =$ ______ b) $\sqrt{\frac{9p^3}{64}} =$ ______

Answers: a) $\frac{2m^2\sqrt{7}}{9}$ b) $\frac{3p\sqrt{p}}{8}$

18-8 RATIONALIZING DENOMINATORS

When a fraction contains a radical in its denominator, it is usually converted to an equivalent form without a radical in the denominator. For example:

$$\frac{3}{\sqrt{7}} \text{ is converted to } \frac{3\sqrt{7}}{7} \qquad \frac{2}{5\sqrt{x}} \text{ is converted to } \frac{2\sqrt{x}}{5x}$$

The conversion process above is called "rationalizing the denominator". We will discuss the "rationalizing" process in this section.

---

57. When the numerator and denominator of a fraction are identical radicals, the fraction equals "1". For example:

$$\frac{\sqrt{7}}{\sqrt{7}} = \sqrt{\frac{7}{7}} = \sqrt{1} = 1 \qquad \frac{\sqrt{2x}}{\sqrt{2x}} = \sqrt{\frac{2x}{2x}} = \sqrt{\quad} = \_\_\_\_$$

$\sqrt{1} = 1$

58. To obtain equivalent forms of fractions, we multiply by a fraction that equals "1". For example:

$$\frac{2}{3} = \frac{2}{3}(1) = \frac{2}{3}\left(\frac{5}{5}\right) = \frac{10}{15} \qquad \frac{x}{5} = \frac{x}{5}(1) = \frac{x}{5}\left(\frac{2}{2}\right) = \frac{2x}{10}$$

When multiplying by a fraction that equals "1", we can multiply by a fraction whose terms are the same radical. That is:

$$\frac{2}{3} = \frac{2}{3}(1) = \frac{2}{3}\left(\frac{\sqrt{5}}{\sqrt{5}}\right) = \frac{2\sqrt{5}}{3\sqrt{5}} \qquad \frac{x}{5} = \frac{x}{5}(1) = \frac{x}{5}\left(\frac{\sqrt{2}}{\sqrt{2}}\right) = \_\_\_\_\_\_$$

$\frac{x\sqrt{2}}{5\sqrt{2}}$

59. The denominator of a fraction is "rationalized" if it does not contain a radical. For example:

$$\frac{\sqrt{2}}{3} \qquad \frac{3\sqrt{5}}{7} \qquad \frac{\sqrt{x}}{11} \qquad \frac{2y\sqrt{2y}}{5}$$

The denominator of a fraction is "not rationalized" if it contains a radical. For example:

$$\frac{2}{\sqrt{3}} \qquad \frac{\sqrt{3}}{2\sqrt{7}} \qquad \frac{\sqrt{x}}{\sqrt{11}} \qquad \frac{1}{3\sqrt{5y}}$$

Which of the fractions below have denominators that are rationalized? ________

a) $\frac{\sqrt{5}}{7}$ b) $\frac{5}{\sqrt{7}}$ c) $\frac{3\sqrt{x}}{4}$ d) $\frac{\sqrt{3}}{4\sqrt{x}}$

(a) and (c)

60. In the process of rationalizing a denominator, we must square a radical. Remember that the square of a radical is the radicand. For example:

$$\left(\sqrt{5}\right)^2 = \sqrt{5}\cdot\sqrt{5} = \sqrt{25} = 5$$

$$\left(\sqrt{3y}\right)^2 = \sqrt{3y}\cdot\sqrt{3y} = \sqrt{9y^2} =$$

Following the examples, complete these:

$\left(\sqrt{7}\right)^2 = 7$ $\left(\sqrt{x}\right)^2 = x$ a) $\left(\sqrt{13}\right)^2 = \_\_\_\_$ b) $\left(\sqrt{2m}\right)^2 = \_\_\_\_$

---

a) 13 b) 2m

61. When the denominator of a fraction is "not rationalized", we can rationalize it by multiplying by a fraction whose terms are the same as the radical in the denominator. For example:

$$\frac{5}{\sqrt{7}} = \frac{5}{\sqrt{7}}\left(\frac{\sqrt{7}}{\sqrt{7}}\right) = \frac{5\sqrt{7}}{(\sqrt{7})^2} = \frac{5\sqrt{7}}{7}$$

$$\frac{1}{\sqrt{2y}} = \frac{1}{\sqrt{2y}}\left(\frac{\sqrt{2y}}{\sqrt{2y}}\right) = \frac{\sqrt{2y}}{(\sqrt{2y})^2} = \frac{\sqrt{2y}}{2y}$$

Notice that we got rid of the radical in each denominator by squaring it. Rationalize each denominator below.

a) $\frac{1}{\sqrt{15}} =$ ____________________

b) $\frac{3}{\sqrt{x}} =$ ____________________

---

62. Following the examples, rationalize the other denominators.

Answers: a) $\frac{\sqrt{15}}{15}$ b) $\frac{3\sqrt{x}}{x}$

$$\frac{3}{2\sqrt{5}} = \frac{3}{2\sqrt{5}}\left(\frac{\sqrt{5}}{\sqrt{5}}\right) = \frac{3\sqrt{5}}{2(\sqrt{5})^2} = \frac{3\sqrt{5}}{2(5)} = \frac{3\sqrt{5}}{10}$$

$$\frac{1}{7\sqrt{x}} = \frac{1}{7\sqrt{x}}\left(\frac{\sqrt{x}}{\sqrt{x}}\right) = \frac{\sqrt{x}}{7(\sqrt{x})^2} = \frac{\sqrt{x}}{7x}$$

a) $\frac{1}{5\sqrt{7}} =$ ____________________

b) $\frac{2}{3\sqrt{y}} =$ ____________________

---

63. Following the example, rationalize the other denominator.

Answers: a) $\frac{\sqrt{7}}{35}$ b) $\frac{2\sqrt{y}}{3y}$

$$\frac{5}{3x\sqrt{2x}} = \frac{5}{3x\sqrt{2x}}\left(\frac{\sqrt{2x}}{\sqrt{2x}}\right) = \frac{5\sqrt{2x}}{3x(\sqrt{2x})^2} = \frac{5\sqrt{2x}}{(3x)(2x)} = \frac{5\sqrt{2x}}{6x^2}$$

$\frac{1}{y\sqrt{5y}} =$ ____________________

---

64. Following the examples, rationalize the other denominators.

Answer: $\frac{\sqrt{5y}}{5y^2}$

$$\frac{2\sqrt{3}}{\sqrt{5}} = \frac{2\sqrt{3}}{\sqrt{5}}\left(\frac{\sqrt{5}}{\sqrt{5}}\right) = \frac{2\sqrt{(3)(5)}}{(\sqrt{5})^2} = \frac{2\sqrt{15}}{5}$$

$$\frac{\sqrt{7}}{\sqrt{3x}} = \frac{\sqrt{7}}{\sqrt{3x}}\left(\frac{\sqrt{3x}}{\sqrt{3x}}\right) = \frac{\sqrt{(7)(3x)}}{(\sqrt{3x})^2} = \frac{\sqrt{21x}}{3x}$$

a) $\frac{\sqrt{11}}{\sqrt{7}} =$ ____________________

b) $\frac{3\sqrt{5}}{\sqrt{y}} =$ ____________________

Answers: a) $\frac{\sqrt{77}}{7}$ b) $\frac{3\sqrt{5y}}{y}$

65. Following the example, rationalize the other denominator.

$$\frac{3\sqrt{7}}{2\sqrt{x}} = \frac{3\sqrt{7}}{2\sqrt{x}}\left(\frac{\sqrt{x}}{\sqrt{x}}\right) = \frac{3\sqrt{7x}}{2(\sqrt{x})^2} = \frac{3\sqrt{7x}}{2x}$$

$$\frac{\sqrt{5}}{y\sqrt{7y}} = \_\_\_\_\_\_\_\_\_\_\_\_\_\_\_\_$$

$\frac{\sqrt{35y}}{7y^2}$

---

66. Following the example, rationalize the other denominators.

$$\frac{2\sqrt{5}}{x^2\sqrt{2x}} = \frac{2\sqrt{5}}{x^2\sqrt{2x}}\left(\frac{\sqrt{2x}}{\sqrt{2x}}\right) = \frac{2\sqrt{(5)(2x)}}{x^2(\sqrt{2x})^2} = \frac{2\sqrt{10x}}{x^2(2x)} = \frac{2\sqrt{10x}}{2x^3}$$

a) $\frac{3}{y^4\sqrt{5y}} = \_\_\_\_\_\_\_\_\_\_\_\_\_\_\_\_$

b) $\frac{\sqrt{7}}{2t^3\sqrt{t}} = \_\_\_\_\_\_\_\_\_\_\_\_\_\_\_\_$

a) $\frac{3\sqrt{5y}}{5y^5}$

b) $\frac{\sqrt{7t}}{2t^4}$

---

67. When a radical containing a fraction has been simplified by factoring out perfect squares, we sometimes obtain a denominator containing a radical. Denominators of that type are usually rationalized. For example:

$$\sqrt{\frac{7}{20}} = \frac{\sqrt{7}}{\sqrt{20}} = \frac{\sqrt{7}}{2\sqrt{5}} = \frac{\sqrt{7}}{2\sqrt{5}}\left(\frac{\sqrt{5}}{\sqrt{5}}\right) = \frac{\sqrt{35}}{10}$$

$$\sqrt{\frac{12}{x^3}} = \frac{\sqrt{12}}{\sqrt{x^3}} = \frac{2\sqrt{3}}{x\sqrt{x}} = \frac{2\sqrt{3}}{x\sqrt{x}}\left(\frac{\sqrt{x}}{\sqrt{x}}\right) = \frac{2\sqrt{3x}}{x^2}$$

Factor out perfect squares and then rationalize the denominator of the resulting fraction.

a) $\sqrt{\frac{8}{27}} = \_\_\_\_\_\_\_\_\_\_\_\_\_\_\_\_$

b) $\sqrt{\frac{3}{4y^5}} = \_\_\_\_\_\_\_\_\_\_\_\_\_\_\_\_$

a) $\frac{2\sqrt{6}}{9}$

b) $\frac{\sqrt{3y}}{2y^3}$

---

68. Even when we cannot factor out perfect squares from a radical containing a fraction, we can still rationalize its denominator. For example:

$$\sqrt{\frac{5}{7}} = \frac{\sqrt{5}}{\sqrt{7}} = \frac{\sqrt{5}}{\sqrt{7}}\left(\frac{\sqrt{7}}{\sqrt{7}}\right) = \frac{\sqrt{35}}{7}$$

$$\sqrt{\frac{3}{2x}} = \frac{\sqrt{3}}{\sqrt{2x}} = \frac{\sqrt{3}}{\sqrt{2x}}\left(\frac{\sqrt{2x}}{\sqrt{2x}}\right) = \frac{\sqrt{6x}}{2x}$$

Following the examples, rationalize each denominator below.

a) $\sqrt{\frac{3}{5}} = \_\_\_\_\_\_\_\_\_\_\_\_\_\_\_\_$

b) $\sqrt{\frac{7}{y}} = \_\_\_\_\_\_\_\_\_\_\_\_\_\_\_\_$

a) $\frac{\sqrt{15}}{5}$

b) $\frac{\sqrt{7y}}{y}$

18-9 SIMPLIFYING PRODUCTS WITH FRACTIONAL RADICANDS

In this section, we will multiply radicals with fractional radicands and show various ways in which the products can be simplified.

69. To multiply radicals with fractional radicands, we simply multiply the radicands. That is:

$$\sqrt{\frac{1}{2}}\cdot\sqrt{\frac{3}{4}} = \sqrt{\left(\frac{1}{2}\right)\left(\frac{3}{4}\right)} = \sqrt{\frac{3}{8}} \qquad \sqrt{\frac{2}{3}}\cdot\sqrt{\frac{x}{5}} = \sqrt{\left(\frac{2}{3}\right)\left(\frac{x}{5}\right)} = \underline{\qquad}$$

70. After multiplying, we simplify the product as much as possible. For example:

$$\sqrt{\frac{2x}{15}}$$

$$\sqrt{\frac{5}{2}}\cdot\sqrt{\frac{4}{5}} = \sqrt{\frac{20}{10}} = \sqrt{2} \qquad \sqrt{18}\cdot\sqrt{\frac{1}{2}} = \sqrt{9} = \underline{\qquad}$$

71. Following the example, simplify the other product by factoring out perfect squares.

3

$$\sqrt{\frac{4}{3}}\cdot\sqrt{\frac{x}{3}} = \sqrt{\frac{4x}{9}} = \frac{2\sqrt{x}}{3} \qquad \sqrt{\frac{3}{y}}\cdot\sqrt{\frac{6}{y^3}} = \sqrt{\frac{18}{y^4}} = \underline{\qquad}$$

72. If we get a radical denominator when factoring out perfect squares, the denominator is rationalized. For example:

$$\frac{3\sqrt{2}}{y^2}$$

$$\sqrt{\frac{1}{2}}\cdot\sqrt{\frac{3}{4}} = \sqrt{\frac{3}{8}} = \frac{\sqrt{3}}{2\sqrt{2}} = \frac{\sqrt{3}}{2\sqrt{2}}\left(\frac{\sqrt{2}}{\sqrt{2}}\right) = \frac{\sqrt{6}}{2(2)} = \frac{\sqrt{6}}{4}$$

$$\sqrt{\frac{2}{3}}\cdot\sqrt{\frac{10}{x}} = \sqrt{\frac{20}{3x}} = \frac{2\sqrt{5}}{\sqrt{3x}} = \frac{2\sqrt{5}}{\sqrt{3x}}\left(\frac{\sqrt{3x}}{\sqrt{3x}}\right) = \frac{2\sqrt{15x}}{3x}$$

Following the examples, simplify each product below.

a) $\sqrt{\frac{2}{5}}\cdot\sqrt{\frac{2x}{3}} = \underline{\qquad}$

b) $\sqrt{\frac{7}{8}}\cdot\sqrt{\frac{1}{y}} = \underline{\qquad}$

73. Even when we cannot factor out perfect squares from the product, we rationalize the denominator. For example:

a) $\frac{2\sqrt{15x}}{15}$

b) $\frac{\sqrt{14y}}{4y}$

$$\sqrt{\frac{1}{2}}\cdot\sqrt{\frac{3}{5}} = \sqrt{\frac{3}{10}} = \frac{\sqrt{3}}{\sqrt{10}} = \frac{\sqrt{3}}{\sqrt{10}}\left(\frac{\sqrt{10}}{\sqrt{10}}\right) = \frac{\sqrt{30}}{10}$$

$$\sqrt{\frac{1}{3}}\cdot\sqrt{\frac{5}{x}} = \sqrt{\frac{5}{3x}} = \frac{\sqrt{5}}{\sqrt{3x}} = \underline{\qquad}$$

$$\frac{\sqrt{15x}}{3x}$$

18-10 ADDING AND SUBTRACTING RADICALS

In this section, we will show how the distributive principle can be used to add and subtract like terms containing radicals.

74. Two radical terms are called "like" terms if they contain the same radical. For example:

$2\sqrt{5}$ and $7\sqrt{5}$ are like terms because both contain $\sqrt{5}$.

$3\sqrt{x}$ and $4\sqrt{x}$ are like terms because both contain ________.

$\sqrt{x}$

75. We use the distributive principle to add like radical terms. For example:

$$2\sqrt{5} + 7\sqrt{5} = (2 + 7)\sqrt{5} = 9\sqrt{5}$$
$$3\sqrt{x} + 4\sqrt{x} = (3 + 4)\sqrt{x} = 7\sqrt{x}$$

Notice that using the distributive principle is the same as adding the coefficients of the radicals. Using that method, do these:

a) $5\sqrt{y} + 6\sqrt{y}$ = ________ b) $8\sqrt{2} + 2\sqrt{2}$ = ________

---

a) $11\sqrt{y}$
b) $10\sqrt{2}$

76. If the coefficient of a radical is not explicitly shown, its coefficient is "1". For example:

$\sqrt{7} = 1\sqrt{7}$ $\sqrt{x} = 1\sqrt{x}$

Therefore: a) $\sqrt{7} + 3\sqrt{7} = 1\sqrt{7} + 3\sqrt{7}$ = ________

b) $5\sqrt{x} + \sqrt{x} = 5\sqrt{x} + 1\sqrt{x}$ = ________

c) $\sqrt{3} + \sqrt{3} = 1\sqrt{3} + 1\sqrt{3}$ = ________

---

a) $4\sqrt{7}$
b) $6\sqrt{x}$
c) $2\sqrt{3}$

77. Complete each addition:

a) $9\sqrt{t} + \sqrt{t}$ = ________ c) $\sqrt{15} + 6\sqrt{15}$ = ________

b) $\sqrt{y} + \sqrt{y}$ = ________ d) $\sqrt{23} + \sqrt{23}$ = ________

---

a) $10\sqrt{t}$
b) $2\sqrt{y}$
c) $7\sqrt{15}$
d) $2\sqrt{23}$

78. We use the distributive principle to subtract like radical terms. For example:

$$10\sqrt{x} - 7\sqrt{x} = (10 - 7)\sqrt{x} = 3\sqrt{x}$$
$$2\sqrt{5} - 6\sqrt{5} = (2 - 6)\sqrt{5} = -4\sqrt{5}$$

Notice that using the distributive principle is the same as subtracting the coefficients of the radicals. Using that method, do these:

a) $15\sqrt{y} - 5\sqrt{y}$ = ________ b) $4\sqrt{3} - 9\sqrt{3}$ = ________

---

a) $10\sqrt{y}$
b) $-5\sqrt{3}$

79. To perform the subtractions below, it is helpful to write the "1" coefficients explicitly. That is:

a) $4\sqrt{7} - \sqrt{7} = 4\sqrt{7} - 1\sqrt{7} =$ ________

b) $\sqrt{x} - 3\sqrt{x} = 1\sqrt{x} - 3\sqrt{x} =$ ________

c) $\sqrt{2} - \sqrt{2} = 1\sqrt{2} - 1\sqrt{2} =$ ________

---

a) $3\sqrt{7}$
b) $-2\sqrt{x}$
c) $0\sqrt{2}$ or 0

80. Two radical terms cannot be combined into one term by addition or subtraction if they do not contain the same radical. For example:

$4\sqrt{3} + 5\sqrt{7}$ cannot be combined into one term.

$8\sqrt{x} + 3\sqrt{y}$ cannot be combined into one term.

Perform these additions and subtractions if possible.

a) $\sqrt{y} + 2\sqrt{y} =$ ________

b) $\sqrt{3} + 5\sqrt{2} =$ ________

c) $2\sqrt{7} - \sqrt{7} =$ ________

d) $4\sqrt{t} - \sqrt{5} =$ ________

---

a) $3\sqrt{y}$
b) Not possible
c) $1\sqrt{7}$ or $\sqrt{7}$
d) Not possible

81. Each expression below contains one radical term and one non-radical term.

$3 + \sqrt{2}$ $\quad$ $5 - \sqrt{x}$

Since the terms are "unlike", they cannot be combined into one. That is:

$3 + \sqrt{2}$ does not equal $3\sqrt{2}$

$5 - \sqrt{x}$ does not equal $-5\sqrt{x}$

Perform the following additions and subtractions if possible.

a) $5 + \sqrt{5} =$ ________

b) $\sqrt{5} + \sqrt{5} =$ ________

c) $y - \sqrt{y} =$ ________

d) $\sqrt{y} - \sqrt{y} =$ ________

---

a) Not possible
b) $2\sqrt{5}$
c) Not possible
d) $0\sqrt{y}$ or 0

82. To multiply radicals, we simply multiply their radicands. That is:

$$\sqrt{7} \cdot \sqrt{3} = \sqrt{(7)(3)} = \sqrt{21}$$

To add or subtract radicals, however, we do not add or subtract their radicands. That is:

$\sqrt{7} + \sqrt{3}$ does not equal $\sqrt{7 + 3}$

$\sqrt{7} - \sqrt{3}$ does not equal $\sqrt{7 - 3}$

Which of the following are true? ________

a) $\sqrt{2} + \sqrt{5} = \sqrt{2 + 5}$

b) $\sqrt{2} \cdot \sqrt{5} = \sqrt{(2)(5)}$

c) $\sqrt{8} - \sqrt{3} = \sqrt{8 - 3}$

d) $\sqrt{8} \cdot \sqrt{3} = \sqrt{(8)(3)}$

Only (b) and (d)

SELF-TEST 54 (pages 468-477)

Divide and simplify each quotient.

1. $\dfrac{\sqrt{40}}{\sqrt{5}}$ = ________	2. $\dfrac{\sqrt{18x^6}}{\sqrt{2x^2}}$ = ________	3. $\dfrac{\sqrt{24t^4}}{\sqrt{2t}}$ = ________

Simplify by factoring out perfect squares.

4. $\sqrt{\dfrac{7}{9}}$ = ________	5. $\sqrt{\dfrac{20}{y^4}}$ = ________	6. $\sqrt{\dfrac{16p^3}{25}}$ = ________

Rationalize each denominator.

7. $\dfrac{1}{\sqrt{3}}$ = ________	8. $\dfrac{\sqrt{7}}{x^2\sqrt{x}}$ = ________	9. $\dfrac{3\sqrt{5}}{m\sqrt{2m}}$ = ________

Multiply and simplify each product. Rationalize all denominators.

10. $\sqrt{\dfrac{1}{5}} \cdot \sqrt{\dfrac{3}{5}}$ = ________	11. $\sqrt{\dfrac{1}{2}} \cdot \sqrt{\dfrac{7}{x}}$ = ________

Perform each addition and subtraction if possible.

12. $2\sqrt{2} + 3\sqrt{2}$ = ________	14. $\sqrt{p} + \sqrt{p}$ = ________	16. $\sqrt{y} - 2\sqrt{y}$ = ________
13. $5\sqrt{3} - \sqrt{3}$ = ________	15. $7 + \sqrt{3}$ = ________	17. $d - \sqrt{d}$ = ________

ANSWERS:

1. $2\sqrt{2}$
2. $3x^2$
3. $2t\sqrt{3t}$
4. $\dfrac{\sqrt{7}}{3}$
5. $\dfrac{2\sqrt{5}}{y^2}$
6. $\dfrac{4p\sqrt{p}}{5}$
7. $\dfrac{\sqrt{3}}{3}$
8. $\dfrac{\sqrt{7x}}{x^3}$
9. $\dfrac{3\sqrt{10m}}{2m^2}$
10. $\dfrac{\sqrt{3}}{5}$
11. $\dfrac{\sqrt{14x}}{2x}$
12. $5\sqrt{2}$
13. $4\sqrt{3}$
14. $2\sqrt{p}$
15. Not possible
16. $-1\sqrt{y}$ or $-\sqrt{y}$
17. Not possible

SUPPLEMENTARY PROBLEMS AND ANSWERS

SUPPLEMENTARY PROBLEMS

UNIT 1: SIGNED WHOLE NUMBERS

Assignment 1

Use a signed number to represent each of these.

1. a gain of 3 yards
2. an increase of 10°
3. a loss of $50
4. 5° below zero
5. a loss of 7 yards
6. a profit of $25
7. a deposit of $50
8. 20° above zero
9. a withdrawal of $100

Write either ">" or "<" in each blank.

10. +4 ______ -3
11. -2 ______ -4
12. 7 ______ 5
13. -5 ______ 0
14. 0 ______ 3
15. -9 ______ -4
16. +17 ______ +31
17. -1 ______ -4
18. +1 ______ -1
19. -15 ______ -20
20. -10 ______ +10
21. 89 ______ -12

Complete.

22. $|+6| =$ ______
23. $|-5| =$ ______
24. $|17| =$ ______
25. $|-17| =$ ______
26. $|0| =$ ______
27. $|-1| =$ ______
28. $|+1| =$ ______
29. $|-35| =$ ______
30. $|47| =$ ______
31. $|-63| =$ ______

Assignment 2

Find each sum.

1. $(-2) + (-4) =$ ______
2. $4 + (-7) =$ ______
3. $8 + (-1) =$ ______
4. $(-9) + 5 =$ ______
5. $(-7) + (-1) =$ ______
6. $(-8) + 0 =$ ______
7. $(-6) + 6 =$ ______
8. $10 + (-5) =$ ______
9. $20 + (-30) =$ ______
10. $(-10) + 15 =$ ______
11. $40 + (-40) =$ ______
12. $(-50) + 20 =$ ______

Find each sum.

13. $6 + (-7) + 5 =$ ______
14. $(-3) + (-8) + 1 =$ ______
15. $(-6) + (-9) + (-2) =$ ______
16. $(-9) + 8 + (-3) =$ ______
17. $2 + (-3) + 9 + (-5) =$ ______
18. $(-7) + 4 + (-3) + 0 =$ ______
19. $(-4) + 9 + (-7) + (-6) + 9 =$ ______
20. $50 + (-10) + 30 + 10 + (-60) =$ ______

UNIT 1: SIGNED WHOLE NUMBERS (continued)

Assignment 3

Complete.

1. $5 - 13 =$ _____ 2. $6 - (-4) =$ _____ 3. $(-1) - 2 =$ _____

4. $(-7) - (-8) =$ _____ 5. $4 - (-4) =$ _____ 6. $(-9) - (-9) =$ _____

7. $0 - 3 =$ _____ 8. $(-1) - 1 =$ _____ 9. $11 - 22 =$ _____

10. $15 - (-17) =$ _____ 11. $(-22) - 33 =$ _____ 12. $(-40) - (-60) =$ _____

Complete.

13. $7 - 8 - (-2) =$ _____ 14. $9 + (-3) - 6 =$ _____

15. $6 + 8 - 15 =$ _____ 16. $2 - 7 - 1 =$ _____

17. $(-8) - (-1) - 6 =$ _____ 18. $(-2) + 7 - 5 =$ _____

19. $9 - (-1) + (-6) - 1 =$ _____ 20. $4 - 9 + 0 - 3 - (-7) =$ _____

Assignment 4

Find each product.

1. $8(-7) =$ _____ 2. $(-5)(4) =$ _____ 3. $0(-3) =$ _____ 4. $(-1)(-10) =$ _____

5. $1(-12) =$ _____ 6. $(-9)(-6) =$ _____ 7. $(-7)(-1)(4) =$ _____ 8. $(5)(-6)(0) =$ _____

9. $(-3)(-7)(-2) =$ _____ 10. $(-2)(1)(-3)(2) =$ _____ 11. $(-9)(0)(-1)(-1) =$ _____ 12. $(1)(-8)(-7)(-1) =$ _____

Complete.

13. $20 + 3(-2) =$ _____ 14. $50 - 10(-1) =$ _____ 15. $8(7) - 40 =$ _____

16. $5(-3) + 15 =$ _____ 17. $6(-4) + 3(-1) =$ _____ 18. $(-3)(-5) - (-2)(-9) =$ _____

Find each quotient.

19. $\frac{42}{-7} =$ _____ 20. $\frac{-20}{4} =$ _____ 21. $\frac{0}{-6} =$ _____ 22. $\frac{15}{-1} =$ _____ 23. $\frac{-9}{1} =$ _____

24. $\frac{-32}{-4} =$ _____ 25. $\frac{-64}{2} =$ _____ 26. $\frac{-40}{-10} =$ _____ 27. $\frac{-5}{-1} =$ _____ 28. $\frac{0}{7} =$ _____

UNIT 2: SOLVING EQUATIONS

Assignment 5

Solve each equation.

1. $5 + x = 7$
2. $t + 6 = 12$
3. $11 - y = 7$
4. $R - 1 = 3$
5. $3 + x = 13$
6. $40 + m = 49$

Solve each equation.

7. $8x = 56$
8. $-8 = 2k$
9. $-4t = 16$
10. $-12 = -2m$
11. $3R = -9$
12. $50 = 10x$
13. $-9d = -72$
14. $26 = -2V$
15. $8q = -8$
16. $12 = 12y$
17. $2x = 0$
18. $0 = -10S$

Assignment 6

Solve each equation.

1. $4m + 6m = 80$
2. $28 = 5x + (-3x)$
3. $(-9p) + 2p = 42$
4. $-18 = (-2y) + (-4y)$
5. $8R - 10R = 12$
6. $40 = 7V - 3V$

Solve each equation.

7. $y + 4y = 55$
8. $48 = (-9x) + x$
9. $3x - x = 40$
10. $-24 = x - 7x$
11. $10B - 9B = 4$
12. $0 = 6F - F$

Solve each equation.

13. $(-9d) + (-d) = 30$
14. $-t = 8$
15. $-12 = -x$
16. $y - 2y = -5$
17. $(-7H) + 6H = 8$
18. $0 = x - 2x$

UNIT 2: SOLVING EQUATIONS (continued)

Assignment 7

Solve each equation.

1. $x + 9 = 4$
2. $12 = 3 + y$
3. $m - 6 = 1$
4. $-5 = m - 10$
5. $F + 4 = 0$
6. $d - 1 = 0$

Solve each equation.

7. $2t + 14 = 24$
8. $11 = 17 + 3S$
9. $25 = 10x - 15$
10. $6 - 5x = 16$
11. $4 - y = 1$
12. $15 - 3p = 0$

Solve each equation.

13. $3t + 20 = 5t$
14. $2p - 3 = 5p$
15. $Q - 10 = 2Q$
16. $10V + 18 = 8V$
17. $12 - 3x = 9x$
18. $d = 14 - 6d$

Assignment 8

Solve each equation.

1. $5(x + 4) = 40$
2. $30 = 3(7 + y)$
3. $7(2d + 1) = 35$
4. $24 = 6(8 + 2p)$
5. $4(x + 9) = 7x$
6. $8y = 2(5y + 7)$

Solve each equation.

7. $2(R - 8) = 4$
8. $20 = 5(3 - k)$
9. $3(4t - 1) = 9$
10. $44 = 4(5 - 2V)$
11. $8(p - 4) = 7p$
12. $3p = 6(5 - 2p)$

Solve each equation.

13. $2x + 7 = x + 12$
14. $15 + 2t = 9 + 5t$
15. $4b - 7 = 5b + 1$
16. $20 - 2y = 30 - 7y$
17. $2(F - 4) = 5F + 1$
18. $3(2x + 5) = 5(x - 1)$

UNIT 3: FRACTIONS AND EQUATIONS: MULTIPLICATION AND DIVISION

Assignment 9

Reduce each fraction to lowest terms!

1. $\frac{4}{8} =$
2. $\frac{6}{4} =$
3. $\frac{10}{25} =$
4. $\frac{12}{10} =$
5. $\frac{12}{16} =$
6. $\frac{10}{4} =$
7. $\frac{9}{15} =$
8. $\frac{35}{28} =$
9. $\frac{6}{18} =$
10. $\frac{24}{20} =$
11. $\frac{12}{48} =$
12. $\frac{36}{24} =$
13. $\frac{20}{32} =$
14. $\frac{35}{10} =$
15. $\frac{25}{100} =$
16. $\frac{48}{18} =$

Perform each multiplication. Reduce each product to lowest terms.

17. $\left(\frac{1}{3}\right)\left(\frac{5}{2}\right) =$
18. $\left(\frac{1}{5}\right)\left(\frac{1}{6}\right) =$
19. $\left(\frac{10}{3}\right)\left(\frac{4}{5}\right) =$
20. $\left(\frac{7}{4}\right)\left(\frac{2}{3}\right) =$
21. $\left(\frac{2}{9}\right)\left(\frac{3}{10}\right) =$
22. $\left(\frac{4}{5}\right)\left(\frac{15}{2}\right) =$
23. $\left(\frac{7}{8}\right)\left(\frac{8}{7}\right) =$
24. $\left(\frac{4}{3}\right)\left(\frac{9}{16}\right) =$
25. $8\left(\frac{1}{2}\right) =$
26. $\frac{2}{3}(15) =$
27. $6\left(\frac{7}{8}\right) =$
28. $\frac{5}{14}(7) =$
29. $(0)\left(\frac{9}{7}\right) =$
30. $1\left(\frac{1}{3}\right) =$
31. $\frac{1}{9}(9) =$
32. $\frac{11}{12}(1) =$

Assignment 10

Perform each multiplication. Reduce each product to lowest terms.

1. $\left(\frac{2}{3}\right)\left(-\frac{1}{5}\right) =$
2. $\left(-\frac{1}{3}\right)\left(-\frac{1}{4}\right) =$
3. $\left(-\frac{10}{4}\right)\left(\frac{12}{5}\right) =$
4. $\left(-\frac{12}{25}\right)\left(-\frac{5}{8}\right) =$
5. $\frac{1}{5}(-20) =$
6. $(-7)\left(-\frac{9}{14}\right) =$
7. $1\left(-\frac{1}{10}\right) =$
8. $\left(-\frac{6}{5}\right)(0) =$

Perform each possible division.

9. $\frac{-6}{3} =$
10. $\frac{20}{-5} =$
11. $\frac{-56}{-7} =$
12. $\frac{-72}{8} =$
13. $\frac{50}{-10} =$
14. $\frac{-12}{-12} =$
15. $\frac{-8}{1} =$
16. $\frac{5}{-1} =$
17. $\frac{0}{-3} =$
18. $\frac{-6}{0} =$

UNIT 3: FRACTIONS AND EQUATIONS: MULTIPLICATION AND DIVISION (continued)

Assignment 10 (continued)

Solve. Reduce each root to lowest terms.

19. $5t = 9$
20. $7 = 12m$
21. $4x = -3$
22. $11 = -2y$
23. $-4R = -7$
24. $-1 = -6p$
25. $6x = 14$
26. $-8 = 12q$
27. $-20x = -16$
28. $-32 = -12d$
29. $5F = -1$
30. $1 = 10v$

Assignment 11

Perform each division. Reduce each quotient to lowest terms.

1. $\dfrac{\frac{1}{2}}{\frac{6}{5}} =$
2. $\dfrac{10}{\frac{2}{3}} =$
3. $\dfrac{\frac{1}{3}}{6} =$
4. $\dfrac{\frac{1}{5}}{\frac{1}{10}} =$
5. $\dfrac{-\frac{2}{7}}{-\frac{5}{3}} =$
6. $\dfrac{-\frac{3}{4}}{6} =$
7. $\dfrac{-12}{-\frac{8}{7}} =$
8. $\dfrac{-\frac{1}{9}}{\frac{1}{3}} =$
9. $\dfrac{1}{-\frac{10}{11}} =$
10. $\dfrac{-\frac{8}{5}}{1} =$
11. $\dfrac{0}{\frac{1}{7}} =$
12. $\dfrac{-\frac{4}{5}}{0} =$

Solve. Reduce each root to lowest terms.

13. $\frac{3}{4}x = 12$
14. $1 = \frac{8}{3}y$
15. $\frac{1}{4}p = -2$
16. $6 = -\frac{12}{5}R$
17. $\frac{1}{2}d = \frac{1}{8}$
18. $-\frac{2}{3} = \frac{1}{5}t$
19. $-\frac{4}{5}V = \frac{2}{15}$
20. $-\frac{9}{8} = -\frac{3}{2}m$
21. $-\frac{1}{12}F = \frac{1}{3}$
22. $-1 = -\frac{1}{4}a$
23. $-\frac{5}{2}y = 1$
24. $\frac{1}{4}x = 0$

UNIT 4: FRACTIONS AND EQUATIONS: ADDITION AND SUBTRACTION

Assignment 12

Perform each addition. Reduce each answer to lowest terms.

1. $\frac{2}{5} + \frac{1}{5} =$ 2. $\frac{1}{6} + \frac{7}{6} =$ 3. $\frac{3}{10} + \frac{7}{10} =$ 4. $\frac{3}{8} + \frac{1}{2} =$

5. $\frac{7}{20} + \frac{3}{10} =$ 6. $\frac{1}{2} + \frac{5}{6} =$ 7. $\frac{1}{3} + \frac{1}{4} =$ 8. $\frac{5}{6} + \frac{3}{8} =$

9. $\frac{3}{4} + \frac{1}{10} =$ 10. $1 + \frac{1}{4} =$ 11. $\frac{2}{3} + 2 =$ 12. $\frac{3}{2} + 1 =$

Perform each addition. Reduce each answer to lowest terms.

13. $\frac{4}{9} + \left(-\frac{5}{9}\right) =$ 14. $\left(-\frac{1}{6}\right) + \frac{5}{6} =$ 15. $\left(-\frac{3}{4}\right) + \left(-\frac{5}{4}\right) =$ 16. $\frac{4}{5} + \left(-\frac{1}{10}\right) =$

17. $\left(-\frac{7}{8}\right) + \frac{1}{2} =$ 18. $\left(-\frac{2}{3}\right) + \left(-\frac{7}{12}\right) =$ 19. $\frac{1}{3} + \left(-\frac{1}{2}\right) =$ 20. $\left(-\frac{3}{4}\right) + \frac{5}{6} =$

21. $\left(-\frac{9}{8}\right) + \left(-\frac{1}{20}\right) =$ 22. $3 + \left(-\frac{11}{4}\right) =$ 23. $\left(-\frac{5}{2}\right) + 1 =$ 24. $(-1) + \left(-\frac{2}{3}\right) =$

Assignment 13

Perform each subtraction. Reduce each answer to lowest terms.

1. $\frac{6}{7} - \frac{6}{7} =$ 2. $\frac{7}{8} - \frac{3}{8} =$ 3. $\frac{5}{4} - \frac{1}{2} =$ 4. $\frac{4}{3} - \frac{7}{12} =$

5. $\frac{1}{4} - \frac{1}{5} =$ 6. $\frac{11}{8} - \frac{5}{6} =$ 7. $2 - \frac{1}{5} =$ 8. $1 - \frac{3}{4} =$

Perform each subtraction. Reduce each answer to lowest terms.

9. $\frac{1}{5} - \frac{3}{5} =$ 10. $\frac{3}{10} - \left(-\frac{7}{10}\right) =$ 11. $\left(-\frac{1}{4}\right) - \frac{1}{8} =$ 12. $\left(-\frac{1}{2}\right) - \left(-\frac{7}{10}\right) =$

13. $\frac{1}{7} - \frac{1}{3} =$ 14. $\left(-\frac{5}{6}\right) - \left(-\frac{5}{4}\right) =$ 15. $2 - \left(-\frac{1}{2}\right) =$ 16. $\left(-\frac{3}{5}\right) - 1 =$

UNIT 4: FRACTIONS AND EQUATIONS: ADDITION AND SUBTRACTION (continued)

Assignment 13 (continued)

Solve each equation. Reduce each root to lowest terms.

17. $2x + x = 1$
18. $m - 7m = 4$
19. $2 = 5R - R$
20. $4y + 1 = 0$
21. $3 - 5q = 7$
22. $9 = 8t - 3$
23. $3F + 2 = 6F$
24. $5d - 7 = d$
25. $8b = 2b - 9$
26. $4x + 5 = x + 9$
27. $2y - 3 = 6y - 1$
28. $1 - V = 10 - 11V$
29. $3(t + 2) = 8$
30. $4(1 - 2m) = 2m$
31. $3p - 8 = 2(5p + 1)$

Assignment 14

Solve each equation. Reduce each root to lowest terms.

1. $\frac{4}{9}x + \frac{1}{9}x = \frac{2}{9}$
2. $12 = y + \frac{1}{2}y$
3. $\frac{5}{8}d + d = 0$
4. $\frac{1}{2}V - \frac{2}{5}V = \frac{1}{3}$
5. $H - \frac{7}{4}H = 1$
6. $\frac{3}{10} = R - 1$

Solve each equation. Reduce each root to lowest terms.

7. $x + \frac{3}{4} = \frac{7}{8}$
8. $\frac{4}{7} = y + 1$
9. $K + \frac{2}{3} = 0$
10. $t - \frac{1}{3} = \frac{3}{5}$
11. $1 = d - \frac{1}{4}$
12. $m - 2 = -\frac{5}{2}$

Solve each equation. Reduce each root to lowest terms.

13. $\frac{3}{8}x + 1 = \frac{1}{2}$
14. $2d + \frac{1}{4} = \frac{5}{6}$
15. $\frac{4}{3} = \frac{2}{3}p - 1$
16. $\frac{1}{4} - 2x = \frac{5}{8}$
17. $0 = 6 - \frac{2}{3}R$
18. $\frac{2}{5} - q = \frac{3}{5}$
19. $\frac{6}{7} = 1 - t$
20. $\frac{2}{3}x - \frac{1}{2} = x$
21. $\frac{5}{8}y + 1 = \frac{3}{8}y + 5$

UNIT 5: DECIMAL NUMBERS AND EQUATIONS

Assignment 15

Perform each addition.

1. \$6.95 + \$1.99	2. 7.4 + 12	3. .5 + 4 + .08	4. .167 + 5.3 + .29
5. (-.08) + (-.07)	6. 5.4 + (-3.7)	7. (-.4) + .5	8. (-.27) + (-.3)
9. 3 + (-.56)	10. (-2.6) + 1	11. (-.86) + (-2)	12. .21 + (-.8)

Perform each subtraction.

13. \$19.95 - \$6.40	14. 7.1 - 4.4	15. \$10 - \$4.35	16. .05 - .029
17. 6.4 - 7.2	18. .25 - (-.36)	19. (-.3) - .8	20. (-1.9) - (-1.3)
21. 3 - 4.88	22. 1 - (-.5)	23. (-.65) - 1	24. (-2) - (-.3)

Solve each equation.

25. x + 1.2 = 7.7	26. .5 + t = .33	27. 10 = R + 12.8
28. y - 3.4 = 6.7	29. p - .87 = 1	30. 0 = T - 3.6

Assignment 16

Perform each multiplication.

1. (.1)(66)	2. 5(.03)	3. (.4)(.08)	4. 1(29.5)
5. 0(-7.5)	6. (-1)(.414)	7. (-.02)(-9)	8. .7(-.6)
9. .4(8.8)	10. (-9)(.61)	11. (54)(-.01)	12. (-3.3)(-4.4)

Perform each division.

13. $\frac{.16}{2}$	14. $\frac{5.5}{5}$	15. $\frac{1.2}{.3}$	16. $\frac{.032}{.08}$
17. $\frac{-6.87}{1}$	18. $\frac{.4}{-1}$	19. $\frac{0}{15.6}$	20. $\frac{-48.2}{-48.2}$
21. $\frac{-.048}{.24}$	22. $\frac{15.6}{-4}$	23. $\frac{-8.8}{-1.1}$	24. $\frac{-98.8}{1.3}$

UNIT 5: DECIMAL NUMBERS AND EQUATIONS (continued)

Assignment 16 (continued)

Solve each equation.

25. $5x = .75$
26. $1 = .4y$
27. $.06R = .54$
28. $-2.6b = 0$
29. $-2.1t = 9.45$
30. $-4.8d = 96$

Assignment 17

Solve each equation.

1. $.5x + .15 = .75$
2. $.9 - .2y = .1$
3. $.08 = .03 - p$
4. $3.7d - 2.5d = 60$
5. $.1(V - .7) = .34$
6. $7.8m - 2.5 = 7.4m + 1.1$

Round to the nearest tenth.

7. 19.53
8. .399

Round to the nearest thousandth.

9. .00701
10. 2.67061

Round to the nearest whole number.

11. $25.60
12. 412.3

Round to the nearest hundredth.

13. .6954
14. 2.1033

Find each quotient to the indicated place.

15. nearest tenth $\frac{-65.3}{7}$
16. nearest whole number $\frac{27.5}{-.3}$
17. nearest thousandth $\frac{-10.6}{-24}$

Solve. Report each root to the indicated place.

18. nearest tenth
$6x = 23.1$
19. nearest hundredth
$.09t - .01 = .03$
20. nearest thousandth
$1.1(y + 1) = .2$

By memory, convert each fraction to a decimal number.

21. $\frac{1}{2}$
22. $\frac{2}{3}$
23. $\frac{1}{4}$
24. $\frac{4}{5}$

Convert each fraction to a decimal number to the nearest hundredth.

25. $\frac{6}{7}$
26. $\frac{10}{9}$
27. $\frac{1}{12}$
28. $\frac{83}{30}$

UNIT 6: FRACTIONAL EQUATIONS AND PROPORTIONS

Assignment 18

Solve each equation.

1. $\frac{x}{2} = 5$
2. $1 = \frac{7}{y}$
3. $\frac{4p}{5} = 0$
4. $3 = \frac{12}{5d}$
5. $\frac{3}{V + 1} = 2$
6. $\frac{m - 7}{9} = 0$
7. $1 = \frac{2b}{b + 1}$
8. $9 = \frac{10 - x}{2}$

Assignment 19

Solve each proportion.

1. $\frac{x}{3} = \frac{2}{5}$
2. $\frac{1}{2} = \frac{4}{y}$
3. $\frac{3}{4} = \frac{2p}{7}$
4. $\frac{1}{4R} = \frac{3}{2}$
5. $\frac{t + 2}{4} = \frac{5}{2}$
6. $\frac{4}{3} = \frac{d - 1}{d}$
7. $\frac{1}{m + 3} = \frac{5}{7}$
8. $\frac{1}{4} = \frac{x}{10 - x}$

Assignment 20

Solve each proportion.

1. $\frac{\frac{1}{3}}{10} = \frac{2}{x}$
2. $\frac{1.5}{6} = \frac{y}{8}$
3. $\frac{1.4}{t} = \frac{.7}{40}$
4. $\frac{5}{\$.75} = \frac{d}{\$1.20}$

Use a proportion to solve each problem.

5. There are 16 ounces in 1 pound. How many pounds are there in 192 ounces?

6. If a machine can drill 20 holes in 4 minutes, how long will it take to drill 150 holes?

7. In a small town, the ratio of registered Democrats to registered Republicans is 2:3. If there are 50 registered Democrats, how many registered Republicans are there?

8. The ratio of "Yes" votes to "No" votes on a referendum was 3:1. If 720 people voted, how many "Yes" votes were there?

9. 1 kilogram equals 2.2 pounds. Therefore, 7 kilograms equals how many pounds?

10. 1 liter equals 2.1 pints. To the nearest tenth, how many liters are there in 25 pints?

11. A small airplane uses 4 gallons of gasoline every $\frac{1}{4}$ hour. How many gallons would it need to fly 3 hours?

12. If 2.2 pounds of meat cost \$3.30, how much would 3.2 pounds of the same meat cost?

UNIT 7: SQUARES, SQUARE ROOTS, AND EQUATIONS

Assignment 21

Find each square.

1. 7^2
2. $(-1)^2$
3. $\left(\frac{1}{5}\right)^2$
4. $\left(-\frac{3}{2}\right)^2$
5. $(32)^2$
6. $(-400)^2$
7. $(.02)^2$
8. $(-1.5)^2$

Find each square root. Use the table when necessary. Approximate the square roots of decimal numbers by rounding to the nearest whole number before using the table.

9. $\sqrt{16}$
10. $-\sqrt{121}$
11. $\pm\sqrt{100}$
12. $\sqrt{3}$
13. $-\sqrt{69}$
14. $\pm\sqrt{191}$
15. $\sqrt{27.38}$
16. $-\sqrt{148.5}$

Assignment 22

Solve each equation.

1. $x^2 = 1$
2. $y^2 - 100 = 0$
3. $46 = 30 + t^2$
4. $25d^2 - 1 = 0$
5. $\frac{1}{4}p^2 = 16$
6. $49 = \frac{81}{r^2}$
7. $\frac{2V^2}{3} = 24$
8. $\frac{3}{100} = \frac{x^2}{48}$

Using the table, solve each equation. Round decimal numbers to whole numbers before using the table.

9. $4y^2 = 180$
10. $27 = \frac{t^2}{5}$
11. $57.6 = 2d^2 - 52.2$
12. $\frac{5}{6} = \frac{h^2}{200}$

Assignment 23

Solve each equation.

1. $\sqrt{R} = \frac{5}{7}$
2. $1 = \sqrt{3p}$
3. $\sqrt{4t} = \frac{4}{3}$
4. $\sqrt{y - 4} = 6$
5. $\sqrt{\frac{x}{2}} = 4$
6. $2 = \sqrt{\frac{13}{5d}}$
7. $\sqrt{\frac{10}{V + 1}} = 5$
8. $\frac{6}{5} = \sqrt{\frac{2}{R}}$
9. $\sqrt{F^2 - 4} = 0$
10. $0 = \sqrt{9m^2 - 1}$
11. $\sqrt{3x^2 - 26} = 1$
12. $5 = \sqrt{100 - y^2}$

UNIT 8: FORMULA EVALUATION

Assignment 24

1. In $\boxed{L = N + D}$, find L if N = \$32 and D = \$8 .

2. In $\boxed{A = 360 - (B + C + D)}$, find A if B = 70 , C = 95 , and D = 90 .

3. In $\boxed{v = 20t}$, find "v" if t = 5 .

4. In $\boxed{A = LW}$, find A if L = 10 and W = 8 .

5. In $\boxed{A = \frac{1}{2}bh}$, find A if b = 10 and h = 2.5 .

6. In $\boxed{y = mx + b}$, find "y" if $m = \frac{3}{4}$, x = 12 , and b = 1 .

7. In $\boxed{C = \frac{5}{9}(F - 32)}$, find C if F = 59 .

8. In $\boxed{I = \frac{E}{R}}$, find I if E = 3.6 and R = 1.2 .

9. In $\boxed{t = \frac{1}{f}}$, find "t" if f = 25 .

10. In $\boxed{V = \frac{4st}{a}}$, find V if s = 3 , t = 8 , and a = 4 .

11. In $\boxed{R = \frac{p + q}{2}}$, find R if p = 2.4 and q = 1.4 .

12. In $\boxed{F = \frac{9C}{5} + 32}$, find F if C = 20 .

13. In $\boxed{A = 6s^2}$, find A if s = 10 .

14. In $\boxed{P = I^2R}$, find P if I = 5 and R = 2 .

15. In $\boxed{P = \frac{E^2}{R}}$, find P if E = 10 and R = 20 .

16. In $\boxed{H = \frac{D^2N}{2.5}}$, find H if D = 5 and N = 6 .

17. In $\boxed{E = \sqrt{PR}}$, find E if P = 2 and R = 18 .

18. In $\boxed{t = \sqrt{\frac{2s}{a}}}$, find "t" if s = 40 and a = 10 .

19. In $\boxed{R = \sqrt{Z^2 - X^2}}$, find R if Z = 7 and X = 5 .

20. In $\boxed{N = K\sqrt{\frac{V}{H}}}$, find N if K = 10 , V = 100 , and H = 4 .

UNIT 8: FORMULA EVALUATION (continued)

Assignment 25

1. In $F_T = F_1 + F_2 + F_3$, find F_T if $F_1 = 2$, $F_2 = 3$, and $F_3 = 4$.

2. In $E = \frac{P_o}{P_i}$, find E if $P_o = 20$ and $P_i = 10$.

3. In $R = \frac{R_1R_2}{R_1 + R_2}$, find R if $R_1 = 40$ and $R_2 = 60$.

4. In $m = \frac{y_2 - y_1}{x_2 - x_1}$, find "m" if $y_2 = 10$, $y_1 = 4$, $x_2 = 5$, and $x_1 = 3$.

5. In $E = \frac{1}{2}(m_1 + m_2)v^2$, find E if $m_1 = 4$, $m_2 = 6$, and $v = 2$.

6. In $d = \sqrt{\frac{I_a - I_c}{m}}$, find "d" if $I_a = 100$, $I_c = 20$, and $m = 5$.

7. In $R = C + M$, find M if $R = \$69.95$ and $C = \$47.50$.

8. In $V = LWH$, find W if $V = 40$, $L = 5$, and $H = 4$.

9. In $V_1 = \frac{P_2V_2}{P_1}$, find V_2 if $V_1 = 10$, $P_2 = 8$, and $P_1 = 4$.

10. In $v = \frac{s_2 - s_1}{t_2 - t_1}$, find t_2 if $v = 20$, $s_2 = 100$, $s_1 = 40$, and $t_1 = 1$.

11. In $P = I^2R$, find I if $P = 128$ and $R = 2$.

12. In $s = \frac{1}{2}at^2$, find "t" if $s = 100$ and $a = 20$.

13. In $F = \frac{m_1m_2}{rd^2}$, find "d" if $F = 2$, $m_1 = 6$, $m_2 = 12$, and $r = 4$.

14. In $v = \sqrt{2as}$, find "s" if $v = 12$ and $a = 6$.

15. In $I = \sqrt{\frac{P}{R}}$, find P if $I = 7$ and $R = 2$.

16. In $c = \sqrt{a^2 + b^2}$, find "b" if $c = 6$ and $a = 4$.

17. In $\frac{F_1}{F_2} = \frac{d_1}{d_2}$, find F_2 if $F_1 = 10$, $d_1 = 60$, and $d_2 = 72$.

18. In $\frac{I_1}{I_2} = \frac{(d_2)^2}{(d_1)^2}$, find d_1 if $I_1 = 3$, $I_2 = 4$ and $d_2 = 6$.

UNIT 9: FORMULA REARRANGEMENT

Assignment 26

1. Solve for "m": $p = 2m$
2. Solve for "t": $v = at$
3. Solve for "c": $a = b^2c$
4. Solve for H: $V = LWH$
5. Solve for V_1: $C_1V_1 = C_2V_2$
6. Solve for "q": $D = 3.5m^2pq$
7. Solve for F: $C = FT$
8. Solve for "b": $a^2 = 2bd$
9. Solve for D: $M = CDE$
10. Solve for P_2: $P_1V_1 = P_2V_2$
11. Solve for T: $b^2Tr = q_1q_2$
12. Solve for Q: $F = .3V^2Qt$
13. Solve for "t" $3bt = m(p - q)$
14. Solve for M: $B = M(S - T)$
15. Solve for "b": $V = ab(r_1 - r_2)$
16. Solve for "c": $F = cd(t_2 - t_1)$

Assignment 27

1. Solve for C: $B = \frac{C}{D}$
2. Solve for "t": $m = \frac{at}{r}$
3. Solve for "v_1": $H = \frac{v_1v_2}{1.8s^2}$
4. Solve for F: $A = \frac{T}{F}$
5. Solve for "d_2": $F_1 = \frac{d_1F_2}{d_2}$
6. Solve for "p": $h = \frac{t_2 - t_1}{2p}$
7. Solve for K: $F = \frac{ACK(v_2 - v_1)}{M}$
8. Solve for "c": $B = \frac{1}{2}cd$
9. Solve for "h": $A = \frac{1}{2}h(m_1 + m_2)$
10. Solve for C_1: $\frac{A_1}{A_2} = \frac{C_1}{C_2}$
11. Solve for T_1: $\frac{P_1Q_1}{T_1} = \frac{P_2Q_2}{T_2}$
12. Solve for "m": $m^2 = st$
13. Solve for "b": $b^2 = \frac{c - d}{t}$
14. Solve for "p": $p^2 = q^2 + r^2$
15. Solve for "v": $H = \frac{cv^2}{2.5}$
16. Solve for "t": $F = \frac{1}{2}dt^2$

UNIT 9: FORMULA REARRANGEMENT (continued)

Assignment 28

1. Solve for V: $P = V + 50$
2. Solve for C: $A = B + C + D$
3. Solve for Q: $L = Q - F$
4. Solve for B: $b = B - cH$
5. Solve for V: $I = M + TV$
6. Solve for "t_2": $k = \frac{t_2 - t_1}{p}$
7. Solve for C_i: $v = \frac{C_o + C_i}{t}$
8. Solve for "d": $I_a = I_c + md^2$
9. Solve for "q": $t^2 = p^2 + q^2$
10. Solve for "b": $c^2 = b^2 - a^2$
11. Solve for T: $R = S - T$
12. Solve for "t_1": $p = \frac{t_2 - t_1}{m}$
13. Solve for F: $d = D - Fp$
14. Solve for "v": $t_o^2 = t_f^2 - 2av$
15. Solve for "t": $h^2 = s^2 - t^2$
16. Solve for M_B: $M = M_A - M_B$

Assignment 29

1. Solve for S: $M = P(S + T)$
2. Solve for "d": $a = b(c + d)$
3. Solve for V_1: $R = F(V_1 - V_2)$
4. Solve for X: $C = P(L - X)$
5. Solve for "t_1": $D = \frac{1}{2}m(t_1 + t_2)$
6. Solve for "p": $h = ap + bp$
7. Solve for T: $CT + FT = S$
8. Solve for I: $Q = IX + IY$
9. Solve for D: $T = \sqrt{D}$
10. Solve for "t": $m = \sqrt{2at}$
11. Solve for "h": $p = \sqrt{\frac{h}{q}}$
12. Solve for "v": $r = \sqrt{\frac{2b}{v}}$
13. Solve for S: $\sqrt{ST} = \frac{P}{R}$
14. Solve for "x": $e = \sqrt{x^2 + y^2}$
15. Solve for "t": $c = \sqrt{k^2 + t^2}$
16. Solve for E: $P = \sqrt{D^2 - E^2}$

UNIT 10: GEOMETRIC FORMULAS

Assignment 30

Find the perimeter and area of each rectangle or square.

1. 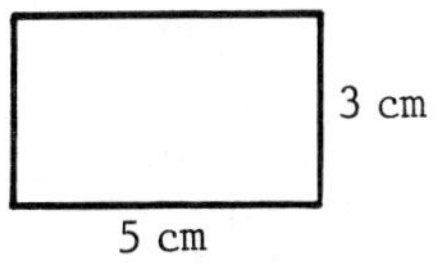

2. 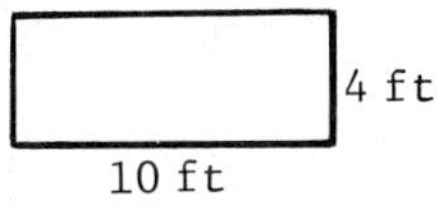

3. 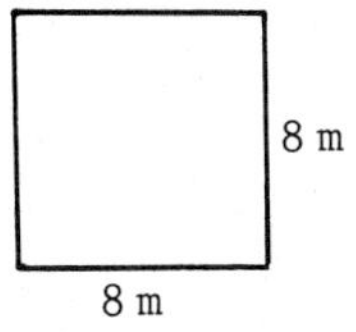

Find the area of each parallelogram and triangle.

4. 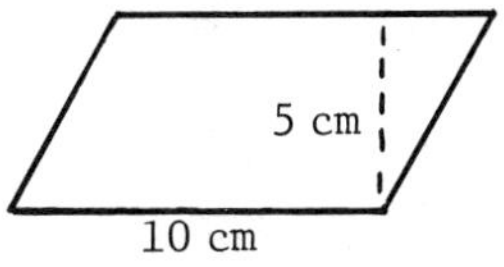

5. 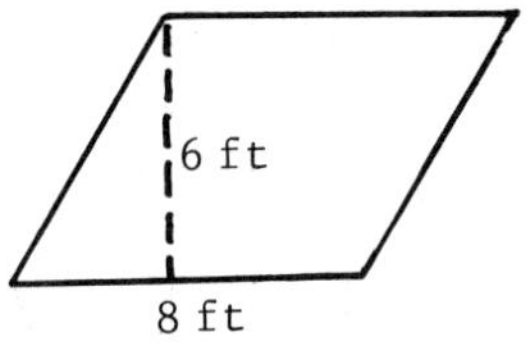

6. 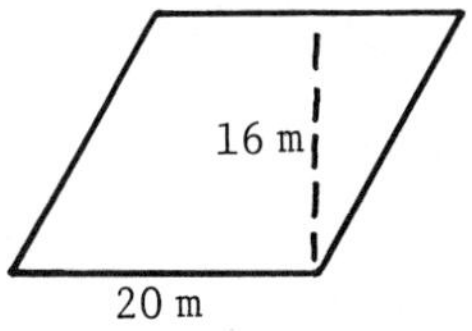

7. 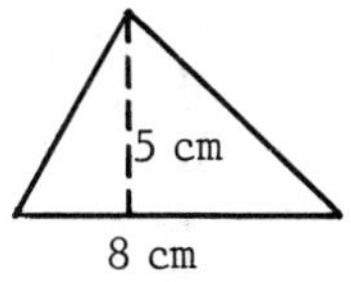

8. 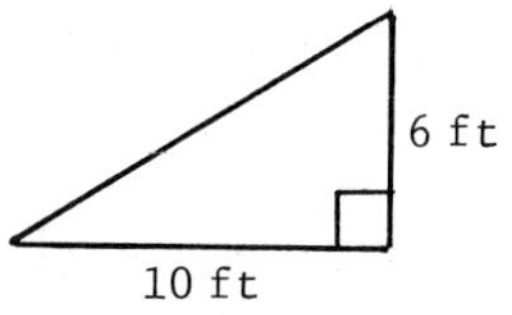

9. 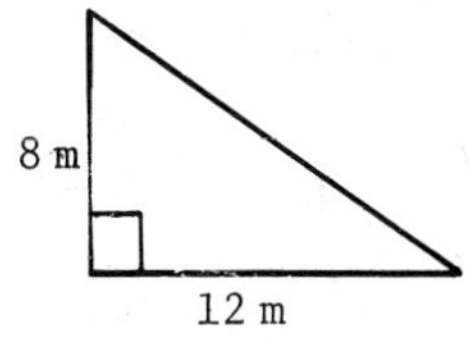

Find the unknown angle in each triangle.

10. 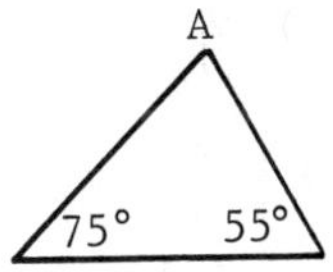

11. 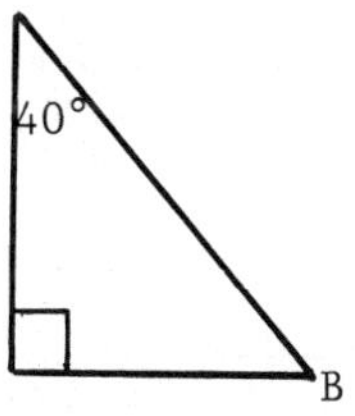

12. 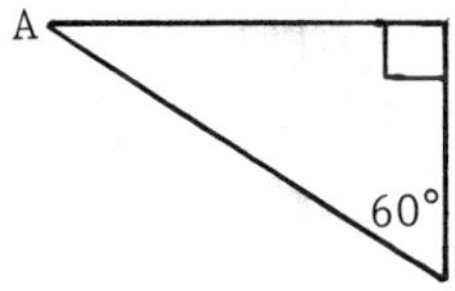

13. Find the width of a rectangle if its perimeter is 30 cm and its length is 8 cm.
14. Find the length of a rectangle if its area is 80 ft^2 and its width is 4 ft.
15. Find the side of a square whose area is 49 m^2.
16. Find the height of a parallelogram if its area is 100 cm^2 and its base is 20 cm.
17. Find the base of a triangle if its area is 40 yd^2 and its height is 8 yd.
18. Find the height of a triangle if its area is 50 cm^2 and its base is 10 cm.
19. Find the area of a right triangle if $l_1 = 16$ in and $l_2 = 8$ in.
20. Find l_2 in a right triangle if its area is 120 m^2 and l_1 is 12 m.

UNIT 10: GEOMETRIC FORMULAS (continued)

Assignment 31

Use the Pythagorean Theorem to find the unknown dimension.

1. 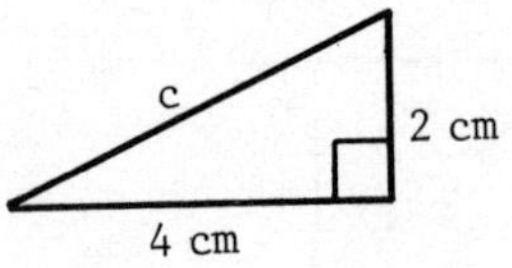

2. 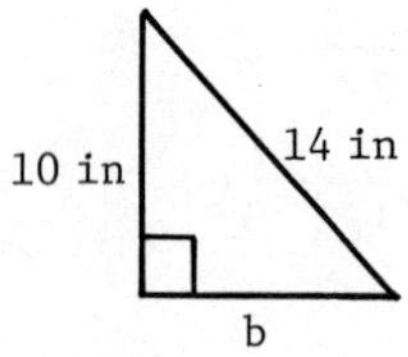

3. 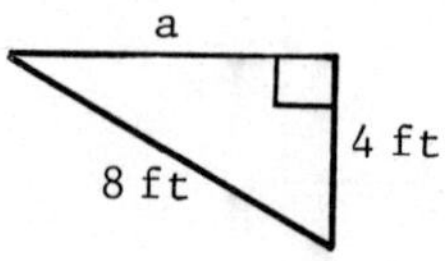

4. 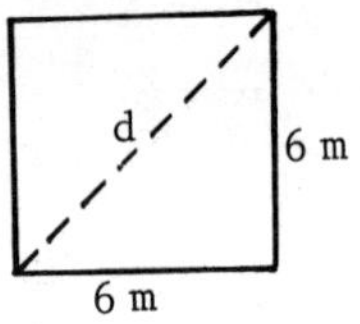

5. 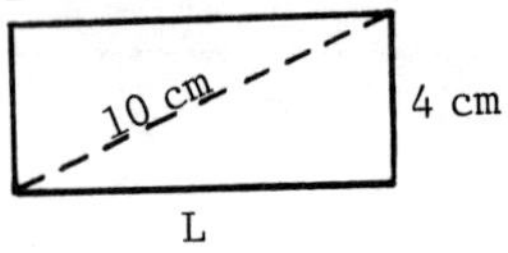

6. 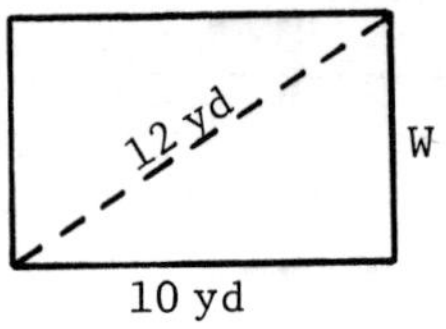

Find the circumference and area of each circle.

7. 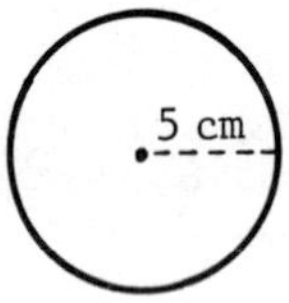

8. 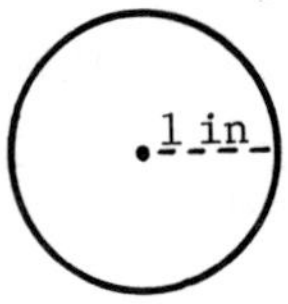

9. 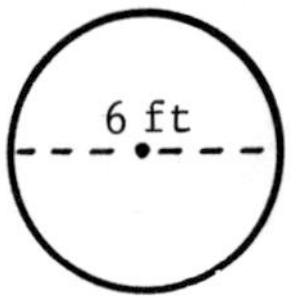

Find the volume of each figure.

10. 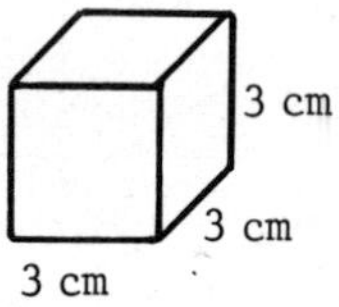

11. 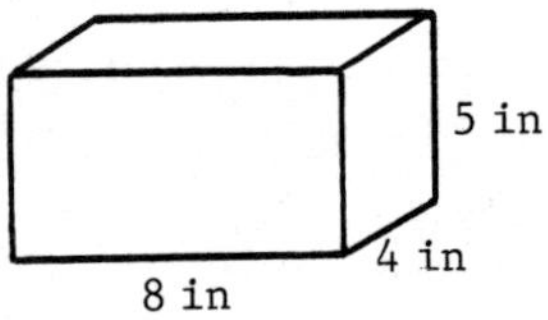

12. 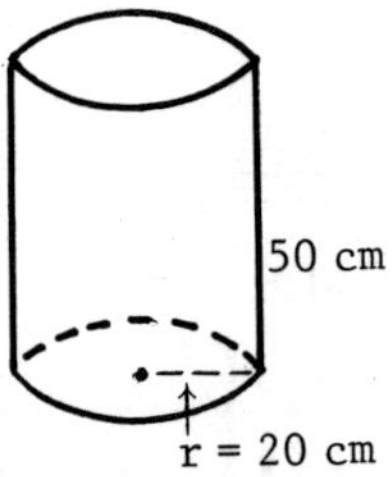

13. Find the hypotenuse of a right triangle if its legs are 6 cm and 2 cm.
14. Find the second leg of a right triangle if the first leg is 6 ft and the hypotenuse is 10 ft.
15. Find the diagonal of a square whose side is 3 yd.
16. Find the length of a rectangle if its diagonal is 5m and its width is 3 m.
17. Find the radius of a circle if its diameter is 14 cm.
18. Find the diameter of a circle if its circumference is 25.12 in.
19. Find the radius of a circle if its area is 12.56 yd^2.
20. Find the width of a rectangular solid if $V = 100\ ft^2$, $L = 5$ ft, and $H = 5$ ft.

UNIT 11: SCIENTIFIC NOTATION

Assignment 32

Convert each power of ten to an ordinary number.

1. 10^2 2. 10^{-3} 3. 10^1 4. 10^0 5. 10^{-1} 6. 10^4 7. 10^{-5}

8. 10^{-2} 9. 10^6 10. 10^{-6} 11. 10^7 12. 10^{-8} 13. 10^9 14. 10^{-10}

Convert each ordinary number to a power of ten.

15. 10 16. 100 17. 10,000 18. 1,000,000 19. 100,000,000

20. 1 21. .1 22. .01 23. .00001 24. .00000001

Perform each multiplication and division.

25. $10^3 \times 10^3$ 26. $10^{-1} \times 10^4$ 27. $10^5 \times 10^{-5}$ 28. $10^{-2} \times 10^{-1}$ 29. $10^0 \times 10^{-7}$

30. $\frac{10^5}{10^2}$ 31. $\frac{10^9}{10^8}$ 32. $\frac{10^4}{10^4}$ 33. $\frac{10^3}{10^6}$ 34. $\frac{10^{-1}}{10^3}$ 35. $\frac{10^{-4}}{10}$ 36. $\frac{10^{-2}}{10^{-9}}$

Assignment 33

Perform each multiplication.

1. $10^2 \times .84$ 2. $10^{-1} \times 527$ 3. $10^4 \times 16$ 4. $10^{-5} \times .9$

5. 91.3×10^0 6. $2{,}000 \times 10^{-3}$ 7. $.75 \times 10^6$ 8. 425×10^{-6}

Convert from scientific notation to an ordinary number.

9. 2.8×10^1 10. 6.56×10^4 11. 7×10^7 12. 9.4×10^9

13. 4.1×10^{-1} 14. 1.14×10^{-3} 15. 2×10^{-6} 16. 8.8×10^{-8}

Convert to scientific notation.

17. 27.7 18. 519 19. 49,600 20. 6,000,000 21. 425,000,000

22. .314 23. .075 24. .000004 25. .0000639 26. .000000011

Write each calculator answer in scientific notation and then convert it to an ordinary number.

	Left Side	Right Side					
27.	1.56	7	________	x	________	=	________________
28.	8.03	-6	________	x	________	=	________________
29.	7.22	06	________	x	________	=	________________
30.	4.73	-08	________	x	________	=	________________

UNIT 11: SCIENTIFIC NOTATION (continued)

Assignment 34

Convert each multiplication to scientific notation.

1. 256×10^3
2. 47.8×10^6
3. 366×10^{-4}
4. 15.9×10^{-6}
5. $.66 \times 10^8$
6. $.949 \times 10^{-7}$
7. $.0025 \times 10^8$
8. $3{,}700 \times 10^{-9}$

Write each product in scientific notation.

9. $3 \times (2 \times 10^6)$
10. $2 \times (4 \times 10^{-7})$
11. $50 \times (7 \times 10^5)$
12. $.04 \times (8 \times 10^{-6})$
13. $(4 \times 10^2) \times (2 \times 10^7)$
14. $(7 \times 10^{-7}) \times (3.4 \times 10^5)$
15. $(6.6 \times 10^{-1}) \times (2 \times 10^{-5})$

Write each quotient in scientific notation.

16. $\dfrac{8 \times 10^6}{2}$
17. $\dfrac{6 \times 10^{-7}}{3}$
18. $\dfrac{2.46 \times 10^{-5}}{6}$
19. $\dfrac{8.8 \times 10^7}{40}$
20. $\dfrac{6 \times 10^8}{2 \times 10^5}$
21. $\dfrac{8 \times 10^{-7}}{4 \times 10^{-2}}$
22. $\dfrac{6.33 \times 10^6}{2.11 \times 10^{12}}$
23. $\dfrac{4.56 \times 10^{-10}}{1.14 \times 10^{-15}}$

Express each answer in scientific notation.

24. If 1 mole of carbon dioxide contains 6.02×10^{23} molecules, how many molecules are there in 10 moles of carbon dioxide?

25. A star is 20 light years from earth. If one light year is 5.87×10^{12} miles, how far is that star from earth?

26. If 22.4 liters of a gas contain 6.02×10^{23} molecules, how many molecules are there in 10 liters of the same gas?

27. In $V_2 = \dfrac{C_1V_1}{C_2}$, find V_2 when $C_1 = 8 \times 10^{-3}$, $V_1 = 200$, and $C_2 = 2 \times 10^{-4}$.

UNIT 12: PERCENTS AND EQUATIONS

Assignment 35

From memory, convert each percent to a fraction and each fraction to a percent.

1. 50% 2. 25% 3. $33\frac{1}{3}\%$ 4. 80% 5. 30% 6. 20% 7. 90%

8. $\frac{2}{3}$ 9. $\frac{1}{2}$ 10. $\frac{3}{4}$ 11. $\frac{2}{5}$ 12. $\frac{7}{10}$ 13. $\frac{3}{5}$ 14. $\frac{1}{10}$ 15. $\frac{1}{4}$

Convert each percent to a decimal number or whole number.

16. 57% 17. 3.5% 18. $7\frac{1}{4}\%$ 19. 400% 20. .15% 21. 269% 22. .01%

Convert each decimal number or whole number to a percent.

23. .85 24. .041 25. 1 26. 3.08 27. 5 28. .0007 29. .0325

Convert to the nearest whole-number percent.

30. $\frac{25}{50}$ 31. $\frac{197}{300}$ 32. $\frac{675,000}{230,000}$

Convert to the nearest tenth of a percent.

33. $\frac{11}{12}$ 34. $\frac{1}{40}$ 35. $\frac{16}{300}$

Perform each operation.

36. 35% + 52% 37. 200% + 17% 38. 100% - 69% 39. 48% - 25%

40. 10% x 77 41. 60% x 50 42. .3% x 1,000 43. 1.2% x 225

44. $\frac{375}{10\%}$ 45. $\frac{54}{20\%}$ 46. $\frac{66}{300\%}$ 47. $\frac{228}{4\%}$

Assignment 36

Solve each problem.

1. What is 50% of 80? 2. 30% of 40 is what? 3. Find $66\frac{2}{3}\%$ of 90.

4. What is 81% of 1,500? 5. 122% of 2,000 is what? 6. Find .15% of 5,000.

Solve each problem.

7. 20 is what percent of 80? 8. What percent of 200 is 140? 9. 50 is what percent of 150?

10. What percent of 500 is 45? 11. 35 is what percent of 20? 12. What percent of 1,000 is 8?

Solve each problem.

13. 8 is 40% of what? 14. 90% of what is 45? 15. 34 is 17% of what?

16. 200% of what is 80? 17. 600 is 150% of what? 18. 12% of what is 180?

UNIT 12: PERCENTS AND EQUATIONS (continued)

Assignment 36 (continued)

19. If 80% of the 30 students in a class can swim, how many can swim?

20. If 12 of 40 people attend movies weekly, what percent attend weekly?

21. A student has 35 items correct on a test. If her grade was 70%, how many items were there on the test?

22. If the sales tax for a $350 sofa is $17.50, what is the sales tax rate?

23. An electric power company found that .92% of its 150,000 customers were late in paying their March bills. How many were late?

24. Only 20 of 10,000 entries in a contest received a prize. What percent received a prize?

25. A family spends $300 per month for housing. If that amount is 25% of their monthly income, what is their monthly income?

26. Find the monthly finance charge for an unpaid balance of $156 at a rate of $1\frac{1}{2}\%$.

Assignment 37

1. The price of a $5,000 car is increased 8%. What is the new price?

2. In a 5-year period, the enrollment at an elementary school decreased from 600 to 480. What was the percent decrease?

3. The price of a swimming suit was reduced $33\frac{1}{3}\%$ for a sale. If the amount of the reduction was $10, what was the original price?

4. Find the sale price of a $450 washer-dryer if it is reduced 10%.

5. In a seven-year period, a woman's weekly salary increased from $80 to $180. What was the percent increase?

6. The taxes for a home increased 25% in 5 years. If the amount of the increase was $200, what are the present taxes?

7. Find the interest on a $1,000 loan at 7% for 1 year.

8. If the interest for a $500 loan for 1 year is $30, what is the interest rate?

9. A woman can invest some money at 8%. In order to earn $400 interest in 1 year, how much must she invest?

10. Find the interest on a $200 loan at $10\frac{1}{2}\%$ for 3 years.

11. If $220 interest is paid for a $1,000 loan for 2 years, what is the interest rate?

12. If $25 interest is paid for a 10% loan for 6 months, what is the principal of the loan?

13. Find the interest for a $2,000 savings deposit at 6.28% for 3 months.

14. If the interest for a $1,200 investment for 4 months is $30, what is the interest rate?

UNIT 13: GRAPHING

Assignment 38

Complete the table of solutions for each equation below.

1. $y = 4x$

x	y
5	
3	
0	
-3	
-5	

2. $y = 3x - 2$

x	y
3	
1	
0	
-1	
-3	

3. $y = 5 - x$

x	y
2	
1	
0	
-1	
-2	

4. $2x - y = 4$

x	y
4	
2	
0	
-2	
-4	

Assignment 39

1. Write the coordinates of the following points on the coordinate system at the right.

A ______ E ______

B ______ F ______

C ______ G ______

D ______ H ______

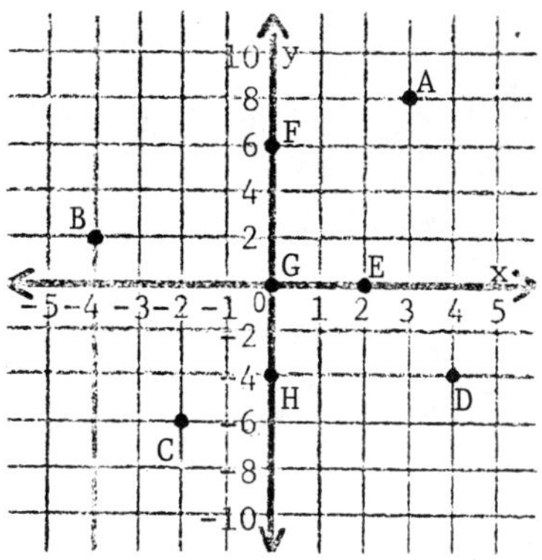

2. List the quadrant (1, 2, 3, or 4) in which each point below lies.

B ______ C ______ A ______ D ______

3. Plot and label the following points on the coordinate system at the right.

A (-2, 6) E (9, 2)

B (5, -3) F (-4, -7)

C (0, 8) G (0, 0)

D (-7, 0) H (0, -5)

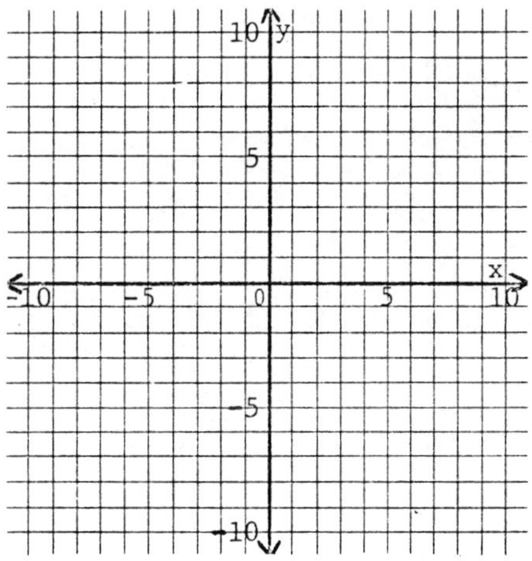

UNIT 13: GRAPHING (continued)

Assignment 40

Complete each table and then graph each equation.

1. $y = x + 2$

x	y
2	
1	
0	
-1	
-2	

2. $x + y = 0$

x	y
2	
1	
0	
-1	
-2	

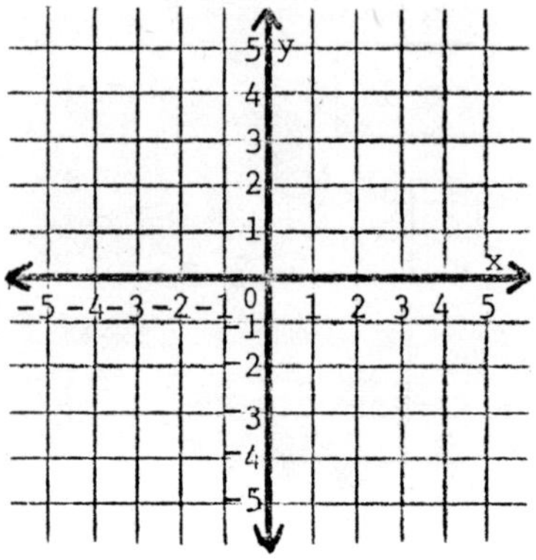

Make up a solution table and then graph each equation.

3. $y = x + 5$

x	y

4. $2x + y = 3$

x	y

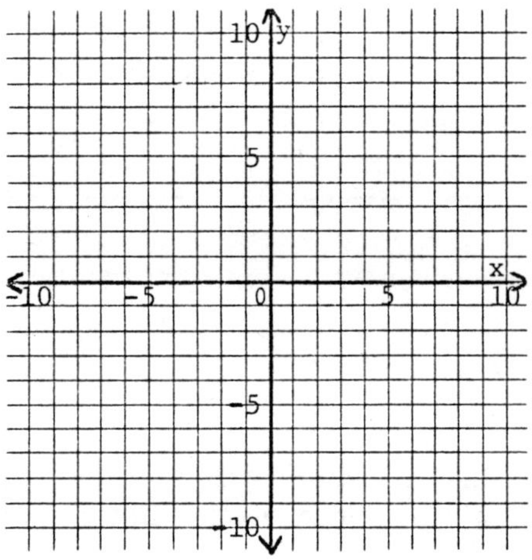

The graph below shows the relationship between "distance traveled" and "time" for an airplane.

5. How far can the plane travel in 3 hours?

6. How many hours would it take to travel 1,000 miles?

7. How far can the plane travel in 6 hours?

8. How many hours would it take to travel 200 miles?

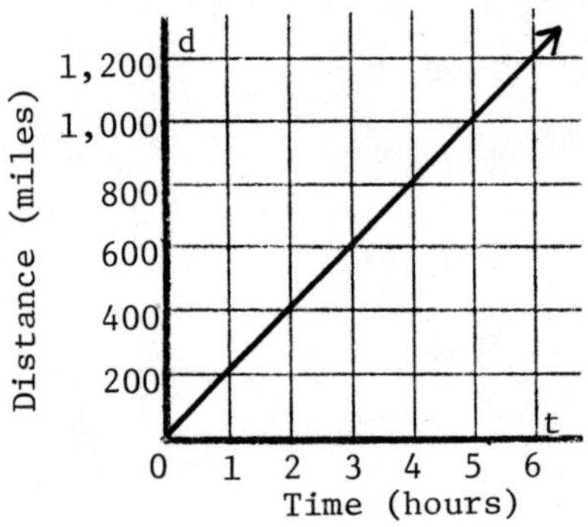

UNIT 14: SYSTEMS OF EQUATIONS

Assignment 41

Use the graphing method to solve each system below.

1. $x + y = 2$
   $y = x$

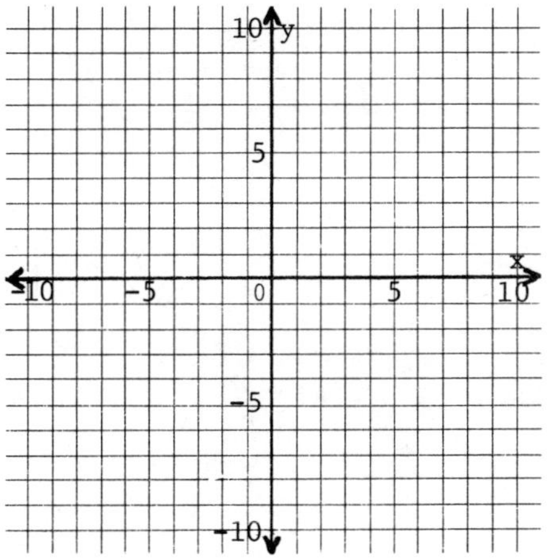

2. $x - y = 2$
   $x + y = 6$

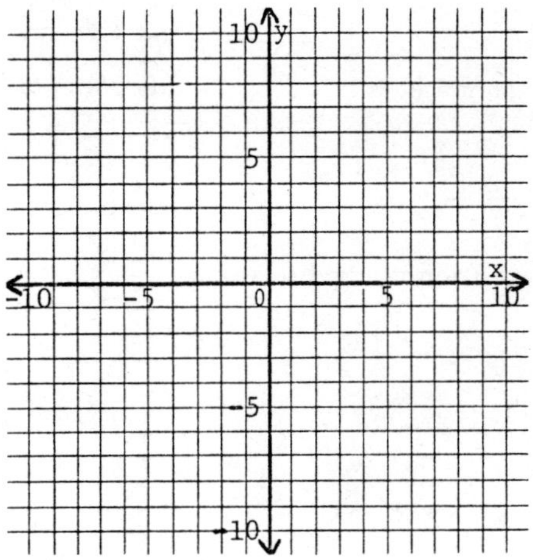

3. $x - y = 6$
   $2x + y = 3$

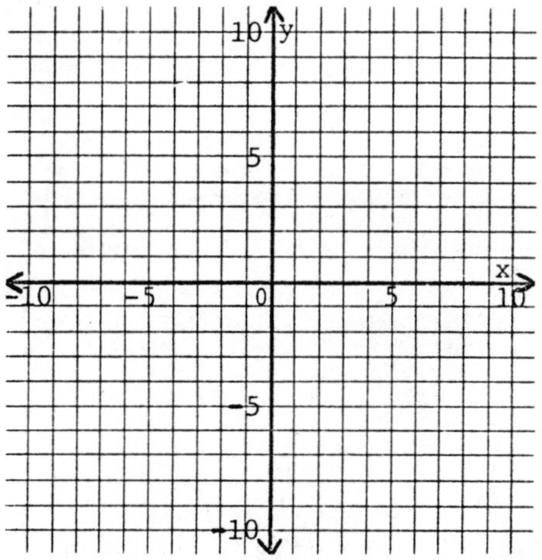

4. $y = 2x$
   $x + y = 9$

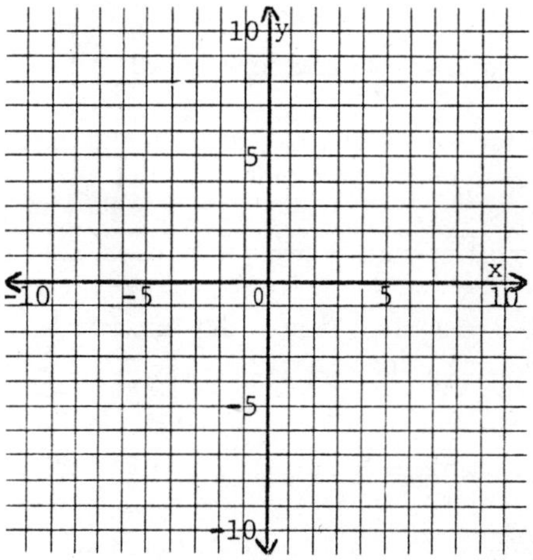

Use the addition method to solve each system.

5. $x + y = 7$
   $x - y = 3$

6. $2p + q = 5$
   $p - q = 1$

7. $s - 2t = 7$
   $3s + 2t = 13$

8. $4a + 3b = 8$
   $a - 3b = 2$

UNIT 14: SYSTEMS OF EQUATIONS (continued)

Assignment 42

Use the addition method to solve each system.

1. $x + 2y = 9$
   $2x - y = 8$

2. $a - 3b = 0$
   $3a + b = 10$

3. $s + 4t = 14$
   $3s - 2t = 0$

4. $p + q = 1$
   $3p - 4q = 10$

5. $3x + 2y = 8$
   $x + y = 2$

6. $a + 2b = 7$
   $3a + 4b = 15$

7. $2m + t = 3$
   $4m + t = 1$

8. $p - 2q = 0$
   $3p - 2q = 8$

Assignment 43

Use the addition method to solve each system.

1. $x - y = 3$
   $y = 2x - 7$

2. $d = h + 5$
   $d + h = 1$

3. $2p + q = -5$
   $p = 2q$

4. $m = 3t$
   $m - t = 4$

5. $2x - y = 1$
   $y = x$

6. $3a + b = 1$
   $a = b - 9$

Solve each problem.

7. There are 29 students in a class. If there are 7 more females than males, how many females and males are there?

8. The sum of two numbers is 48. The larger is three times the smaller. Find the two numbers.

9. A husband weighs 40 pounds more than his wife. Together they weigh 280 pounds. Find the weight of each.

10. The difference of two numbers is 10. The larger is twice the smaller. Find the two numbers.

11. A board 18 feet long is cut into two parts. If the larger part is 2 feet longer than the smaller, how long is each part?

12. The perimeter of a rectangle is 50 meters. If the length is four times the width, find the length and width.

13. A piece of wire 36 centimeters long is cut into two parts. If the larger part is five times the smaller, how long is each part?

14. The perimeter of a rectangle is 36 feet. If the length is 4 feet longer than the width, find the length and width.

UNIT 15: SPECIAL PRODUCTS AND FACTORING

Assignment 44

Perform each multiplication.

1. $(x)(3)$ 2. $(5)(4y)$ 3. $(4p)(1)$ 4. $(a)(a)$ 5. $(x)(8x)$

6. $(3y)(y)$ 7. $(2m)(3m)$ 8. $(6x)(5x)$ 9. $(3y)(7y)$ 10. $(8t)(8t)$

Perform each multiplication.

11. $3(2x + 5)$ 12. $4(3y - 1)$ 13. $p(p - 1)$ 14. $h(h + 1)$

15. $x(x + 8)$ 16. $y(3y - 4)$ 17. $5a(a + 7)$ 18. $6b(2b - 1)$

Write the missing factor.

19. $6x = (3)(\quad)$ 20. $8y = (8y)(\quad)$ 21. $b^2 = (b)(\quad)$ 22. $2m^2 = (\quad)(2m)$

23. $9x^2 = (x)(\quad)$ 24. $4t^2 = (\quad)(2t)$ 25. $30p^2 = (5p)(\quad)$ 26. $32b^2 = (\quad)(4b)$

Factor completely.

27. $2x + 8$ 28. $16y - 12$ 29. $8t + 4$ 30. $a^2 + 4a$

31. $t^2 - 9t$ 32. $h^2 + h$ 33. $s^2 - s$ 34. $2p^2 + 3p$

35. $5d^2 - 15d$ 36. $8x^2 + 6x$ 37. $6y^2 - 12y$ 38. $16t^2 + 4t$

Assignment 45

Find each trinomial product.

1. $(x + 1)(x + 5)$ 2. $(y + 6)(y + 4)$ 3. $(m + 2)(m + 2)$ 4. $(t + 1)(5t + 2)$

5. $(3p + 4)(p + 2)$ 6. $(2x + 1)(3x + 5)$ 7. $(3y + 2)(4y + 1)$ 8. $(2a + 3)(2a + 3)$

Find each trinomial product.

9. $(x + 7)(x - 1)$ 10. $(y - 1)(y + 2)$ 11. $(t + 4)(t - 5)$ 12. $(m + 8)(m - 2)$

13. $(2a + 7)(a - 3)$ 14. $(b - 4)(3b + 1)$ 15. $(4x + 1)(2x - 5)$ 16. $(3y - 1)(5y + 4)$

17. $(x - 3)(x - 2)$ 18. $(y - 8)(y - 9)$ 19. $(b - 3)(b - 3)$ 20. $(2b - 1)(b - 1)$

21. $(t - 2)(4t - 3)$ 22. $(3y - 2)(3y - 4)$ 23. $(5p - 4)(2p - 1)$ 24. $(2t - 3)(2t - 3)$

UNIT 15: SPECIAL PRODUCTS AND FACTORING (continued)

Assignment 45 (continued)

Find each binomial product.

25. $(x + 5)(x - 5)$ 26. $(y - 7)(y + 7)$ 27. $(m + 1)(m - 1)$ 28. $(a - 2)(a + 2)$

29. $(2t + 3)(2t - 3)$ 30. $(4x - 1)(4x + 1)$ 31. $(5b + 3)(5b - 3)$ 32. $(6y - 2)(6y + 2)$

Assignment 46

Factor each binomial.

1. $x^2 - 9$ 2. $y^2 - 100$ 3. $b^2 - 1$ 4. $h^2 - 81$

5. $9m^2 - 4$ 6. $4y^2 - 1$ 7. $49t^2 - 36$ 8. $64x^2 - 25$

Factor each trinomial.

9. $x^2 + 7x + 6$ 10. $y^2 + 6y + 9$ 11. $t^2 + 9t + 20$ 12. $m^2 - 4m + 3$

13. $b^2 - 11b + 18$ 14. $p^2 - 10p + 25$ 15. $x^2 + 2x - 15$ 16. $y^2 - 3y - 10$

17. $t^2 + t - 6$ 18. $a^2 - a - 12$ 19. $p^2 + 6p - 16$ 20. $m^2 - 7m - 18$

Factor each trinomial.

21. $2x^2 + 5x + 3$ 22. $7t^2 + 17t + 6$ 23. $3x^2 - 8x + 5$ 24. $5y^2 - 14y + 8$

25. $2a^2 + 5a - 3$ 26. $5h^2 - 3h - 2$ 27. $3y^2 + y - 2$ 28. $5y^2 - 2y - 7$

UNIT 16: QUADRATIC EQUATIONS

Assignment 47

Solve each equation. Use the square-root table when necessary.

1. $x^2 = 25$
2. $y^2 = 144$
3. $t^2 = 37$
4. $m^2 = 185$
5. $p^2 = \frac{9}{49}$
6. $d^2 = \frac{81}{25}$
7. $x^2 = \frac{1}{9}$
8. $y^2 = \frac{1}{36}$
9. $b^2 - 1 = 0$
10. $h^2 - 64 = 0$
11. $t^2 - 22 = 0$
12. $m^2 - 130 = 0$
13. $4x^2 = 9$
14. $25y^2 = 75$
15. $4a^2 - 1 = 0$
16. $25k^2 - 49 = 0$
17. $\frac{1}{4}t^2 = 16$
18. $30 = \frac{1}{3}m^2$
19. $\frac{1}{2}x^2 - 50 = 0$
20. $\frac{1}{5}y^2 - 30 = 0$

Solve each equation.

21. $(x - 5)(x - 7) = 0$
22. $(y + 10)(y - 1) = 0$
23. $(t + 1)(t + 3) = 0$
24. $(4m - 3)(2m - 9) = 0$
25. $(3a - 5)(2a + 1) = 0$
26. $(7a + 1)(4a + 5) = 0$
27. $x(x - 8) = 0$
28. $y(y + 1) = 0$
29. $m(5m - 2) = 0$
30. $2t(t + 4) = 0$
31. $5d(d - 1) = 0$
32. $8p(2p + 3) = 0$

Assignment 48

Use the factoring method to solve each equation.

1. $x^2 - 3x = 0$
2. $y^2 + 10y = 0$
3. $p^2 - p = 0$
4. $d^2 + d = 0$
5. $3t^2 - 2t = 0$
6. $5m^2 + 7m = 0$
7. $4a^2 - a = 0$
8. $9b^2 + b = 0$
9. $5x^2 - 10x = 0$
10. $8y^2 + 2y = 0$
11. $4h^2 - 8h = 0$
12. $24c^2 + 18c = 0$
13. $y^2 = 5y$
14. $7x^2 = x$
15. $3p^2 = 12p$
16. $2t^2 = -5t$

UNIT 16: QUADRATIC EQUATIONS (continued)

Assignment 48 (continued)

Use the factoring method to solve each equation.

17. $x^2 + 6x + 5 = 0$

18. $y^2 + 5y + 6 = 0$

19. $t^2 + 10t + 25 = 0$

20. $m^2 - 7m + 12 = 0$

21. $a^2 - 11a + 10 = 0$

22. $d^2 - 8d + 16 = 0$

23. $x^2 + 2x - 15 = 0$

24. $y^2 + y - 12 = 0$

25. $t^2 - 4t - 12 = 0$

26. $p^2 - p = 20$

27. $h^2 + 1 = 2h$

28. $x^2 = 5x + 14$

Use the factoring method to solve each equation.

29. $2x^2 + 5x + 3 = 0$

30. $5y^2 + 14y + 8 = 0$

31. $3t^2 - 8t + 5 = 0$

32. $7y^2 - 9y + 2 = 0$

33. $2p^2 + p - 6 = 0$

34. $3d^2 - 2d - 5 = 0$

35. $5m^2 + 2 = 7m$

36. $2x^2 - x = 3$

37. $7t^2 = 10t - 3$

Assignment 49

Use the quadratic formula to solve each equation.

1. $3x^2 - 8x - 3 = 0$

2. $2y^2 + 7y - 15 = 0$

3. $5d^2 - d - 4 = 0$

4. $3t^2 + 5t - 12 = 0$

5. $4x^2 + 11x - 3 = 0$

6. $10y^2 - 7y + 1 = 0$

Use the quadratic formula to solve each equation. Round each root to the nearest hundredth.

7. $x^2 - 5x + 3 = 0$

8. $5y^2 - 4y - 2 = 0$

9. $2t^2 + 8t + 5 = 0$

10. $m^2 - 9m + 10 = 0$

11. $3d^2 + 10d + 1 = 0$

12. $6h^2 + h - 4 = 0$

UNIT 17: EXPONENTS AND POLYNOMIALS

Assignment 50

Convert each power to an ordinary number.

1. 8^2 2. 2^3 3. 4^4 4. 1^8 5. 9^1 6. 8^0

Write each product as a power.

7. $3^2 \cdot 3^3$ 8. $t^5 \cdot t^4$ 9. $4^3 \cdot 4^3$ 10. $m^6 \cdot m^6$

11. $8 \cdot 8^2$ 12. $b^4 \cdot b$ 13. $7^5 \cdot 7^0$ 14. $a^0 \cdot a$

Write each quotient as a power.

15. $\frac{5^4}{5^2}$ 16. $\frac{y^{10}}{y^6}$ 17. $\frac{7^4}{7^3}$ 18. $\frac{p^2}{p}$

19. $\frac{2^5}{2^0}$ 20. $\frac{t}{t^0}$ 21. $\frac{9^2}{9^2}$ 22. $\frac{b^{10}}{b^{10}}$

Raise each power to the indicated power.

23. $(4^2)^5$ 24. $(x^6)^3$ 25. $(7^4)^2$ 26. $(y^5)^5$

27. $(2^3)^1$ 28. $(2^3)^0$ 29. $(t^9)^1$ 30. $(t^9)^0$

Assignment 51

Perform each multiplication.

1. $3(4x^5)$ 2. $-2(7y^2)$ 3. $(b)(-b^2)$ 4. $(3t^4)(t)$

5. $(4m)(2m)$ 6. $(-x^4)(3x^3)$ 7. $(-2d)(d^5)$ 8. $(-7p^3)(-3p^3)$

Perform each division.

9. $\frac{8x^2}{2}$ 10. $\frac{-12y^5}{6}$ 11. $\frac{15t^4}{-5t^2}$ 12. $\frac{-10p^6}{-2p^5}$

13. $\frac{7m^6}{7m}$ 14. $\frac{-2t^3}{-t^3}$ 15. $\frac{-5d^4}{d}$ 16. $\frac{-3y}{3y}$

UNIT 17: EXPONENTS AND POLYNOMIALS (continued)

Assignment 51 (continued)

Perform each multiplication.

17. $x(2x^2 - 1)$
18. $m^3(m^4 + 5m^2)$
19. $2y(y^3 - y)$
20. $3p^2(2p + 7)$
21. $t^3(3t^2 - t + 1)$
22. $5d(d^5 + d^3 - d)$

Perform each division.

23. $\frac{14x - 7}{7}$
24. $\frac{12y^3 + 8y}{4y}$
25. $\frac{m^{10} - 5m^6}{m^2}$
26. $\frac{3b^5 + 6b^3}{3b^2}$
27. $\frac{d^4 - 2d^2 - d}{d}$
28. $\frac{4p^6 + 6p^5 - 8p^4}{2p^3}$

Assignment 52

Perform each addition and subtraction.

1. $5t + 3t$
2. $2x^3 + 7x^3$
3. $m^2 + 2m^2$
4. $8d^4 + d^4$
5. $p^3 + p^3$
6. $9q - 6q$
7. $3t^3 - 7t^3$
8. $a^2 - 2a^2$
9. $7x^4 - x^4$
10. $b^3 - b^3$

Perform each addition horizontally.

11. $(x^2 + 3) + (2x^2 + 7)$
12. $(3y + 9) + (y - 6)$
13. $(t^4 - 8) + (5t^4 + 7)$
14. $(5p^3 - p^2) + (2p^2 - p)$
15. $(m^2 - 5m + 3) + (m^2 + 5m - 10)$
16. $(a^3 - 3a^2 - 1) + (2a^5 + a^4 - a^3)$

Perform each subtraction horizontally.

17. $(3b^2 + b) - (b^2 + 5b)$
18. $(5x - 7) - (4x - 8)$
19. $(t^4 - t^2) - (2t^4 - 3t^2)$
20. $(2b^2 + b) - (2b^2 + b)$
21. $(3y^2 - 7y + 8) - (y^2 - 7y - 1)$
22. $(p^2 + 3p - 6) - (p^3 + p^2 - 6p)$

Perform each addition in columns.

23. Add $5b^2 - 8$ and $b^2 + 5$.
24. Add $t^3 - 3t^2$ and $-4t^3 + 3t^2$.
25. Add $y^2 - 5y + 6$, $y - 5$, and $2y^2 + 3y - 3$.
26. Add $2x + 9$, $x^2 - x - 1$, and $4x^3 + 3x^2 - 6$.

Perform each subtraction in columns.

27. Subtract $3y + 2$ from $7y + 9$.
28. Subtract $x^2 - x$ from $4x^2 - 3x$.
29. Subtract $a^2 + 2a + 1$ from $2a^2 + a + 2$
30. Subtract $t^3 + 3t^2 - 5$ from $t^4 + 3t^2 - 5$.

UNIT 18: RADICALS

Assignment 53

Perform each multiplication.

1. $\sqrt{2} \cdot \sqrt{7}$
2. $\sqrt{5} \cdot \sqrt{x}$
3. $\sqrt{3t} \cdot \sqrt{10}$
4. $\sqrt{7} \cdot \sqrt{x} \cdot \sqrt{3}$
5. $\sqrt{2d} \cdot \sqrt{3} \cdot \sqrt{11}$
6. $5 \cdot 4\sqrt{3}$
7. $10\sqrt{5} \cdot \sqrt{3}$
8. $\sqrt{2} \cdot 8\sqrt{3}$
9. $10\sqrt{3} \cdot 5\sqrt{7}$

Simplify by factoring out perfect squares.

10. $\sqrt{16}$
11. $\sqrt{45}$
12. $\sqrt{200}$
13. $\sqrt{49y}$
14. $\sqrt{12x}$
15. $\sqrt{32t}$
16. $\sqrt{16d^4}$
17. $\sqrt{5x^2}$
18. $\sqrt{18p^6}$
19. $\sqrt{4y^3}$
20. $\sqrt{7y^5}$
21. $\sqrt{50m^7}$

Multiply and simplify each product.

22. $\sqrt{3} \cdot \sqrt{12}$
23. $\sqrt{2} \cdot \sqrt{10}$
24. $\sqrt{8} \cdot \sqrt{2x}$
25. $\sqrt{5y} \cdot \sqrt{15}$
26. $\sqrt{2t} \cdot \sqrt{3t}$
27. $\sqrt{5b} \cdot \sqrt{5b}$
28. $\sqrt{t^3} \cdot \sqrt{4t}$
29. $\sqrt{2x^2} \cdot \sqrt{50x}$
30. $\sqrt{y^3} \cdot \sqrt{8y^2}$

Assignment 54

Divide and simplify each quotient.

1. $\frac{\sqrt{32}}{\sqrt{2}}$
2. $\frac{\sqrt{21}}{\sqrt{3}}$
3. $\frac{\sqrt{50x}}{\sqrt{2x}}$
4. $\frac{\sqrt{y^6}}{\sqrt{y^4}}$
5. $\frac{\sqrt{p^4}}{\sqrt{p}}$
6. $\frac{\sqrt{6x^3}}{\sqrt{2x}}$
7. $\frac{\sqrt{24t^5}}{\sqrt{3t^2}}$
8. $\frac{\sqrt{14d^7}}{\sqrt{2d^2}}$

Simplify by factoring out perfect squares.

9. $\sqrt{\frac{1}{25}}$
10. $\sqrt{\frac{13}{16}}$
11. $\sqrt{\frac{18}{49}}$
12. $\sqrt{\frac{3x}{4}}$
13. $\sqrt{\frac{y^4}{36}}$
14. $\sqrt{\frac{12x^2}{25}}$
15. $\sqrt{\frac{24}{t^6}}$
16. $\sqrt{\frac{16p^3}{9}}$

UNIT 18: RADICALS (continued)

Assignment 54 (continued)

Rationalize each denominator.

17. $\dfrac{3}{\sqrt{5}}$ 18. $\dfrac{1}{\sqrt{x}}$ 19. $\dfrac{7}{2\sqrt{3}}$ 20. $\dfrac{1}{5\sqrt{2y}}$

21. $\dfrac{2\sqrt{5}}{3\sqrt{7}}$ 22. $\dfrac{7\sqrt{2}}{t^2\sqrt{3t}}$ 23. $\sqrt{\dfrac{2}{3}}$ 24. $\sqrt{\dfrac{18}{m^3}}$

Multiply and simplify. Rationalize all denominators.

25. $\sqrt{\dfrac{1}{2}} \cdot \sqrt{\dfrac{3}{2}}$ 26. $\sqrt{\dfrac{7}{x}} \cdot \sqrt{\dfrac{5}{x}}$ 27. $\sqrt{\dfrac{y}{6}} \cdot \sqrt{\dfrac{y^4}{6}}$

28. $\sqrt{\dfrac{1}{3}} \cdot \sqrt{\dfrac{1}{2}}$ 29. $\sqrt{\dfrac{3}{5}} \cdot \sqrt{\dfrac{4}{x}}$ 30. $\sqrt{\dfrac{1}{3}} \cdot \sqrt{\dfrac{8}{3t}}$

Perform each possible addition and subtraction.

31. $2\sqrt{5} + 2\sqrt{5}$ 32. $\sqrt{3} + 4\sqrt{3}$ 33. $5\sqrt{7} + 3\sqrt{2}$ 34. $2\sqrt{x} + \sqrt{x}$

35. $\sqrt{m} + \sqrt{m}$ 36. $3 + \sqrt{6}$ 37. $9\sqrt{2} - 7\sqrt{2}$ 38. $8\sqrt{5} - 6\sqrt{3}$

39. $3\sqrt{t} - \sqrt{t}$ 40. $\sqrt{p} - 2\sqrt{p}$ 41. $x - \sqrt{x}$ 42. $7\sqrt{11} - 7\sqrt{11}$

ANSWERS FOR SUPPLEMENTARY PROBLEMS

UNIT 1: SIGNED WHOLE NUMBERS

Assignment 1

1. +3 yards 2. +10° 3. −$50 4. −5° 5. −7 yards 6. +$25 7. +$50 8. +20°
9. −$100 10. > 11. > 12. > 13. < 14. < 15. < 16. < 17. > 18. >
19. > 20. < 21. > 22. 6 23. 5 24. 17 25. 17 26. 0 27. 1 28. 1
29. 35 30. 47 31. 63

Assignment 2

1. −6 2. −3 3. 7 4. −4 5. −8 6. −8 7. 0 8. 5 9. −10 10. 5 11. 0
12. −30 13. 4 14. −10 15. −17 16. −4 17. 3 18. −6 19. 1 20. 20

Assignment 3

1. −8 2. 10 3. −3 4. 1 5. 8 6. 0 7. −3 8. −2 9. −11 10. 32
11. −55 12. 20 13. 1 14. 0 15. −1 16. −6 17. −13 18. 0 19. 3 20. −1

Assignment 4

1. −56 2. −20 3. 0 4. 10 5. −12 6. 54 7. 28 8. 0 9. −42 10. 12
11. 0 12. −56 13. 14 14. 60 15. 16 16. 0 17. −27 18. −3 19. −6
20. −5 21. 0 22. −15 23. −9 24. 8 25. −32 26. 4 27. 5 28. 0

UNIT 2: SOLVING EQUATIONS

Assignment 5

1. 2 2. 6 3. 4 4. 4 5. 10 6. 9 7. 7 8. −4 9. −4 10. 6 11. −3
12. 5 13. 8 14. −13 15. −1 16. 1 17. 0 18. 0

Assignment 6

1. 8 2. 14 3. −6 4. 3 5. −6 6. 10 7. 11 8. −6 9. 20 10. 4 11. 4
12. 0 13. −3 14. −8 15. 12 16. 5 17. −8 18. 0

Assignment 7

1. −5 2. 9 3. 7 4. 5 5. −4 6. 1 7. 5 8. −2 9. 4 10. −2 11. 3
12. 5 13. 10 14. −1 15. −10 16. −9 17. 1 18. 2

Assignment 8

1. 4 2. 3 3. 2 4. −2 5. 12 6. −7 7. 10 8. −1 9. 1 10. −3 11. 32
12. 2 13. 5 14. 2 15. −8 16. 2 17. −3 18. −20

UNIT 3: FRACTIONS AND EQUATIONS: MULTIPLICATION AND DIVISION

Assignment 9

1. $\frac{1}{2}$ 2. $\frac{3}{2}$ 3. $\frac{2}{5}$ 4. $\frac{6}{5}$ 5. $\frac{3}{4}$ 6. $\frac{5}{2}$ 7. $\frac{3}{5}$ 8. $\frac{5}{4}$ 9. $\frac{1}{3}$ 10. $\frac{6}{5}$ 11. $\frac{1}{4}$ 12. $\frac{3}{2}$

13. $\frac{5}{8}$ 14. $\frac{7}{2}$ 15. $\frac{1}{4}$ 16. $\frac{8}{3}$ 17. $\frac{5}{6}$ 18. $\frac{1}{30}$ 19. $\frac{8}{3}$ 20. $\frac{7}{6}$ 21. $\frac{1}{15}$ 22. 6

UNIT 3: FRACTIONS AND EQUATIONS: MULTIPLICATION AND DIVISION (continued)

Assignment 9 (continued)

23. 1   24. $\frac{3}{4}$ 25. 4   26. 10   27. $\frac{21}{4}$ 28. $\frac{5}{2}$ 29. 0   30. $\frac{1}{3}$ 31. 1   32. $\frac{11}{12}$

Assignment 10

1. $-\frac{2}{15}$ 2. $\frac{1}{12}$ 3. -6   4. $\frac{3}{10}$ 5. -4   6. $\frac{9}{2}$ 7. $-\frac{1}{10}$ 8. 0   9. -2   10. -4

11. 8   12. -9   13. -5   14. 1   15. -8   16. -5   17. 0   18. Impossible   19. $\frac{9}{5}$

20. $\frac{7}{12}$ 21. $-\frac{3}{4}$ 22. $-\frac{11}{2}$ 23. $\frac{7}{4}$ 24. $\frac{1}{6}$ 25. $\frac{7}{3}$ 26. $-\frac{2}{3}$ 27. $\frac{4}{5}$ 28. $\frac{8}{3}$ 29. $-\frac{1}{5}$

30. $\frac{1}{10}$

Assignment 11

1. $\frac{5}{12}$ 2. 15   3. $\frac{1}{18}$ 4. 2   5. $\frac{6}{35}$ 6. $-\frac{1}{8}$ 7. $\frac{21}{2}$ 8. $-\frac{1}{3}$ 9. $-\frac{11}{10}$ 10. $-\frac{8}{5}$

11. 0   12. Impossible   13. 16   14. $\frac{3}{8}$ 15. -8   16. $-\frac{5}{2}$ 17. $\frac{1}{4}$ 18. $-\frac{10}{3}$ 19. $-\frac{1}{6}$

20. $\frac{3}{4}$ 21. -4   22. 4   23. $-\frac{2}{5}$ 24. 0

UNIT 4: FRACTIONS AND EQUATIONS: ADDITION AND SUBTRACTION

Assignment 12

1. $\frac{3}{5}$ 2. $\frac{4}{3}$ 3. 1   4. $\frac{7}{8}$ 5. $\frac{13}{20}$ 6. $\frac{4}{3}$ 7. $\frac{7}{12}$ 8. $\frac{29}{24}$ 9. $\frac{17}{20}$ 10. $\frac{5}{4}$ 11. $\frac{8}{3}$

12. $\frac{5}{2}$ 13. $-\frac{1}{9}$ 14. $\frac{2}{3}$ 15. -2   16. $\frac{7}{10}$ 17. $-\frac{3}{8}$ 18. $-\frac{5}{4}$ 19. $-\frac{1}{6}$ 20. $\frac{1}{12}$

21. $-\frac{47}{40}$ 22. $\frac{1}{4}$ 23. $-\frac{3}{2}$ 24. $-\frac{5}{3}$

Assignment 13

1. 0   2. $\frac{1}{2}$ 3. $\frac{3}{4}$ 4. $\frac{3}{4}$ 5. $\frac{1}{20}$ 6. $\frac{13}{24}$ 7. $\frac{9}{5}$ 8. $\frac{1}{4}$ 9. $-\frac{2}{5}$ 10. 1   11. $-\frac{3}{8}$

12. $\frac{1}{5}$ 13. $-\frac{4}{21}$ 14. $\frac{5}{12}$ 15. $\frac{5}{2}$ 16. $-\frac{8}{5}$ 17. $\frac{1}{3}$ 18. $-\frac{2}{3}$ 19. $\frac{1}{2}$ 20. $-\frac{1}{4}$

21. $-\frac{4}{5}$ 22. $\frac{3}{2}$ 23. $\frac{2}{3}$ 24. $\frac{7}{4}$ 25. $-\frac{3}{2}$ 26. $\frac{4}{3}$ 27. $-\frac{1}{2}$ 28. $\frac{9}{10}$ 29. $\frac{2}{3}$

30. $\frac{2}{5}$ 31. $-\frac{10}{7}$

UNIT 4: FRACTIONS AND EQUATIONS: ADDITION AND SUBTRACTION (continued)

Assignment 14

1. $\frac{2}{5}$ 2. 8 3. 0 4. $\frac{10}{3}$ 5. $-\frac{4}{3}$ 6. $\frac{13}{10}$ 7. $\frac{1}{8}$ 8. $-\frac{3}{7}$ 9. $-\frac{2}{3}$ 10. $\frac{14}{15}$

11. $\frac{5}{4}$ 12. $-\frac{1}{2}$ 13. $-\frac{4}{3}$ 14. $\frac{7}{24}$ 15. $\frac{7}{2}$ 16. $-\frac{3}{16}$ 17. 9 18. $-\frac{1}{5}$ 19. $\frac{1}{7}$

20. $-\frac{3}{2}$ 21. 16

UNIT 5: DECIMAL NUMBERS AND EQUATIONS

Assignment 15

1. $8.94 2. 19.4 3. 4.58 4. 5.757 5. -.15 6. 1.7 7. .1 8. -.57
9. 2.44 10. -1.6 11. -2.86 12. -.59 13. $13.55 14. 2.7 15. $5.65
16. .021 17. -.8 18. .61 19. -1.1 20. -.6 21. -1.88 22. 1.5 23. -1.65
24. -1.7 25. 6.5 26. -.17 27. -2.8 28. 10.1 29. 1.87 30. 3.6

Assignment 16

1. 6.6 2. .15 3. .032 4. 29.5 5. 0 6. -.414 7. .18 8. -.42 9. 3.52
10. -5.49 11. -.54 12. 14.52 13. .08 14. 1.1 15. 4 16. .4 17. -6.87
18. -.4 19. 0 20. 1 21. -.2 22. -3.9 23. 8 24. -76 25. .15 26. 2.5
27. 9 28. 0 29. -4.5 30. -20

Assignment 17

1. 1.2 2. 4 3. -.05 4. 50 5. 4.1 6. 9 7. 19.5 8. .4 9. .007
10. 2.671 11. $26 12. 412 13. .70 (or .7) 14. 2.10 (or 2.1) 15. -9.3 16. -92
17. .442 18. 3.9 19. .44 20. -.818 21. .5 22. .67 23. .25 24. .8
25. .86 26. 1.11 27. .08 28. 2.77

UNIT 6: FRACTIONAL EQUATIONS AND PROPORTIONS

Assignment 18

1. 10 2. 7 3. 0 4. $\frac{4}{5}$ 5. $\frac{1}{2}$ 6. 7 7. 1 8. -8

Assignment 19

1. $\frac{6}{5}$ 2. 8 3. $\frac{21}{8}$ 4. $\frac{1}{6}$ 5. 8 6. -3 7. $-\frac{8}{5}$ 8. 2

Assignment 20

1. 60 2. 2 3. 80 4. 8 5. 12 pounds 6. 30 minutes 7. 75 Republicans

8. 540 "Yes" votes 9. 15.4 pounds 10. 11.9 liters 11. 48 gallons 12. $4.80

UNIT 7: SQUARES, SQUARE ROOTS, AND EQUATIONS

Assignment 21

1. 49 2. 1 3. $\frac{1}{25}$ 4. $\frac{9}{4}$ 5. 1,024 6. 160,000 7. .0004 8. 2.25 9. 4
10. -11 11. ±10 12. 1.732 13. -8.307 14. ±13.82 15. 5.196 16. -12.207

Assignment 22

1. ±1 2. ±10 3. ±4 4. $\pm\frac{1}{5}$ 5. ±8 6. $\pm\frac{9}{7}$ 7. ±6 8. $\pm\frac{6}{5}$ 9. ±6.708
10. ±11.619 11. ±7.416 12. ±12.923

Assignment 23

1. $\frac{25}{49}$ 2. $\frac{1}{3}$ 3. $\frac{4}{9}$ 4. 40 5. 32 6. $\frac{13}{20}$ 7. $-\frac{3}{5}$ 8. $\frac{25}{18}$ 9. ±2 10. $\pm\frac{1}{3}$
11. ±3 12. ±8.66

UNIT 8: FORMULA EVALUATION

Assignment 24

1. L = \$40 2. A = 105 3. v = 100 4. A = 80 5. A = 12.5 6. y = 10 7. C = 15
8. I = 3 9. t = .04 10. V = 24 11. R = 1.9 12. F = 68 13. A = 600 14. P = 50
15. P = 5 16. H = 60 17. E = 6 18. t = 2.828 19. R = 4.899 20. N = 50

Assignment 25

1. $F_T = 9$ 2. E = 2 3. R = 24 4. m = 3 5. E = 20 6. d = 4 7. R = \$22.45
8. W = 2 9. $V_2 = 5$ 10. $t_2 = 4$ 11. I = 8 12. t = 3.162 13. d = 3 14. s = 12
15. P = 98 16. b = 4.472 17. $F_2 = 12$ 18. $d_1 = 6.928$

UNIT 9: FORMULA REARRANGEMENT

Assignment 26

1. $m = \frac{p}{2}$ 2. $t = \frac{v}{a}$ 3. $c = \frac{a}{b^2}$ 4. $H = \frac{V}{LW}$ 5. $V_1 = \frac{C_2V_2}{C_1}$ 6. $q = \frac{D}{3.5m^2p}$ 7. $F = \frac{C}{T}$
8. $b = \frac{a^2}{2d}$ 9. $D = \frac{M}{CE}$ 10. $P_2 = \frac{P_1V_1}{V_2}$ 11. $T = \frac{q_1q_2}{b^2r}$ 12. $Q = \frac{F}{.3V^2t}$ 13. $t = \frac{m(p - q)}{3b}$
14. $M = \frac{B}{S - T}$ 15. $b = \frac{V}{a(r_1 - r_2)}$ 16. $c = \frac{F}{d(t_2 - t_1)}$

Assignment 27

1. C = BD 2. $t = \frac{mr}{a}$ 3. $v_1 = \frac{1.8Hs^2}{v_2}$ 4. $F = \frac{T}{A}$ 5. $d_2 = \frac{d_1F_2}{F_1}$ 6. $p = \frac{t_2 - t_1}{2h}$
7. $K = \frac{FM}{AC(v_2 - v_1)}$ 8. $c = \frac{2B}{d}$ 9. $h = \frac{2A}{m_1 + m_2}$ 10. $C_1 = \frac{A_1C_2}{A_2}$ 11. $T_1 = \frac{P_1Q_1T_2}{P_2Q_2}$
12. $m = \sqrt{st}$ 13. $b = \sqrt{\frac{c - d}{t}}$ 14. $p = \sqrt{q^2 + r^2}$ 15. $v = \sqrt{\frac{2.5H}{c}}$ 16. $t = \sqrt{\frac{2F}{d}}$

UNIT 9: FORMULA REARRANGEMENT (continued)

Assignment 28

1. $V = P - 50$ 2. $C = A - B - D$ 3. $Q = L + F$ 4. $B = b + cH$ 5. $V = \frac{I - M}{T}$

6. $t_2 = kp + t_1$ 7. $C_i = vt - C_o$ 8. $d = \sqrt{\frac{I_a - I_c}{m}}$ 9. $q = \sqrt{t^2 - p^2}$ 10. $b = \sqrt{c^2 + a^2}$

11. $T = S - R$ 12. $t_1 = t_2 - mp$ 13. $F = \frac{D - d}{p}$ 14. $v = \frac{t_f^2 - t_o^2}{2a}$ 15. $t = \sqrt{s^2 - h^2}$

16. $M_B = M_A - M$

Assignment 29

1. $S = \frac{M - PT}{P}$ 2. $d = \frac{a - bc}{b}$ 3. $V_1 = \frac{R + FV_2}{F}$ 4. $X = \frac{PL - C}{P}$ 5. $t_1 = \frac{2D - mt_2}{m}$

6. $p = \frac{h}{a + b}$ 7. $T = \frac{S}{C + F}$ 8. $I = \frac{Q}{X + Y}$ 9. $D = T^2$ 10. $t = \frac{m^2}{2a}$ 11. $h = p^2q$

12. $v = \frac{2b}{r^2}$ 13. $S = \frac{P^2}{R^2T}$ 14. $x = \sqrt{e^2 - y^2}$ 15. $t = \sqrt{c^2 - k^2}$ 16. $E = \sqrt{D^2 - P^2}$

UNIT 10: GEOMETRIC FORMULAS

Assignment 30

1. P = 16 cm, A = 15 cm^2 2. P = 28 ft, A = 40 ft^2 3. P = 32 m, A = 64 m^2 4. A = 50 cm^2
5. A = 48 ft^2 6. A = 320 m^2 7. A = 20 cm^2 8. A = 30 ft^2 9. A = 48 m^2 10. A = 50°
11. B = 50° 12. A = 30° 13. W = 7 cm 14. L = 20 ft 15. s = 7 m 16. h = 5 cm
17. b = 10 yd 18. h = 10 cm 19. A = 64 in^2 20. l_2 = 20 m

Assignment 31

1. c = 4.472 cm 2. b = 9.798 in 3. a = 6.928 ft 4. d = 8.485 m 5. L = 9.165 cm
6. W = 6.633 yd 7. C = 31.4 cm, A = 78.5 cm^2 8. C = 6.28 in, A = 3.14 in^2
9. C = 18.84 ft, A = 28.26 ft^2 10. V = 27 cm^3 11. V = 160 in^3 12. V = 62,800 cm^3
13. c = 6.325 cm 14. l_2 = 8 ft 15. d = 4.243 yd 16. L = 4 m 17. r = 7 cm
18. d = 8 in 19. r = 2 yd 20. W = 4 ft

UNIT 11: SCIENTIFIC NOTATION

Assignment 32

1. 100 2. .001 3. 10 4. 1 5. .1 6. 10,000 7. .00001 8. .01
9. 1,000,000 10. .000001 11. 10,000,000 12. .00000001 13. 1,000,000,000
14. .0000000001 15. 10^1 16. 10^2 17. 10^4 18. 10^6 19. 10^8 20. 10^0 21. 10^{-1}
22. 10^{-2} 23. 10^{-5} 24. 10^{-8} 25. 10^6 26. 10^3 27. 10^0 or 1 28. 10^{-3} 29. 10^{-7}
30. 10^3 31. 10^1 or 10 32. 10^0 or 1 33. 10^{-3} 34. 10^{-4} 35. 10^{-5} 36. 10^7

Assignment 33

1. 84 2. 52.7 3. 160,000 4. .000009 5. 91.3 6. 2 7. 750,000 8. .000425
9. 28 10. 65,600 11. 70,000,000 12. 9,400,000,000 13. .41 14. .00114
15. .000002 16. .000000088 17. 2.77×10^1 18. 5.19×10^2 19. 4.96×10^4 20. 6×10^6
21. 4.25×10^8 22. 3.14×10^{-1} 23. 7.5×10^{-2} 24. 4×10^{-6} 25. 6.39×10^{-5}
26. 1.1×10^{-8} 27. 1.56×10^7 = 15,600,000 28. 8.03×10^{-6} = .00000803
29. 7.22×10^6 = 7,220,000 30. 4.73×10^{-8} = .0000000473

UNIT 11: SCIENTIFIC NOTATION (continued)

Assignment 34

1. 2.56×10^5 2. 4.78×10^7 3. 3.66×10^{-2} 4. 1.59×10^{-5} 5. 6.6×10^7 6. 9.49×10^{-8}
7. 2.5×10^5 8. 3.7×10^{-6} 9. 6×10^6 10. 8×10^{-7} 11. 3.5×10^7 12. 3.2×10^{-7}
13. 8×10^9 14. 2.38×10^{-1} 15. 1.32×10^{-5} 16. 4×10^6 17. 2×10^{-7} 18. 4.1×10^{-6}
19. 2.2×10^6 20. 3×10^3 21. 2×10^{-5} 22. 3×10^{-6} 23. 4×10^5
24. 6.02×10^{24} molecules 25. 1.174×10^{14} miles 26. 2.69×10^{23} molecules 27. $V_2 = 8,000$

UNIT 12: PERCENTS AND EQUATIONS

Assignment 35

1. $\frac{1}{2}$ 2. $\frac{1}{4}$ 3. $\frac{1}{3}$ 4. $\frac{4}{5}$ 5. $\frac{3}{10}$ 6. $\frac{1}{5}$ 7. $\frac{9}{10}$ 8. $66\frac{2}{3}\%$ 9. 50% 10. 75%
11. 40% 12. 70% 13. 60% 14. 10% 15. 25% 16. .57 17. .035 18. .0725
19. 4 20. .0015 21. 2.69 22. .0001 23. 85% 24. 4.1% 25. 100% 26. 308%
27. 500% 28. .07% 29. 3.25% 30. 50% 31. 66% 32. 293% 33. 91.7% 34. 2.5%
35. 5.3% 36. 87% 37. 217% 38. 31% 39. 23% 40. 7.7 41. 30 42. 3
43. 2.7 44. 3,750 45. 270 46. 22 47. 5,700

Assignment 36

1. 40 2. 12 3. 60 4. 1,215 5. 2,440 6. 7.5 7. 25% 8. 70% 9. $33\frac{1}{3}\%$
10. 9% 11. 175% 12. .8% 13. 20 14. 50 15. 200 16. 40 17. 400 18. 1,500
19. 24 students 20. 30% 21. 50 items 22. 5% 23. 1,380 customers 24. .2%
25. $1,200 26. $2.34

Assignment 37

1. $5,400 2. 20% 3. $30 4. $405 5. 125% 6. $1,000 7. $70 8. 6%
9. $5,000 10. $63 11. 11% 12. $500 13. $31.40 14. 7.5%

UNIT 13: GRAPHING

Assignment 38

1.

x	y
5	20
3	12
0	0
-3	-12
-5	-20

2.

x	y
3	7
1	1
0	-2
-1	-5
-3	-11

3.

x	y
2	3
1	4
0	5
-1	6
-2	7

4.

x	y
4	4
2	0
0	-4
-2	-8
-4	-12

Assignment 39

1. A (3, 8) E (2, 0)
   B (-4, 2) F (0, 6)
   C (-2, -6) G (0, 0)
   D (4, -4) H (0, -4)

2. B (2)
   C (3)
   A (1)
   D (4)

3.

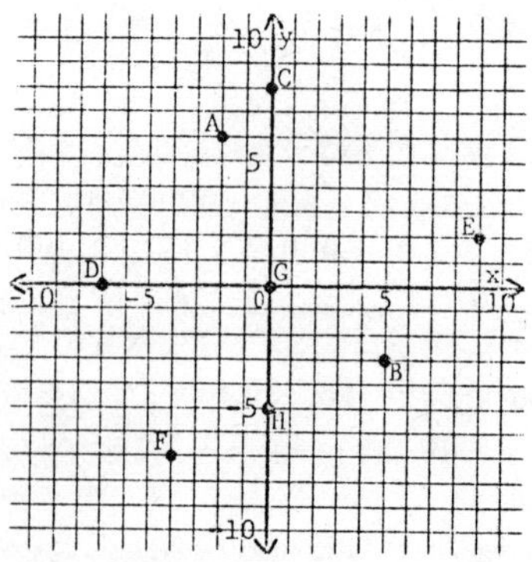

UNIT 13: GRAPHING (continued)

Assignment 40

1 and 2.

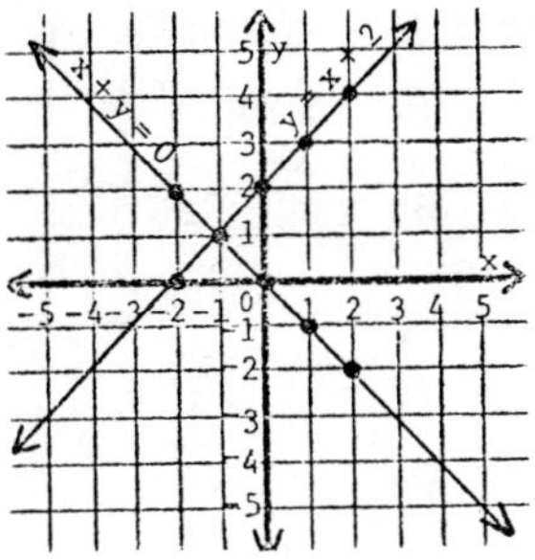

3 and 4.

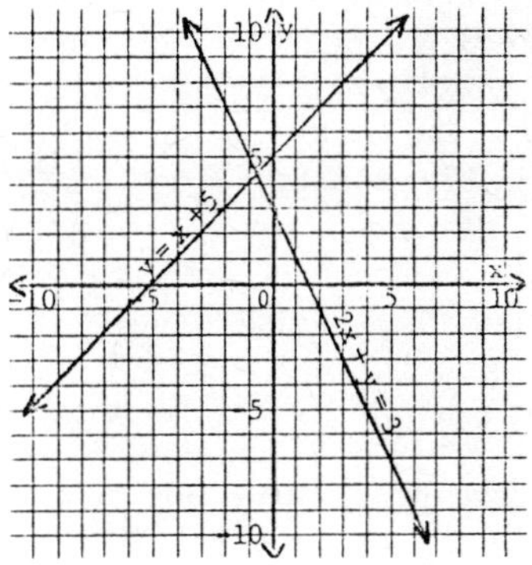

5. 600 miles
6. 5 hours
7. 1,200 miles
8. 1 hour

UNIT 14: SYSTEMS OF EQUATIONS

Assignment 41

1. $x = 1, y = 1$

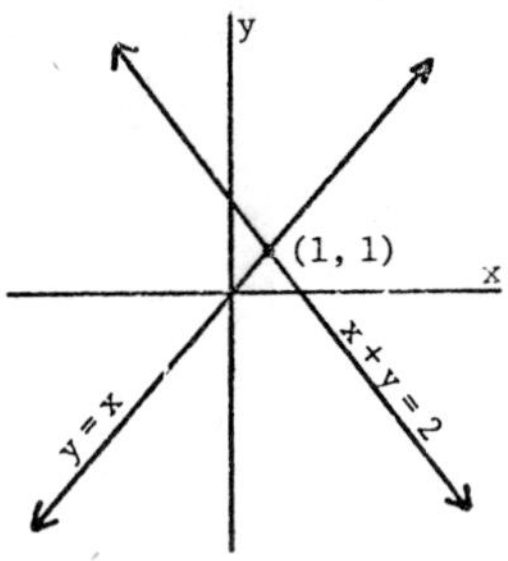

2. $x = 4, y = 2$

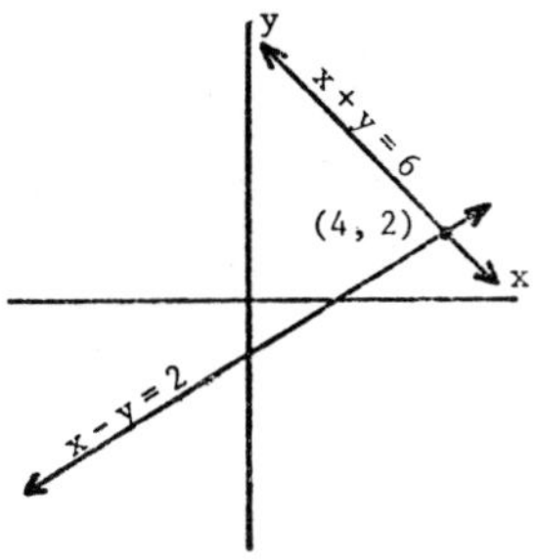

3. $x = 3, y = -3$

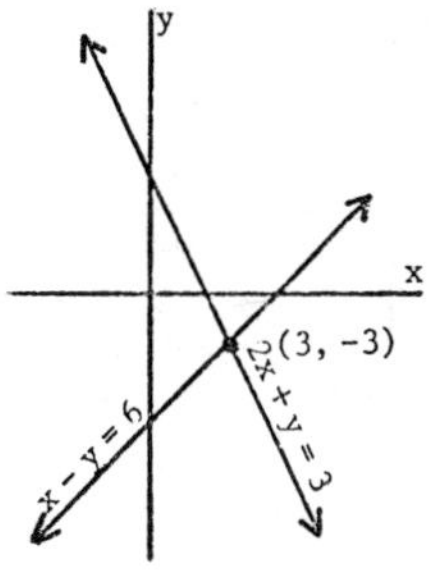

4. $x = 3, y = 6$

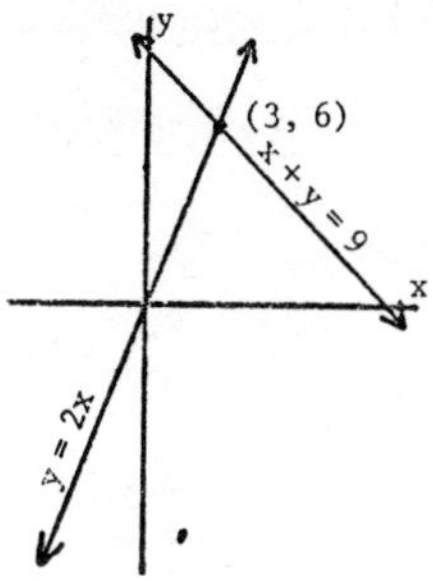

5. $x = 5, y = 2$ 6. $p = 2, q = 1$ 7. $s = 5, t = -1$ 8. $a = 2, b = 0$

Assignment 42

1. $x = 5, y = 2$ 2. $a = 3, b = 1$ 3. $s = 2, t = 3$ 4. $p = 2, q = -1$ 5. $x = 4, y = -2$
6. $a = 1, b = 3$ 7. $m = -1, t = 5$ 8. $p = 4, q = 2$

Assignment 43

1. $x = 4, y = 1$ 2. $d = 3, h = -2$ 3. $p = -2, q = -1$ 4. $m = 6, t = 2$ 5. $x = 1, y = 1$
6. $a = -2, b = 7$ 7. 18 females, 11 males 8. larger is 36, smaller is 12
9. husband is 160 pounds, wife is 120 pounds 10. larger is 20, smaller is 10
11. larger is 10 ft, smaller is 8 ft 12. length is 20 m, width is 5 m
13. larger is 30 cm, smaller is 6 cm 14. length is 11 ft, width is 7 ft

UNIT 15: SPECIAL PRODUCTS AND FACTORING

Assignment 44

1. $3x$ 2. $20y$ 3. $4p$ 4. a^2 5. $8x^2$ 6. $3y^2$ 7. $6m^2$ 8. $30x^2$ 9. $21y^2$
10. $64t^2$ 11. $6x + 15$ 12. $12y - 4$ 13. $p^2 - p$ 14. $h^2 + h$ 15. $x^2 + 8x$ 16. $3y^2 - 4y$
17. $5a^2 + 35a$ 18. $12b^2 - 6b$ 19. $3(2x)$ 20. $(8y)(1)$ 21. $(b)(b)$ 22. $(m)(2m)$
23. $(x)(9x)$ 24. $(2t)(2t)$ 25. $(5p)(6p)$ 26. $(8b)(4b)$ 27. $2(x + 4)$ 28. $4(4y - 3)$
29. $4(2t + 1)$ 30. $a(a + 4)$ 31. $t(t - 9)$ 32. $h(h + 1)$ 33. $s(s - 1)$ 34. $p(2p + 3)$
35. $5d(d - 3)$ 36. $2x(4x + 3)$ 37. $6y(y - 2)$ 38. $4t(4t + 1)$

UNIT 15: SPECIAL PRODUCTS AND FACTORING (continued)

Assignment 45

1. $x^2 + 6x + 5$ 2. $y^2 + 10y + 24$ 3. $m^2 + 4m + 4$ 4. $5t^2 + 7t + 2$ 5. $3p^2 + 10p + 8$
6. $6x^2 + 13x + 5$ 7. $12y^2 + 11y + 2$ 8. $4a^2 + 12a + 9$ 9. $x^2 + 6x - 7$ 10. $y^2 + y - 2$
11. $t^2 - t - 20$ 12. $m^2 + 6m - 16$ 13. $2a^2 + a - 21$ 14. $3b^2 - 11b - 4$ 15. $8x^2 - 18x - 5$
16. $15y^2 + 7y - 4$ 17. $x^2 - 5x + 6$ 18. $y^2 - 17y + 72$ 19. $b^2 - 6b + 9$ 20. $2b^2 - 3b + 1$
21. $4t^2 - 11t + 6$ 22. $9y^2 - 18y + 8$ 23. $10p^2 - 13p + 4$ 24. $4t^2 - 12t + 9$ 25. $x^2 - 25$
26. $y^2 - 49$ 27. $m^2 - 1$ 28. $a^2 - 4$ 29. $4t^2 - 9$ 30. $16x^2 - 1$ 31. $25b^2 - 9$
32. $36y^2 - 4$

Assignment 46

1. $(x + 3)(x - 3)$ 2. $(y + 10)(y - 10)$ 3. $(b + 1)(b - 1)$ 4. $(h + 9)(h - 9)$ 5. $(3m + 2)(3m - 2)$
6. $(2y + 1)(2y - 1)$ 7. $(7t + 6)(7t - 6)$ 8. $(8x + 5)(8x - 5)$ 9. $(x + 1)(x + 6)$ 10. $(y + 3)(y + 3)$
11. $(t + 4)(t + 5)$ 12. $(m - 1)(m - 3)$ 13. $(b - 2)(b - 9)$ 14. $(p - 5)(p - 5)$ 15. $(x - 3)(x + 5)$
16. $(y + 2)(y - 5)$ 17. $(t - 2)(t + 3)$ 18. $(a + 3)(a - 4)$ 19. $(p - 2)(p + 8)$ 20. $(m + 2)(m - 9)$
21. $(2x + 3)(x + 1)$ 22. $(7t + 3)(t + 2)$ 23. $(3x - 5)(x - 1)$ 24. $(5y - 4)(y - 2)$
25. $(2a - 1)(a + 3)$ 26. $(5h + 2)(h - 1)$ 27. $(3y - 2)(y + 1)$ 28. $(5y - 7)(y + 1)$

UNIT 16: QUADRATIC EQUATIONS

Assignment 47

1. ± 5 2. ± 12 3. ± 6.083 4. ± 13.601 5. $\pm\frac{3}{7}$ 6. $\pm\frac{9}{5}$ 7. $\pm\frac{1}{3}$ 8. $\pm\frac{1}{6}$ 9. ± 1
10. ± 8 11. ± 4.69 12. ± 11.402 13. $\pm\frac{3}{2}$ 14. ± 1.732 15. $\pm\frac{1}{2}$ 16. $\pm\frac{7}{5}$ 17. ± 8
18. ± 9.487 19. ± 10 20. ± 12.247 21. 5 and 7 22. -10 and 1 23. -1 and -3
24. $\frac{3}{4}$ and $\frac{9}{2}$ 25. $\frac{5}{3}$ and $-\frac{1}{2}$ 26. $-\frac{1}{7}$ and $-\frac{5}{4}$ 27. 0 and 8 28. 0 and -1 29. 0 and $\frac{2}{5}$
30. 0 and -4 31. 0 and 1 32. 0 and $-\frac{3}{2}$

Assignment 48

1. 0 and 3 2. 0 and -10 3. 0 and 1 4. 0 and -1 5. 0 and $\frac{2}{3}$ 6. 0 and $-\frac{7}{5}$
7. 0 and $\frac{1}{4}$ 8. 0 and $-\frac{1}{9}$ 9. 0 and 2 10. 0 and $-\frac{1}{4}$ 11. 0 and 2 12. 0 and $-\frac{3}{4}$
13. 0 and 5 14. 0 and $\frac{1}{7}$ 15. 0 and 4 16. 0 and $-\frac{5}{2}$ 17. -1 and -5 18. -2 and -3
19. -5 (double root) 20. 3 and 4 21. 1 and 10 22. 4 (double root) 23. 3 and -5
24. 3 and -4 25. -2 and 6 26. -4 and 5 27. 1 (double root) 28. -2 and 7
29. $-\frac{3}{2}$ and -1 30. $-\frac{4}{5}$ and -2 31. $\frac{5}{3}$ and 1 32. $\frac{2}{7}$ and 1 33. $\frac{3}{2}$ and -2 34. $\frac{5}{3}$ and -1
35. $\frac{2}{5}$ and 1 36. $\frac{3}{2}$ and -1 37. $\frac{3}{7}$ and 1

Assignment 49

1. $-\frac{1}{3}$ and 3 2. $\frac{3}{2}$ and -5 3. $-\frac{4}{5}$ and 1 4. $\frac{4}{3}$ and -3 5. $\frac{1}{4}$ and -3 6. $\frac{1}{5}$ and $\frac{1}{2}$
7. 4.30 and .70 8. 1.15 and -.35 9. -.78 and -3.22 10. 7.70 and 1.30
11. -.10 and -3.23 12. .74 and -.90

UNIT 17: EXPONENTS AND POLYNOMIALS

Assignment 50

1. 64 2. 8 3. 256 4. 1 5. 9 6. 1 7. 3^5 8. t^9 9. 4^6 10. m^{12}
11. 8^3 12. b^5 13. 7^5 14. a 15. 5^2 16. y^4 17. 7 18. p 19. 2^5 20. t
21. 1 22. 1 23. 4^{10} 24. x^{18} 25. 7^8 26. y^{25} 27. 2^3 28. 1 29. t^9 30. 1

Assignment 51

1. $12x^5$ 2. $-14y^2$ 3. $-b^3$ 4. $3t^5$ 5. $8m^2$ 6. $-3x^7$ 7. $-2d^6$ 8. $21p^6$ 9. $4x^2$
10. $-2y^5$ 11. $-3t^2$ 12. 5p 13. m^5 14. 2 15. $-5d^3$ 16. -1 17. $2x^3 - x$
18. $m^7 + 5m^5$ 19. $2y^4 - 2y^2$ 20. $6p^3 + 21p^2$ 21. $3t^5 - t^4 + t^3$ 22. $5d^6 + 5d^4 - 5d^2$
23. $2x - 1$ 24. $3y^2 + 2$ 25. $m^8 - 5m^4$ 26. $b^3 + 2b$ 27. $d^3 - 2d - 1$ 28. $2p^3 + 3p^2 - 4p$

Assignment 52

1. 8t 2. $9x^3$ 3. $3m^2$ 4. $9d^4$ 5. $2p^3$ 6. 3q 7. $-4t^3$ 8. $-a^2$ 9. $6x^4$ 10. 0
11. $3x^2 + 10$ 12. $4y + 3$ 13. $6t^4 - 1$ 14. $5p^3 + p^2 - p$ 15. $2m^2 - 7$ 16. $2a^5 + a^4 - 3a^2 - 1$
17. $2b^2 - 4b$ 18. $x + 1$ 19. $-t^4 + 2t^2$ 20. 0 21. $2y^2 + 9$ 22. $-p^3 + 9p - 6$ 23. $6b^2 - 3$
24. $-3t^3$ 25. $3y^2 - y - 2$ 26. $4x^3 + 4x^2 + x + 2$ 27. $4y + 7$ 28. $3x^2 - 2x$ 29. $a^2 - a + 1$
30. $t^4 - t^3$

UNIT 18: RADICALS

Assignment 53

1. $\sqrt{14}$ 2. $\sqrt{5x}$ 3. $\sqrt{30t}$ 4. $\sqrt{21x}$ 5. $\sqrt{66d}$ 6. $20\sqrt{3}$ 7. $10\sqrt{15}$ 8. $8\sqrt{6}$
9. $50\sqrt{21}$ 10. 4 11. $3\sqrt{5}$ 12. $10\sqrt{2}$ 13. $7\sqrt{y}$ 14. $2\sqrt{3x}$ 15. $4\sqrt{2t}$ 16. $4d^2$
17. $x\sqrt{5}$ 18. $3p^3\sqrt{2}$ 19. $2y\sqrt{y}$ 20. $y^2\sqrt{7y}$ 21. $5m^3\sqrt{2m}$ 22. 6 23. $2\sqrt{5}$
24. $4\sqrt{x}$ 25. $5\sqrt{3y}$ 26. $t\sqrt{6}$ 27. 5b 28. $2t^2$ 29. $10x\sqrt{x}$ 30. $2y^2\sqrt{2y}$

Assignment 54

1. 4 2. $\sqrt{7}$ 3. 5 4. y 5. $p\sqrt{p}$ 6. $x\sqrt{3}$ 7. $2t\sqrt{2t}$ 8. $d^2\sqrt{7d}$ 9. $\frac{1}{5}$
10. $\frac{\sqrt{13}}{4}$ 11. $\frac{3\sqrt{2}}{7}$ 12. $\frac{\sqrt{3x}}{2}$ 13. $\frac{y^2}{6}$ 14. $\frac{2x\sqrt{3}}{5}$ 15. $\frac{2\sqrt{6}}{t^3}$ 16. $\frac{4p\sqrt{p}}{3}$
17. $\frac{3\sqrt{5}}{5}$ 18. $\frac{\sqrt{x}}{x}$ 19. $\frac{7\sqrt{3}}{6}$ 20. $\frac{\sqrt{2y}}{10y}$ 21. $\frac{2\sqrt{35}}{21}$ 22. $\frac{7\sqrt{6t}}{3t^3}$ 23. $\frac{\sqrt{6}}{3}$
24. $\frac{3\sqrt{2m}}{m^2}$ 25. $\frac{\sqrt{3}}{2}$ 26. $\frac{\sqrt{35}}{x}$ 27. $\frac{y^2\sqrt{y}}{6}$ 28. $\frac{\sqrt{6}}{6}$ 29. $\frac{2\sqrt{15x}}{5x}$ 30. $\frac{2\sqrt{2t}}{3t}$
31. $4\sqrt{5}$ 32. $5\sqrt{3}$ 33. Not possible 34. $3\sqrt{x}$ 35. $2\sqrt{m}$ 36. Not possible 37. $2\sqrt{2}$
38. Not possible 39. $2\sqrt{t}$ 40. $-\sqrt{p}$ 41. Not possible 42. 0

INDEX